Human Factors Engineering

Human Factors Engineering

Chandler Allen Phillips, M.D., P.E.

John Wiley & Sons, Inc.
New York / Chichester / Weinheim / Brisbane / Singapore / Toronto

Acquisitions Editor	*Wayne Anderson*
Marketing Manager	*Katherine Hepburn*
Production Editor	*Ken Santor*
Designer	*Kevin Murphy*
Illustration Coordinator	*Sigmund Malinowski/Sandra Rigby*
Cover Photo	*provided courtesy of Kaiser Porcelain, Ltd.*

This book was set in Times Ten by PRD Group, Inc. and printed and bound by Hamilton Printing Company, Inc. The cover was printed by Lehigh Press, Inc.

This book is printed on acid-free paper. ♾

The paper in this book was manufactured by a mill whose forest management programs include sustained yield harvesting of its timberlands. Sustained yield harvesting principles ensure that the numbers of trees cut each year does not exceed the amount of new growth.

Library of Congress Cataloging in Publication Data:

Phillips, Chandler A.

Human factors engineering/Chandler Allen Phillips.

p. cm.

ISBN 0-471-24089-3 (cloth: alk. paper)

1. Human engineering. I. Title.

TA166 .P53 2000

620.8—dc21

99-048715

Printed in the United States of America

10 9 8 7 6 5 4 3 2 1

For Janie

To love
Is not to gaze
Steadfastly at each other
But to look together
In the same direction.

And For Sage and Shelby

Beauty without vanity
Strength without insolence
Courage without ferocity
And all the virtues of man
Without his vices.

PREFACE

This book began with a drive my wife, Janie, and I were taking to our local grocery store about four years ago. I was sharing with her a conversation earlier that month between the Department Chair (the late Anthony J. Cacioppo, Ph.D.) and myself. He and I had been discussing (and mildly lamenting) the current state of engineering education that occurred when the "human element" was an important interaction with a device, system, or process that was the subject of an engineering analysis.

Such interactions occurred frequently and across different engineering disciplines. For industrial engineers, the human operator was often an integral element in performing manual labor tasks, production assembly tasks, or technological system control tasks (to name but a few). For mechanical engineers (including manufacturing engineers) there was not only interest in people doing work, but also in people operating the devices that are engineered (e.g. automobiles, power tools, and other consumer products). For aeronautical engineers, above and beyond airplane design, the "human-in-the-loop" problem was an additional responsibility.

Finally, all of these disciplines (I.E., M.E. and A.E.) have a fundamental interest in virtual environment systems and human control. In particular, such engineers have a primary responsibility for "haptic" control devices that provide a "mechanical" sense of what the human is experiencing visually. Additionally, these disciplines are very involved in the environmental control engineering of locations in which human beings are working.

Our concern that day was that the "human factor" was not subjected to the same engineering analysis and rigor that the "inanimate factors" of the device, system, or process were. In essence, the engineering student was not being taught how to apply their engineering fundamentals (mechanics, thermodynamics, electronics, and control theory) to the analysis of this human factor. Rather, engineering faculty often released their students to "human centered courses" in other academic departments. For example, I.E.'s would often take a human factors course taught in a psychology department by psychology faculty, and in which many of the students were social science majors. Another example would be M.E.'s referred to a human anatomy and physiology course taught in a biology department by biology faculty, and in which many of the students would be nursing majors.

Quite frequently, the reaction of the more vocal engineering students would be something rather blunt, such as:

"Where is the engineering in all of this?"

Or something more articulately stated, such as:

"How do I apply this information by using the mathematical and analytical skills in which an engineer is trained?"

My department chair and I concluded that meeting with a clear understanding of the problem, but no obvious or practical solution was evident for the *undergraduate* engineering student (who wanted to graduate and go to work). For these students, an evasive answer to the above questions [such as: "Go to graduate school, and take more courses in this area"] would not be well received.

As my wife and I continued to drive to the grocery store, I bounced an idea off her that had been forming over some period of time. What was needed as a practical solution was a good engineering textbook for these undergraduate students. If properly done, it would allow engineering faculty themselves to teach their own students *human factors engineering* using the same mathematical and analytical methods that are applied to inanimate devices, systems, and processes.

What I then said to Janie was "I can write this book." It would be based upon my twenty-five year university career of teaching, research, and professional practice. Each chapter would be based upon a course (or a part of a course) that I had taught (usually many times). The worked examples and the chapter problems would come from classroom examinations, homework problems, and term paper assignments. Finally, the chapters would progress from engineering fundamentals through professional practice.

Over this twenty-five year period, the types of courses that I taught, the continuing progression of my research, and even the areas of my professional practice had all evolved such that in the last decade I was clearly at a juncture in which *human factors engineering* intertwined with both traditional human factors and also biomedical engineering. This juncture had been both intellectually exciting and philosophically satisfying. There was a real sense of discovery as I interacted both with my students as well as like-minded colleagues.

By then, Janie and I had arrived at the grocery store, stopped the car, and were about to open the doors. I had shared all of these thoughts with her as she quietly listened.

And just then, as a final statement that condensed all of my feelings at that one moment in time, I said, "If I don't write this book, then someone else will."

With a quick instinct and flash of insight that I have grown to admire over our many years together, she replied–

"Then do it!"

And that is how this book began.

ACKNOWLEDGEMENTS

My deepest appreciation and gratitude are extended to the following individuals:

Janie Draper Phillips for that spark of creativity during the many hours we worked together to develop the real world activities used for the various worked examples. A special acknowledgement goes to her idea for the cover design. Engineering practice is both an art and a science and the human focus combines the two.

Anthony J. Cacioppo, Ph.D. (former Department Chair) for being my mentor in human factors and for sharing a common interest in all things classical. He will be missed.

Daniel W. Repperger, Ph.D., P.E., who is Director of the Human Sensory Feedback Laboratory (Wright-Patterson Air Force Base, Dayton, Ohio) for hosting my Sabbatical year leave (1996-1997) during which this book began. We all know Dan as "Mr. Haptic Control" (he holds the original 1984 U.S. Patent in that field). Dan generously donated innumerable hours to provide a detailed technical review and edit of the entire textbook manuscript.

Richard J. Koubek, Ph.D., who is the current Department Chair, for his sage advise and counsel in the development of this text.

Amir Faghri, Ph.D., who is Dean of the University of Connecticut School of Engineering, for his continuing encouragement and for his critical review of the textbook manuscript.

Karen Strickler, a former student of mine who is now a biomedical engineer (B.S.B.M.E.) and also a human factors engineer (M.S.E.), for working countless hours on the typing and technical editing of the successive drafts for the textbook manuscript. Thanks to her skill, dedication, and tireless productivity we were able to meet all of the contractual deadlines set by the publisher. Her work was above and beyond the call of duty and is greatly appreciated.

A. Wayne Anderson, who is my editor at John Wiley, for moving this textbook through the various phases of development with both thoughtful consideration and consummate professionalism.

The Department office staff: Sharon Brannon and Carol Harward for their operational support and Daisy Stieger for the technical editing of the Instructor's Manual that accompanies this text.

CONTENTS

Part I Introduction to Human Factors Engineering

Part II Human Factors Engineering: Fundamentals

Part I

Principles of Analysis and Design

CHAPTER 1

Human Factors Engineering

1.1 INTRODUCTION TO THE STUDENT

Human factors engineering may be defined as follows:

The discipline of engineering concerned with the analysis, design, and development of human-technological systems in which primary emphasis is on the human.

Human factors engineering is based upon the concept of a human-technological system that may be described quantitatively using the standard methods of engineering analysis. The system itself may be simple or complex, but generally consists of four elements as shown in Figure 1.1. By definition, there is always a human subsystem as part of the overall system.

A system is defined as a set of elements which are related by means of their interaction with each other and which collectively act together in order to achieve an objective or purpose. In human factors engineering the human subsystem is the major focus of the system engineering analysis. This analysis may be mechanic, electronic, thermodynamic, biophysical, ergonomic, or informatic (for example), depending upon the specific objectives of the analysis. Some examples are given in the review of the learning objectives that concludes this chapter.

With respect to human factors engineering, the technological subsystem may vary from the very simple to the very complex. A toothbrush or a hammer is an example of a simple technological subsystem. At the opposite extreme, a jet airliner or a nuclear submarine is an example of a very complex technological subsystem. Furthermore, the technological subsystem may be engineering-discipline specific, such as a mechanical drill press, an electronic computer, or an industrial assembly line.

As shown in Figure 1.1, human-technological systems may interact with the environment (the environmental subsystem), in which case we have an *open* human-technological system. A *closed* human-technological system is one in which there is minimal environmental interaction. Some examples are given in Chapter 2.

The task to be performed is the fourth element that interacts with a human-technological system. This final element ensures that the human-technological system has an objective and does something purposeful. The human operator's individual strategy may vary, as well as their physical and mental capabilities. Indeed, even the human operator's motivation and attitude may vary. However, when there is a purposeful task, the human factors engineer has some basis for predictability and human performance measurement. Without a specified task, human-technological systems may be random, reflex responsive, or erratic and are not well suited for quantitative engineering analysis.

As shown in Figure 1.2, human factors engineering may be viewed as a multidisciplinary engineering specialty. Its scientific foundations are based upon a social

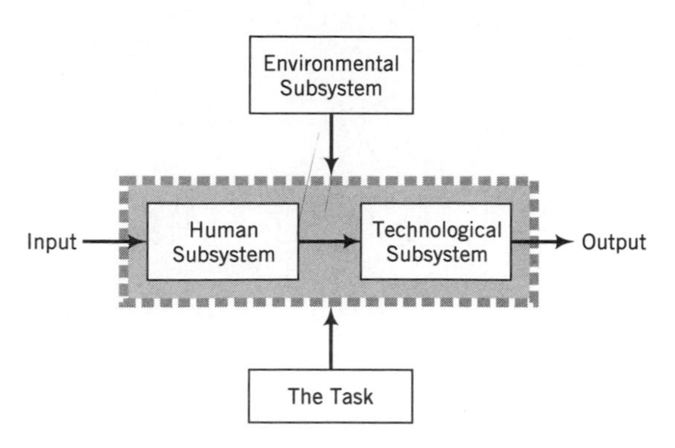

Figure 1.1 A human-technological system (enclosed within dashed lines).

science (psychology), applied science (engineering), and life science (biology/medicine). Figure 1.2 indicates that human factors engineering is directly a discipline of engineering since it is based directly upon the general engineering elements of physics, statics, dynamics, electronics, thermodynamics, and control systems.

Consequently, upper division undergraduate engineering students may be pleasantly surprised to know that much of their general engineering knowledge, which they have already acquired, directly interfaces with the discipline of human factors engineering. The previously described engineering science elements define a set of analytic skills, which (as we shall see in subsequent chapters) can be directly applied to human factors engineering analysis, design, and development.

Figure 1.2 also indicates that human factors engineering is indirectly dependent upon both psychology and biomedicine for its fundamental scientific basis. From

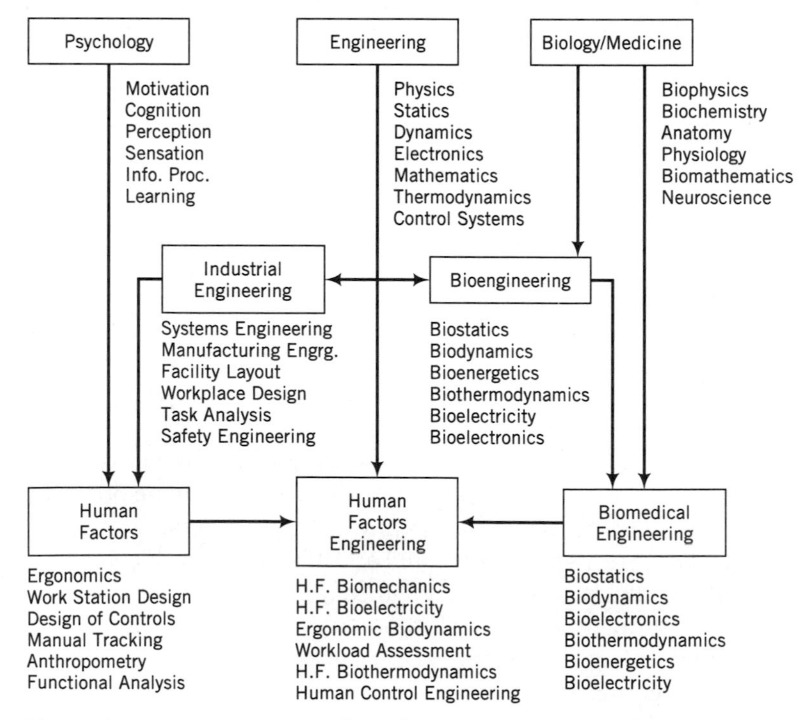

Figure 1.2 Human factors engineering viewed as a multidisciplinary engineering specialty.

psychology, some representative fundamental knowledge elements are motivation, cognition, perception, sensation, information processing, and learning. From biology/medicine, representative fundamental knowledge elements are biomathematics, biophysics, biochemistry, anatomy, physiology, and neuroscience.

Direct interfacing of fundamental knowledge elements from psychology and biology/medicine with human factors engineering is not as straightforward, so it is necessary to preprocess these fundamental knowledge elements and integrate them into subfields of knowledge. Psychology and biomedicine are seldom (if ever) incorporated into traditional engineering fields such as electrical, mechanical, industrial, or chemical engineering. However, human factors is a recognized specialty of industrial engineering and also a specialty of psychology. From industrial engineering, representative applied (human activity) knowledge elements are systems engineering, manufacturing engineering, facility layout, workplace design, task analysis, and safety engineering.

The specialty of human factors can be defined as *design for human use.* In human factors, some representative elements are ergonomics, workstation design, design of controls, manual tracking, design of tools, and functional analysis. The majority of human factors specialists are industrial engineers and psychologists. In this specialty, human beings (with their abilities and limitations) are studied as human operators that interact with a system (e.g., a machine, a task, and/or an environment). Consequently, human factors provides the second interface with the discipline of human factors engineering. It preprocesses fundamental knowledge elements from psychology with the applied (human activity) knowledge elements of industrial engineering. Human factors as a specialty is important as an interface because it provides the context and some content of human factors engineering.

Bioengineering is an interdisciplinary engineering field that interfaces engineering with biology. The resultant knowledge elements are applied to any living system (e.g., plants, animals, microorganisms, human beings, etc.). From bioengineering, some representative applied knowledge elements are biostatics, biodynamics, bioelectronics, biothermodynamics, and bioenergetics.

Biomedical engineering can be defined as *design and development of biomedical devices and systems.* In biomedical engineering, the representative elements are similar to bioengineering, but the specific focus of these elements is the human being. This is because biomedical engineering is a subfield of medical science and also a subfield of bioengineering. It is practiced by engineers, the majority of whom are biomedical engineers both by training and by experience. Consequently, biomedical engineering knowledge elements focus on the application of engineering methods to analyze the human being and so develop devices and systems that can be applied to human beings. As such, it represents the third interface with the discipline of human factors engineering. Biomedical engineering is important as an interface because it translates engineering theory and practice directly into human application.

The convergence of engineering science, human factors, and biomedical engineering defines the discipline of human factors engineering (Figure 1.2). Representative knowledge elements of this engineering discipline (as presented in this text) are human factors biomechanics (Chapter 3 and 4), human factors bioelectricity (Chapter 5), human factors biothermodynamics (Chapter 6), ergonomic biodynamics (Chapter 7), workload assessment (Chapter 8), and human control engineering (Chapters 9 and 10).

This text has been written for undergraduate engineering students majoring in a variety of disciplines. As such, it is anticipated that the upper division undergraduate engineering students have no prior course work in either human factors or biomedical engineering. If such is not the case (e.g., industrial engineering majors or biomedical engineering majors), these students may indeed move through certain chapters of the text at a more accelerated rate than their peers. A few students may have had prior course work in both human factors and biomedical engineering. However, it is very likely that human factors courses were *separate* from biomedical engineering courses. Such students will also find this text useful in integrating these two areas.

In order to develop a single text that presents human factors engineering at the undergraduate level, a three-stage approach has been used. First, it is expected that prior to undertaking the material in this text, students will have completed their engineering fundamentals (calculus, physics, statics and dynamics, electricity/electronics, and thermodynamics).

Second, that part of this text (containing Chapters 3–6) is to be studied first. In these chapters, students will learn the fundamentals of the interface between human factors and biomedical engineering based upon their prior engineering knowledge. This part of the text is entitled "Human Factors Engineering: Fundamentals." The student will learn and understand human factors as it provides the context and some content in which human factors engineering is practiced. The same student will also learn and understand biomedical engineering as it provides the analytical methodology in which human factors engineering is practiced. The specific areas of human factors engineering described in this part of the text (Chapters 3–6) are direct extensions of the engineering fundamentals already mastered by students in prior undergraduate course work.

Third, the subsequent part of this text (Chapters 7–10) is to be studied after the earlier part has been mastered. This part of the text is entitled "Human Factors Engineering: Practice" and is divided into two subparts. In the first subpart, entitled "Ergonomic Engineering," students will apply human factors engineering fundamentals to ergonomic engineering practice. This should reinforce students' integration of material and so enhance their understanding of human factors engineering.

In the second subpart, entitled "Human Control Engineering," students should have some prior computational experience with a mathematical-based computer application package in order to perform system simulation and graphical analysis. MATLAB® is employed in the worked examples (primarily because it is used for undergraduate laboratory and homework assignments by our own faculty and students). Every attempt has been made to make Chapters 9 and 10 self-contained so that a prior course in control systems (although useful) is not required in order to understand and apply the material presented.

The definition of human factors engineering (HFE) has now been presented in the context of a human-technological system. HFE has also been defined with respect to its related scientific and engineering disciplines. The three-step approach to HFE used in this text was then described. This introductory chapter now concludes with a review of the learning objectives toward which this text is directed.

1.2 HFE LEARNING OBJECTIVES

Eight learning objectives should be attained upon completion of this text. Each objective in this section is illustrated by an example that is actually worked out later in the text. Each learning objective correlates with a chapter in the text, so that as students work through the text, there will be numerous worked examples of each objective. The initial four learning objectives refer to the engineering fundamentals of human factors engineering. The final four learning objectives refer to the professional practice of human factors engineering.

Objective 1: Apply the Methods and Equations of Static Mechanics to Analyze Human-Technological Systems. In Chapter 3, biostatic mechanics is applied to various regions of the human operator. One such region is the shoulder and arm. It is noted that as human skeletal muscles perform a demanding task over a prolonged time, the muscle undergoes fatigue. Muscle fatigue is physically characterized as a progressive decrease in its ability to generate force. Modifying the task to reduce muscle force can increase endurance and prolong the time period before the onset of fatigue.

EXAMPLE 1 (adapted from Example 3.1)

a. A volunteer emergency worker must continuously extend an arm straight out holding an 18 N sign that says "DETOUR RIGHT" as shown in Figure 1.3.a. As the person performs this task, determine the shoulder muscle force ($\mathbf{F_M}$) and the shoulder joint reaction forces ($\mathbf{R_y}$ and $\mathbf{R_x}$).
b. As time passes by and fatigue sets in, the extended arm sags to an angle of 65° (from the side of the body) as shown in Figure 1.3.b. Find the new shoulder muscle force ($\mathbf{F_M}$).
c. Realizing the extended arm is sagging, the volunteer bends the elbow joint by 90° (so that the lower arm and hand are vertically upward) and also reextends the upper arm straight outward, holding the sign as shown in Figure 1.3.c. Find the new shoulder muscle force ($\mathbf{F_M}$).

Completion of this example problem will demonstrate that modifying the task (by bending the elbow joint) reduces the shoulder muscle force by 40% (compared to keeping the arm straight out).

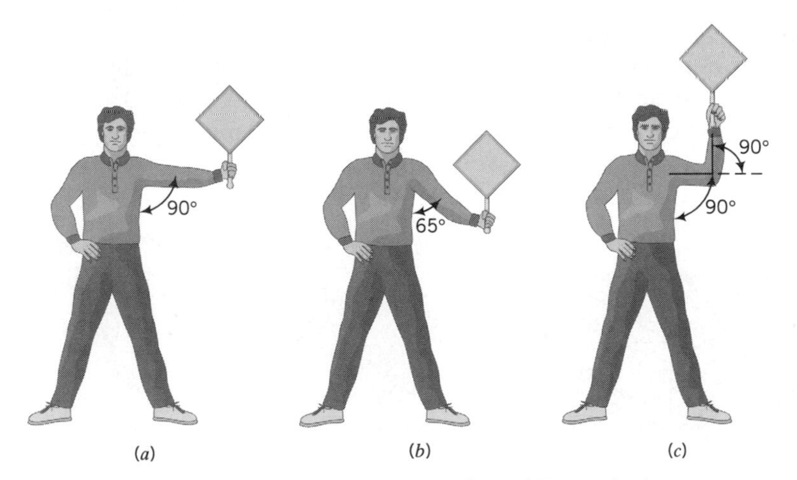

Figure 1.3 The volunteer emergency worker of Example 1 as seen in frontal view.

Objective 2: Apply the Methods and Equations of Dynamic Mechanics to Analyze Human-Technological Systems. In Chapter 4, biodynamic mechanics is applied (in part) to analyze human work. A human factors engineer should be able to analyze a task in which the human (as part of a system) is doing work. Initially, the HFE will analyze the task that the human is currently performing. Subsequent analysis may then indicate how the task might be performed more easily and efficiently.

EXAMPLE 2 (adapted from Example 4.6)

a. Consider a person pulling a crate by means of an attached cable (Figure 1.4.a). The crate (with a mass of 60 kg) is being pulled along a rough cement horizontal surface (with a coefficient of sliding friction, $\mu = 0.3$). If the person exerts a constant force ($\mathbf{T} = 250$ N) at an angle ($\theta = 30°$) from the horizontal, find the acceleration of the loaded crate. You may assume that the entire bottom of the crate remains in continuous contact with the ground.
b. The task described in part (a) is now modified by placing the crate on a wheeled pallet (Figure 1.4.b). This elevates the load and reduces the angle of the rope pull to 15°. If the coefficient of rolling friction is now $0.1(\mu)$, find the new force (**T**) that must be exerted on the rope to achieve the same horizontal acceleration obtained in part (a).

Completion of this example problem will show that the force exerted by the person on the rope is reduced by almost half when using a wheeled platform. It will be noted that *two* contributing factors were present. The coefficient of friction was reduced *and* the load was elevated (decreasing the angle θ).

Objective 3: Apply the Methods and Equations of Electricity and Electronics to Analyze Human-Technological Systems. In Chapter 5, bioelectronic devices are applied to the human operator in order to record the person's physiological responses when performing work. These devices can record, for example, respiration (referred to as electropneumogram), heart rate (the electrocardiogram), and muscle activity (the electromyogram) by means of surface electrodes applied to the person's skin.

The differential amplifier is the first signal conditioning unit that receives input from these surface electrodes. Differential amplifiers are commonly used when very small voltages (such as the electromyogram) are encountered. A major advantage

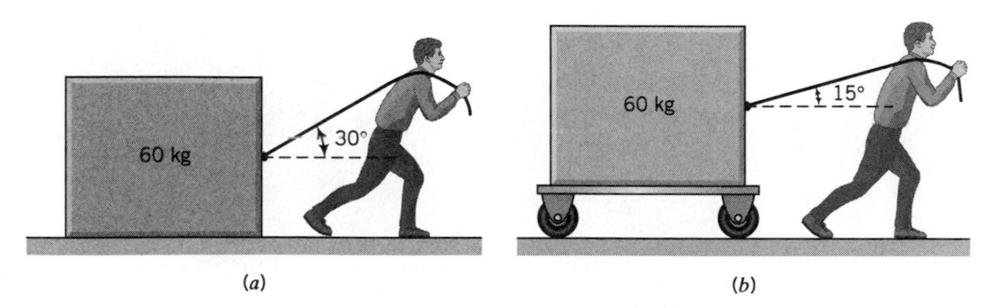

Figure 1.4 A person pulling a crate as described in Example 2 as seen in profile view.

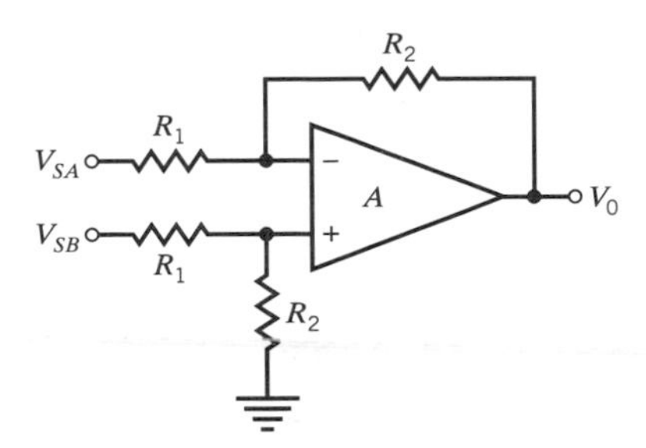

Figure 1.5 The differential amplifier circuit to be analyzed in Example 3.

of a differential amplifier is that any signals common to both channels (e.g., 60 Hz noise) are suppressed (the common mode rejection).

EXAMPLE 3 (adapted from Example 5.4)

For a differential amplifier (Figure 1.5), find the voltage output (V_0) as a function of the differential voltage input ($V_{SB} - V_{SA}$). You may invoke approximations as appropriate.

Completion of this example problem will show that the human factors engineer may adjust the voltage gain of the amplifier by varying the *ratio* of two resistors (R_2 with respect to R_1 of Figure 1.5).

Objective 4: Apply the Methods and Equations of Thermodynamics and Energetics to Analyze Human-Technological Systems. In Chapter 6, biothermodynamics and bioenergetics are applied to analyze the energy requirements and energy exchanges of the human operator when performing work. One application of this analytical approach is the specification of a "neutral" environment in which work is performed. A neutral environment is one in which the internal heat generated by a person's metabolism ($\dot{M}$) is exactly equal to the person's heat loss, i.e., the heat that is transferred ($\dot{Q}$) from the person to the environment.

EXAMPLE 4 (adapted from Example 6.5)

A human factors engineer has been asked to provide input on the environmental design of a business office. Light physical activity will be performed (typing, filing, sorting, telephoning) by a group of office workers. The mean individual body mass for an office staff member is 65 kg. A standard deviation of 7.5 kg is common for this office population (mixed men and women).

a. For these clothed workers, the HFE is requested to define a neutral environment with respect to human operator heat transfer (see Figure 1.6). It may be approximated that essentially the entire human operator heat transfer is by free convection (still air) and radiation. Heat transfer by conduction and evaporation is negligible.
b. Repeat the analysis performed in part (a), with the exception that forced convection (air moving at velocity, v) replaces free convection (still air).

Figure 1.6 Office worker in a "neutral" environment as described in Example 4.

Completion of this example problem will result in a *combination* of free air temperatures and surrounding surface (e.g., office wall) temperatures that will provide the desired working environment.

Objective 5: Understand Ergonomic Biodynamics as Applied to the Practice of Ergonomic Engineering. In Chapter 7, the student's prior exposure to biodynamic mechanics (Chapter 4) will be extended to describe (in part) the motion of the particular body area of the human operator as the person performs a task.

EXAMPLE 5 (adapted from Example 7.7)

An industrial firm is developing a wrist brace for vocational bowlers to obtain better wrist support and control when swinging and releasing their bowling ball. A professional bowler is performing this task and his position in time is being filmed in right profile view. Forearm position coordinate data (as a function of time) are acquired from an elbow marker (E), a wrist marker (W), and a hand marker (H) as shown in Figure 1.7.

You are the human factors engineer who had previously performed the data analysis in Examples 7.2, 7.4, 7.5, and 7.6. You have now been assigned to develop the preliminary engineering requirements for the bowler's wrist brace as part of a Phase I systems engineering effort (you may wish to review Section 2.1 at this point).

a. Interpret the kinematic data of the forearm and hand segment from Examples 7.2 and 7.4.

Figure 1.7 Forearm position markers for the elbow (E), wrist (W), and hand (H) as described in Example 5.

b. Interpret the joint reaction forces and net muscle moment data at the wrist and elbow from Example 7.5.
c. Interpret the muscle mechanical power and work performed on the hand and forearm segments from Example 7.6.

Successful completion of the example will provide a quantitative description of human performance during an important phase of this task. This understanding will translate into a set of engineering specifications that will quantitatively address both wrist support and arm-hand-ball control when swinging and releasing the bowling ball.

Objective 6: Understand Quantitative Workload Analysis as Applied to the Practice of Ergonomic Engineering. In Chapter 8, quantitative methods are presented that will allow the human factors engineer to analyze the workload experienced by a human operator when performing a task. Workload is the amount of effort that a human operator experiences (i.e., that occurs *inside* the person's body) in response to a specific amount of external work (i.e., that occurs *outside* of the person's body).

The material presented in Chapter 8 will extend the student's prior exposure to bioelectronics (Chapter 5) since quantitative workload analysis is dependent (in part) upon electronic devices that measure a human operator's physiological reaction to the external work being performed.

EXAMPLE 6 (adapted from Example 8.6)

A field engineer is assisting in the evaluation of a manual "crank" generator, which can be used to power a portable communication set. The engineer sits in a chair with the box-like generator at table (chest) height and holds a crank handle (at each side of the box) in each hand. That person then proceeds to apply tangential force ($\mathbf{F_L}$) to rotate each crank handle (offset 180° from each other). The radius of handle rotation is 0.2 M ($\mathbf{F_L}$ = 20 N) and the rotation rate is 360° per 1.25 s. In order to evaluate the required muscular strength and aerobic endurance for this dynamic work, the following are measured with an electropneumogram (see Chapter 5): respiratory cycle = 3.75 s, and peak air flow rate = 0.545 L/s. The maximal oxygen uptake = 2.1 L/min was measured at a prior time.

a. Calculate the external work and power, the customary pulmonary parameters, and the aerobic index.
b. Regarding energetic assessment of this dynamic workload [and using the results from part (a)], calculate the customary energetic parameters.

Successful completion of this worked example will allow the human factors engineer to specify this human operator task as mild intensity dynamic work.

Objective 7: Understand Neuromuscular Control Systems as Applied to the Practice of Human Control Engineering. In Chapter 9, the human subsystem (of the human-technological system presented in Figure 1.1) will be characterized at the

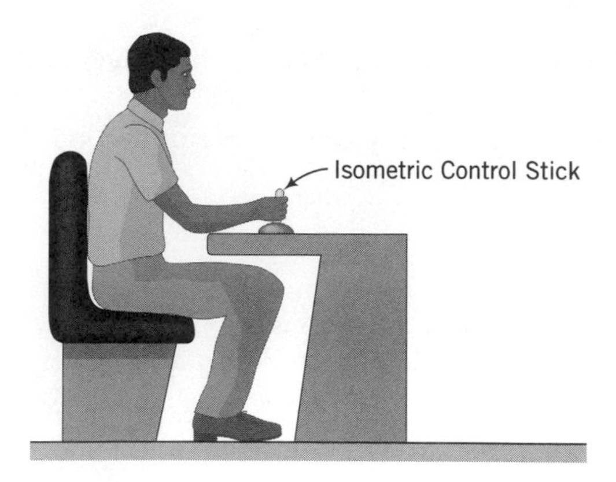

Figure 1.8 Forklift operator using an isometric control stick as described in Example 7.

individual element level. By understanding human operator control at the elemental level, the human factors engineer will realize *both* the capabilities *and* the limitations of central nervous system control of human skeletal muscle.

Worked examples using a computer simulation approach will allow the human factors engineer to evaluate, specify, and design specific types of control devices by which the human operator subsystem interacts with (i.e., controls) the technological subsystem. One such control device is an "isometric" control stick. Such a device responds to an increasing amount of applied force, but the control stick itself remains rigid (i.e., does not respond with any amount of stick movement).

EXAMPLE 7 (adapted from Example 9.6)

A forklift operator is seated in the cab and is controlling the forklift position with an isometric control stick as shown in Figure 1.8. The operator has the upper arm extended downward and the lower arm extended straight ahead (of the person's body). The operator grips the control stick handle and pushes (isometrically) straight forward using the anterior shoulder muscle. By doing so, the forklift operator is able to raise the fork with the isometric control stick. The operator has just engaged a pallet (with a load on it) at ground level using the forks of the lift. The operator now wants to lift the pallet and load an upward vertical distance, $\Delta x = 2$ M.

a. Write a computer simulation program of this human-technological control system when performing a unit displacement task using an isometric control stick (as depicted in Figure 9.22).
b. Plot the time course of the displacement (x_0) of the load on the pallet ($x_0 = 0$ at ground level), and the displacement error (x_e) during the performance of this task.
c. Plot the time course of the central nervous processor force command ($\mathbf{P_R}$) and the neuromuscular actuator output force ($\mathbf{P'}$) during the performance of this task.

Successful completion of this computer simulation example will indicate that the human operator control strategy is a *reciprocal function* of both the control device characteristics and the technological device characteristics.

Objective 8. Understand Human Operator Control as Applied to the Practice of Human Control Engineering. In Chapter 10, the human operator will be characterized as an information processor. This will allow the human factors engineer to evaluate alternative equipment design configurations in the context of the human operator's mental and physical capabilities.

EXAMPLE 8 (adapted from Example 10.4)

A ship's sonar operator is sitting at a console viewing a sonar screen with an information content of $H_s = 5$ bits. The human factors engineer is evaluating two types of operators (Operator A and Operator B) that are being used in order to evaluate two different sonar station-console switch configurations (State 1 and State 2). When a suspect object suddenly appears on the sonar screen, the operator will immediately alert the computer system (and the watch supervisor) by manually depressing an alert button.

Operator A is a well-trained, experienced, and fast-reacting individual (β = 9.2/s), and Operator B is a recently trained, less experienced, and slower-reacting individual (β = 5.0/s). In the first proposed switch configuration (State 1), the center of a square button (of height H_1) is to be located at a vertical height (A_1) of 0.5 M above the operator's hand and will require a coordinated hand-arm-shoulder movement (α_M = 4.5/s) to reach and depress the push button switch. In the second proposed switch configuration (State 2), the center of the square button (of height H_2) is to be located at a vertical height (A_2) of 0.1 M and will require only a hand-wrist movement (α_M = 10/s) to reach and depress the push-button switch.

a. If the average diameter of an operator's index finger height (h) is .01 M, find H_1 and H_2 so that both switch configurations have an index of difficulty (I.D.) of 5 bits.
b. Develop the relevant feed-forward transfer function (human operator informatic model) for each human operator-state combination.
c. Solve for the action time that the human operator informatic model [from part (b)] would predict for each of the four operator-state combinations.

Successful completion of this worked example will demonstrate a very interesting trade-off in equipment design and human operator capabilities. The *less* experienced operator (when using the *second* switch configuration) will have an action time comparable to a *more* experienced operator, who is using the *first* switch configuration.

FURTHER INFORMATION

J.D. Bronzino (ed.): *The Biomedical Engineering Handbook.* CRC Press. Boca Raton, FL. 1995.

D.G. Newnan (ed.): *Engineer-In-Training License Review.* 14th Edition. Engineering Press. Austin, TX. 1997.

C.A. Phillips: *Human Factors Bioengineering*. ClassNote Publications (Wright State University). Dayton, OH. 1995.

C.A. Phillips and D.W. Repperger (eds.): *Human Interaction with Technological Systems* (AUTOMEDICA 16:4). Gordon and Breach. New York. 1998.

G. Salvendy (ed.): *Handbook of Human Factors and Ergonomics*. 2nd Edition. John Wiley. New York. 1997.

CHAPTER 2

Human Systems Design and Modeling

The human-factors-engineering systems approach examines a system from the top down instead of from the bottom up. The human factors engineer first directs attention to the human-technological system as a black box that interacts with the environment. Subsequently, the human factors engineer considers how the subsystem black boxes combine to create the overall system. The final level of analysis is with respect to the individual units that combine to form the subsystems. This process is applicable when bringing a system into being as well as when improving already existing systems. By approaching the total system in this manner, the human factors engineer will be able to achieve a more thorough analysis when interfacing human subsystems with technological subsystems. This structured approach of defining the entire system, constituent subsystems, and finally the individual units, will ensure that the human factors engineer has considered all of the relevant functional relationships. The individual units are important, and perhaps initially more interesting to the traditional engineer. However, for the human factors engineer the highest priority is given to the operational purpose of the complete human-technological system, which will be achieved by means of a thorough understanding of all the functional relationships interconnecting the human subsystem with the technological subsystem.

2.1 HUMAN SYSTEMS DESIGN AND DEVELOPMENT

It may be initially convenient for the human factors engineer to classify the particular human-technological system as either static or dynamic, as well as either closed or open. A *static* system is one in which the technological subsystem has structure and activity, but the human subsystem is without activity. For example, many biomedical engineering systems that involve medical imaging of the human anatomy or physiological data recording are commonly performed with the human in a quiet, passive mode. In these human-technological systems, the human subsystem frequently is viewed as a signal generator, and the major engineering focus is upon the technological subsystem.

A *dynamic* system involves a technological subsystem (with structure and activity) and a human subsystem (with corresponding structure and activity). For example, if the medical imaging system were to require human operator interaction to achieve the optimal medical imaging, the human (operator) subsystem and the technological (medical imaging) subsystem would collectively represent a dynamic system. A large-scale human-technological system might be a production plant, combining a building, workers, supervisors and administrators, production parts, and assembly tasks.

A *closed* human-technological system is one in which there is negligible interaction between the system and its environment. Such closed systems are self-contained, and their operational performance is not affected by environmental conditions or alterations. An example might be a human operator inside a diving bell. The technological (diving bell) subsystem is designed to be closed in the sense that it must protect the vulnerable human being from a very hostile and adverse environment. When operating in steady state, the human subsystem is isolated from changing environmental factors such as external water pressure, ambient luminance, and thermal variations. The technological (diving bell) subsystem must interact with the environment since its function is to isolate the human subsystem from a wide range of adverse environmental variables.

An *open* human-technological system will allow information, energy, and matter to transition across the system boundaries. Consequently, open systems will interact with the environment, in the sense that the human operator subsystem will have its performance modified as various environmental factors are altered. An example would be a human driving an automobile, in which case the human operator performance would be modified by such environmental factors as tortuosity of the road, rain and fog, and traffic density.

The preceding discussion indicates that it is not always a simple distinction whether to classify a system as static or dynamic and/or closed or open. However, with respect to human-technological systems, it is convenient to apply these terms with respect to the human subsystem.

The requirement for human systems design and development is due to the fact that many of the design engineers in one or more of the discipline-specific engineering fields (e.g., industrial engineering, mechanical engineering, or aeronautical engineering) are not adequately experienced with respect to the human subsystem so that all of the functional relationships of the human-technological system are evaluated in a correct and expeditious manner. Human systems design and development is based upon the standard concept of systems engineering as defined in numerous texts. This chapter will emphasize the specific aspect of human systems modeling.

a. First Stage: Conceptual Design

Conceptual design is the first stage of human systems design and development. Human systems modeling is a significant element during this stage. Three substages constitute conceptual design. First is the definition of system requirements. Second is the preliminary systems analysis. Third is the development of a set of system specifications.

The human factors engineer must begin by defining the basic requirements for the human technological system with respect to the input criteria for design. Six key questions should be asked:

1. What is the system to achieve with respect to operational performance criteria (speed, accuracy, length of performance, product output, etc.)?
2. When will the system actually be required? Under what circumstances? What is the anticipated operational lifetime?

3. How often is the system to be used? Frequency, duration, duty cycle, etc.?
4. How is the system to be distributed in space? Where will the various subsystems of the overall system be located?
5. What are the requirements for effectiveness that the system should demonstrate? Effectiveness may be quantitatively defined by such factors as cost effectiveness, system dependability, and overall system reliability.
6. What are the environmental requirements (pressure, temperature, humidity, vibration, etc.)?

Conceptual design continues with the preliminary systems analysis. With the definition of the system requirements, the human factors engineer must then consider how the human-technological system is to be made operational if it is to satisfy the required need. With a set of requirements as a guideline, the human factors engineer will investigate and define the different technical methods by which the system requirements can be satisfied. This is a crucial phase since the decision must be made concerning whether it is feasible to proceed further with the human-technological system development. This preliminary system analysis involves six constituent elements:

1. Definition of the problem. A *precise* definition of the problem must be formulated. A problem that may appear obvious at first glance may still be rather difficult to define clearly and precisely. Subsequent analysis will defend this definition of the problem.
2. Identification of the feasible alternatives. It is then necessary to formulate all possible alternative solutions that will address the problem. Initially, all conceivable design alternatives should be considered for completeness sake, with subsequent elimination of solutions not desirable. The remainder will then be subjected to evaluation.
3. Selection of the evaluation criteria. These will be the quantitative parameters used in selecting among alternative solutions. With respect to the overall human-technological system, such parameters might include the operational performance and the cost effectiveness among others.
4. Application of the modeling methods. The human factors engineer will then apply analytical techniques in the form of a systems model. A model is a simplified representation of certain aspects of the real system (e.g., a doll is a model of the human body).

A mathematical model is a model created using a set of mathematical equations, functions, and logic rules that describe the system (e.g., $F = ma$, for an object in motion). A computer model is a computer program that embodies the equations, functions, logic rules, and solution methods to the equations of the mathematical model. An example of where computer models have been effective in human systems design are the neuromuscular control system models described in Chapter 9. Model development, in general, is described in Section 2.2.a.

5. The generation of input data. The human factors engineer must now specify the system requirements for appropriate input data. These data requirements can be identified from both the evaluation criteria and the parameter requirements of the model.
6. Model simulation. Simulation is the exercising of the model and the obtaining of results. When one takes all of the modeling processes and runs the model

for a number of different values, the results obtained are the simulation. Once the required data are input into the model, the output can be analyzed with respect to feasibility of the design alternative. Mathematical models will also allow for identifying the system robustness with respect to individual parameter sensitivity.

At the completion of the preliminary system analysis, a major decision must be made regarding the feasibility of the proposed system. A recommendation to proceed should also be accompanied by an identification of areas of potential risk.

The final substage of conceptual design is the development of a set of system specifications that are generated and compiled as the governing technical document.

b. Second Stage: Advanced Development

The second major phase of human system design and development is the advanced development stage, which basically represents the preliminary system design. Advanced development is composed of three substages. The first is the allocation of requirements. The second is the trade-off and optimization. The third is a set of detailed design specifications.

The first substage of advanced development is the allocation of the human-technological system factors to the various subsystems and their constituent units. For example, in a human-technological system, if the allowable operator response time of the human subsystem is X, what should be the allowable central nervous processor unit response time, and what should be the allowable neuromuscular actuator unit response time? As an alternative example, if the required force output of the human operator (human subsystem) is $\mathbf{F}$, what is the allowable upper extremity neuromuscular actuator unit force ($\mathbf{G}$), and what is the allowable lower extremity neuromuscular actuator unit force ($\mathbf{H}$)? In conclusion, the allocation of requirements will provide the initial guideline to the human factors engineer for developing subsystem performance and its constituent unit performance, which is compatible with the overall system requirements.

The second substage of advanced development is trade-off and optimization. The allocation of requirements to the various units and subsystems will establish the limitations and constraints for their design (the maximum and/or minimum values to which subsystem and unit design must comply). Various design configurations may satisfy the boundary requirements, and the human factors engineer must identify the optimal configuration by using various analytical methods. The evaluation process of various design alternatives relies on the use of mathematical techniques in the form of human operator and technological system models. Models are of central importance in this second substage of advanced development.

Modeling allows experimentation that otherwise would place humans at risk (e.g., simulating car crashes). Modeling allows for the study of systems that are difficult to observe (e.g., neuromuscular phenomena). It allows for the study of systems that are difficult to experiment with directly (e.g., aircraft collisions). Modeling permits nondestructive testing and allows many experiments to be performed in a short time merely by changing parameters. Modeling also allows a better understanding of the system and produces sensitivity tests to answer "what if" questions about the system in question. Finally, modeling may replace ethically sensitive experimentation (e.g., in human performance under very hazardous environmental parameters or operational circumstances).

The third substage of advanced development is the construction of a set of detailed design specifications. The human factors engineer will define a set of detailed specifications (e.g., developmental human operator and/or task-process specifications). This set of detailed specifications will then represent the input required for the detailed design and development of the various subsystems and individual elements.

c. Third Stage: Detail Design and Development

The final stage of human system design and implementation is detail design and development. This is a sequential process with four substages:

1. The description of the systems, units, assemblies, subassemblies, and parts (Figure 2.1) of the technological subsystem and the human subsystem (e.g., technical data, human operator requirements and training).
2. The preparation of design documentation (e.g., detail production drawings, human specification, and related technological databases).
3. The development of the engineering model and/or a prototype model of the human-technological system for the purpose of verification of the detailed design.
4. Tests and evaluation of the physical model of the human-technological system, and redesign, retesting, and reevaluation of the system as necessary.

The necessary and sufficient element in the detailed design activities is the establishment of the design team. Even the simplest project design requires an integrated team representing management, engineering, and supporting personnel. This team will transform the second-stage document (consisting of a detailed set of specifications) into a system that is ready for production. The project engineering team should include a balance of engineering technical expertise, engineering technical support, and administrative support. For less complex projects, this would be represented by the human factors engineer (focusing on the human subsystem) and a discipline-specific engineer (focusing upon the technological system). Engineering technical support might be anything from a secretary/librarian to a partial effort from the appropriate electrical/mechanical/laboratory technician. In the case of a

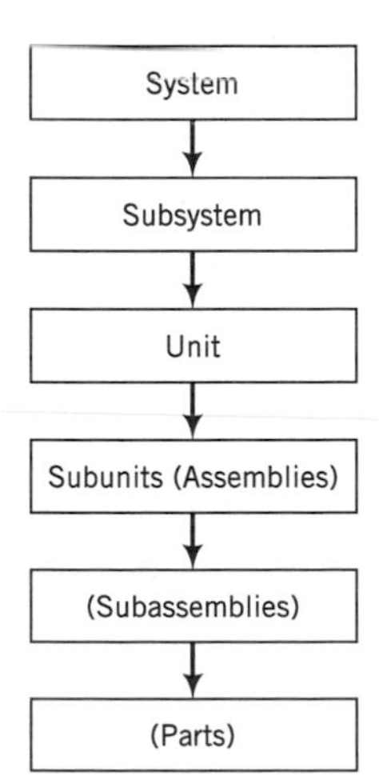

Figure 2.1 Levels of systems design and development [parenthetical terms denote production levels].

less complex human-machine system, the administrative support might represent a marketing individual who would transition the system into an existing product.

For the more complex system, the engineering technical expertise would reside with a group of professional engineers, consisting of one or more industrial engineers, mechanical engineers, human factors engineers, aeronautical engineers, system engineers, and/or other engineers as appropriate. The engineering technical support would be provided by the design draftsman, a technical librarian, and/or technical writer, appropriate laboratory technicians, test technicians, computer programmers, and others as appropriate. The administrative support for a more complex human-technological system would require representatives from marketing, budgeting, purchasing, and industrial/public relations as necessary.

As the detailed design and development proceeds, the engineering documentation will progress from simple stages to the required level of detail necessary for the production group to successfully manufacture the system. The final goal of the assembled design team is to provide detailed production design layouts that specify supplier data and manufacturing tolerances. These detailed production design drawings must be prepared for the entire system proceeding hierarchically from subsystems, units, subunits (assemblies), subassemblies, and parts (Figure 2.1).

Recall that human systems modeling is an essential element of the various stages of human system design. This chapter now proceeds with a description of the developmental steps in the formulation of a model and a review of model applications.

2.2 HUMAN SYSTEMS MODELING

Biomathematical modeling is a technique by which the human factors engineer applies the physical equations of engineering science to describe the human-technological system. Biomathematical modeling is useful to the human factors engineer because the equations of the physical sciences and engineering sciences are traditionally applied to describe the behavior of inanimate objects. Biomathematical modeling, therefore, is a transitional technique that allows the human factors engineer to apply the same equations to animate objects as well. As we shall see in this section, biomathematical models are of various types and occur at various levels. In order to perform biomathematical modeling effectively, the human factors engineer must exercise skill and judgment so as to achieve an optimal solution. Some of this judgment and skill will result from the actual practice and experience of the human factors engineer as that person works with real-world human-technological systems. The purpose of this section is to overview the various categories and levels of application of biomathematical models and present a brief example at each level of application.

It was once said that a good engineer makes a thousand different mistakes while a poor engineer makes the same mistake a thousand times. You will begin building your experience with biomathematical models as you proceed through the remaining chapters of this text. In each chapter you will observe different applications of biomathematical models depending upon the particular aspect of the human-technological system to be analyzed. As you commence and conclude each chapter, briefly review the types of biomathematical models used and the level at which they were applied in that chapter. This will develop an understanding of model application in a systematic manner.

a. Introduction to Human Systems Modeling

Models can be broadly classified as either deterministic or stochastic. *Deterministic models* have outcomes and output that are a direct consequence of their initial conditions. With the same initial conditions, the same output is always produced. Examples would be Newton's second law, $F = ma$ (Chapter 3), and Ohm's law, $\mathbf{E} = IR$ (Chapter 5). *Stochastic models* are those whose outcome is dependent on random effects, such as rolling a die. The input is the same, but the result is not uniquely determined.

Model types may further be subdivided by the type of mathematical assumptions used in construction of the model. As most models contain several variables, the grouping of the variables is widely used as a description of the type of model. Models in which multiple parameters are grouped together into a smaller number of variables are called *lumped parameter models*. An example of a lumped parameter model is the spring-mass model to represent the elastic nature of biological tissue. In the spring-mass model, many factors influencing the elastic response of the tissue are grouped into two broad mechanical types, springs and masses. Using this model, simple, well-defined mathematical functions can be used to express the complex nature of the tissue's elastic response (Chapter 4). Such models have the advantage of well-defined mathematical precedents but may suffer from lack of realism, especially in complex systems.

In a *distributed parameter model,* variables are not collected and generalized as in the lumped parameter model. In this type of model, variables are allowed to influence changes in the system as they would in a practical event. These types of models are often much more complex than lumped parameter models and are usually implemented as a computer model, involving software solutions. Specific models see wide application in vehicle biodynamics and have been used to re-create the impact response of the human body in car-pedestrian accidents.

Models may be further classified as to the type of time duration allowed for in the model. In a *discrete model,* model parameters are allowed to change over a finite, defined time interval. In a *continuous model,* time changes are continual and require simulation using real-time parameters for the model. *Discrete models* are often used for several reasons. In many situations, change really is discrete; it occurs at well-defined time intervals. Continuous change can often be approximated very well by a discrete change. Discrete change is often easier to work with when using digital computer computations. Finally, data are usually discrete rather than continuous, even when the underlying model is continuous. An example is analog-to-digital conversion in the processing of an electromyogram (EMG). The use of a discrete model is necessary because of the limitations of the available data (Chapter 5). *Continuous models* are widely found in automatic control theory. For example, human operator control may be represented as a physically realizable, linear and time invariant system, which is mathematically modeled by transfer functions (Chapter 10).

Development of a model and simulation requires a structured and sequential approach. All modeling is an approximation and not the actual system. Consequently, a four-stage process is followed.

The first stage is the identification of the real problem. Development of the model is started by stating exactly what must be known. The objectives and purpose for the model are defined and the model is then classified as either deterministic or stochastic. Further refinement of the model type is accomplished by classifying the model as presented above. At this point, constants and variables for the model are identified and precisely defined.

The second stage of modeling involves formulation of the mathematical model itself. A simple system is first defined and the objective and bounds of the system are considered. Most models are begun rather simply and complexity is added as needed. Approximations used in the model are then invoked and documented and relevant factors influencing the model are listed. Inputs and outputs of the system are then defined. The relationships between each of the systems variables are developed.

The third stage of model development involves obtaining the mathematical solution. Algebraic and numerical methods are needed to solve the sets of equations that represent the model. Many different resources are available to the engineer. Along with classical mathematical theory, currently many different types of computer application packages are available to help with solving complex mathematical relationships. Spreadsheets, software packages, and programmed models are all widely available to aid in the solution. As an example, MATLAB has been used to solve various mathematical models developed in Chapters 9 and 10.

The fourth stage of model development is the model validation. The model results must be compared to the real world, as validation of the model results is of paramount importance. Results obtained from various validation tests (see Example 2.1) must be compared with the mathematical model to determine if the mathematical model is appropriately accurate. By checking the signs of variables and observing increase and/or decrease of variables, a real-world representation of the mathematical model can be assessed. Results from the model should be compared with actual data from physical experiments to determine the reliability of the model. Statistical analysis may also be performed to gain confidence in the results predicted by the model.

EXAMPLE 2.1

The validity of system equations may be confirmed by performing a unit check after the final numerical solution is obtained. A unit check is one method for testing the validity of a derived equation. Outline the standard methods available to the human factors engineer for evaluating equation validity. Arrange these methods in a progressive order for validity checking purposes.

SOLUTION 2.1

A progressive sequence of methods for checking the validity of derived equations consists of three stages: 1. checking for the law of parsimony; 2. testing for internal consistency; and 3. testing for external consistency.

1. Checking that the equation has been formatted to represent the simplest equation that describes the system. This is known as the "law of parsimony."
   a. Check that the equation is written in its simplest form. Evaluate all equation terms in order to determine whether some could be collected and/or cancelled. Check for complex functions that might be stated more simply.
   b. Check that the equation is written using only primary parameters. If an equation term is not a primary parameter but a composite of two or more other primary variables, then the equation could be concealing an identity. Alternatively, the relationship between the dependent and independent variable could be an artificial correlation.

   c. Check that the equation cannot be further simplified. Are there one or more simplifying approximations that may still be invoked? Check that further simplification will not result in significant loss of accuracy.
2. Testing for internal consistency.
   a. Check that the equation is dimensionally correct. This will require that each equation variable be expressed in terms of mass (kilogram), length (meter), and time (second).
   b. Perform a unit check on the equation. Determine that the same set of units has been used consistently throughout the equation. When variables have different sets of units, check that appropriate numerical conversion factors are included in the equation.
   c. Check the equation at its limits. The equation is checked by first letting each appropriate variable approach zero. The equation is then checked by letting each appropriate variable approach infinity. This will result in a series of special cases and two outcomes are possible. One is that the equation appears reasonable at these limits for these special cases. Otherwise, there are other limits for which the equation appears reasonable, and these limits should be clearly stated.
   d. Test the equation by numerical substitution. This test should be performed in duplicate or triplicate by developing sets of arbitrary parameter values that individually would appear reasonable. When appropriate numerical substitution has been performed, does the equation predict a correct value?
3. Testing for external consistency.
   a. The equation is compared with other similar equations that have been previously derived. Determine whether observed differences might be reasonably anticipated from the use of different models. Also determine whether there is at least proportional agreement based upon the variation of different factors.
   b. The equation should be compared with real experimental data. Check the degree of correlation by means of the root-mean-squared error method (for example). Determine whether the equation predicts data trends within a reasonable degree by calculating the coefficient of correlation (for example). Where deviations are observed between equation prediction and actual data, consider whether this might be acceptable with respect to the simplifying approximations that have been used.

---

b. Model Applications: Lower System Level

Beginning at the *molecular level*, models describing the structure and function of microscopic chemical structures are many and varied. There are models that describe the structure of individual atoms, such as the Bohr model, which describes electron configurations as discrete energy levels. Models may further be constructed to simulate the properties of larger molecules, both organic and inorganic.

At the molecular/subcellular level, molecules interact with one another by means of various chemical reactions. The most basic of these reactions is Substance A being converted into Substance B with no back reaction. This is an example of a first-order reaction resulting in a rate of change (of concentration) characterized by an exponential decaying function (for Substance A). This particular type of model also predicts a monotonic curve that rises with ever-decreasing slope to an asymptote (for Substance B). The overall reaction is characterized by a first-order rate constant. Second-order reactions occur when Substance A combines with Substance B to form Product C and Product D. The second-order rate constant determines the rate of the concentration change of the system. An important example is the combustion of a sugar (glucose) molecule:

$$C_6H_{12}O_6 + 6O_2 = 6CO_2 + 6H_2O \tag{1}$$

This important example of human operator bioenergetics is presented in Chapter 6.

With the addition of enzymes, chemical reaction dynamics began to change. The catalyzed reaction is governed by a rate constant that is related to the initial rate constant. To model this time-varying concentration rate, several models were historically proposed. The earliest model considered four differential equations of the individual substances' concentrations. This treatment was not adequate. Later models gave more rigorous derivations, but they did not provide a good predictive model. Ultimately, a fairly good model, the Michaelis-Menten model was devised. This model defines the original rate constant (which is accounted for as the Michaelis constant of the equation) of the uncatalyzed reaction. This model then predicts a curve with the following characteristics: An intermediate asymptote is reached, and the half-way point of the reaction can be identified with the proper parameter substitutions.

The next level, or organization, is the *cellular level*. Models relating the mechanical properties of cells and tissues, transport properties, chemical properties, and rheological properties are all well documented in the relevant literature. One example is the relationship between electrical activity of skeletal muscle (EMG) and the isometric force developed by the muscle as described in Chapter 8. It is the release of the calcium ion (Ca^{++}) from the sarcoplasmic reticulum (SR) that is responsible for this mechanism known as electromechanical coupling.

A specific model has been developed to model this mechanism using a computer simulation. This model describes muscle fiber calcium cycling based upon the kinetics of sarcoplasmic reticulum Ca^{++} release channels. The SR channel regulating mechanism includes two types of Ca^{++} binding sites: 1. low affinity sites with high binding rates, which regulate the opening of Ca^{++} channels and 2. high affinity sites with low binding rates, which regulate their closing. The model developed describes intracellular phenomena related to the release of Ca^{++} from the SR, the recovery mechanism of Ca^{++} availability, and their interval dependence. The results of the model support the basic hypothesis that Ca^{++} release in the SR channels is controlled by Ca^{++}-sensitive activating and inactivating sites. This model does suffer from several approximations, the most basic of which is that there is no physical barrier which can prevent the free movement of the calcium ions in the cytoplasm. This model does correlate well with the available experimental data.

Tissue and organ is the next level of organization for modeling. When modeling a local region or organ of the body, the substantial inward and outward fluid flows of blood (lymph flow is usually neglected) across the compartment boundaries must be accounted for. The model of particular interest to the human factors engineer is the distribution and concentration (availability) of oxygen transported by the circulation (blood) to specific organ tissues (e.g., brain, skeletal muscle, etc.). The simplest example is that of the concentration of a solute (e.g., oxygen) across a two-compartment organ model. In this model, the organ model is considered to consist of blood and tissue subregions, each of which is assumed to be homogeneous.

Blood and tissue are also assumed to be in equilibrium. A mass balance for this system is:

$$\dot{Q}(C_{Bi} - C_{Bo}) = V_T \frac{dC_T}{dt} + V_B \frac{dC_B}{dt} \tag{2}$$

where $\dot{Q}$ is the volumetric blood flow rate; C_{Bi}, C_{Bo} are the concentrations of solute in the inlet and outlet bloodstreams; C_B, C_T are the concentrations of the solute in the blood and tissue regions; and V_B, V_T are the volumes of the blood and tissue regions.

In many regions of the body, the volume of blood is much less than the volume of tissue and hence the total compartment volume is roughly equal to V_T. Introducing these simplifying approximations, we can reduce the right-hand side of equation (2) so that:

$$\dot{Q}(C_{Bi} - C_{Bo}) = V_T \frac{dC_T}{dt} \tag{3}$$

Also, because equilibrium exists between blood and tissue, C_B and C_T can be related by the expression:

$$C_B = \alpha C_T \tag{4a}$$

or alternatively:

$$C_T = \frac{1}{\alpha} \cdot C_B \tag{4b}$$

where α is a dimensionless "partition coefficient" and represents the ratio of C_B to C_T at equilibrium, and we have approximated this as a linear equilibrium relation. Define the outlet concentration in the blood as C_B, so that:

$$C_B = C_{Bo} \tag{5}$$

Upon substituting equations (4b) and (5) into equation (3), the mass balance is then:

$$\frac{V_T}{\alpha} \frac{dC_B}{dt} = \dot{Q}(C_{Bi} - C_B) \tag{6}$$

For the case where a solute (e.g., oxygen) is suddenly introduced at a constant amount into the blood, C_{Bi} goes from zero to a constant value C_{Bi}^0 at $t = 0$, and equation (6) has the solution:

$$C_B = C_{Bi}^0 \left[1 - \exp\left(-\frac{\dot{Q}\alpha t}{V_T} \right) \right] \tag{7}$$

This model demonstrates the concentration of a solute in an organ (e.g., brain or muscle) as a function of time predicted by the two compartment model. This model is a reasonable approximation of experimental data but suffers because of the approximation that the concentrations of solute in the blood and tissues follow a linear relationship, which is rarely the case.

EXAMPLE 2.2

Equation (6) may be used to solve for the concentration of a solute in the blood (C_B) as a function of time:

$$C_B = f(t)$$

where C_B is the dependent variable and time is the independent variable. Check the validity of equation (6) using the three-stage process defined in Example 2.1. Your solution should include: (a) checking that the equation is in its simplest and most concise form; (b) testing for internal consistency; and (c) testing for external consistency.

SOLUTION 2.2

a. Check for the law of parsimony. Begin by restating equation (6) and inspecting it carefully:

$$\frac{V_T}{\alpha}\frac{dC_B}{dt} = \dot{Q}(C_{Bi} - C_B) \tag{6}$$

Careful inspection will then reveal that equation (6) is actually a first-order, linear differential equation with constant coefficients. First, rearrange equation (6) in a form more consistent with this type of differential equation:

$$\frac{dC_B}{dt} + \frac{\dot{Q}\alpha}{V_T}C_B = \frac{\dot{Q}\alpha}{V_T}C_{Bi} \tag{i}$$

Second, from inspection of equation (i), it is apparent that the three constant coefficients may be further collected into a single term representing a reciprocal time constant:

$$\frac{1}{\tau} = \frac{\dot{Q}\alpha}{V_T} \tag{ii}$$

Third, substitution of equation (ii) into equation (i) results in the final simplified and condensed form of equation (6):

$$\frac{dC_B}{dt} + \left(\frac{1}{\tau}\right)C_B = \left(\frac{1}{\tau}\right)C_{Bi} \tag{iii}$$

b. Test for internal consistency. First, perform a unit check of the final simplified form of the equation. Simple inspection of equation (iii) indicates that each term in the differential equation is the ratio of a concentration with respect to time. However, a unit check should also be performed on equation (ii). Recall that blood flow ($\dot{Q}$) has the units of volume per time and that α is dimensionless. Making these substitutions into equation (ii) results in:

$$\frac{1}{T} = \frac{V(1)}{TV} \tag{iv}$$

So both equation (ii) and equation (iii) pass the unit check test. Second, test the equation by letting each appropriate variable first approach zero, and then approach infinity. Inspection of equation (iii) indicates that there are only two relevant variables: the concentration of the solute in the blood at the inflow to the tissue volume (C_{Bi}) and the system time constant (τ). The result is four limit checks which proceed as follows.

When $C_{Bi} \to 0$ (at t_0), equation (iii) reduces to:

$$\frac{dC_B}{dt} = -\frac{1}{\tau}C_B \tag{v}$$

which is solved as a function of time:

$$C_B(t) = C_{B\cdot 0}e^{-t/\tau} \tag{vi}$$

where $C_{B\cdot 0}$ is the concentration of solute already in the blood flowing through the tissue at t_0 (when the concentration of solute in the blood at the inflow point goes to zero).

When $C_{Bi} \rightarrow \infty$ (at t_0), equation (iii) is solved for a unit step of magnitude, $C_{B\infty}$ (at t_0). This results in a solution analogous to equation (7):

$$C_B(t) = C_{B\cdot\infty}[1 - e^{-t/\tau}] \tag{vii}$$

Equation (vii) goes to infinity as $C_{B\cdot\infty} \rightarrow \infty$, which is the expected result.

When $\tau \rightarrow 0$, first rearrange equation (iii) as:

$$\tau\frac{dC_B}{dt} + C_B = C_{Bi} \tag{viii}$$

Equation (viii) then reduces to:

$$C_B = C_{Bi} \tag{ix}$$

which is reasonable and expected for a very small (near zero) time constant [or when time (t) approaches infinity].

When $\tau \rightarrow \infty$, equation (iii) reduces to:

$$\frac{dC_B}{dt} = 0 \tag{x}$$

which is reasonable and expected for a very large (near infinity) time constant.

Third, equation (iii) should be checked at least twice by numerical substitution.

c. Test for external consistency. First, identify a potentially similar equation. Equation (63) of Chapter 5 is the Fick equation, which may be described as follows. The Fick equation is a steady-state solution of a two-compartment model derived to calculate the cardiac output ($\dot{V}_{CO}$), which is the blood flow through the heart and lungs. One compartment is the "blood" volume and the other compartment is the "lung" volume through which a steady-state oxygen flow ($\dot{V}_{O_2}$) occurs as the person breathes in and out. The solute carried by the blood is "oxygen" and this solute (oxygen) is at a concentration difference ($C_{O_2a} - C_{O_2v}$) between the blood inflow and outflow from the lung.

Equation (63) of Chapter 5 is:

$$\dot{V}_{CO} = \frac{\dot{V}_{O_2}}{(C_{O_2a} - C_{O_2v})} \tag{8}$$

Second, check to see if an equation analogous to equation (8) can be written for the steady-state blood flow ($\dot{Q}$) of equation (6):

$$\dot{Q} \approx \frac{\dot{V}_{O_2}}{(C_{O_2a} - C_{O_2v})} \tag{xi}$$

Begin by substituting equation (ii) into equation (viii):

$$\left(\frac{V_T}{\dot{Q}\alpha}\right)\frac{dC_B}{dt} + C_B = C_{Bi} \tag{xii}$$

Recall that $C_B = C_{Bo}$, and rearrange equation (xii):

$$\dot{Q} = \left(\frac{V_T}{\alpha}\right)\left[\frac{\frac{dC_B}{dt}}{(C_{Bi} - C_{Bo})}\right] \tag{xiii}$$

which, when compared to equation (xi), indicates that the steady-state dC_B/dt for the oxygen entering the blood (from the lung) as the blood flows through the lung, is analogous and proportional to the steady-state $\dot{V}_{O_2}$ for volume of oxygen flowing into the lung (during a person's breathing cycle) via equation (xi).

Second, the equation should be compared with actual experimental data. Goodness of fit could then be tested by either the root-mean-squared error method or by the calculation of the coefficient of correlation.

In the organizational hierarchy of the human body, the different specialized organs are combined into an *anatomical and physiological system*, which is the next level of organization for modeling. "Physiological" refers to the functional elements of a system, and "anatomical" refers to the structural elements of the system. By their very nature, physiologic and anatomic systems are complex and subject to a multitude of variables and influences. This complexity of both structure and function makes these systems challenging targets for modeling efforts.

A very useful series of models has been developed for the analysis of the human musculoskeletal system. For example, the biomechanical (biostatic) model employed in Chapter 3 may be subdivided into three levels of models for the purpose of sequential analysis. The first level is an *anatomical* model. This model represents a rendering of the musculoskeletal system with complete anatomical fidelity. In addition to skeletal and muscular tissue, nerves, blood vessels, subcutaneous fat, skin, and all other tissues are represented.

The second and subsequent level is an *approximate* model. Using suitable simplifying approximations, only those elements of anatomical detail that are most important to the understanding of the interaction of the organism with respect to external and internal forces are retained. Customarily, the specific muscles (or muscle groups) and specific skeletal bones will be the essential anatomical elements that are used at this second (approximate) model level. Chapter 3 will consider these particular elements in significantly more detail.

The third and final level is the *analytical* model. The approximate model is further simplified so that it may be represented by an engineering diagram. For the biostatic mechanical models presented in Chapter 3, the free-body diagram is the analytical model that behaves as a rigid body and obeys Newton's equations (Newtonian mechanics).

c. Model Applications: Higher System Level

Having considered models at the lower system level, models then progress to the higher system level. Higher-system modeling begins at the human/individual level. Models of the entire human body have many advantages, mostly due to ethical and availability issues associated with testing and experimentation on humans. For example, the availability of cadavers has never been adequate to supply the needs for "destructive" testing. Therefore, models of the human body have received much attention and focused resources. Primary contributors to human/individual modeling are the aerospace and automotive research laboratories.

One example of human models is in the area of pedestrian safety research. Ideally, experimentation with cadavers would result in the most desirable data, but due to the limited availability of cadavers, research with models is prevalent. One approach involves the modeling of a human using numerical computational software. In order to complete the model, identification of the following parameters is required: (1) the location, axis of rotation, and ranges of motion of each joint; (2) the joint's stiffness and viscous properties; (3) the inertial characteristics of each body segment; (4) the contact stiffness and energy recovery properties of each body segment; and (5) the external profile of each body segment.

The body is then modeled as a series of ellipsoids with one ellipsoid each for the head, neck, thorax, abdomen, pelvis, upper leg, lower leg, foot, upper arm, lower arm, and hand. Using parameters just described, the dynamic relationships between these segments are input into the computer as are the physical and inertial properties of the segment. When such a model is constructed, impacts with various objects can be simulated and the resulting effects on the body segments recorded. This data is then compared to injury criteria previously established. Using this method, variations in vehicle front end geometry and structural compliance can be explored to determine the optimal design in order to minimize pedestrian injuries.

The modeling of a human/individual and the interaction with the environment while performing a task is the next level of modeling. Much research in the field of human factors engineering is directed toward this *human operator–technological system* model. Models in these cases find their greatest usefulness when substituted for testing large numbers of subjects of varying physical dimensions and varying physical and mental capacities. It is simply not possible to test for every type of human operator that may interact with a device or system. Consequently, a model simulating as much of the interaction as possible is desirable and has a high degree of usefulness.

Biomathematical models used by the human factors engineer must account for biological variability. Individuals within a population group have widely varying physical dimensions as well as varying physical capacities and mental capacities. Based upon the acquisition and compilation of a large volume of measurement data on human physical dimensions, an extensive set of tables and charts has been prepared describing the physical characteristics (e.g., width and length of the hand), and functional characteristics (e.g., range-of-motion of the fingers) of various populations of human beings. From these sets of data, very useful models may be developed.

One specific application is the design of horizontal work surfaces. Horizontal work surfaces are normally used by seated and "sit-stand" workers and should provide for manual activities that are within convenient arm's reach. Certain normal and maximum areas have been proposed and are used rather widely. These two areas are defined as: 1. the normal area, or the area that can be conveniently reached with a sweep of the forearm while the upper arm remains in a natural position at the side; and 2. the maximum area, or the area that could be reached by extending the arm from the shoulder. Using these principles and available anthropometric data, models in an "ideal" workplace can be constructed. Following development of the model, differing size individuals can be "fit" theoretically into the workspace and the quality of the fit determined. Using this method, differing designs can be explored theoretically without the need for a full-size mockup and without the need to obtain subjects representative of a particular population. This model does suffer from the shortcoming that individuals rarely conform to the statistical data and unforeseen parameters such as jewelry or clothing may affect the quality of the design.

The next level of organization is the *human social–societal group* model. Some of the earliest and most numerous of the various models developed have described populations. An example of a population model is one that is age dependent. In an age-dependent population model, data are obtained as to the age of the population expressed as a percentage of the total population. This model uses discrete dynamic systems to model population growth, such as human populations, for which age is important. In some ways, continuous dynamic models might be more desirable because human populations are changing continuously. In practice, however, the data are discrete and so discrete models are often used. Moreover, human populations do exhibit some of the kinds of behavior (that have already been seen for discrete dynamic systems) that are not seen with continuous models. A possible reason for this is that humans can have a persistent influence on our environment.

Given trend data for some time in the past, a mathematical model can be developed to describe the age of the population at any point in time. After verification of model with past data, the model can then be used to extrapolate the age of the population at some point in the future. In this example, a model is the only way to arrive at this data as there is no experimental method to determine a value in the future. These models tend to be good predictors of future trends, but they do suffer from the quality of the available data and do not take into account human influences on the populations such as war, famine, and other catastrophes.

Finally, the highest organizational level is the level involving a human group interfacing or interacting with a large technological system. This represents a *human group–technological system* model. Such interactions occur commonly in our modern society, with production and assembly activities being the most obvious and most numerous examples. An important theoretical model is that of a production system design. When designing such a system, several steps must be pursued so that the model can be constructed. The first stage is the determination of the objectives and performance specifications for the model. Exact objectives must be defined and simulation restrictions generated. The second stage of this production system design model is the definition of the system. Boundaries as to what technologies define the system and what tasks will be required of the operators must be precisely defined.

The third stage is the theoretical design and layout of the technological system. As any practical system is constrained by the physical space available, budget, and construction time, all of these parameters must be synthesized into a practical design model. The experience of the engineers and the design team, and available literature play an important part in this step of the design. Also at the third stage, the interface system (between human and technological) must be designed and modeled. The fourth stage is practical testing and simulation to examine various components of the system before actual construction of the system begins. Computer simulation is of considerable use at this fourth stage. Upon completion of various task scenarios, the evaluation of the effectiveness of the current system and any redesigns may take place. Using this four-stage approach, complete human-technological systems can be modeled and evaluated without the expense and time of completing a final system and then having to redesign or reengineer that system.

In conclusion, models are most useful tools when dealing with unknown or undesirable (even hazardous) situations. Many types of models are available, and each has unique advantages and disadvantages. Modeling can decrease design time and avoid the expense and delay of practical testing in a variety of situations. Models can be employed with structures as small as single atoms or as large as the world population.

FURTHER INFORMATION

B.S. Blanchard and W.J. Fabrycky: *Systems Engineering and Analysis*. 3rd Edition. Prentice-Hall. Upper Saddle River, NJ. 1998.

G.R. McMillan, et al. (eds.): *Applications of Human Performance Models to System Design*. Plenum Press. New York. 1989.

C.A. Phillips (ed.): *Effective Upper and Lower Extremity Prostheses* (AUTOMEDICA 11:1-3). Gordon and Breach. New York. 1989.

D.S. Riggs: *The Mathematical Approach to Physiological Problems*. M.I.T. Press. Cambridge, MA. 1970.

G. Salvendy (ed.): *Handbook of Industrial Engineering*. 2nd Edition. John Wiley. New York. 1992.

Part II

Human Factors Engineering: Fundamentals

CHAPTER 3

Biostatic Mechanics

Biomechanics is the study of mechanics as it applies to biological systems. Table 3.1 indicates that *biomechanics* is a subdiscipline of *biophysics* and *biomedical engineering*. This is as it should be, since much of the work of *biomechanics* needs to be done without continually justifying why it is done. Science, or *biophysics* in this case, seeks the reasons why, and thus collects much important information. *Biomedical engineering* and human factors engineering (in particular) then seek to apply this knowledge to the betterment of human beings.

Table 3.1 also indicates that *biomechanics* can be subdivided into three sub-sub-disciplines. *Biostatics* is the science of the structure of living organisms in relation to the forces with which they interact. *Biodynamics* studies the nature of and determinants of the movement (and associated forces) of living organisms. *Bioenergetics* is the study of the energy transformation in living organisms. *Bioenergetics* includes (in part) biothermodynamic processes.

With regard to the subdisciplines identified in Table 3.1, *biodynamic* mechanics (biomechanics) will be presented in Chapter 4, *bioelectricity and bioelectronics* will be presented in Chapter 5, and *biothermodynamics and bioenergetics* will be presented in Chapter 6.

This chapter presents those elements of *biostatic* mechanics that represent the engineering fundamentals (statics) necessary for the human factors engineer. In order to do this, we must: 1. review the basics of static equilibrium; 2. define the specific system to be studied and the analytical model to be used; and 3. apply 1. and 2. in order to obtain useful information for the human factors engineer.

Consequently, this chapter has been divided into four sections. Section 3.1 reviews the basics and defines the system and model. Section 3.2 addresses the upper extremity, including the hand. Section 3.3 addresses the lower extremity, including the foot. Section 3.4 addresses the human back with respect to bending, lifting, and carrying.

As already noted in human systems modeling (Section 2.2), human factors engineering models must account for biological variability. Human beings demonstrate a wide range of physical characteristics, of which height (H) and weight (W) are two of the basic elements. Consequently, as a free-body diagram (FBD) is developed for biostatic mechanics analysis, we shall use the parameter convention (summarized in Table 3.2) where H represents the erect standing body height (in meters) and W represents the total body weight (in newtons). In the worked examples, W and H are often assigned an arbitrary value (for didactic

Table 3.1 HFE Fundamentals: Summary of Relationships

Discipline	Subdiscipline	Subsubdiscipline
Biophysics		
	Biomechanics	
		Biostatics
		Biodynamics
		Bioenergetics
	Bioelectricity	
	Biothermodynamics	
Biochemistry		
	Physical Biochemistry	
		Bioenergetics
Biomedical Engineering		
	Biomechanics	
		Biostatics
		Biodynamics
		Bioenergetics
	Bioelectronics	
	Biothermodynamics	

purposes only). Alternatively, the human operator may vary in size so that the results are proportionately related to the *W* and *H* of the human being.

Upper division (third and fourth year) undergraduate engineering students may be surprised to realize that they already know a great deal of biomechanics. Actually, the information and skills that were obtained in the first course in static and dynamic mechanics have provided the tools to understand much of bio-

Table 3.2 Anthropometric Modeling Data

Body Segment (Used in the Worked Examples)	Segment Length (Fraction of H[a])	Segment Weight (Fraction of W[b])
Head and neck	.17	.08
Forearm and hand	.20	.02
Upper arm	.20	.03
Arm	.40	.05
Head, neck, and both arms	—	.18
Thorax and abdomen	.30	.36
Pelvis	—	.16
Foot and foreleg	.29	.05
Upper leg	.24	.10
Leg	.53	.15
Head, neck, both arms, thorax, abdomen, and three-eighths pelvis	—	.60
One leg and five-eighths pelvis	—	.25

[a] H = Total body height, erect and standing (meters)
[b] W = Total body weight (newtons)

solid mechanics. No where is this more apparent than in the study of biostatic mechanics of the human musculoskeletal system.

3.1 STATICS OF RIGID BODIES

a. Static Equilibrium Equations

This chapter will consider the statics of rigid bodies in two dimensions. In general, a body may be treated as a combination of a large number of particles. The analysis will have to consider the size of the body as well as the forces that will act upon different particles so that the external forces will be applied at different points. The bodies considered in this chapter are assumed to be rigid. A rigid body is one that does not deform. Although some amount of deformation is always present in real physical systems, the rigid-body approximation does not appreciably affect the equilibrium state of the body under consideration.

The forces that act upon rigid bodies are conveniently divided into two types: external forces and internal forces. External forces represent the action of those forces outside the rigid body system being evaluated. External forces are completely responsible for the external behavior of the rigid body. This chapter will only consider external forces. Internal forces are those inside the rigid body that hold the constituent particles together. As we shall see in Chapter 7, when the rigid body is a structural element of a system composed of several such elements, the forces that hold the component elements together may also be considered as internal forces.

During the seventeenth century, Sir Isaac Newton developed the three fundamental laws, which we today recognize as the science of mechanics. Newton's first law can be stated as follows: When the resultant force acting upon a particle is zero, that particle will remain at rest if it was originally at rest or will move with a constant velocity in a straight line if it was originally in motion. This statement refers to a particle and defines the conditions necessary for translational equilibrium. In order for a rigid body to be in static equilibrium, that body must be not only in translational equilibrium but in rotational equilibrium.

According to Newton's first law, a body is said to be in translational equilibrium if the net (or resultant) external force acting upon it is equal to zero. For two-dimensional motion in x y coordinate space, this represents:

$$\sum F_y = 0 \quad (1)$$

$$\sum F_x = 0 \quad (2)$$

When using the condition for translational equilibrium, for example, in the y direction, the forces acting in the positive y direction carry a positive sign and the forces acting in the negative y direction must carry a negative sign.

In order for a rigid body to be in rotational equilibrium, the net external moment (due to the externally applied forces) about an arbitrary axis located at a point in the body must be equal to zero. For a system of external forces acting in the x-y plane, and designating M_0 as the origin, then the condition for rotational equilibrium is:

$$\sum M_0 = 0 \quad (3)$$

Some of the external moments about the origin (caused by the externally applied forces) may act in the clockwise direction, and others may act in a counterclockwise direction. Based upon the assignment of positive direction as either clockwise or counterclockwise, some of the moments will be of positive sign, while other moments will be of negative sign.

The general procedure for analyzing the forces and moments acting upon a rigid body in two dimensions for the condition of static equilibrium is as follows:

1. Draw the free-body diagram of the elements of the system and indicate all of the known and unknown external forces on it.
2. Establish an x- and y-coordinate system and indicate positive directionality for translational and rotational movement. Resolve all external forces into their components along these orthogonal axes.
3. For each free-body diagram, apply the required condition for translational and rotational equilibrium. For two-dimensional static equilibrium problems, the governing equations (1), (2), and (3) are applicable.
4. Solve the above equations simultaneously for the unknown parameters. Be sure to include the correct directions and units of both forces and moments when solving these equations.

As a final note, be aware that it may be more convenient in some cases to apply the condition for rotational equilibrium more than once. The condition for rotational equilibrium may be applied twice, for example, by considering the moments of the externally applied forces at two different points within the rigid body. Consequently, the third independent equation would be the condition for translational equilibrium (either in the x or the y direction). Since this equation must also be independent, the direction chosen should not be in parallel to a straight line connecting the two moment points. Finally, the condition for rotational equilibrium may be applied three times in a given problem. The moments generated by the externally applied forces are then considered with respect to three points within the rigid body. Be aware that the three points selected must not form a straight line.

b. Simply Supported Structures

Let us now consider some specific applications by examining the forces on beams and cables. A beam may be defined as the structural member of a system that is designed to support various loads applied at different points along that member. In some cases, the applied forces are orthogonal to the central axis of the beam so that only shear and bending occur in the beam. In other cases where the applied loads are not perpendicular to the beam, these loads will also produce axial forces within the beam. A beam is an element in a mechanical system. A beam may be connected to the ground or to other beams by various supports or connection devices. These include rollers, knife-edges, hinges, pivots, or cables. Each connection or support device experiences different reactions that occur at the specific support device or connection. In Sections 3.2, 3.3, and 3.4, which follow, the human skeletal elements (specifically the various bones) will be treated as functionally analogous to the beams of mechanical systems.

Two specific connection and support devices used in mechanical systems are of particular interest when performing a static equilibrium analysis of the human musculoskeletal system. These are the hinge joint and the cable. The hinge joint is a connection device and the cable is a support device. Both of these mechanical-system elements are useful in the analysis of the human musculoskeletal system.

A hinge joint may connect a beam to another beam or to the ground. A hinge joint will constrain a beam both in the x and y directions. Consequently, it permits no translational movement of the beam. With respect to translational forces, recall that those forces denoted by $\mathbf{R}_x$ and $\mathbf{R}_y$ represent the components of the reaction forces applied by the ground (or another beam) through the hinge joint. However, a hinge joint cannot constrain rotational forces about its axis so a hinge joint will rotate if it experiences a net moment. Consequently, in order to satisfy the condition of static equilibrium, the net moment about a hinge joint must be zero. In Sections 3.2, 3.3, and 3.4, the various joints of the human skeletal system will be treated as functionally analogous to the hinge joint of mechanical systems.

Cables are used to connect various members of a mechanical system together or to the ground. Cables transmit forces from one member to the other or to the ground. Recall that cables are flexible elements of the mechanical system and can sustain only tensile forces. A cable cannot sustain a compressive force, and if one is experienced, the cable will collapse and the condition of static equilibrium will not be maintained. With respect to tensile forces, the tension in cables will be uniform throughout that element. As will be seen in subsequent chapters, when skeletal muscle contracts, it pulls inward upon itself and generates tension at its two ends. Consequently, in the remaining sections of this chapter, the cable of a mechanical system will be treated as functionally analogous to human skeletal muscle.

The following example is presented as a brief review of these concepts. Note that the mechanical system consists of a beam, a cable, and a joint. In this example the hinge joint connects the beam to the ground and the cable supports the beam with respect to the ground. To make it interesting, the problem is presented as an engineering dynamics question which requires an analysis of the system and two different conditions of static equilibrium. As you proceed to the various examples of Sections 3.2, 3.3, and 3.4, you will appreciate the analogy of the *mechanical* beam-joint-cable-ground system to the *human* bone-joint-muscle-body system.

EXAMPLE 3.0

A horizontal lamp pole, which is a uniform horizontal beam of weight, W_p = 100 N, is hinged to the wall at point A. The other end of the lamp pole (B) is supported by a cable that makes an angle, θ = 50° with the horizontal, as shown in Figure 3.0. The length of the lamp pole (beam) is 1.33 M. The lamp itself has a weight,

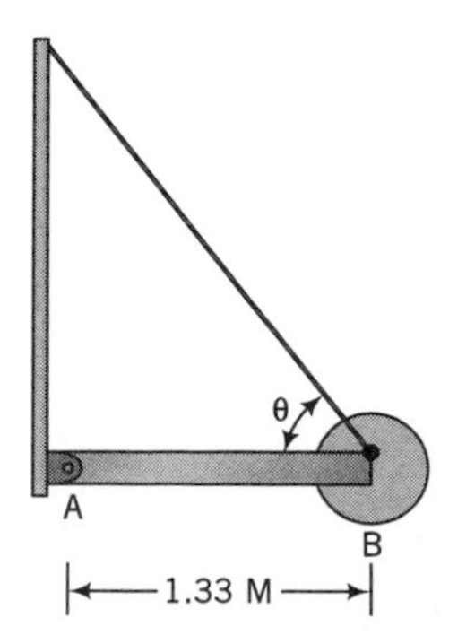

Figure 3.0 A simply supported structure representing a beam-joint-cable mechanical system.

$W_L = 25$ N, and its center of mass is located at B. Solve for the force in the cable (**F**) and for the horizontal and vertical reaction forces at the hinge joint ($\mathbf{R}_x$ and $\mathbf{R}_y$).

SOLUTION 3.0

From Figure 3.0, draw the free-body diagram (FBD) of the mechanical system:

At this point, note the following: 1. we have assigned a sign convention and

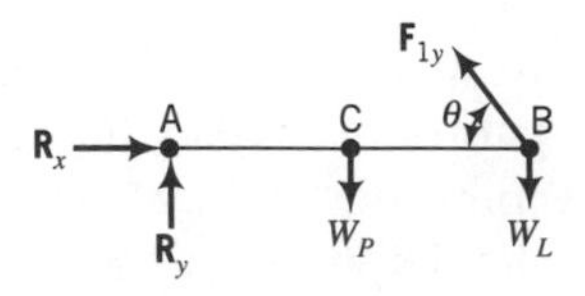

arbitrarily made both reaction forces positive; 2. since A is a hinge joint, it cannot sustain a rotational force, so we will sum the moments about this point; 3. we can approximate the pole (beam) as having a uniform mass distribution *and* uniform shape, so its center of mass is located at its midpoint; 4. The lamp is approximated as having a point center of mass located at the end of the pole (beam); 5. The mass of the cable is considered to be negligible (i.e., more than a magnitude of order less than the combined mass of pole and lamp); 6. there are three unknowns (**F**, $\mathbf{R}_x$, and $\mathbf{R}_y$), so we will have to write and solve three equations in order to have a statically determinate system.

Refine the FBD to indicate all known parameters:

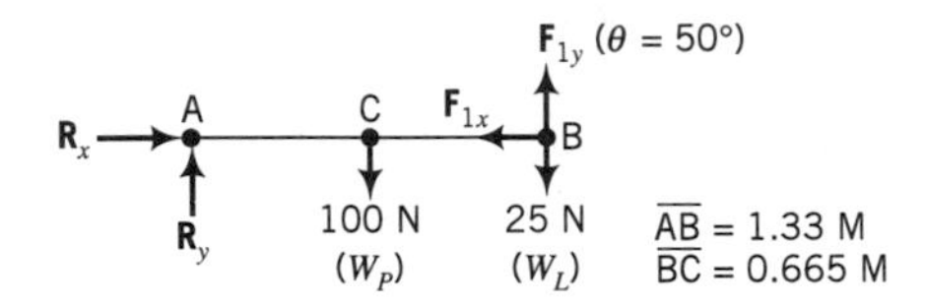

Note at this point, **F** is resolved into its orthogonal components, and all known parameters (forces and lengths) depicted.

Write and solve the governing equations for the condition of static equilibrium:

$$\sum M_A = 0:$$

$$-W_p(AC) - W_L(AB) + \mathbf{F}_y(AB) = 0 \qquad \text{(i)}$$

Note: 1. moments are summed about point A (the hinge); 2. counterclockwise positive sign convention is observed; 3. $\mathbf{F}_x$, $\mathbf{R}_x$, and $\mathbf{R}_y$ exert no moment at point A since they act directly through point A (and so there is zero moment arm); and 4. *all* moments about point A have been identified.

$$\sum \mathbf{F}_x = 0$$

$$\mathbf{R}_x - \mathbf{F}_x = 0$$

$$\mathbf{R}_x - \mathbf{F} \cdot \cos\vartheta = 0 \qquad \text{(ii)}$$

Note that rightward positive sign convention is observed and that we have accounted for *all* horizontal forces.

$$\sum \mathbf{F}_y = 0$$

$$\mathbf{R}_y - W_p - W_L + \mathbf{F}_y = 0$$

$$\mathbf{R}_y - W_p - W_L + F \cdot \sin\vartheta = 0 \qquad \text{(iii)}$$

Note that upward positive sign convention is observed and *all* vertical forces are considered. Substitute knowns into equations (i), (ii), and (iii) to solve for **F**, $\mathbf{R}_x$, and $\mathbf{R}_y$:

$$\sum M_A = 0\text{: [equation (i)]:}$$

$$-(100)(.665) - (25)(1.33) + \mathbf{F}(.766)(1.33) = 0$$

$$\mathbf{F} = \frac{99.8}{1.02} = 97.8\text{ N} \tag{iv}$$

$$\sum \mathbf{F}_x = 0\text{: [equation (ii)]:}$$

$$\mathbf{R}_x - (97.8)(.643) = 0$$

$$\mathbf{R}_x = 62.9\text{ N } (\rightarrow)$$

($\mathbf{R}_x$ acts rightward)

$$\sum \mathbf{F}_y = 0\text{: [equation (iii)]:}$$

$$\mathbf{R}_y - (100) - (25) + (97.8)(.766) = 0$$

$$\mathbf{R}_y = 125 - 74.9 = 50.1\text{ N } (\uparrow)$$

($\mathbf{R}_y$ acts upward)

---

c. The Musculoskeletal System

For the purposes of a human factors engineering application, the specific anatomical system which biostatic mechanics focuses upon is the musculoskeletal system. This system is characterized by five essential elements. Two of these are the *proximal* segment and the *distal* segment. If one thinks of the navel (or umbilicus) as the center of the body with the arms stretched straight out to the side and the legs in a standing position spread somewhat apart, then the proximal segment is that anatomical structure nearest the navel and the distal segment is its adjoining structure, but farther from the navel. A *joint* is defined as the junction of the proximal segment with the distal segment. The *agonist muscle* is an internal force generator that crosses the joint. One end (the origin) is usually connected to the proximal segment, and the other end (the insertion) is usually connected to the distal segment. This muscle is referred to as the "agonist" when it is the prime mover of the anatomical system. The *antagonist muscle* is an internal force generator that also crosses the joint and usually has its origin at the proximal segment and its insertion at the distal segment. An antagonist muscle develops an opposing force with respect to its agonist partner. In many biostatic mechanic models of the musculoskeletal system, only one muscle will be represented. This will be a *functional* agonist muscle (hereafter identified as muscle) in which the internal force generated will be the difference between the *structural* agonist muscle and the *structural* antagonist muscle. In real (not idealized) biological systems, simultaneous contraction of both the agonist and antagonist muscle is necessary to stabilize joints.

In order to analyze this anatomical system, we will use a biomechanical (biostatic) model. As noted in Section 2.2.b, such a model proceeds through three levels of development. The anatomical model and the approximate model are the first two models. Generally speaking, at least four of the five essential elements of the system are included in each model. The third and final level is the *analytical* model. The approximate model is further simplified so that it may be represented by an engineering diagram. For the biostatic mechanical models presented in this chapter, the free-body diagram is the analytical model that behaves as a rigid body and obeys Newton's first and third laws.

Human factors engineers employ biostatic analytical models in two different ways. In the first (and most common) application, external forces act upon the human body. These forces represent inputs to the biostatic model of the musculoskeletal system of interest. The outputs of the biostatic model are the internal forces that are reactive in nature. In actuality, the biostatic model represents the *inverse* solution of what actually happens. In the second (less common) application, a more realistic scenario is constructed. Internal forces are actively generated by contracting muscle (as well as internal reactive forces at joints). These internal forces are the inputs to the biostatic model of the musculoskeletal system of interest. The outputs of the model are the forces that act upon the external environment. In this case, the model represents a *forward* solution of what the human being is actually doing.

Having defined the anatomical system of interest and having developed a suitable analytical model, the human factors engineer must then identify a suitable application in order to obtain useful information for the design and development process. HFE system design and development was described in Section 2.1. However, for the purposes of demonstrating the development and utility of biostatic mechanical models, we shall use a simple three-step approach in this chapter:

1. Input. A specific HFE application is stated, which requires analysis. Here, we identify the HFE application.
2. Modeling and analysis. In this step, we construct both an approximate and analytic model of the system. This chapter will then proceed with the mathematical analysis and provide an answer to a specific problem.
3. Assessment and interpretation. What does the answer mean? What useful information about the system does the answer provide? Can we satisfy a system specification? Do we need more or different measurement and descriptive data? What new problem (if any) has been uncovered? In this chapter, we shall limit assessment and interpretation to how the answer is useful in telling us something about the system.

3.2 UPPER EXTREMITY AND HAND

a. Shoulder and Arm

The *anatomical system* for the shoulder and arm is as follows:

Proximal segment: Scapula (shoulder bone)

Distal segment: Humerus (arm bone)

Joint: Shoulder joint

Muscle (action): Deltoid (abduction of the arm: when viewing a person from the front, raising the arm from the side of the body outward from the side of the body as in flapping the arms).

The *approximate* model for the shoulder and arm is presented in Figure 3.1.a, and the *free-body diagram* for the shoulder and arm is presented in Figure 3.1.b.

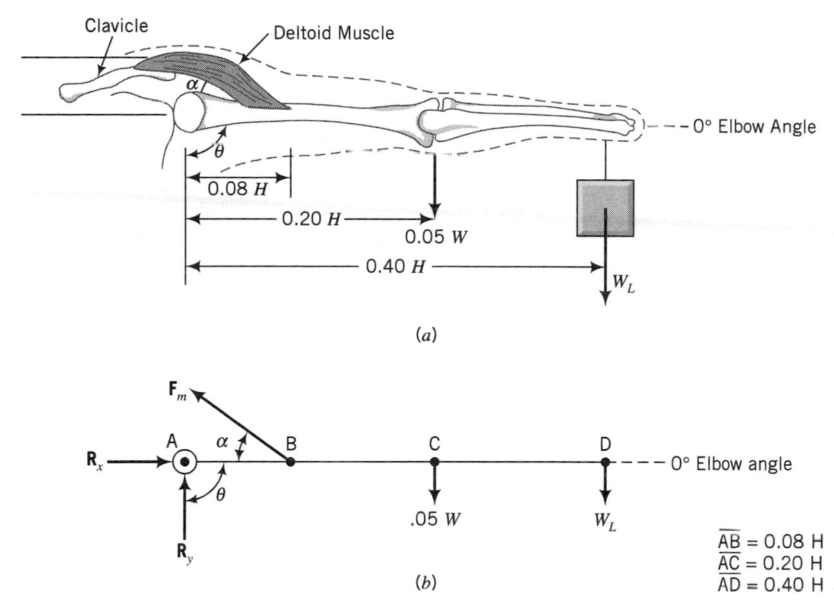

Figure 3.1.a Approximate model of the shoulder and arm. **b** Free-body diagram of the shoulder and arm.

HFE Application As human skeletal muscles perform a demanding task over a prolonged time, the muscle undergoes fatigue. The biochemical and/or biophysical basis for skeletal muscle fatigue remains uncertain. However, muscle fatigue is physically characterized as a progressive decrease in its ability to generate force. Modifying the task to reduce muscle force can increase endurance and prolong the time period before onset of fatigue.

EXAMPLE 3.1

a. A volunteer emergency worker (of height, $H = 1.75$ M and weight, $W = 700$ N) must continuously extend his arm straight out so that the arm is at a right angle ($\theta = 90°$) to the side of the body (Figure 3.1.a). You may neglect the elbow angle. The worker is holding a "DETOUR RIGHT" sign that represents a load weight, $W_L = 18$ N. Determine the force exerted by the deltoid muscle ($\mathbf{F}_M$) assuming that it inserts on the arm bone at an angle, $\alpha = 30°$. Also determine the vertical reaction force of the body ($\mathbf{R}_y$) and the horizontal reaction force of the body ($\mathbf{R}_x$) at the shoulder joint.

b. As time passes by and fatigue sets in, the extended arm sags (to an angle, $\theta = 65°$, with respect to the side of the body) as shown in Figure 1.3.b. Find the new $\mathbf{F}_M$ (assuming now that $\alpha = 25°$).

c. Realizing the extended arm is sagging, the volunteer bends the elbow and also reextends the upper arm straight outward so that $\theta = 90°$. The sign, hand, and lower arm are now held vertically upward at a 90° right angle to the upper arm as shown in Figure 1.3.c. Find the new $\mathbf{F}_M$ (assuming now that $\alpha = 30°$).

SOLUTION 3.1(a)

Given: $H = 1.75$ M, $W = 700$ N
Given: $W_L = 18$ N, $\theta = 90°$, $\propto = 30°$
Find: $\mathbf{F}_w$, $\mathbf{R}_x$, and $\mathbf{R}_y$:
Using the FBD of Figure 3.1.b, and the given data:

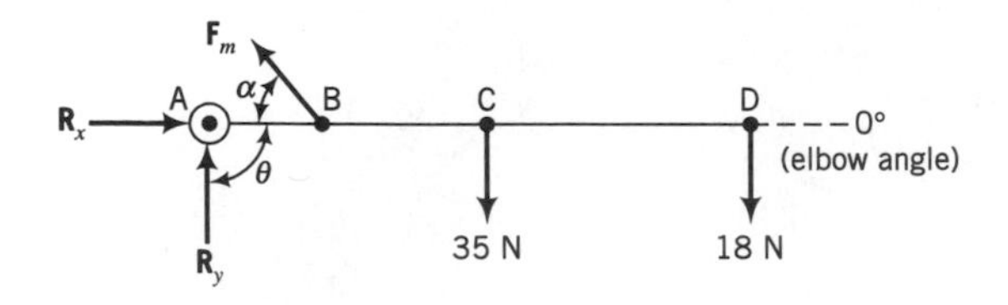

AB = 0.14 M
AC = 0.35 M
AD = 0.70 M

Sign convention:

$$\sum \mathbf{F}_y = 0:$$

$$\mathbf{F}_m \cdot \sin(30°) + \mathbf{R}_y - 35 - 18 = 0 \quad \text{(i)}$$

$$\sum F_x = 0:$$

$$-\mathbf{F}_M \cdot \cos(30°) + \mathbf{R}_x = 0 \quad \text{(ii)}$$

$$\sum M_A = 0:$$

$$[\mathbf{F}_M \cdot \sin(30°)](.14) - (35)(.35) - (18)(.70) = 0 \quad \text{(iii)}$$

$$(.5)(.14)\mathbf{F}_M - (12.25) - (12.6) = 0 \quad \text{(iv)}$$

$$\mathbf{F}_M = \frac{24.85}{0.07} = 355 \text{ N} \quad \text{(v)}$$

The force exerted by the deltoid muscle is *one-half* the body weight!
Substituting equation (v) into equation (ii):

$$-(355)(0.866) + \mathbf{R}_x = 0 \quad \text{(vi)}$$

$$\mathbf{R}_x = 307 \text{ N (rightward)} \quad \text{(vii)}$$

Substituting equation (v) into equation (i):

$$(355)(0.5) + \mathbf{R}_y - 52 = 0 \quad \text{(viii)}$$

$$\mathbf{R}_y = 53 - 177.5 = 125.5 \text{ N (downward)} \quad \text{(ix)}$$

SOLUTION 3.1(b)

Given: $W_L = 18$ N, $\theta = 65°$, $\alpha = 25°$
Find: $\mathbf{F}_M$
Modify the FBD of Example 3.1(a):

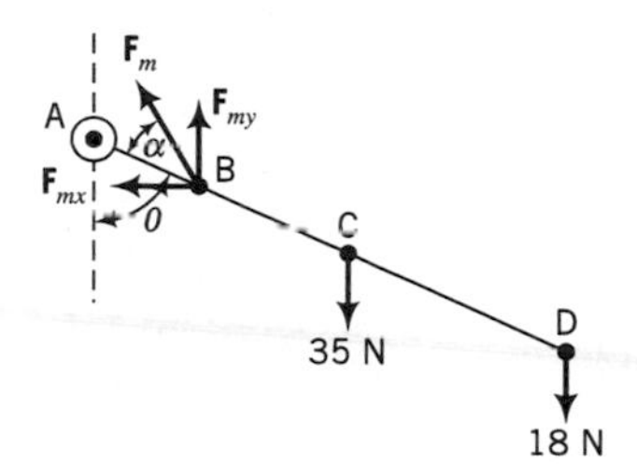

AB =.14 M

AC = .35 M

AD =.70 M

Since we are solving *only* for $\mathbf{F}_M$ (not $\mathbf{R}_x$ and $\mathbf{R}_Y$); sum the moments about A:

$$\sum M_A = 0:$$

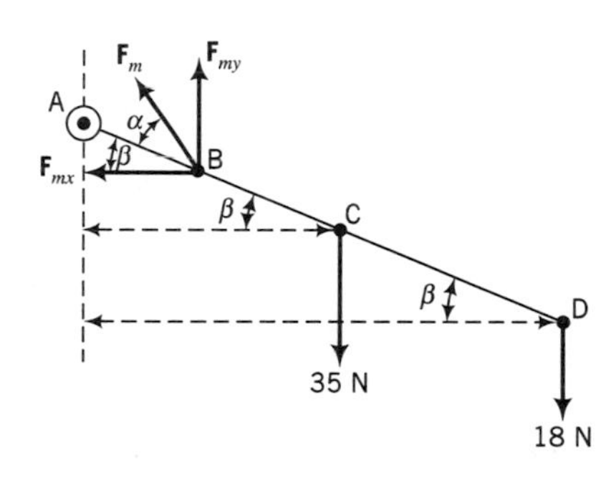

$$\beta = 90° - \theta = 90° - 65° = 25°$$

$$(\mathbf{F}_{MY})[AB \cdot \cos\beta] - \mathbf{F}_{MX}[AB \cdot \sin\beta] - (35)[AC \cdot \cos\beta] - (18)[AD \cdot \cos\beta] = 0$$

$$(\mathbf{F}_M)[\sin(\alpha+\beta)](.14)(.906) - \mathbf{F}_M[\cos(\alpha+\beta)](.14)(.423) - (35)(.35)(.906) - (18)(.70)(.906) = 0$$

$$\mathbf{F}_M(.766)(.127) - \mathbf{F}_M(.643)(.059) - 11.1 - 11.4 = 0$$

$$\mathbf{F}_M = \frac{22.5}{.0973 - .0379} = 379 \text{ N}$$

As the extended arm sags, the deltoid muscle must exert *even more* force!

SOLUTION 3.1(c)

Given: $\theta = 90°$, $\propto = 30°$

Find: $\mathbf{F}_M$ (at $\hat{W}_L$)

$\hat{W}_L$ is the new arm load with the elbow bent upward at 90°. From anthropometry data (Table 3.2), we divide the total arm weight into:

$$W_{UA} = W(\text{upper arm}) = 0.03 \text{ W}$$

$$W_{LAH} = W(\text{lower arm and hand}) = 0.02 \text{ W}$$

The new load weight ($\hat{W}_{LX}$):

$$\hat{W}_L = W_{LAH} + W_L$$
$$\hat{W}_L = 14\text{ N} + 18\text{ N} = 32\text{ N}$$

and

$$W_{UA} = 21\text{ N}$$

The FBD for Example 3.1(c):

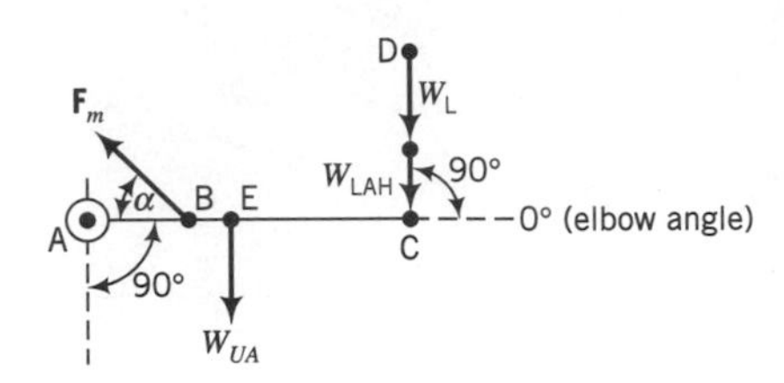

AB = .14 M
AC = .35 M
Approximate AE = 0.175 M
The simplified FBD:

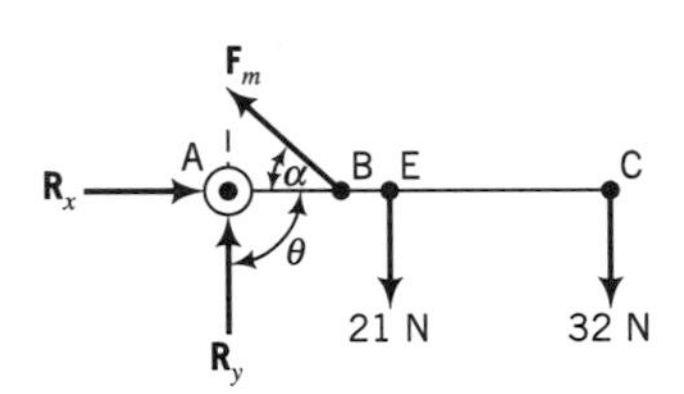

Since we are solving *only* for $\mathbf{F}_M$ (*not* $\mathbf{R}_x$ and $\mathbf{R}_Y$):

$$\sum M_A = 0:$$
$$[\mathbf{F}_M \cdot \sin(30°)](.14) - (21)(.175) - (32)(.35) = 0$$
$$(.5)(.14)\mathbf{F}_M - 3.67 - 11.20 = 0$$
$$\mathbf{F}_M = \frac{14.88}{0.07} = 213\text{ N}$$

Modifying the task by bending the elbow upright, reduces the deltoid muscle force by 40% (compared to the arm straight and extended).

---

b. Elbow and Forearm

The *anatomical* system for the elbow and forearm is:

Proximal segment: Humerus (arm bone)
Distal segment: Radius and ulna (forearm bones)
Joint: Elbow joint
Agonist muscle (action): Triceps (extension of forearm: lowering the forearm away from the upper arm, widening the elbow angle).

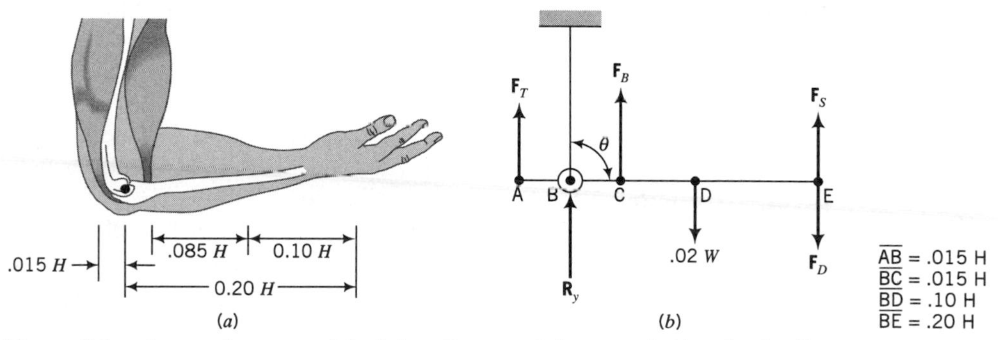

Figure 3.2.a Approximate model of the elbow and forearm. **b** Free-body diagram of the elbow and forearm.

Antagonist muscle (action): Biceps (flexion of forearm: raising the forearm toward the upper arm, decreasing the elbow angle).

The *approximate* model for the elbow and forearm is represented in Figure 3.2.a, and the FBD for the elbow and forearm is shown in Figure 3.2.b.

HFE Application Although many biomechanical models of the musculoskeletal system depict only a *functional* agonist, real muscles always act in agonist-antagonist pairs. Many tasks require the interplay of these agonist-antagonist muscle pairs. During motion in *one* direction, the agonist is (by definition) the prime mover in *that* direction. However, the muscle generating the predominant force may alternate between the actual agonist and its opposing antagonist depending on the nature of the task performed.

EXAMPLE 3.2

An overhead roller door has two opposing forces acting upon it. As the door is progressively lowered, there is an upward spring force ($\mathbf{F}_S$) that progressively decreases (Figure 3.3.a). Also, during progressive lowering, there is a downward door force, due to its weight ($\mathbf{F}_D$), that progressively increases (Figure 3.3.b). A person (H = 1.67 M and W = 550 N) is standing directly in front of the overhead roller

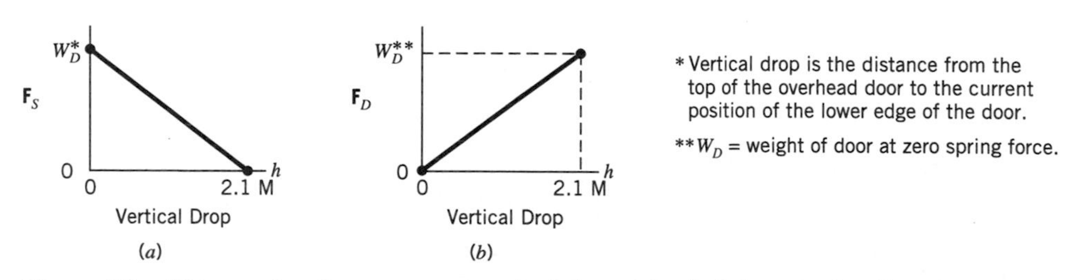

Figure 3.3.a Door spring forces versus vertical drop. Vertical drop is the distance from the top of the overhead door to the current position of the lower edge of the door. W_D = weight of the door at zero spring force. **b** Door force (weight) versus vertical drop.

door and pulling its lower edge downward with the forearm and hand (while the upper arm is straight along the person's side). Using Figures 3.2.a and 3.2.b for reference (and then modifying accordingly):

a. If the angle at the elbow (between the upper arm and the forearm) is $\theta = 25°$ when the vertical door drop (h) is 0.78 M, then for a door weight (W_D) of 120 N: find the force exerted by the triceps muscle ($\mathbf{F}_T$), the force exerted by the biceps muscle ($\mathbf{F}_B$) and the reaction force at the elbow ($\mathbf{R}_e$).
b. If the angle at the elbow (between the upper arm and the forearm) is $\theta = 142°$ when the vertical drop (h) is 1.41 M, find $\mathbf{F}_T$, $\mathbf{F}_B$, and $\mathbf{R}_e$ for a 120-N door (W_D).

SOLUTION 3.2(a)

Given: $H = 1.67$ M, $W = 550$ N
Given: $\theta = 25°$, $h = 0.78$ M, $W_D = 120$ N
Find: $\mathbf{F}_T$, $\mathbf{F}_B$, $\mathbf{R}_e$

Using the FBD of Figure 3.2.b, and the given θ and assigned H, W:

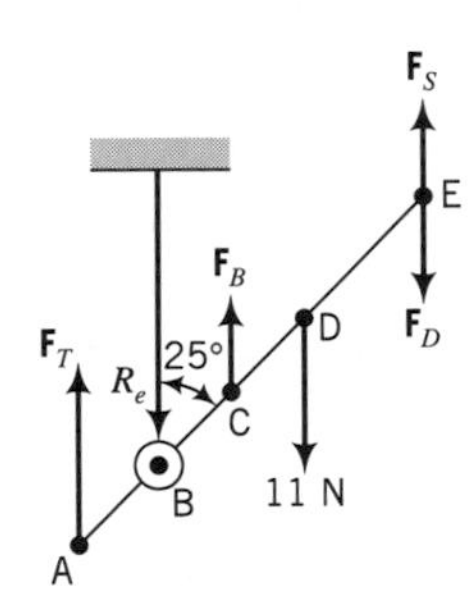

AB = .025 M
BC = .025 M
BD = .167 M
BE = .334 M

From Figure 3.3.a, find F_s:
By inspection:

$$\mathbf{F}_S = -\left(\frac{120\text{ N}}{2.1\text{ M}}\right)h + 120\text{ N}$$

For $h = 0.78$ M,

$$\mathbf{F}_S = -(57.1)(0.78) + 120 = 75.4\text{ N}$$

From Figure 3.3.b, find $\mathbf{F}_D$:

By inspection:

$$\mathbf{F}_D = \left(\frac{120\ \text{N}}{2.1\ \text{M}}\right)h$$

$$\mathbf{F}_D = (57.1)(0.78) = 44.5\ \text{N}$$

Calculate *net* force:

$$\mathbf{F}_N = \mathbf{F}_S - \mathbf{F}_D = 75.4 - 44.5 = 309\ \text{N (upward)}$$

Modify the FBD for F_N upward:

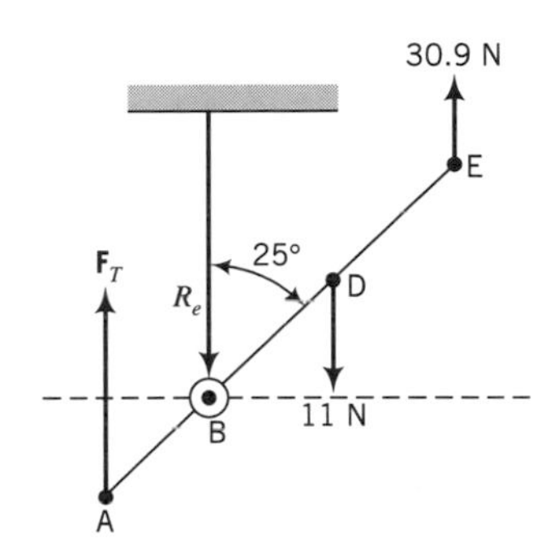

$\sum \mathbf{F}_y = 0$:

$$\mathbf{F}_T - \mathbf{R}_e - 11 + 30.9 = 0$$

$$\mathbf{R}_e - \mathbf{F}_T + 19.9$$

$\sum M_B = 0$:

$$-(\mathbf{F}_T)[AB \cdot \sin(25°)] - (11)[BD \cdot \sin(25°)] + (30.9)[BE \cdot \sin(25°)] = 0$$

$$-(\mathbf{F}_T)(.025)(.423) - (11)(.167)(.423) + (30.9)(.334)(.423) = 0$$

$$-(\mathbf{F}_T)(.0106) - (.7771) + 4.3656 = 0$$

$$\mathbf{F}_T = \frac{3.5885}{.0106} = 339\ \text{N (upward)}$$

The net force is exerted by the *triceps* muscle and is slightly over 60% of the body weight.

Find $\mathbf{R}_e$:

$$\mathbf{R}_e = 339 + 19.9 = 359\ \text{N (downward)}$$

SOLUTION 3.2(b)

Given: $\theta = 142°$, $h = 1.41$ M, $W_D = 120$ N
Find: $\mathbf{F}_T$, $\mathbf{F}_B$, $\mathbf{R}_e$

Using the FBD of Example 3.2(a), *except* $\theta = 142°$:

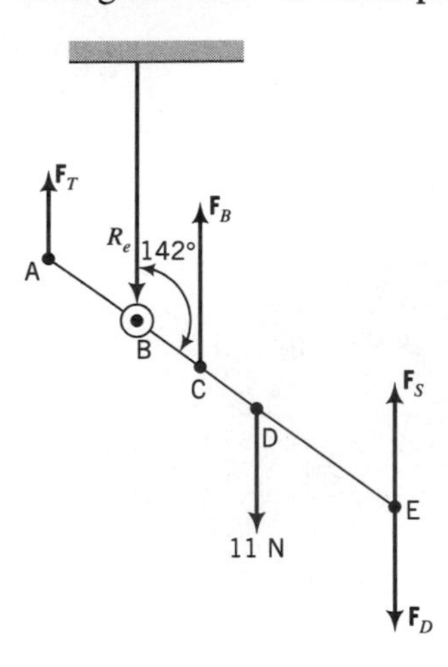

Find $\mathbf{F}_S$ (Figure 3.3.a):

$$\mathbf{F}_S = -(57.1)h + 120$$
$$\mathbf{F}_S = -(57.1)(1.41) + 120 = 39.5\ \text{N}$$

Find $\mathbf{F}_D$ (Figure 3.3.a):

$$\mathbf{F}_D = (57.1)h$$
$$\mathbf{F}_D = (57.1)(1.41) = 80.5\ \text{N}$$

Find net force:

$$\mathbf{F}_N = \mathbf{F}_S - \mathbf{F}_D = 39.5 - 80.5 = -41\ \text{N}$$

Modify FBD for $\mathbf{F}_N$ downward:

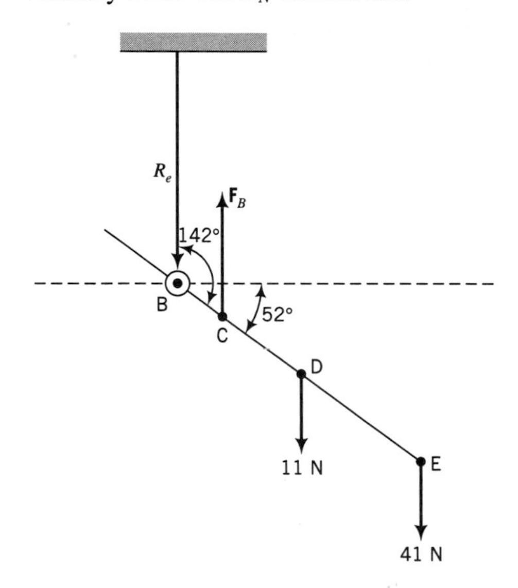

$\sum \mathbf{F}_y = 0$:

$$\mathbf{F}_B - \mathbf{R}_e - 11 - 41 = 0$$
$$\mathbf{R}_e = \mathbf{F}_B - 52$$

$\sum M_B = 0$:

$$(\mathbf{F}_B)[BC \cdot \cos(52°)] - (11)[BD \cdot \cos(25°)] - (41)[BE \cdot \cos(25°)] = 0$$
$$(\mathbf{F}_B)(.025)(.616) - (11)(.167)(.616) - (41)(.334)(.616) = 0$$
$$(.0154)(F_B) - 1.1316 - 8.4355 = 0$$
$$\mathbf{F}_B = \frac{9.5671}{.0154} = 621 \text{ N}$$

The net force is exerted by the *biceps* muscle and is slightly over 110% of the body weight
Find R_e:

$$\mathbf{R}_e = 621 - 52 = 569 \text{ N (downward)}$$

---

c. Wrist and Hand

The *anatomical system* for the wrist and hand is:

Proximal segment: Radius and ulna (forearm bones)
Distal segment: Scaphoid and lunate (and other wrist bones) and the hand bones
Joint: Wrist joint
Muscle (action): Flexor digitorum (flexion of the fingers toward the palm of the hand).

The *approximate* model for the wrist and hand (while gripping a round object) interacting with the *free-body diagram* for the wrist and hand is depicted in Figure 3.4.

HFE Application Gripping strength of the hand is a function of the length of the finger flexor muscles (which run the length of the forearm). The optimal length for these muscles (strongest gripping force) is when the wrist is straight, as when we shake hands. Bending the wrist either toward the palm (wrist flexion) or away from the palm (wrist extension) lowers grip strength by under-lengthening or over-lengthening the finger flexor muscles from their optimal length.

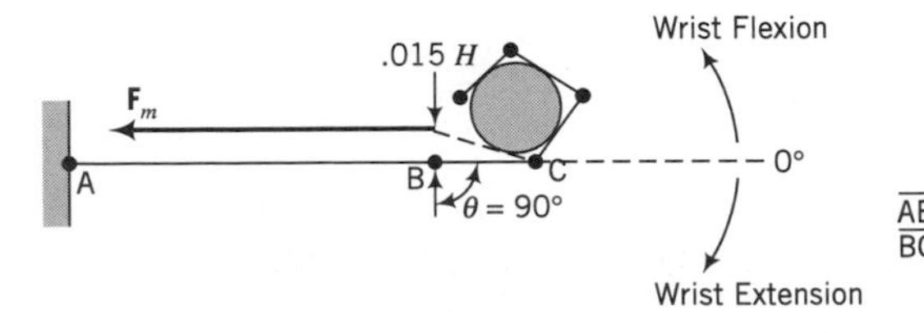

Figure 3.4 Approximate model and free-body diagram of the wrist and hand.

EXAMPLE 3.3

Referring to Figure 3.5, the gripping force ($\mathbf{F}_g$) is shown to be a function of wrist joint angle (θ). For a person (for whom $\mathbf{F}_g = 150$ N when $\theta = 0°$) who then proceeds to lift a floor lamp (of weight, W_L) by gripping its polished brass pole (with coefficient of sliding friction, $\rho = 0.3$):

a. If the wrist is flexed inward toward the palm at an angle of 60° (with respect to the wrist straight), find the maximum lamp weight (W_L) that can be held before the lamp will slip in the person's hand.
b. Repeat Example 3.3(a) when the wrist is in straight alignment with the forearm (zero degrees of wrist flexion or extension).
c. Repeat Example 3.3(a) when the wrist is extended outward away from the palm at an angle of 40° (with respect to the wrist straight).

SOLUTION 3.3(a)

Given: $\rho = 0.3$

Find: $W_L\,(\theta = 60°)$ Wrist flexion

FBD of systems:

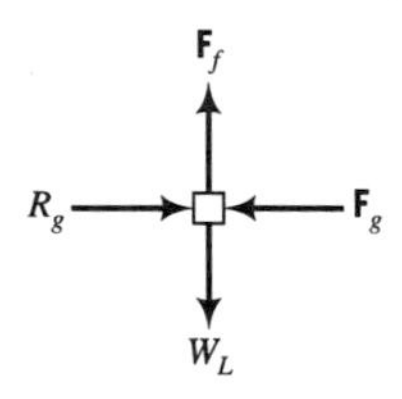

$$\sum \mathbf{F}_x = 0:$$

$$\mathbf{R}_g - \mathbf{F}_g = 0 \quad \text{(i)}$$

$$\sum \mathbf{F}_y = 0:$$

$$\mathbf{F}_f - W_L = 0 \quad \text{(ii)}$$

where

$\mathbf{F}_g$ = human grip force (around pole)
$\mathbf{R}_g$ = lamp pole reactions force
W_L = weight of lamp
$\mathbf{F}_f$ = gripping frictional force:

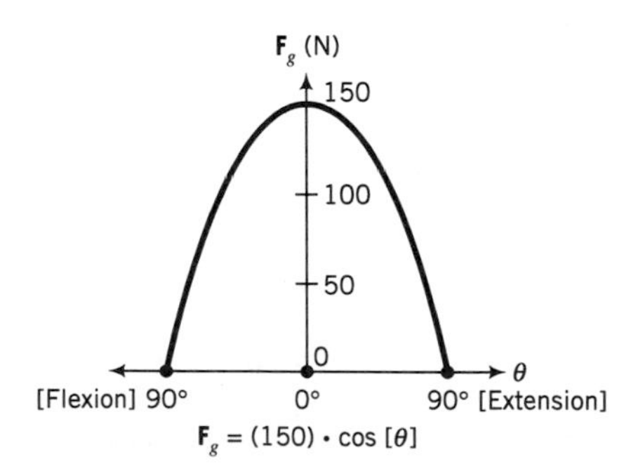

Figure 3.5 Gripping force (F_g) as a function of wrist joint angle.

$$\mathbf{F}_f = \rho \cdot \mathbf{F}_g \qquad \text{(iii)}$$

Referring to Figure 3.5 when $\theta = 60°$:

$$\mathbf{F}_g = (150\ \text{N}) \cdot \cos[60°] = 75\ \text{N}$$
$$\mathbf{F}_f = \rho \cdot \mathbf{F}_g = (0.3)(75) = 22.5\ \text{N}$$

Rearrange:

$$W_L = \mathbf{F}_f = 22.5\ \text{N} \qquad \text{(ii)}$$

SOLUTION 3.3(b)

Given: Same as Example 3.3(a)
Find: W_L ($\theta = 0°$): wrist straight
FBD: as per Example 3.3(a)

Rearrange:

$$W_L = \mathbf{F}_f \qquad \text{(ii)}$$
$$\mathbf{F}_f = \rho \cdot \mathbf{F}_g \qquad \text{(iii)}$$

Referring to Figure 3.5 when $\theta = 0°$:

$$\mathbf{F}_g = (150\ \text{N}) \cos[0°] = 150\ \text{N}$$
$$\mathbf{F}_f = \rho \cdot \mathbf{F}_g = (0.3)(150) = 45\ \text{N}$$
$$W_L = \mathbf{F}_f = 45\ \text{N}$$

The gripping force doubled, so a lamp twice as heavy could be held with a straight wrist (compared to 60° of wrist flexion).

SOLUTION 3.3(c)

Given: Same as per Example 3.3(a)
Find: $W_L(\theta = 40°)$: wrist extension
FBD as per Example 3.3(a)

Rearrange:

$$W_L = \mathbf{F}_f \qquad \text{(ii)}$$
$$\mathbf{F}_f = \rho \cdot \mathbf{F}_g \qquad \text{(iii)}$$

Referring to Figure 3.5 when $\theta = 40°$

$$\mathbf{F}_g = (150\ \text{N}) \cos[40°] = 115\ \text{N}$$
$$\mathbf{F}_f = \rho \cdot \mathbf{F}_g = (0.3)(115) = 34.5\ \text{N}$$
$$W_L = \mathbf{F}_f = 34.5\ \text{N}$$

Gripping force again decreases with the wrist in extensions (compared to a straight wrist).

3.3 LOWER EXTREMITY AND FOOT

a. Hip and Leg

The *anatomical* system for the hip and leg is:

Proximal segment: Pelvis (pelvic bones)
Distal segment: Femur (thigh bone)
Joint: Hip joint
Muscle (action): Gluteus medius and minimus, and tensor fascia lata (abduction of leg: extending the straightened leg out from the side of the body, as seen in from the front of a person when kicking a ball sideways with the side of the foot).

The *approximate* model for the hip and leg (when a person is standing on one leg only) is presented in Figure 3.6.a. The *free-body diagram* for the hip and leg is presented in Figure 3.6.b.

HFE Application Common tasks such as walking or climbing stairs require the alternate shifting of body weight onto one leg and then the other as that leg contacts the ground (while the other leg swings through the air). This generates high forces at the hip joint and can be physically demanding for people with weak hip muscles (such as the elderly or debilitated) or actually painful for people with hip joint disorders (such as arthritis or recent hip surgery). Designing assistive aids such as canes, crutches, handrails, and banisters will benefit these people by lowering forces at the hip joint.

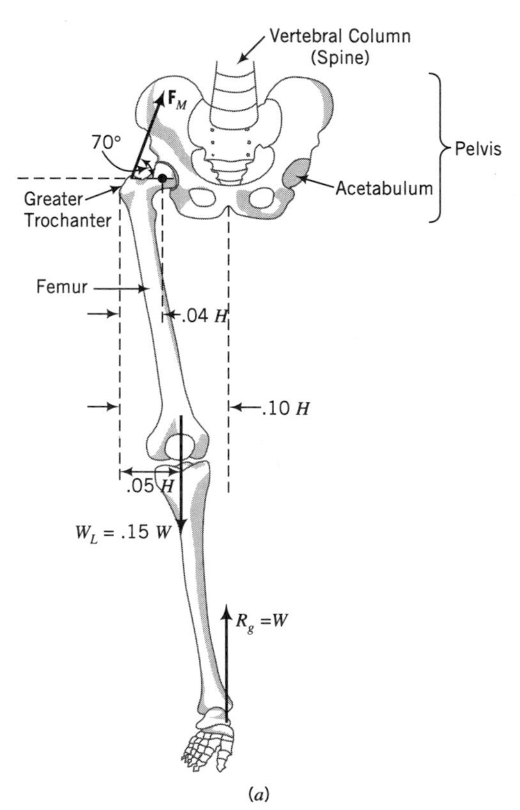

Figure 3.6.a Approximate model of the hip and leg.

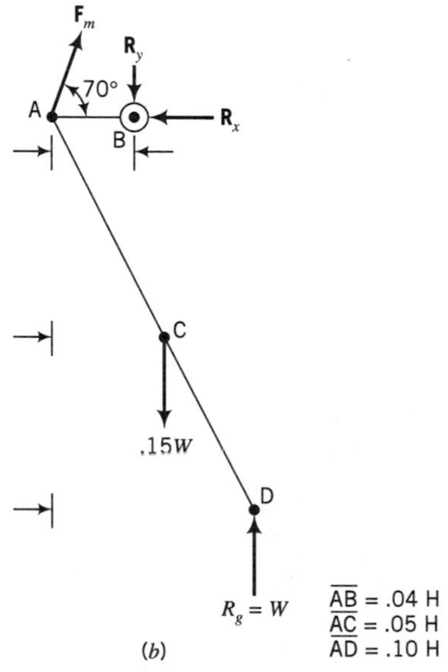

Figure 3.6.b Free-body diagram of the hip and leg.

EXAMPLE 3.4

a. A person (H = 1.60 M and W = 500 N) shown in Figure 3.7.a is descending a series of steps that do not have a handrail. (This is quite common in many sports arenas and concert hall balconies!) The person has the right leg fully extended and in ground contact while the left leg swings through the air. Referring to Figures 3.6.a and 3.6.b, find $\mathbf{F}_M$ (the hip abductor muscle force) and $\mathbf{R}_H$ (the resultant hip joint reaction force).

b. A person shown in Figure 3.7.b is now descending a series of steps for which a handrail has been provided. The person now grips this handrail with the left hand and offloads 15% of body weight ($\mathbf{R}_{HR}$). The FBD for someone gripping this handrail is shown in Figure 3.7.c. Repeat Example 3.4(a) for the person in Figure 3.7.b and find $\mathbf{F}_M$ and $\mathbf{R}_H$.

SOLUTION 3.4(a)

Given: H = 1.60 M, W = 500 N

Find: $\mathbf{F}_M$, $\mathbf{R}_H$

Place the *body* in static equilibrium:
FBD:

$$\sum \mathbf{F}_y = 0\text{: [Figure 3.7.a]:}$$

$$\mathbf{R}_g - W = 0$$

$$\mathbf{R}_g = W$$

The center of mass (weight) of the body must be aligned directly over the right foot (which, in turn experiences a ground reaction force equal to the weight of the body).
Referring to Figure 3.6.b, modify the FBD with the assigned values:

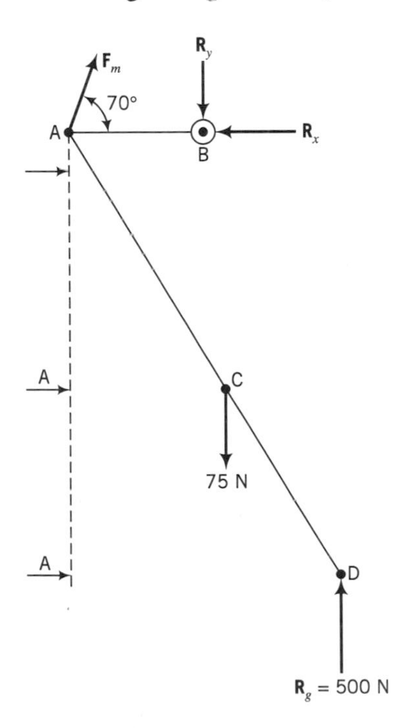

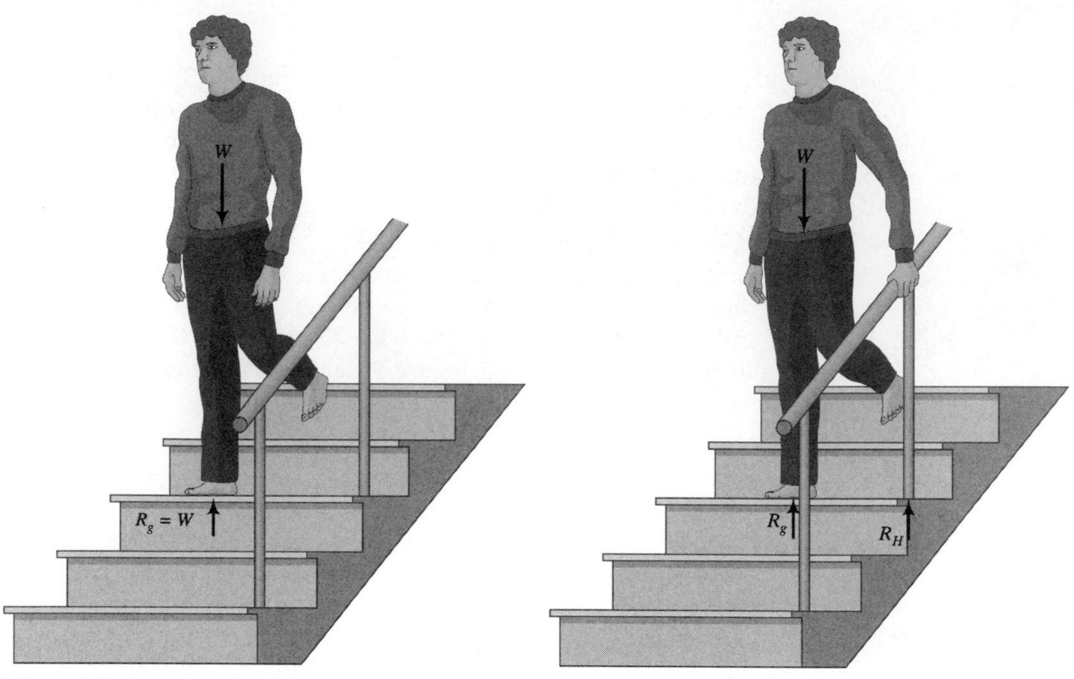

Figure 3.7.a Person descending stairs without a hand rail.

Figure 3.7.b Person descending stairs using a hand rail.

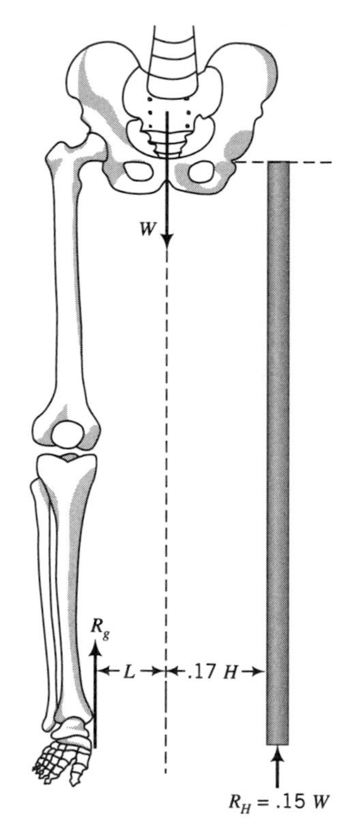

Figure 3.7.c Approximate diagram of person pictured in Figure 3.7.b.

AB = .064 M
AC = .080 M
AD = .16 M

$$\sum \mathbf{F}_y = 0:$$

$$\text{FM} \cdot \sin(70°) - \mathbf{R}_y - 75 + 500 = 0 \tag{i}$$

$$\mathbf{R}_y = (.940)\mathbf{F}_M + 425 \tag{ii}$$

$$\sum \mathbf{F}_x = 0:$$

$$-\mathbf{R}_x + \mathbf{F}_M \cdot \cos(70°) = 0 \tag{iii}$$

$$\mathbf{R}_x = (.342)\mathbf{F}_M \tag{iv}$$

$$\sum M_B = 0:$$

$$-\mathbf{F}_M \cdot \sin(70°) \cdot [.064] - (75)[.080 - .064](500) \tag{v}$$

$$(\mathbf{F}_M)(.940)(.064) = (500)(.096) - (75)(.016) \tag{vi}$$

$$\mathbf{F}_M = \frac{46.8}{.0602} = 777 \text{ N} \tag{vii}$$

Thc force exerted by the hip abductor muscle group is one and a half times the body weight!
Substituting (vii) into (ii):

$$\mathbf{R}_y = (.940)(777) + 425$$

$$\mathbf{R}_y = 1155 \text{ N}$$

Substituting (vii) into (iv):

$$\mathbf{R}_x = (.342)(777) = 266 \text{ N}$$

The hip reaction force ($\mathbf{R}_H$) is the resultant force vector at the hip:

$$R_H = \sqrt{R_y^2 + R_x^2} \;\Big/ \underline{\tan^{-1}\left(\frac{R_y}{R_x}\right)}$$

$$R_H = 1185 \text{ N } \underline{/77.0°}$$

The hip reaction force is two and a third times the body weight! It is at an angle of 77° from the horizontal.

SOLUTION 3.4(b)

Given: $\mathbf{R}_{HR} = 0.15\ W$, $D = .17\ H$ (from Figure 3.7.c)
Find: $\mathbf{F}_M$, $\mathbf{R}_H$

Place the *body* in static equilibrium, using the approximate diagram of Figure 3.7.c, so that the FBD (with the assigned values):

L .27 M
O
500 N
R_g R_{HR} = .75 N

The unknowns (at this point) are the ground reaction force ($\mathbf{R}_g$) directed upward through the right foot and the distance (L) that the right foot is offset from the midline of the body. $\mathbf{R}_{HR}$ is the ground reaction force to the 15% body weight off-loaded to the handrail post [at a distance (D) of 0.27 M from the body center of mass (midline)].

$$\sum \mathbf{F}_y = 0:$$

$$\mathbf{R}_g + 75 - 500 = 0$$

$$\mathbf{R}_g = 425 \text{ N}$$

$$\sum M_0 = 0:$$

$$-(\mathbf{R}_g)(L) + (\mathbf{R}_{HR})(D) = 0$$

$$-(425)) + (75)(.27) = 0$$

$$L = \frac{20.25}{425} = .048 \text{ M}$$

Referring to Figure 3.6.b, we now modify the FBD to account for the new $\mathbf{R}_g$ and L required to keep the body in static equilibrium.

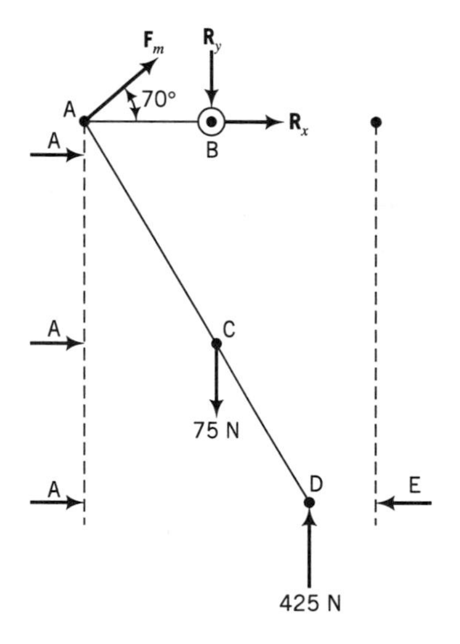

AE = 0.16 M
DE = .048 M
AD = .112 M

$$\text{AC} = .056 \text{ M}$$
$$\text{AB} = .064 \text{ M}$$

where E is the midline of the body.

Note that the AB distance is approximated as constant, and that the center of mass of the leg is still approximated at one-half the AD distance.

$$\sum \mathbf{F}_y = 0:$$

$$\mathbf{F}_M \cdot \sin(70°) - \mathbf{R}_y - 75 + 425 = 0 \quad \text{(i)}$$

$$\mathbf{R}_y = (.940)\mathbf{F}_M + 350 \quad \text{(ii)}$$

$$\sum \mathbf{F}_x = 0:$$

$$-\mathbf{R}_x + \mathbf{F}_M \cdot \cos(70°) = 0 \quad \text{(iii)}$$

$$\mathbf{R}_x = (.342)\mathbf{F}_M \quad \text{(iv)}$$

$$\sum M_B = 0:$$

$$-(\mathbf{F}_M)\sin(70°)[.064] + (75)(.064 - .056) + (425)[.112 - .064] = 0 \quad \text{(v)}$$

$$(\mathbf{F}_M)(.940)(.064) = (425)(.048) + (75)(.008) \quad \text{(vi)}$$

$$\mathbf{F}_M = \frac{21.0}{.0602} = 349 \text{ N} \quad \text{(vii)}$$

The force extended by the hip abductor muscle group (when the handrail is used) is now reduced to less than half the force required when no handrail was used! Substituting (vii) into (ii):

$$\mathbf{R}_y = (.940)(439)350$$
$$\mathbf{R}_y = 678 \text{ N}$$

Substituting (vii) into (iv):

$$\mathbf{R}_x = (.342)(349) = 119 \text{ N}$$

The hip reaction force ($\mathbf{R}_H$) is:

$$R_H = \sqrt{R_y^2 + R_x^2} \;\underline{/\tan^{-1}\left(\frac{R_y}{R_x}\right)}$$

$$R_H = 688 \text{ N } \underline{/80.0°}$$

The hip reaction force is reduced by 42% (when using a hand rail) compared to the force at the hip when no handrail is used.

---

b. Knee and Foreleg

The *anatomical* system for the knee and foreleg is as follows:

Proximal segment: Femur (thigh bone)
Distal segment: Tibia (shin bone) and fibula
Joint: Knee joint

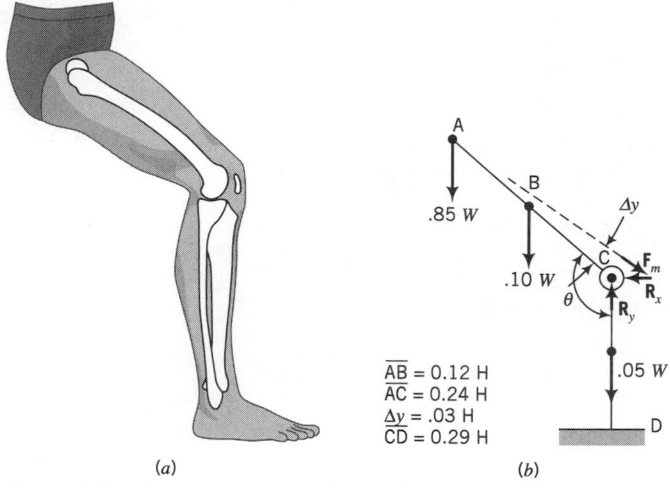

Figure 3.8.a Approximate model of the knee and foreleg.
b Analytical model of the knee and foreleg.

Muscle (action): Quadriceps group (with the knee flexed [back of heel toward the back of the thigh], extends the foreleg (shin bone) forward, as in kicking a football).

The *approximate* model for the knee and foreleg is depicted in Figure 3.8.a, and the *FBD* for the knee and foreleg is presented in Figure 3.8.b.

HFE Application As previously noted regarding the essential elements of the anatomical system, the agonist (and antagonist) muscles cross over the joint and (usually) have an origin on the proximal segment and an insertion on the distal segment. In all of the tasks considered so far, the proximal segment (muscle origin) has been reactive (generating the body's reaction forces) and the distal segment (muscle insertion) has been active (generating the body's active forces). In many tasks, however, this arrangement is reversed, and the distal segment (muscle insertion) is reactive while the proximal segment (muscle origin) is active. Knowing this, some tasks (and the task design parameters) must be analyzed with biostatic mechanical models that reverse the usually reactive and active segments.

EXAMPLE 3.5

a. Some older homes and buildings have a rather high step rise (R). The person ($H = 1.83$ M) shown in Figure 3.9.a has just shifted all of his weight, $W = 670$ N, onto his right foot at the point where the left foot is $h = 1.0$ cm off the ground. If the step rise (R) is 22 cm, find the force exerted by the quadriceps muscle ($\mathbf{F}_M$), and find the vertical and horizontal reaction forces at the knee joint, $\mathbf{R}_y$ and $\mathbf{R}_x$. Use the analytical model of the knee and foreleg shown in Figure 3.8.b.
b. The person now modifies his step climbing by springing off with the left foot and bending forward at the waist as shown in Figure 3.9.b. Repeat Example 3.5(a) and find the $\mathbf{F}_M$, $\mathbf{R}_y$, and $\mathbf{R}_X$.

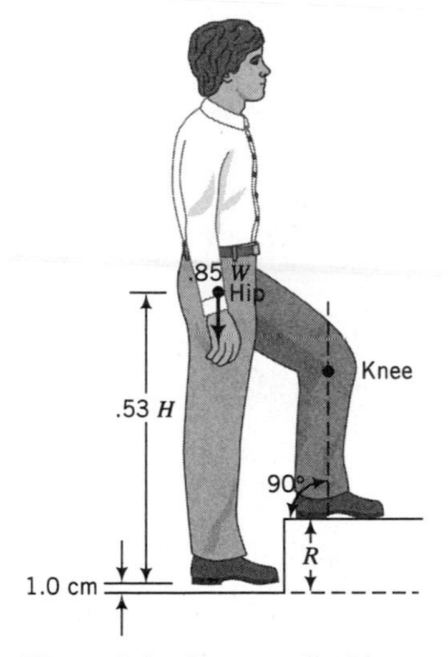

Figure 3.9.a Person climbing a tall step.

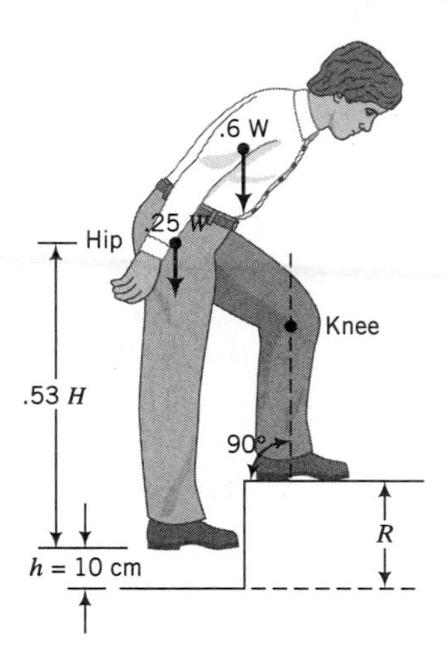

Figure 3.9.b Alternate method for person climbing a tall step.

SOLUTION 3.5(a)

Given: $H = 1.83$ M, $W = 670$ N
Given: $h = 1.0$ cm, $R = 22$ cm
Find: $\mathbf{F}_m$, $\mathbf{R}_x$, $\mathbf{R}_y$

Using the assigned and given values and from Figure 3.9.a, modify the FBD of Figure 3.8.b to account for the location of hip and knee joint:

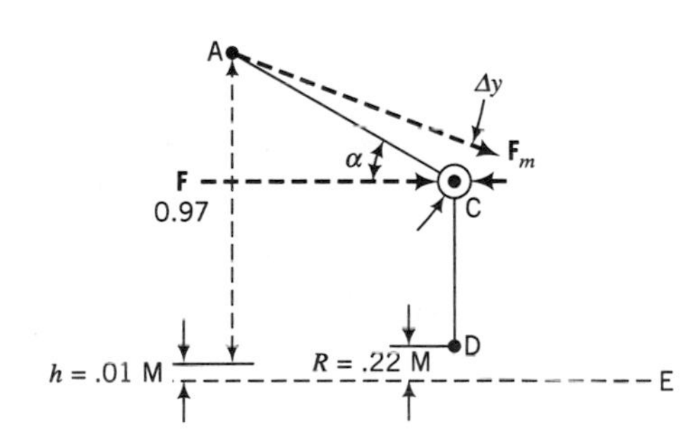

AC = 0.439 M
CD = 0.531 M

Using E as a ground plane:

AE = .980 M
CE = .751 M

Defining F as the datum line through the knee joint:

$$AF = .980 - .751 = .229 \text{ M}$$

Solve for θ, ϕ, and α:

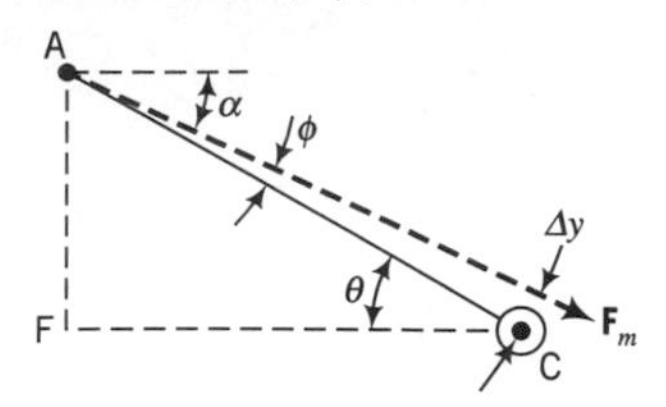

$$\theta = \sin^{-1}\left(\frac{AF}{AC}\right) = \sin^{-1}\left(\frac{.229}{.439}\right) = 31.4°$$

$$\phi = \tan^{-1}\left(\frac{\Delta y}{AC}\right) = \tan^{-1}\left(\frac{.055}{.439}\right) = 7.1°$$

$$\alpha = \theta - \phi = 31.4 - 7.1 = 24.3°$$

Given the assigned values, and θ and α, use the FBD of Figure 3.8.b:

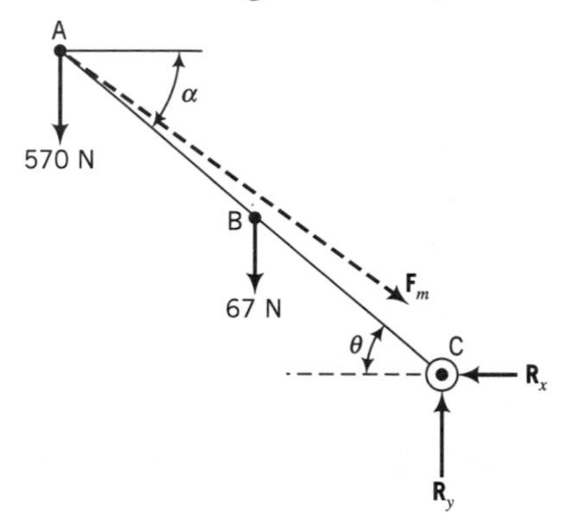

AB = .220 M

AC = .439 M

Note: $\mathbf{R}_y = \mathbf{R}_g - (.05)\,W$, that is, $\mathbf{R}_y$ is the ground reaction force under the right foot minus the weight of the right foreleg and foot.

$$\sum \mathbf{F}_y = 0:$$

$$-\mathbf{F}_M \cdot \sin\alpha - 570 - 67 + \mathbf{R}_y = 0$$

$$\mathbf{R}_y = (.412)\mathbf{F}_M\, 637$$

$$\sum \mathbf{F}_x = 0:$$

$$\mathbf{F}_M \cdot \cos\alpha - \mathbf{R}_x = 0$$

$$\mathbf{R}_x = (.911)\mathbf{F}_M$$

$$\sum M_C = 0$$

$$\mathbf{F}_M \cdot \sin\alpha[.439] \cdot \cos\theta + (570) \cdot [.439] \cdot \cos\theta$$
$$-\mathbf{F}_M \cdot \cos\alpha[.439] \cdot \sin\theta + (67) \cdot [.220]\cos\theta = 0$$

$$(\mathbf{F}_M)(.412)(.439)(.854) + (570)(.439)(.854)$$
$$-(\mathbf{F}_M)(.911)(.439)(.521)(67)(.220)(.854) = 0$$

$$(.208)\mathbf{F}_M - (.154)\mathbf{F}_M = 213.7 + 12.6$$

$$\mathbf{F}_M = \frac{226.3}{.054} = 4191 \text{ N}$$

This is six times the body weight. Most middle-aged and older people could not perform this task in the manner in which it is being attempted.
Solve for $\mathbf{R}_y$:

$$\mathbf{R}_y = (.412)(4191) + 637$$
$$\mathbf{R}_y = 2364 \text{ N}$$

Solve for $\mathbf{R}_X$:

$$\mathbf{R}_x = (.911)(2191) = 3818 \text{ N}$$

Note the *very* high horizontal reaction force at the knee. This force is applied between the patella (knee cap) and the knee joint itself.

SOLUTION 3.5(b)

Given: h = 10 cm, R = 22 cm
Find: $\mathbf{F}_M$, $\mathbf{R}_x$, $\mathbf{R}_y$

Using the assigned and given values, and from Figure 3.9.b, modify the FBD of Figure 3.8.b to account for location of hip and knee joint:

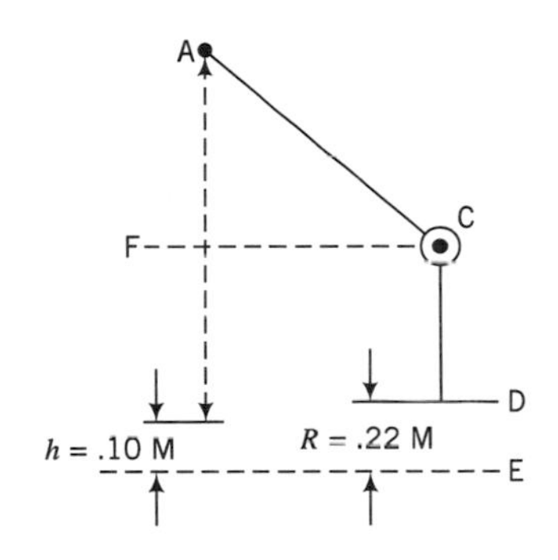

AC = 0.439 M
CD = 0.531 M
AE = 1.07 M
CE = .751 M
AF = 1.07 − .751 = .319 M

Solve for θ, ϕ, and $\propto$:

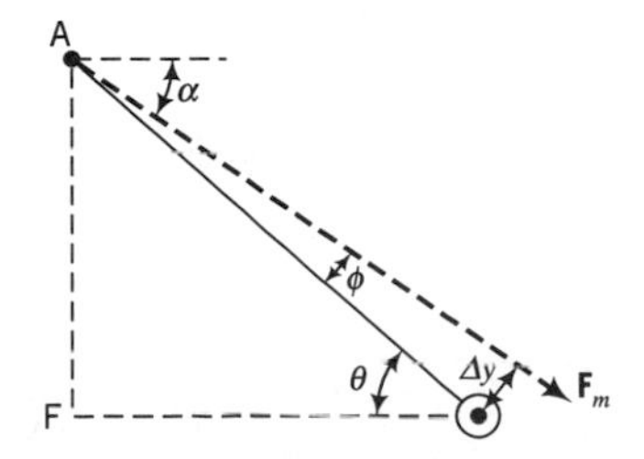

$$\theta = \sin^{-1}\left(\frac{AF}{AC}\right) = \sin^{-1}\left(\frac{.319}{.439}\right) = 46.6°$$

$$\phi = \tan^{-1}\left(\frac{\Delta y}{AC}\right) = \tan^{-1}\left(\frac{.055}{.439}\right) = 7.2°$$

$$\alpha = \theta - \phi = 46.6 - 7.1 = 39.4°$$

Modify the FBD of Figure 3.8.b to represent the person in Figure 3.9.b:

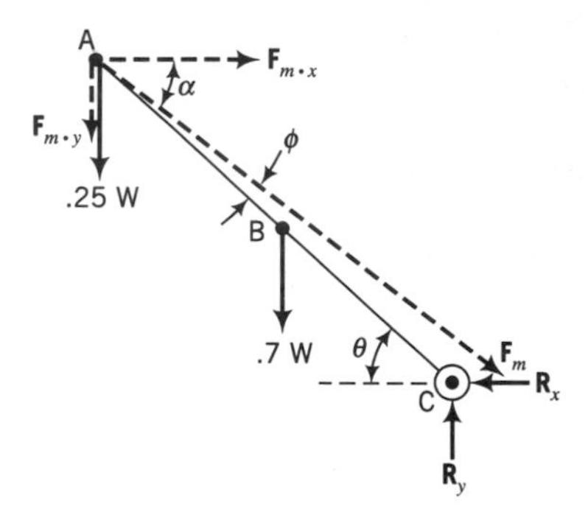

AB = .220 M
AC = .439 M

Given the assigned values, and θ and $\propto$, define the FBD for the person in Figure 3.9.b:

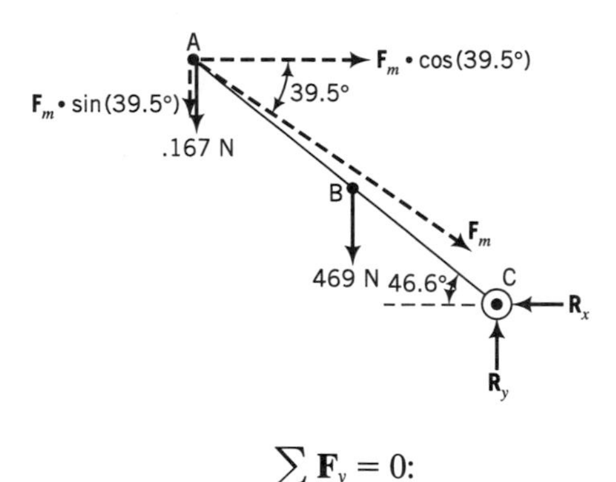

$$\sum \mathbf{F}_y = 0:$$

$$-\mathbf{F}_M(.636) - 167 - 469 + \mathbf{R}_y = 0$$

$$\mathbf{R}_y = (.636)\mathbf{F}_M + 636$$

$$\sum \mathbf{F}_x = 0:$$

$$\mathbf{F}_M(.772) - \mathbf{R}_x = 0$$

$$\mathbf{R}_x = (.772)\mathbf{F}_M$$

$$\sum M_c = 0:$$

$$\mathbf{F}_M(.636)[.439](.687) + (167)[439](.687)$$

$$-\mathbf{F}_M(772)[.439](.727) + (469)[.220](.687) = 0$$

$$(.246)\mathbf{F}_M - (.192)\mathbf{F}_M = 70.9 + 50.4$$

$$\mathbf{F}_M = \frac{121.3}{.054} = 2250 \text{ N}$$

This is now $3\frac{1}{3}$ times the body weight, since the quadriceps are a very powerful muscle group, and our alternate method has approximately halved the required force. Hence, the posture depicted in Figure 3.9.b is the better way to perform this task.
Solve for $\mathbf{R}_y$:

$$\mathbf{R}_y = (.636)(2250) + 636$$
$$\mathbf{R}_y = 2070 \text{ N}$$

Solve for $\mathbf{R}_x$:

$$\mathbf{R}_x = (.772)(2250) = 1740 \text{ N}$$

Note that $\mathbf{R}_y$ did not change that much, but that $\mathbf{R}_X$ was reduced by more that half!

c. Ankle and Foot

The *anatomical* system for the ankle and foot is as follows:

Proximal segment: Tibia (shin bone) and fibula
Distal segment: Talus (ankle bone) and other foot bones
Joint: Ankle joint
Muscle (action): Gastrocnemius muscle (plantar flexes the foot at the ankle, as the downward motion when tapping with the foot. After the ball and toes of the foot have been raised off the floor, plantar flexion results in the ball of the foot being brought down to contact the ground).

The *approximate model* for the ankle and foot is presented in Figure 3.10.a and the *free-body diagram* is shown in Figure 3.10.b.

HFE Application Plantar flexion strength of the foot is a function of the length of the gastrocnemius muscle (which runs the length of the foreleg and inserts on the back of the calcaneus (heel bone). As was the case for the finger flexors (see Example 3.3), the force-length relationship is frequently expressed as a force-joint

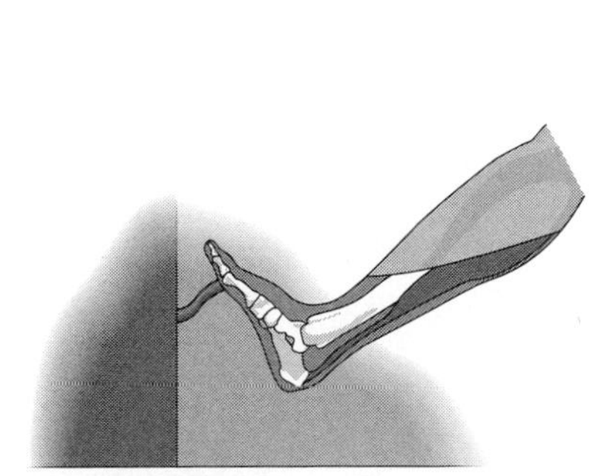

Figure 3.10.a Approximate model of ankle and foot.

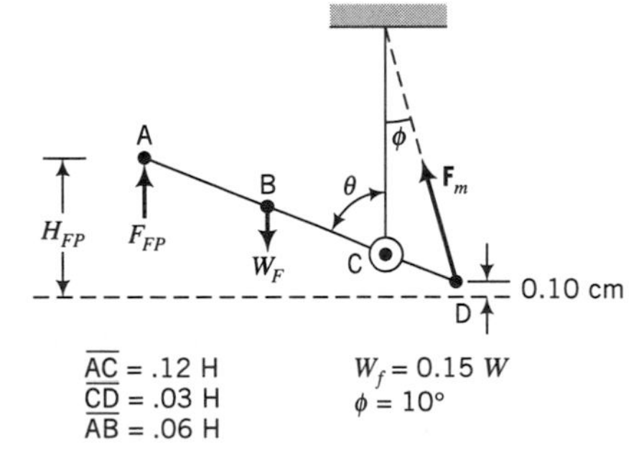

Figure 3.10.b Analytic model of ankle and foot.

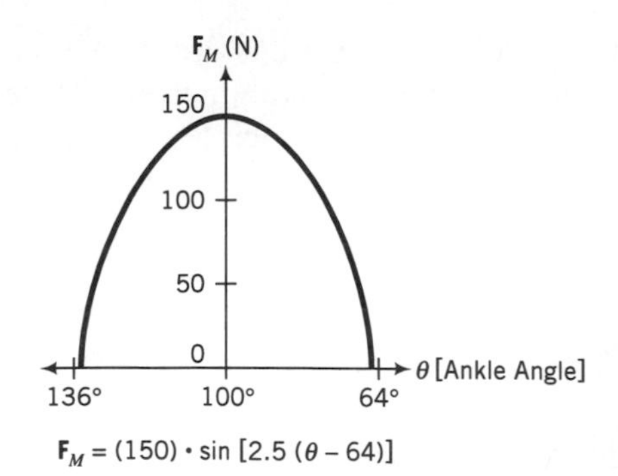

Figure 3.11 Plantar flexion force of the foot as a function of ankle angle (θ). The approximation is made that both heel and toe are off the ground and that ankle joint reaction forces at F_M = 150 N are less than or greater than the body weight W.

angle relationship. This is shown for the gastrocnemius muscle in Figure 3.11. The presentation of data in this format is useful to the HFE design engineer.

EXAMPLE 3.6

a. A foot pedal activated trash basket requires 50 N of foot pedal force ($\mathbf{F}_{FP}$) in order to activate the lid raising operation. The system shown in Figure 3.10.a and 3.10.b is being operated by a person of height, H = 1.64 M and weight, W = 600 N. Use the FBD of Figure 3.10.b to identify the weight of the foot (W_F), the angle at the ankle (θ), and the force exerted by the plantar flexor muscle ($\mathbf{F}_M$) acting at an angle (ϕ) to the lower leg. Also use the muscle force ($\mathbf{F}_M$) versus ankle joint angle (θ) relationship of Figure 3.11. If the lowest point of the foot (at the heel) is h = 0.1 cm above ground level when the front of the foot pushes down on the pedal, find the maximum height above ground level (H_{FP}) that the pedal could be placed in order to assure the minimum required 50 N of foot pedal force.
b. For the joint angle θ and F_{FP} from Example 3.6(a), proceed to find F_M and the joint reaction forces, ($\mathbf{R}_y$ and $\mathbf{R}_x$) at the ankle.

SOLUTION 3.6(a)

Given: H = 1.64 M, W = 600 N
Given: h = 0.1 cm, $\mathbf{F}_{FP}$ = 50 N
Find: $\mathbf{H}_{FP}$

Using the force-angle data of Figure 3.11, find θ for the ankle:

$$50 = (150) \cdot \sin[2.5(\theta - 64)]$$

$$2.5\,\theta - 160 = \sin^{-1}(.333) = 19.5$$

$$\theta = \frac{179.5}{2.5} = 71.8°$$

Using the assigned and given data and the θ, define the FBD of Figure 3.10.b to solve for the height of the foot pedal ($\mathbf{H}_{FP}$).

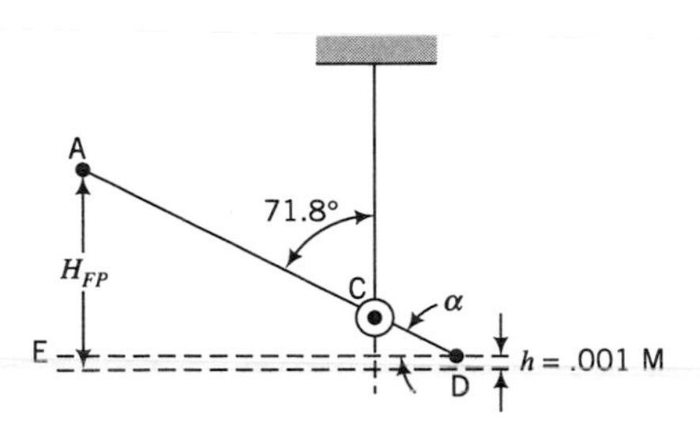

AC = .197 M
CD = .049 M

$$\mathbf{H}_{FP} = AE + .001$$
$$\alpha = 90° - 71.8° = 18.2°$$
$$AD = .197 + .049 = .246 \text{ M}$$
$$AE = AD \cdot \sin(\alpha)$$
$$AE = (.246)(.312) = .077 \text{ M}$$
$$\mathbf{H}_{FP} = .077 + .001 = .078 \text{ M}$$

The height of the foot pedal above ground level should be no more than 3 inches!

SOLUTION 3.6(b)

Given: $\mathbf{F}_{FP} = 50$ N, $\theta = 71.8°$ [from Solution 3.6(a)]
Find: $\mathbf{F}_M$, $\mathbf{R}_x$, $\mathbf{R}_y$

Using the assigned and given data, modify the FBD of Figure 3.10.b:

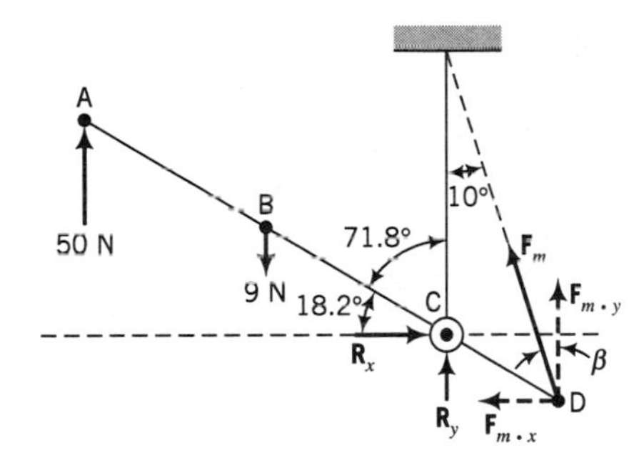

As per Example 3.6(a):

BC = .098
AC = .197
CD = .049
α = 18.2°.

Find β: β may be approximated as 10° since Figure 3.10.b states that ϕ may be approximated as 10° regardless of ankle angle θ.

$$\sum \mathbf{F}_y = 0:$$

$$\mathbf{R}_y + 50 - 9 + \mathbf{F}_M \cdot \cos(10°) = 0$$

$$\mathbf{R}_y = -(.985)\mathbf{F}_M - 41$$

$$\sum \mathbf{F}_x = 0:$$

$$\mathbf{R}_x - \mathbf{F}_M \cdot \sin(10°) = 0$$

$$\mathbf{R}_x = (.174)\mathbf{F}_M$$

$$\sum M_C = 0:$$

$$-(50)(AC)\cos[18.2°] + (9)(BC)\cos(18.2°) + (\mathbf{F}_M)\cdot\cos[10°](CD)\cdot\cos(18.2°)$$

$$-(\mathbf{F}_M)\cdot\sin[10°](CD)\cdot\sin(18.2°) = 0$$

$$-(50)[.197](.950) + (9)(.098)(.950) + (\mathbf{F}_M)(.985)(.049)(.950) - (\mathbf{F}_M)(.174)(.049)(.312) = 0$$

$$(.046)\mathbf{F}_M - (.003)\mathbf{F}_M = 9.36 - 0.84$$

$$\mathbf{F}_M = \frac{8.52}{.043} = 198\text{ N}$$

Find $\mathbf{R}_y$:

$$\mathbf{R}_y = -(.985)(198) - 41$$

$$\mathbf{R}_y = 236\text{ N (downwards)}$$

Find $\mathbf{R}_X$:

$$\mathbf{R}_x = (.174)(198) = 34.5\text{ N (rightward)}$$

3.4 BENDING, LIFTING, AND CARRYING

When bending over, lifting or carrying objects, the *anatomical system* of interest is the same for all three activities. This section will first define the essential elements of that system, and then describe the relevant model. Subsequently, bending, lifting, and carrying will be treated as three separate activities with respect to the HFE applications.

The worked examples that illustrate the HFE applications are solved in terms of a nonspecific body height (H) and a nonspecific body weight (W). This will allow the student to compare and contrast forces between the three types of activities. It will also demonstrate that the system model is independent of body height (H).

The *anatomical system* for the back is:

Proximal segment: Sacrum/pelvis (although the sacrum is a continuation of the spine, it is a functional unit with the pelvis).

Distal segment: Thoracolumbar spine (consisting of twelve thoracic vertebral bones and five lumbar vertebral bones).

Joint: Lumbo-sacral joint (defined as the intervertebral space between the fifth lumbar vertebral bone and first sacral vertebral bone within which the lumbosacral disc is contained).

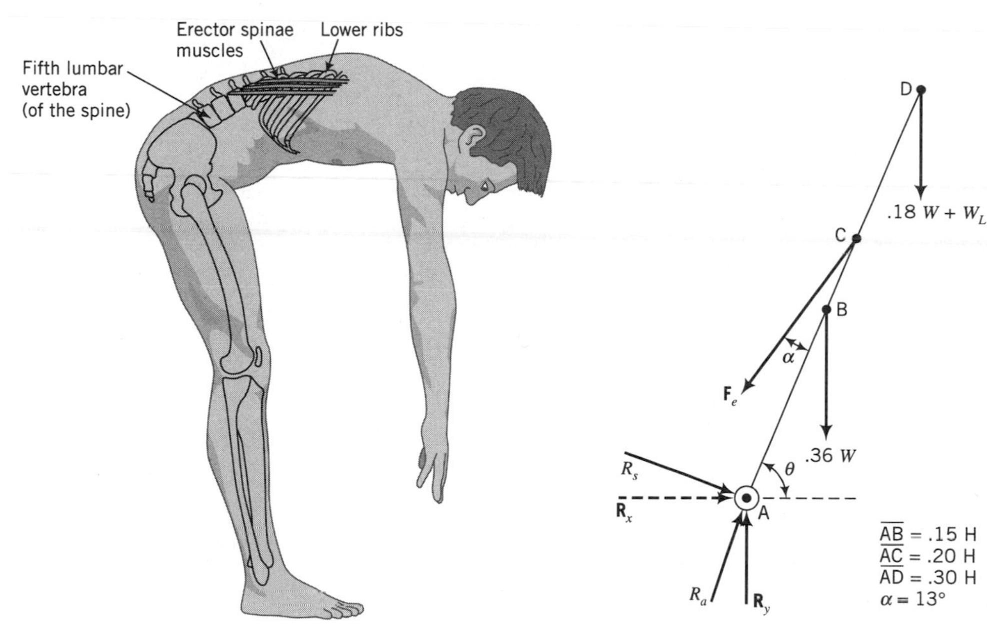

Figure 3.12.a Approximate model of the back.

Figure 3.12.b Free-body diagram of the thoraco-lumbar spine.

Muscle (action): Erector spinae and sacrospinalis muscles (extension of the spine, as when we stand straight up, at attention, during a military inspection).

The *approximate model* for the back is presented in Figure 3.12.a, and the *free-body diagram* for the back is presented in Figure 3.12.b.

HFE Application Note that in the FBD of the thoracolumbar spine [Figure 3.12.b], the vertical and horizontal reaction forces ($\mathbf{R}_y$ and $\mathbf{R}_x$) at the lumbosacral joint are translated to $\mathbf{R}_a$ (axial reaction force along the central axis of the spine) and $\mathbf{R}_s$ (shear reaction force perpendicular to the axis of the spine). This is because these are the forces that cause many of the injuries to the low back. Excessive axial forces can result in the fracture of vertebral bones and/or injury of intervertebral disc. Excessive shear forces can result in the dislocation of adjacent vertebral bones and also injury to the intervertebral disc.

EXAMPLE 3.7

a. For the vertical and horizontal reaction forces ($\mathbf{R}_y$ and $\mathbf{R}_x$) in Figure 3.12.b, derive the equations that translate these forces to the axial and shear reaction forces ($\mathbf{R}_a$ and $\mathbf{R}_s$).

b. In Example 3.7(a), both $\mathbf{R}_a$ and $\mathbf{R}_s$ were derived in terms of $\mathbf{R}_y$ and $\mathbf{R}_x$ that followed positive sign convention. Indicate how the $\mathbf{R}_a$ and $\mathbf{R}_s$ equations from Example 3.7(a) would be modified if $\mathbf{R}_y$ or $\mathbf{R}_x$ or both were negative.

SOLUTION 3.7(a)

For $\mathbf{R}_a$:

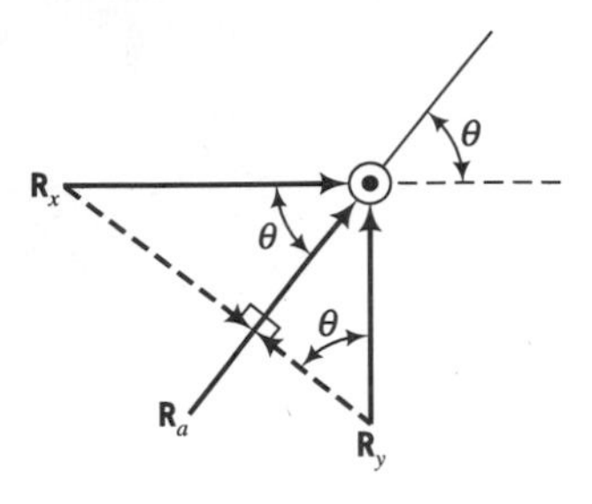

$$\mathbf{R}_a = \mathbf{R}_y \cdot \sin\theta + \mathbf{R}_x \cdot \cos\theta \qquad \text{(i)}$$

For $\mathbf{R}_s$:

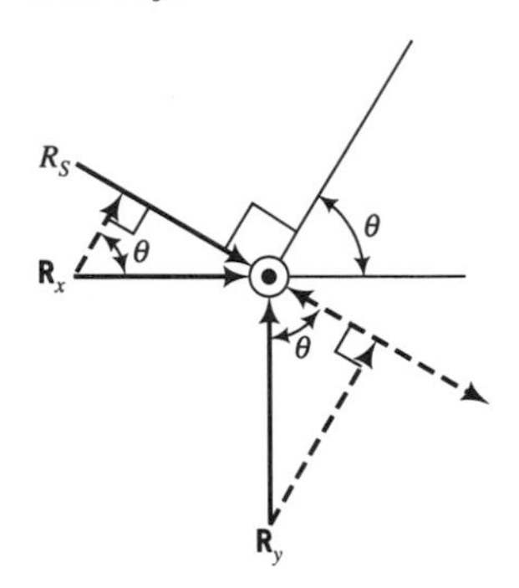

$$\mathbf{R}_s = -\mathbf{R}_y \cdot \cos\theta + \mathbf{R}_x \cdot \sin\theta \qquad \text{(ii)}$$

SOLUTION 3.7(b)

When $\mathbf{R}_y$ and/or $\mathbf{R}_x$ are negative (by sign convention), equations 3.7(a)(i) and 3.7(a)(ii)are modified by inserting a $-\mathbf{R}_y$ and/or $-\mathbf{R}_x$ in place of the $\mathbf{R}_y$ and $\mathbf{R}_x$ terms.

a. Bending

HFE Application Workplace design is an important function of the human factors engineer.

One aspect of this is the design of worktable height and working surface area so that people do not experience back discomfort or unnecessary back fatigue.

When working in the erect position, the muscles of the back should not be subjected to undue stress.

EXAMPLE 3.8

a. The person (of body height, H_B and body weight, W_B) pictured in Figure 3.13.a is working in the erect position with a worktable at a correct (midthigh)

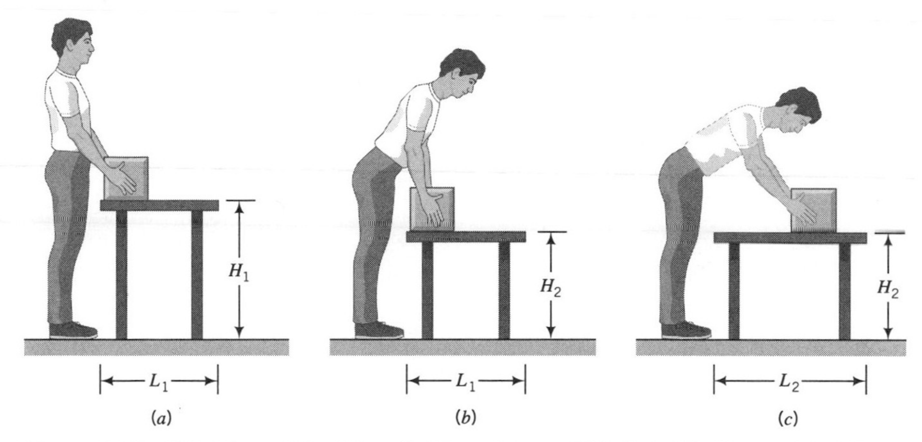

Figure 3.13.a Working table (H_1 = 0.4H and L_1 = 0.22H) **b** Working table (H_1 = 0.28H and L_1 = 0.33H) **c** Working table (H_1 = 0.28H and L_1 = 0.33H)

height (H_1) and a work area of correct arm reach length (L_1) so that the thoracolumbar spine is at an angle, $\theta = 85°$ (see Figure 3.12.b). Per Table 3.2 and Figure 3.12.b, the weight of the thorax and abdomen is 36% of W_B acting through the center (point B) of the thoracolumbar spine. The weight of the head, neck and both arms is 18% of W_B acting through the end (point D) of the thoracolumbar spine (Table 3.2 and Figure 3.12.b). The weight of any additional load (W_L) carried at point D is zero. Calculate the extersor muscle force ($\mathbf{F}_e$) directed at an angle, $\alpha = 13°$ (point C) to the axis of the spine, and calculate the axial reaction force ($\mathbf{R}_a$) and shear reaction force ($\mathbf{R}_s$) at the base of the spine (point A).

b. The person pictured in Figure 3.13.b is now working in the erect position, but the worktable is at a too low (knee-high) height (H_2) even though the work area is of the correct arm-reach length (L_1), so that the thoracolumbar spine is now inclined at an angle, $\theta = 75°$. Keeping all other parameters the same as per Example 3.8(a), calculate $\mathbf{F}_e$, $\mathbf{R}_a$, and $\mathbf{R}_s$.

c. The person pictured in Figure 3.13.c is now working in the erect position, but the worktable is at a too low (knee-high) height (H_2), and also the work area is excessively long at an extended arm-reach length (L_2) that requires both reaching out and down. In this case, the thoracolumbar spine is now inclined at an angle of 60°. Keeping all other parameters the same as per Example 3.8(a), calculate $\mathbf{F}_e$, $\mathbf{R}_a$, and $\mathbf{R}_s$.

SOLUTION 3.8(a)

Given: $H = H_B$, $W = W_B$

Given: $\theta = 85°$

Find: $\mathbf{F}_e$, $\mathbf{R}_a$, $\mathbf{R}_s$

Note that we are now making our solutions more applicable to a general (worker) population. H is *any* body height (H_B), and W is *any* body weight (W_B).

Use the FBD of Figure 3.12.b, where $\theta = 85°$, $H = H_B$, $W = W_B$, and $W_L = 0$: First resolve $\mathbf{F}_e$ into $\mathbf{F}_{ex}$ and $\mathbf{F}_{ey}$ as follows:

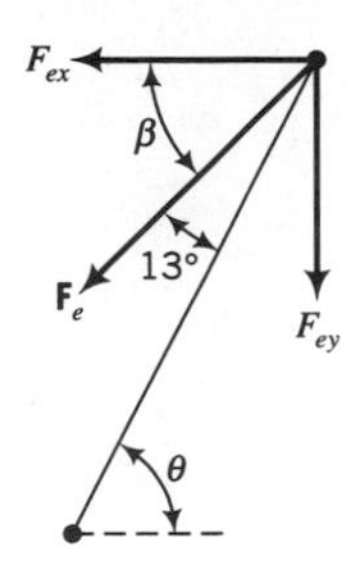

$$\theta = \beta + 13° \quad \text{(i)}$$

For this example: $\beta = \theta - 13° = 85° - 13° = 72°$

$$\mathbf{F}_{ex} = \mathbf{F}_e \cdot \cos(\beta) = (.309)\mathbf{F}_e \quad \text{(ii)}$$

$$\mathbf{F}_{ey} = \mathbf{F}_e \cdot \sin(\beta) = (.951)\mathbf{F}_e \quad \text{(iii)}$$

$$\sum F_y = 0:$$

$$\mathbf{R}_y - .36W_B - (.951)\mathbf{F}_e - .18W_B = 0$$

$$\mathbf{R}_y = (.951)F_e + .54W_B$$

$$\sum \mathbf{F}_x = 0:$$

$$\mathbf{R}_x - (.309)\mathbf{F}_e = 0$$

$$\mathbf{R}_x = (.309)\mathbf{F}_e$$

$$\sum M_A = 0:$$

$$\mathbf{F}_{ex}(.20H_B)\sin(85°) - \mathbf{F}_{ey}(.20H_B)\cos(85°) - (.36)W_B(.15H_B)\cos(85°) - (.18)W_B(.30H_B)\cos(85°) = 0$$

Note that H_B is common to all four terms and divides out:

$$\mathbf{F}_{ex}(.20)\sin(85°) - \mathbf{F}_{ey}(.20)\cos(85°) - (.36)W_B(.15)\cos(85°) - (.18)W_B(.30)\cos(85°) = 0$$

Substituting equations (ii) and (iii) into the above and separating variables:

$$(.309)\mathbf{F}_e(.20)(.996) - (.951)\mathbf{F}_e(.20)(.087) = (.36)W_B(.15)(.087) + (.18)W_B(.30)(.087)$$

And proceed accordingly:

$$\mathbf{F}_e[.0616 - .0165] = W_B[.0047 + .0047]$$

$$\mathbf{F}_e = \frac{.0094}{.0451} = .208W_B$$

$$\mathbf{R}_y = (.951)(.208W_B) + .54W_B$$

$$\mathbf{R}_y = .738W_B$$

$$\mathbf{R}_x = (.309)(.208W_B) = .064W_B$$

Substituting the above into equations 3.7(a)(i) and 3.7(a)(ii):

$$\mathbf{R}_a = (.738)W_B(.996) + (.064)W_B(.087)$$

$$\mathbf{R}_a = 0.741W_B$$

$$\mathbf{R}_s = -(.738)W_B(0.87) + (.064)W_B(.996)$$

$$\mathbf{R}_s = -0.578W_B$$

In an erect position, and only body weight forces being experienced, the lumbosacral joint is subjected to an axial force of three fourths of the body weight.

SOLUTION 3.8(b)

Given: as per Example 3.8(a)

Given: $\theta = 75°$

Find: $\mathbf{F}_e$, $\mathbf{R}_a$, $\mathbf{R}_s$

Use the FBD of Figure 3.12.b where $\theta = 75°$ and $W_L = 0$:
From equation 3.8(a)(i): For $\theta = 75°$, then $\beta = 62°$
From equation 3.8(a)(ii): $\mathbf{F}_{ex} = \mathbf{F}_e \cdot \cos(62°) = .469\mathbf{F}_e$
From equation 3.8(a)(iii): $\mathbf{F}_{ey} = \mathbf{F}_e \cdot \sin(62°) = .883\mathbf{F}_e$

$$\sum \mathbf{F}_y = 0:$$

$$\mathbf{R}_y - .36W_B - .883F_e - .18W_B = 0$$

$$\mathbf{R}_y = (.883)F_e + .54W_B$$

$$\sum \mathbf{F}_x = 0:$$

$$\mathbf{R}_x = -(.469)\mathbf{F}_e = 0$$

$$\mathbf{R}_x = (.469)F_e$$

$$\sum M_A = 0:$$

$$\mathbf{F}_{ex}(.20H_B)\sin(75°) - \mathbf{F}_{ey}(.20H_B)\cos(75°) - (.36)W_B(.15H_B)\cos(75°) - (.18)W_B(.30H_B)\cos(75°) = 0$$

Dividing through by H_B and substituting equation 3.8(a)(ii) and 3.8(a)(iii):

$$(.469)\mathbf{F}_e(.20)(.966) - (.883)\mathbf{F}_e(.20)(.25) = (.36)W_B(.15)(.259) + (.18)W_B(.30)(.259)$$

$$\mathbf{F}_e[.0906 - .0457] = W_B[.0140 + .0140]$$

$$F_e = \left(\frac{.0280}{449}\right)W_B = .624W_B$$

$$\mathbf{R}_y = (.883)(.624W_B) + .54W_B$$

$$\mathbf{R}_y = 1.09W_B$$

$$\mathbf{R}_x = (.469)(.624W_B) = 0.29W_B$$

Substituting the above into equations 3.7(a)(i) and 3.7(a)(ii):

$$\mathbf{R}_a = (1.09)W_B(.966) + (.29)W_B(.259)$$

$$\mathbf{R}_a = 1.13W_B$$

$$\mathbf{R}_s = -(1.09)W_B(.259) + (.29)W_B(.966)$$

$$\mathbf{R}_s = .002W_B \text{ (leftward)}$$

Poor workplace design, requiring the person to bend at the waist by an additional 10°, *increases* lumbosacral axial forces by 65%!

SOLUTION 3.8(c)

Given: as per Example 3.8(a)

Given: $\theta = 60°$

Find: $\mathbf{F}_e$, $\mathbf{R}_a$, $\mathbf{R}_s$

Once again, use the FBD of Figure 3.12.b, where $\theta = 60°$ and $W_L = 0$:
From equation 3.8(a)(i): For $\theta = 60°$, $\beta = 47°$
From equation 3.8(a)(ii): $\mathbf{F}_{ex} = \mathbf{F}_e \cdot \cos(47°) = .682\mathbf{F}_e$
From equation 3.8(a)(iii): $\mathbf{F}_{ey} = \mathbf{F}_e \cdot \sin(47°) = .731\mathbf{F}_e$

$$\sum F_y = 0:$$

$$\mathbf{R}_y - .36W_B - .731F_e - .18W_B = 0$$

$$\mathbf{R}_y = (.731)\mathbf{F}_e + .54W_B$$

$$\sum F_x = 0:$$

$$\mathbf{R}_x = -(.682)\mathbf{F}_e = 0$$

$$\mathbf{R}_x = (.682)\mathbf{F}_e$$

$$\sum M_A = 0:$$

$$\mathbf{F}_{ex}(.20H_B)\sin(60°) - \mathbf{F}_{ey}(.20H_B)\cos(60°) - (.36)W_B(.15H_B)\cos(60°) - (.18)W_B(.30H_B)\cos(60°) = 0$$

Dividing through by H_B and substituting equation 3.8(a)(ii) and 3.8(a)(iii):

$$(682)\mathbf{F}_e(.20)(.866) - (.731)\mathbf{F}_e(.20)(.5) = (.36)W_B(.15)(.5) + (.18)W_B(.30)(.5)$$

$$\mathbf{F}_e[.1181 - .0731] = W_B[.0270 - .0270]$$

$$\mathbf{F}_e = \left(\frac{.054}{.045}\right)W_B = 1.20W_B$$

$$\mathbf{R}_y = (.731)(1.20)W_B + .54W_B$$

$$\mathbf{R}_y = 1.42W_B$$

$$\mathbf{R}_x = (.682)(1.20)W_B = 0.82W_B$$

Substituting the above into equation 3.7(a)(i) and 3.7(a)(ii):

$$\mathbf{R}_a = (1.42)W_B(.866) + (.82)W_B(.5)$$

$$\mathbf{R}_a = 1.64W_B$$

$$\mathbf{R}_s = -(1.42)W_B(.5) + (.82)W_B(.866)$$

$$\mathbf{R}_s = .0001W_B \text{ (rightward)}$$

Extremely poor workplace design, requiring the person to bend at the waist by 30°, *increases* the lumbosacral axial forces by 120%!! Imagine that you had to work an eight-hour shift in this manner (five days per week)!!

---

b. Lifting

HFE Application Manual work often requires people to lift rather heavy loads with their arms and hands. The human factors engineer should understand the forces placed upon the back muscles and the spine itself in order to avoid manual work methods that are excessively stressful and to design manual lifting tasks that minimize these stressful forces. The design process involves not only the technique (posture) with which the person lifts but also the load itself that is lifted. With respect to technique or posture, industrial hygiene signs instruct "lift with the legs,

Figure 3.14.a Person lifting with the back.

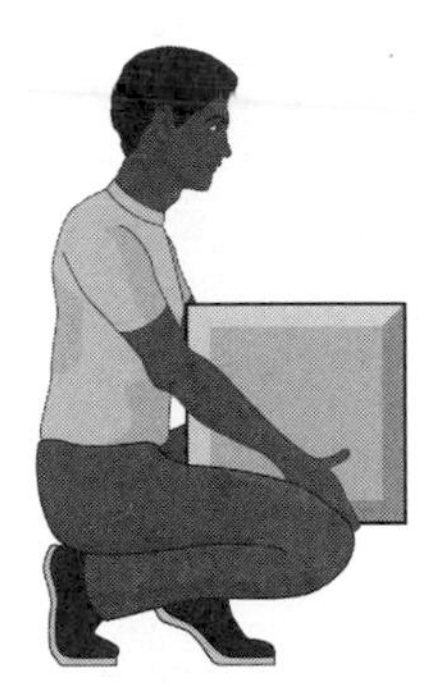

Figure 3.14.b Person lifting with the legs.

Figure 3.14.c Person lifting a divided load with handle (profile view: one side only).

not with the back," that is, keep the spine as vertical as possible. With respect to load design, the HFE should consider (when practicable) dividing the load into two equal halves *and* providing gripping handles on the top and center of each load.

EXAMPLE 3.9

a. A person (of body height, H_B and body weight, W_B) is lifting a load (W_L) of 0.2W with the back (as shown in Figure 3.14.a) so that the thoracolumbar spine is inclined at an angle, $\theta = 35°$. Using the free body diagram FBD of Figure 3.12.b, calculate $\mathbf{F}_e$, $\mathbf{R}_a$, and $\mathbf{R}_s$.
b. A person is now lifting a load (W_L) of 0.2W with the legs (as shown in Figure 3.14.b) so that the thoracolumbar spine is now inclined at an angle, $\theta = 75°$. Using the FBD of Figure 3.12.b, calculate $\mathbf{F}_e$, $\mathbf{R}_a$, and $\mathbf{R}_s$.
c. The person is now lifting a load (W_L) of 0.2W, which has been divided into two loads of 0.1 W each, and each load is provided with handles (as shown in Figure 3.14.c). In this case the thoracolumbar spine is now inclined at an angle, $\theta = 85°$. Using the FBD of Figure 3.12.b, calculate $\mathbf{F}_e$, $\mathbf{R}_a$, and $\mathbf{R}_s$.

SOLUTION 3.9(a)

Given: $H = H_B$, $W = W_B$
Given: $\theta = 35°$, $W_L = 0.2W_B$
Find: $\mathbf{F}_e$, $\mathbf{R}_a$, $\mathbf{R}_s$

Using the FBD of Figure 3.12.b where $\theta = 35°$ and $W_L = 0.2W_B$:
For

$$\theta = 35°, \beta = 22° \quad \text{(i)}$$

$$\mathbf{F}_{ex} = \mathbf{F}_e \cdot \cos(22°) = .927\mathbf{F}_e \quad \text{(ii)}$$

$$\mathbf{F}_{ey} = \mathbf{F}_e \cdot \sin(22°) = .375\mathbf{F}_e \quad \text{(iii)}$$

$$\sum \mathbf{F}_y = 0:$$

$$\mathbf{R}_y - .36W_B - .375F_e - .38W_B = 0$$

$$\mathbf{R}_y = (.75)\mathbf{F}_e + .74W_B$$

$$\sum \mathbf{F}_x = 0:$$

$$\mathbf{R}_x = -(.927)F_e = 0$$

$$\mathbf{R}_x = (.927)F_e$$

$$\sum M_A = 0:$$

$$\mathbf{F}_{ex}(.20H_B)\sin(35°) - \mathbf{F}_{ey}(.20H_B)\cos(35°)$$
$$- (.36)W_B(.15H_B)\cos(35°) - (.38)W_B(.30H_B)\cos(35°) = 0$$

Eliminating H_B and substituting equations (ii) and (iii):

$$(.927)F_e(.20)(.574) - (.375)F_e(.20)(.819) = (.36)W_B(.15)(.819) + (.38)W_B(.30)(.819)$$

$$\mathbf{F}_e[.1064 - .0614] = W_B[.0442 + .0934]$$

$$\mathbf{F}_e = \frac{.1376}{.045} = 3.06W_B$$

$$\mathbf{R}_y = (.731)(1.20)W_B + .54W_B$$

$$\mathbf{R}_y = 1.89W_B$$

$$\mathbf{R}_x = (.927)(3.06)W_B = 2.84W_B$$

Substituting the above into equation 3.7(a)(i) and 3.7(a)(ii):

$$\mathbf{R}_a = (1.89)W_B(.0574) + (2.84)W_B(.819)$$

$$\mathbf{R}_a = 3.41W_B$$

$$\mathbf{R}_s = -(1.89)W_B(.819) + (2.84)W_B(.574)$$

$$\mathbf{R}_s = 0.08W_B \text{ (leftward)}$$

Lifting a load of $0.2W_B$ *with the back* results in an axial force at the lumbosacral joint of 340% of the body weight! A rather high shearing force of about 50% of the body weight is also experienced!

SOLUTION 3.9(b)

Given: $H = H_B$, $W = W_B$
Given: $\theta = 75°$, $W_L = 0.2W_B$
Find: $\mathbf{F}_e$, $\mathbf{R}_a$, $\mathbf{R}_s$

Use the FBD of Figure 3.12.b where $\theta = 75°$ and $W_L = 0.2W_B$:
The solution proceeds exactly as in Example 3.9(a) except $\theta = 75°$. Note that $W_L = 0.2W_B$ is added to $.18W_B$ (the weight of the head and arms) so that $W_T = 0.38W_B$ is now acting at the distal end of the thoracolumbar spine (point D).

The results are:

$$\mathbf{F}_e = 0.97W_B$$
$$\mathbf{R}_y = 1.60W_B$$
$$\mathbf{R}_x = 0.45W_B$$
$$\mathbf{R}_a = 1.66W_B$$
$$\mathbf{R}_s = .024W_B$$

Lifting a load of $0.2W_B$ *with the legs* reduces the axial lumbosacral forces by 53% as compared to lifting the same load with the back.
Note also that the shearing force is now only 2.5% of body weight, a reduction of 95%. With respect to $\mathbf{R}_a$ the lifting task is now equivalent to a bending task where $\theta = 60°$ [e.g., see Example 3.8(c)].

SOLUTION 3.9(c)

Given: $H = H_B$, $W = W_B$
Given: $\theta = 85°$, $W_L = 0.2W_B$
Find: $\mathbf{F}_e$, $\mathbf{R}_a$, $\mathbf{R}_s$

Use the FBD of Figure 3.12.b where $\theta = 85°$ and $W_L = 0.2W_B$.
The solution proceeds exactly as in Example 3.9(a) except $\theta = 85°$. As before $W_L = 0.2\ W_B$ is added to $.18\ W_B$ (the weight of the head and arms) so that $W_T = 0.38W_B$ is now acting at the distal end of the thoracolumbar spine (point D).
The results are:

$$\mathbf{F}_e = .324W_B$$
$$\mathbf{R}_y = 1.05W_B$$
$$\mathbf{R}_x = 0.10W_B$$
$$\mathbf{R}_a = 1.05W_B$$
$$\mathbf{R}_s = .008W_B$$

Lifting a load of $0.2W_B$ with the *legs* and using *handles* and *load redistributions*, reduces the axial lumbosacral forces by about 71% compared to lifting the same load with the back. With respect to $\mathbf{R}_a$, the lifting task is now equivalent to a bending task where $\theta = 75°$ [i.e. see Example 3.8 (b)].

---

c. Carrying

HFE Application Another aspect of manual work is the carrying of the load once it is lifted. Again, good human factors engineering design requires an understanding of biostatic mechanics. And once again, the person's carrying posture and the spatial distribution of the load are two important design parameters. We conclude this chapter with two final examples that indicate that these two variables (posture and load distribution) are not necessarily independent. Rather, the optimal task design involves an interplay between these two variables.

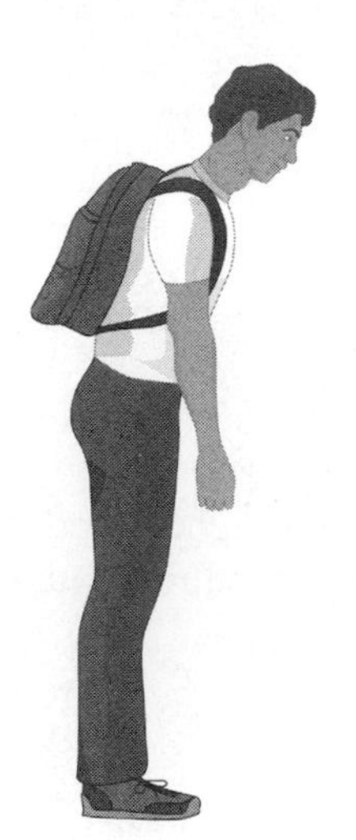

Figure 3.15.a Person carrying a load on the back.

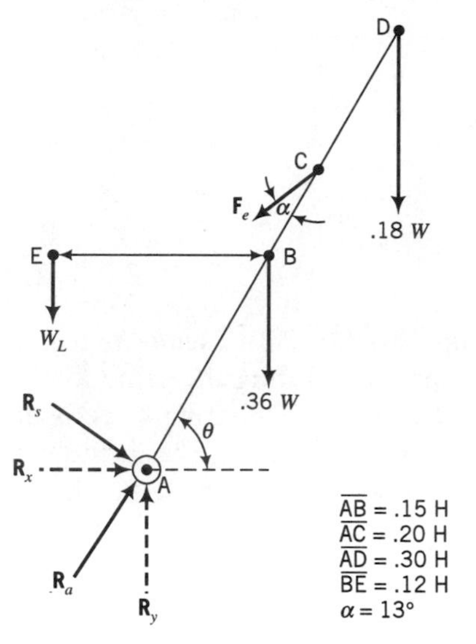

Figure 3.15.b Free-body diagram of a person carrying a load on the back.

EXAMPLE 3.10

a. The person (of height, H_B and weight, W_B) in Figure 3.14.c is now carrying the 0.2W load (W_L), equally distributed into a 0.1W load in each hand. Referring to the FBD of Figure 3.12.b, determine analytically the angle θ of the thoracolumbar spine at which $\mathbf{F}_e = 0$. (What are $\mathbf{R}_a$ and $\mathbf{R}_s$?)

b. The person is now carrying a single load ($W_L = 0.2W$) on the back (a knapsack) as shown in Figure 3.15.a. Referring to the FBD of Figure 3.15.b, determine analytically the angle θ of the thoracolumbar spine at which $\mathbf{F}_e = 0$. Use center of mass location for W_L described in Figure 3.15.b. (What are $\mathbf{R}_a$ and $\mathbf{R}_s$?)

SOLUTION 3.10(a)

Given: $H = H_B$, $W = W_B$
Given: $W_L = 0.2W_B$
Find: θ ($\mathbf{F}_e = 0$)

Use the FBD of Figure 3.12.b, where $W_L = 0.2W_B$:

$$\sum \mathbf{F}_y = 0:$$

$$\mathbf{R}_y - .36W_B - \mathbf{F}_e \cdot \sin(\theta - 13°) - .38W_B = 0$$

$$\mathbf{R}_y = \mathbf{F}_e \cdot \sin(\theta - 13°) + .74W_B$$

$$\sum F_x = 0:$$

$$\mathbf{R}_x - \mathbf{F}_e \cdot \cos(\theta - 13°)$$

$$\mathbf{R}_x = \mathbf{F}_e \cdot \cos(\theta - 13°)$$

$$\sum M_A = 0:$$

$$\mathbf{F}_e \cdot \cos(\theta - 13°)[.20H_B] \cdot \sin\theta - \mathbf{F}_e \cdot \sin(\theta - 13°)[.20H_B]$$

$$\cdot \cos\theta - (.36W_B)(.15H_B)\cos\theta - (.38W_B)(.30H_B)\cos\theta = 0$$

$$\mathbf{F}_e[(.20)\cos(\theta - 13°) \cdot \sin\theta - (.20) \cdot \sin(\theta - 13°) \cdot \cos\theta] = \cos\theta[.054W_B + .114W_B]$$

$$\mathbf{F}_e = [0.84W_B]\left[\frac{\cos\theta}{\cos(\theta - 13°) \cdot \sin\theta - \sin(\theta - 13°)\cos\theta}\right]$$

For $\mathbf{F}_e = 0$, then $\cos\theta = 0$

$$\theta = \cos^{-1}(0) = 90°$$

Also check whether the solution is indeterminate (i.e., denominator equals zero) at $\theta = 90°$.
The denominator is $\cos(77°) \cdot \sin(90°) = 0.225$, so the solution is determinate.
What are $\mathbf{R}_a$ and $\mathbf{R}_s$?

$$\mathbf{R}_y = 0.74W_B$$

$$\mathbf{R}_x = 0$$

Substituting the above into equation 3.7(a)(i) and 3.7(a)(ii):

$$\mathbf{R}_a = 0.74W_B$$

$$\mathbf{R}_s = 0$$

SOLUTION 3.10(b)

Given: $H = H_B$, $W = W_B$
Given: W_L(on the back) $= .2$W
Find: θ ($\mathbf{F}_e = 0$), $\mathbf{R}_a$, $\mathbf{R}_s$,

Use the FBD of Figure. 3.15.b, where $W_L = 0.2W_B$:
First derive an expression for Δx [the distance rearward of A that W_L(on the back) acts]:

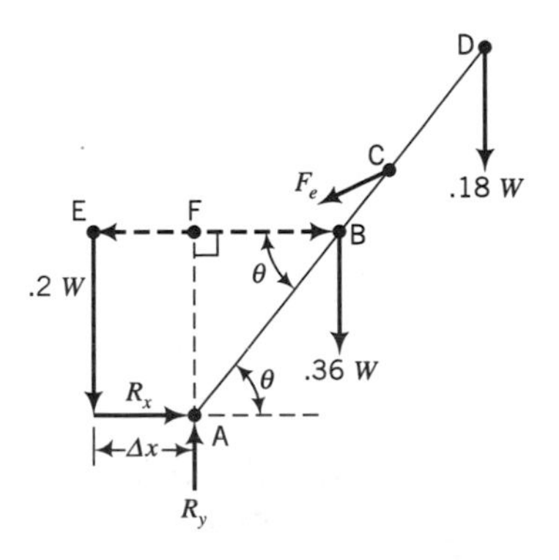

$$\text{BE} = .12H$$
$$\text{EF} = \Delta x$$
$$\text{BF} = (.15H)\cos\theta$$
$$\text{EF} = \text{BE} - \text{BF}$$
$$\Delta x = .12H - (.15H)\cos\theta$$

Second, proceed with the analysis:

$$\sum \mathbf{F}_y = 0:$$
$$\mathbf{R}_y - .2W_B - .36W_B - \mathbf{F}_e \cdot \sin(\theta - 13°) - .18W_B = 0$$
$$\mathbf{R}_y = F_e \cdot \sin(\theta - 13°) + .74W_B$$
$$\sum F_X = 0:$$
$$\mathbf{R}_x - \mathbf{F}_e \cdot \cos(\theta - 13°) = 0$$
$$\mathbf{R}_x = \mathbf{F}_e \cdot \cos(\theta - 13°)$$
$$\sum M_A = 0:$$
$$(.2W_B)[.12H_B - (.15H_B)\cdot\cos\theta] - .36W_B(.15H_B)\cdot\cos\theta + \mathbf{F}_e \cdot \cos(\theta - 13°)[.20H_B]\cdot\sin\theta$$
$$- .18W_B(.30H_B)\cdot\cos\theta - \mathbf{F}_e \cdot \sin(\theta - 13°)[.20H_B]\cos\theta = 0$$
$$\mathbf{F}_e[(.20H_B)\cdot\cos(\theta - 13°)\cdot\sin\theta - (.20H_B)\cdot\sin(\theta - 13°)\cdot\cos\theta]$$
$$= .2W_B(.15H_B)\cos\theta + .36W_B).15H_B(\cos\theta + .18W_B(.30H_B)\cdot\cos\theta - .2W_B(.12H_B)$$
$$\mathbf{F}_e = 5W_B\left[\frac{(.03)\cos\theta + (.054)\cos\theta + (.054)\cos\theta - (.024)}{\cos(\theta - 13°)\cdot\sin\theta - \sin(\theta - 13°)\cdot\cos\theta}\right]$$

For $\mathbf{F}_e = 0$, then:

$$(.03)\cos\theta + (.054)\cos\theta + (.054)\cos\theta = .024$$
$$\cos\theta = \frac{.024}{.138} = .174$$
$$\theta = \cos^{-1}(.174) = 80°$$

Also check whether the solution is indeterminate when $\theta = 80°$. This would occur if the denominator is zero.
However,

$$\cos(67°)\cdot\sin(80°) - \sin(67°)\cdot\cos(80°) = .385 - .160 = .225$$

So the solution is determinate.
What are $\mathbf{R}_a$ and $\mathbf{R}_s$?

$$\mathbf{R}_y = .74W_B$$
$$\mathbf{R}_x = 0$$

Substituting the above into equations 3.7(a)(i) and 3.7(a)(ii):

$$\mathbf{R}_a = .74W_B$$
$$\mathbf{R}_s = .13W_B \text{ (leftward)}$$

Note the interplay between posture and load. In Example 3.10(a), shifting the shoulders somewhat rearward would result in the desired spinal angle.

In Example 3.10(b), placing the knapsack at the midback (rather than high up) and placing heavier items at the bottom of the sack (water bottle and ax) and lighter items at the top (e.g., the bedroll) will result in the desired center-of-gravity.

FURTHER INFORMATION

G.R. Benedek and F.M.H. Villars: *Physics with Illustrative Examples from Medicine and Biology: Mechanics.* Addison-Wesley. Reading, MA. 1981.

D.B. Chaffin, G.B.J. Andersson and B.J. Martin: *Occupational Biomechanics.* 3rd Edition. John Wiley. New York. 1999.

M. Nordin and V.H. Frankel: *Basic Biomechanics of the Musculoskeletal System.* 2nd Edition. Lea and Febiger. Philadelphia. 1989.

N. Ozkaya and M. Nordin: *Fundamentals of Biomechanics.* 2nd Edition. Springer-Verlag. New York. 1999.

C.A. Phillips and J.S. Petrofsky (eds.): *Mechanics of Skeletal and Cardiac Muscle.* Charles C. Thomas. Springfield, IL. 1983.

PROBLEMS

3.1. In your first job as a human factors engineer, you are asked to evaluate the effects of a new exercise device on the shoulder area of the person exercising. The system of interest is shown in Diagram 3.1.

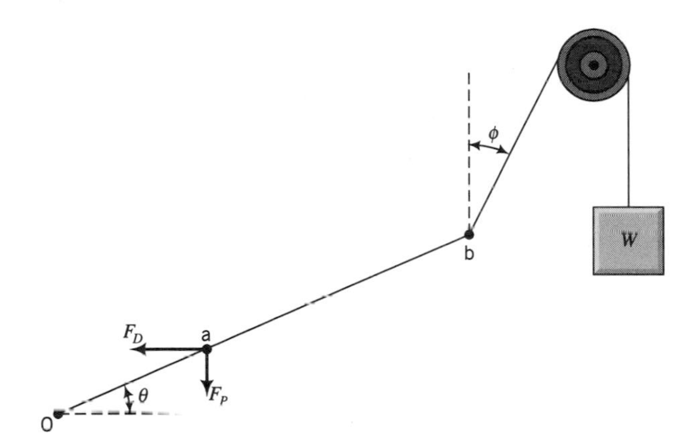

Diagram 3.1 The system of interest for Problem 1 (where $a = 0.1L$ and $b = 1.0L$)

A person extends the arm of axial length L straight out from the shoulder (point O) at an angle $\theta = 30°$ to the horizontal. The person holds a massless cord at point B (suspended over a frictionless pulley and connected to a weight, $W = 100$ N). The cord is at an angle, $\phi = 30°$ to the vertical. The force of the pectoral muscle ($\mathbf{F}_P$) is directed vertically downward from the axis of the arm at point A. The force of the deltoid muscle ($\mathbf{F}_D$) is directed horizontally leftward from the axis of the arm at point A, and is one-half the force of the pectoral muscle. Assume the arm itself is massless.

What are the values (and directions) of the horizontal reaction force $\mathbf{R}_x$ and the vertical reaction force $\mathbf{R}_y$ exerted at the shoulder (point O)? Also, what are the values of the pectoral muscle force ($\mathbf{F}_P$) and the deltoid muscle force ($\mathbf{F}_D$)?

3.2. A 65 kg patient, who just left the hospital after right hip surgery, was incorrectly instructed on how to use his cane. As shown in Diagram 3.2, he held the cane in his right hand while weight bearing with his right leg (the left leg was swinging through the air). Also, per Diagram 3.2, he found that in order to maintain static equilibrium, he had to shift the center of mass of his body *between* his right foot and the cane (in his right hand). In doing this his right leg tilted at angle β, and his cane tilted at angle α.

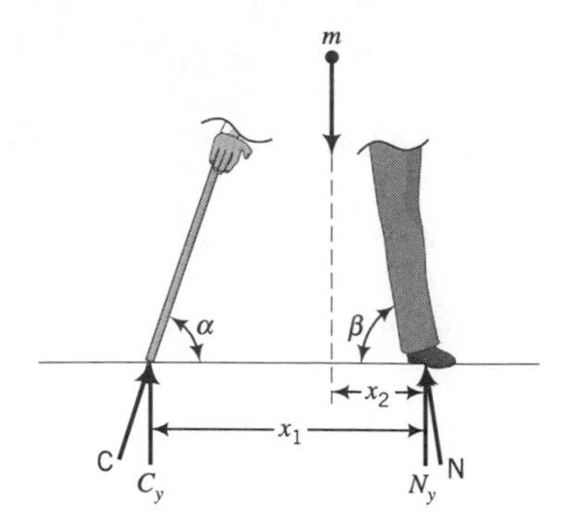

Diagram 3.2 (x_1 = 25 cm, x_2 = 5 cm, β = 80°).

Given the above, and the data in the diagram legend:

1. What is the value of the vertical component of the right leg ground reaction force N_y and what is the value of the vertical component of the cane ground reaction force C_y?
2. In order to maintain static equilibrium, what is the cane's angle of tilt (α)?

3.3. A postal worker is walking along straight and level ground and leaning forward in this human factors biomechanics problem. His thoraco-lumbar spine is of axial length L and is tilted at some "optimal" angle θ. The back packer is of total body weight W and the weight of this thorax ($.4W$) acts through a center of mass of $0.5L$ axial length. The force of the erector spinae muscle ($\mathbf{F}_e$) acts at $.67L$ downward at an axial angle of 12°. The weight of his head/neck and arms ($0.2W$) acts through a center of mass at $1.0L$ axial length. The mail bag he is carrying has a weight of $0.2W$ and its center of mass is located as shown in Diagram 3.3.

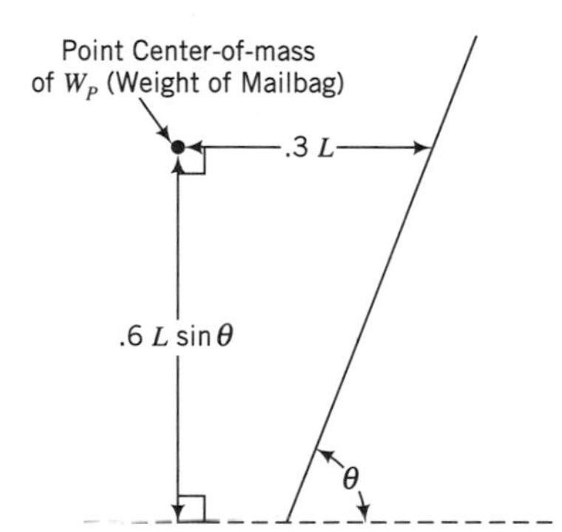

Diagram 3.3 Thoracolumbar spine tilted at angle θ.

If the "optimal" angle θ is defined as that angle at which the force of the erector spinae muscle ($\mathbf{F}_e$) is reduced to zero, what is the value of θ?

3.4. You are requested to design an exercise machine that will maintain a *constant* force on the deltoid muscle ($\mathbf{F}_D$) when the arm is extended outward from the side of the body.

Diagram 3.4.a shows an 80 Newton arm (W_A) of axial length L, which is extended downward at an angle θ_1 of 15° when the exercise machine exerts a force (W_{M1}) of 200 N downward:

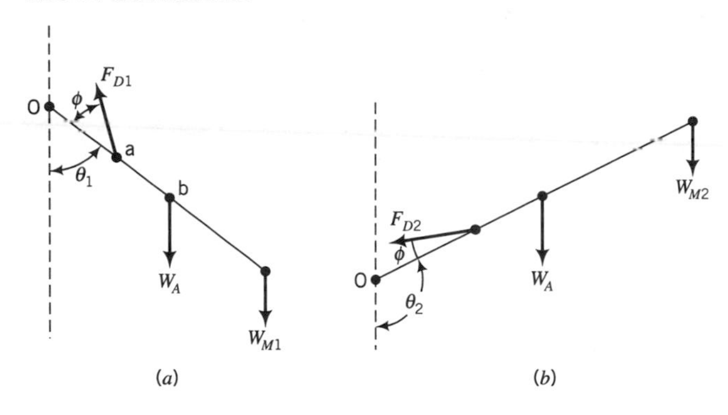

Diagram 3.4a (where $\phi = 15°$, and $a = 0.2L$, $b = 0.5L$, $c = 1.0L$). **b** (where ϕ, a, b, and c are as per Diagram 3.4.a).

The person then raises the extended arm upward at an angle θ_2 of 105° as shown in Diagram 3.4.b.

In order for $\mathbf{F}_D$ to *remain* constant ($\mathbf{F}_{D1} = \mathbf{F}_{D2}$), to what value must your machine adjust W_{M2}?

3.5. A disabled person (wearing knee braces and weighing 1000 N) passes out and leans over at the hip (Diagram 3.5), so that the thoraco-lumbar spine makes an angle of $\theta = 60°$ (as measured counterclockwise from the horizontal).

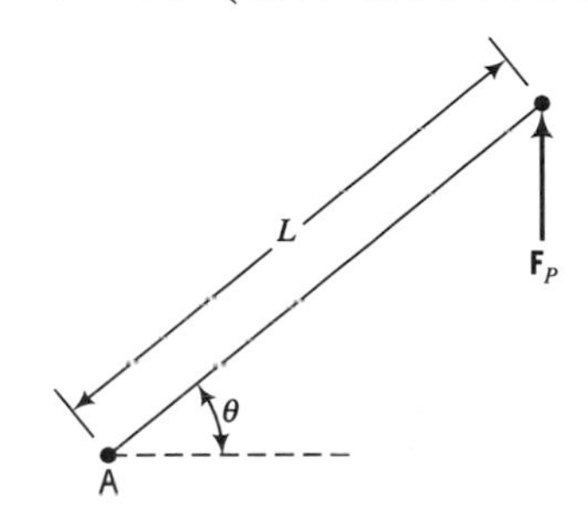

Diagram 3.5 Thoracolumbar spine tilted at angle θ.

The axial length of the thoraco-lumbar spine is L (as measured from the base of the spine, point A). The weight of the torso W_T is uniformly distributed over the length of the thoraco-lumbar spine so that its center of mass is located at an axial distance $L/2$. The weight of the head/neck and arms acts as a point center of mass located at the end of the spine (W_{HA}) at axial distance L.

An attendant catches the person by exerting a force $\mathbf{F}_P$ directed vertically upward at the chin and which can be approximately located at the end of the thoraco-lumbar spine at an axial distance L. Since the person has passed out, this effectively relaxes the erector spinae muscles, so that $\mathbf{F}_e = 0$.

Find the following three forces:

1. The force exerted by the attendant ($\mathbf{F}_P$) that will effectively stop the person at this point (i.e., place the system is in static equilibrium).
2. The magnitudes and directions of the axial reaction force $\mathbf{R}_a$ and shear reaction force R_s that occur at point A.

3.6. A stock clerk of weight W balances equally on the toes of both feet, at an angle θ to the horizontal in order to reach a high shelf. Shown in Diagram 3.6 is one foot where the axial distance from the balance point (toes) to the heel is length L. The applied force at the heel is from the gastrocnemius muscle $\mathbf{F}_g$ and is vertically upward. The joint reaction force $\mathbf{R}_y$ is applied vertically at an axial distance $0.8L$ from the toes. The weight of the foot itself is negligible.

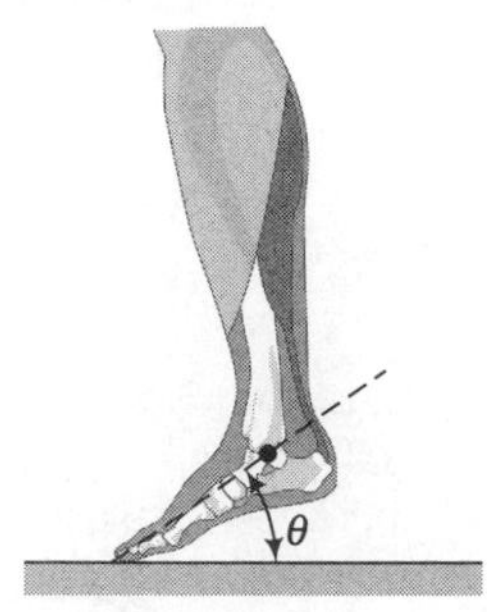

Diagram 3.6

1. What is the muscle force $\mathbf{F}_g$ for this foot?
2. What is the joint reaction force $\mathbf{R}_y$ for this foot?

3.7. A person holds a container of weight W at the end of the forearm (inclined at an angle α to the horizontal) as shown in Diagram 3.7. The axial distance from the center-of-rotation of the elbow joint to the center-of-gravity of the ball is distance L. The weight of the forearm itself is $2W$ and its center-of-force is located at the axial distance $L/2$. Finally, the force of the biceps muscle ($\mathbf{F}_b$) is $8W$ and applied at an axial distance ($L/5$) from the center-of-rotation of the elbow joint at an angle θ to the horizontal.

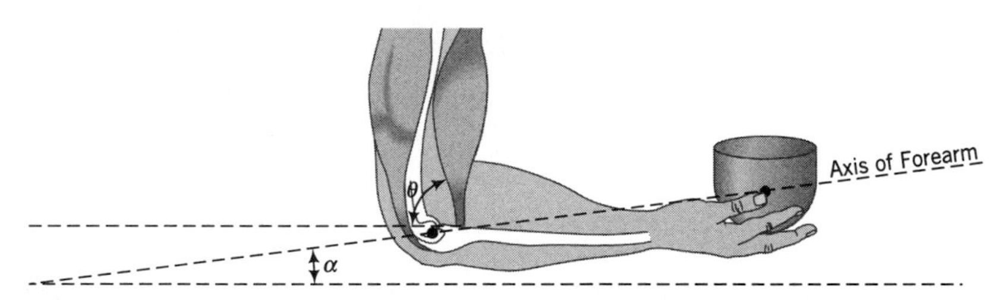

Diagram 3.7

If $\alpha + \theta = 90°$, what is the elbow joint vertical reaction force $\mathbf{R}_y$?

3.8. The patient in Diagram 3.8 lies in bed with the leg (axial length = L) elevated at an angle, $\theta = 25°$. The leg weighs $W_L = 100$ N, and its center-of-gravity is $L/2$. A cylindrical cast (enclosing the leg) weighs $W_c = 50$ N, and its center-of-gravity is $3L/4$ (measured from the hip, point A).

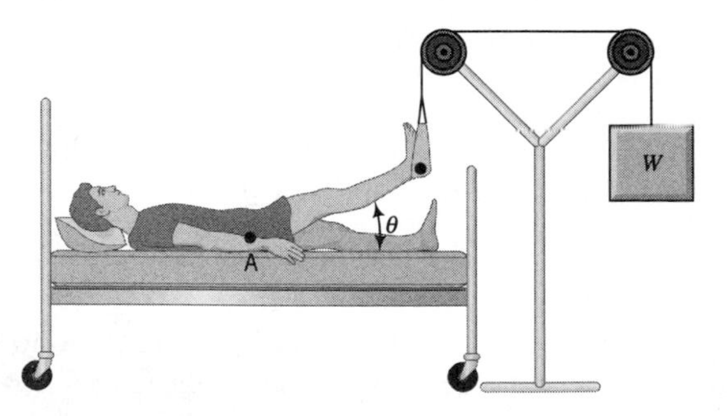

Diagram 3.8

What is the amount of weight W that is needed to put the leg in static equilibrium? What is the value (and direction) of the vertical reaction force $\mathbf{R}_y$ at the hip (point A)?

3.9. A driver's manual recommends that the hands be placed at 10 and 2 o'clock on the steering wheel. However, many people prefer to drive with one hand on the 12 o'clock position. Assuming the driver is maneuvering the auto around a curve, we will look at the forces at the shoulder and the muscles that will be used to hold the wheel for the one-handed driving position. Referring to Diagram 3.9, find the deltoid (shoulder) muscle force (CD) and the vertical (A_y) and horizontal (A_z) reaction forces at the shoulder joint for an 18 N steering wheel force.

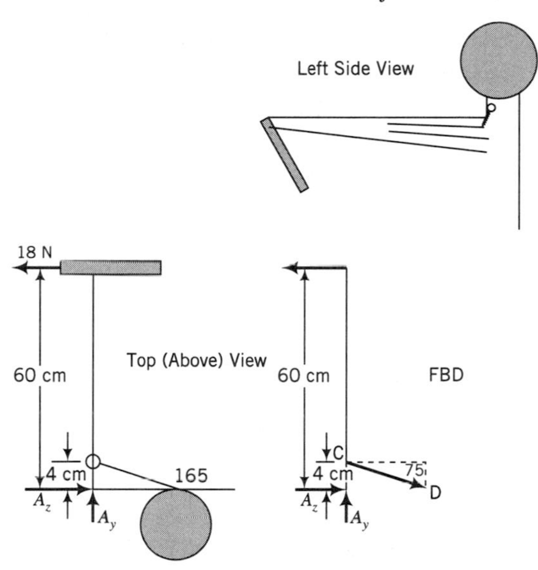

Diagram 3.9

3.10. A gymnast is performing a maneuver (as shown in Diagram 3.10) where he hangs between two cables with his arms extended outward. If his mass is 68 kg, how much force does his latissimus dorsi (back) muscle (EF) need to exert to accomplish this? Find the vertical (A_y) and horizontal (A_x) shoulder joint reaction forces.

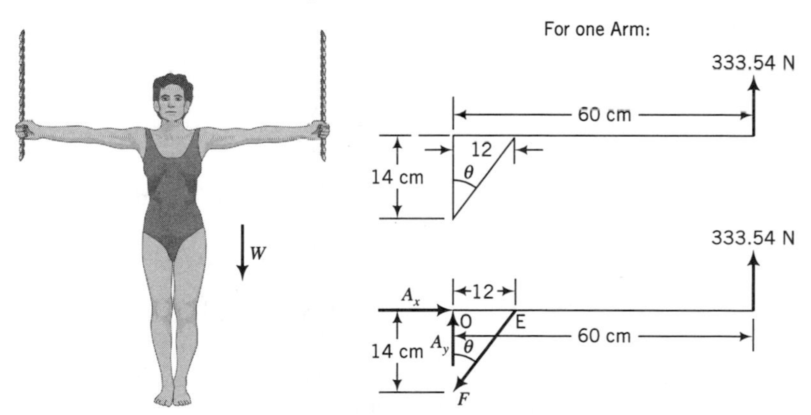

Diagram 3.10

3.11. A man (m = 80 kg) grips a hand drill (m = 5 kg) parallel to the ground, as shown in Diagram 3.11.a. If the extensor carpi radialis (ECR), that is, the wrist muscle, counteracts the weight of the drill in order to hold it parallel to the ground, what is the force muscle ($\mathbf{F}_{ECR}$)? Knowing $\mathbf{F}_{ECR}$, compute the vertical force of the biceps muscle ($\mathbf{F}_B$) as well as the vertical reaction force ($\mathbf{R}_y$) at the elbow in order to hold the forearm parallel to the ground. If this same man grips the same drill but this time with an additional 25-cm-long 0.5 kg drill bit attached (see Diagram 3.11.b), calculate the same parameters as before ($\mathbf{F}'_{ECR}$, $\mathbf{F}'_B$, and $\mathbf{R}'_y$) that are now necessary to keep the drill and forearm parallel to the ground.

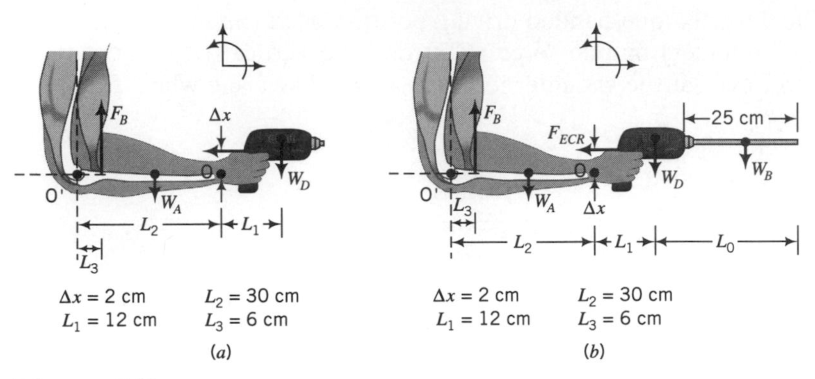

Diagram 3.11

3.12. An elastic cord attached to a support bar tethers a small battery-operated drill suspended above a worktable. A worker stands at the table. Diagram 3.12 is of the worker's arm (at rest) with the elbow (A) bent at 90° and holding the tool at the end of the elastic cord.

In this position there is zero tension in the spring ($\mathbf{F}_S$ = 0). The tether cord elastic spring constant (K) = 25 N/cm, and the forearm now bends clockwise 10° at the elbow, pulling on the spring (stretching it downward) toward the work table.

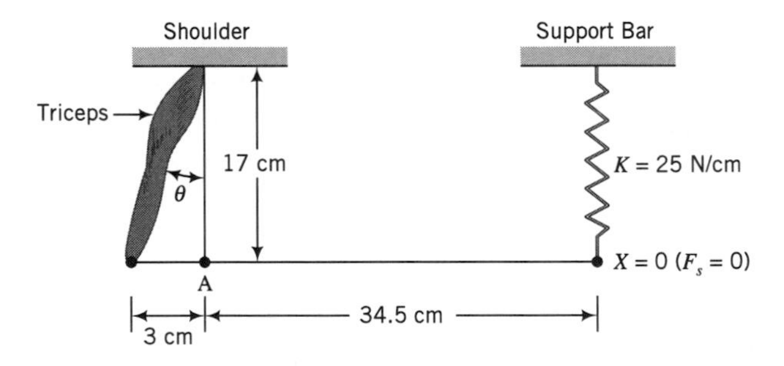

Diagram 3.12

Assuming that the vertical elbow position (point A) remains constant and the angle of the triceps muscle origin at the shoulder (θ) remains constant, and the spring remains purely vertical before and after stretching:

1. How much force does the triceps muscle develop ($\mathbf{F}_T$) by stretching the elastic cord?
2. What are the reactive forces (magnitude and directions) at the elbow in the horizontal and vertical directions ($\mathbf{R}_x$ and $\mathbf{R}_y$)?

CHAPTER 4

Biodynamic Mechanics

4.1 HUMAN BODY KINEMATICS

a. Linear Kinematics

Understanding of the human body in a dynamic state begins with the kinematic equations for uniaxial linear motion. Kinematic analysis is based upon the relationship between position, its first derivative (velocity), and its second derivative (acceleration). Recall that these are vector quantities, and so for uniaxial motion, we define a coordinate axis (such as x) along which the movement occurs. The relevant kinematic parameters are then defined in that direction.

Uniaxial linear movement (or translation) refers to motion which occurs along a straight line. There are many situations in which the motion of objects is only in one direction. One example is that of a car being driven on a straight road as it moves upward along a uniformly inclined hill. Another example might be a skier moving along a straight course as the person proceeds down a uniformly declined hill. These examples illustrate that the direction of motion need not be purely vertical, nor purely horizontal, but may also be at any diagonal to these.

The basic kinematic equations for uniaxial linear (or translational) motion are derived from the definition of velocity and acceleration:

$$\mathbf{v} = \frac{dx}{dt} \tag{1}$$

$$\mathbf{a} = \frac{dv}{dt} \tag{2}$$

One type of translational motion occurs when uniform acceleration is applied to an object so that the acceleration remains a constant with respect to time. If a_0 represents a constant acceleration, and the initial velocity v_0 (at $t_0 = 0$), then the velocity ($\mathbf{v}$) at any point in time (greater than t_0) is obtained by integrating equation (2):

$$\mathbf{v} = v_0 + a_0 t \tag{3}$$

Finally, if the initial position of the object (x_0) is also known, then the position of the object at any point in time (greater than t_0) can be found by substituting equation (1) into equation (3), separating variables, and integrating:

$$x = x_0 + v_0 t + \frac{1}{2} a_0 t^2 \tag{4}$$

When the acceleration is constant, we may also derive a relationship between displacement, velocity and time by rearranging equation (3) for a_0, then substituting this expression into equation (4), and then rearranging for x:

$$x = x_0 + \frac{1}{2}(\mathbf{v} + v_0)t \quad (5)$$

Finally, we may derive the relationship between velocity, acceleration and displacement by rearranging equation (3) for t, then substituting this expression into equation (4), and then rearranging for v^2:

$$\mathbf{v}^2 = v_0^2 + 2a_0(x - x_0) \quad (6)$$

Equations (3)–(6) represent the four kinematic equations that describe uniaxial linear motion of an object when the acceleration is constant. Two common types of motion for which these kinematic expressions are useful are that of *free fall* and also *sliding*.

A very common example of the motion of a body subjected to constant acceleration occurs when the body falls vertically to the ground. The object is said to experience *free fall*. In the absence of air resistance, all objects are subjected to a constant gravitational acceleration so that $\mathbf{a} = \mathbf{g}$ in equations (3), (4), and (6). Recall that $\mathbf{g}$ is a vector quantity, and so must follow the sign convention as applied to the other vector quantities, that is, x (displacement) and $\mathbf{v}$ (velocity).

Sliding is another common example of uniformly accelerated motion. In the absence of the frictional resistance of the sliding surface (and in the absence of air resistance), sliding may be viewed as diagonal free fall (as contrasted to vertical free fall, above). The object is still subjected to the constant gravitational acceleration ($\mathbf{g}$) vertically downward, but the object's linear movement is along a diagonal axis. When that axis is inclined from the horizontal at some inclusive angle θ, the object is subjected to a constant acceleration (a_0) which is a function of that angle.

EXAMPLE 4.1

a. Find an expression for the acceleration ($\mathbf{a}$) of a body sliding down a frictionless diagonal surface inclined at an angle θ. The body has mass m and its center of mass is acted upon vertically downward by the gravitational constant $\mathbf{g}$.

b. Repeat Example 4.1a for a body sliding upward on the diagonal surface.

SOLUTION 4.1a

Draw the free-body diagram of the body moving down a diagonal straight line.

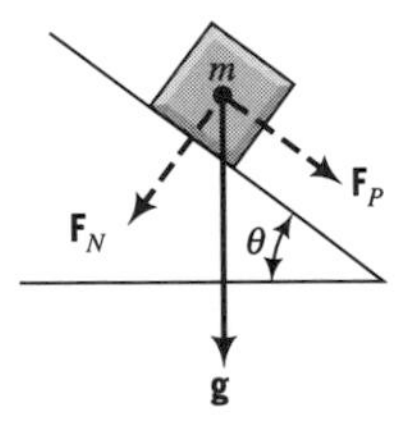

In this case the body is moving along a uniaxial direction (x) oriented rightward and downward. It is therefore useful to make the vertical axis positive downward ($\mathbf{g}$ is positive).

For the net force (F_T):

$$F_T = F_P \text{ (since the surface is frictionless)}$$

$$W = m \cdot \mathbf{g} \quad \text{(i)}$$

$$F_T = W \cdot \sin\theta \quad \text{(ii)}$$

Substituting (i) into (ii):

$$F_T = m \cdot \mathbf{g} \cdot \sin\theta \quad \text{(iii)}$$

From Newton's second law:

$$\mathbf{a} = \frac{F_T}{m} \quad \text{(iv)}$$

Substituting (iii) into (iv):

$$\mathbf{a} = \mathbf{g} \cdot \sin\theta \quad \text{(v)}$$

SOLUTION 4.1b

Draw the free-body diagram, and retain the same sign convention:

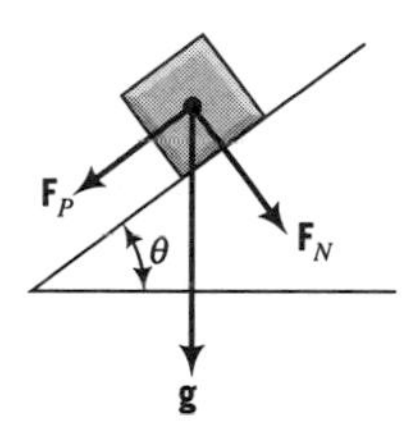

For the net force (F_T)

$$F_T = -F_P$$

$$W = m \cdot \mathbf{g} \quad \text{(i)}$$

$$F_T = -W \cdot \sin\theta \quad \text{(ii)}$$

Substituting (i) into (ii):

$$F_T = -m \cdot g \cdot \sin\theta \quad \text{(iii)}$$

$$\mathbf{a} = \frac{F_T}{m} \quad \text{(iv)}$$

Substituting (iii) into (iv):

$$\mathbf{a} = -\mathbf{g} \cdot \sin\theta \quad \text{(v)}$$

HFE Application: A human factors engineer will often consider various ways for the human subsystem to interact with the technological subsystem. Often this will involve defining and then analyzing an upper limit for performance of the human-technological system.

EXAMPLE 4.2

A piece of playground equipment (a slide) is being considered for installation at a location known for icy winters. The dimensions of the slide are shown in Figure

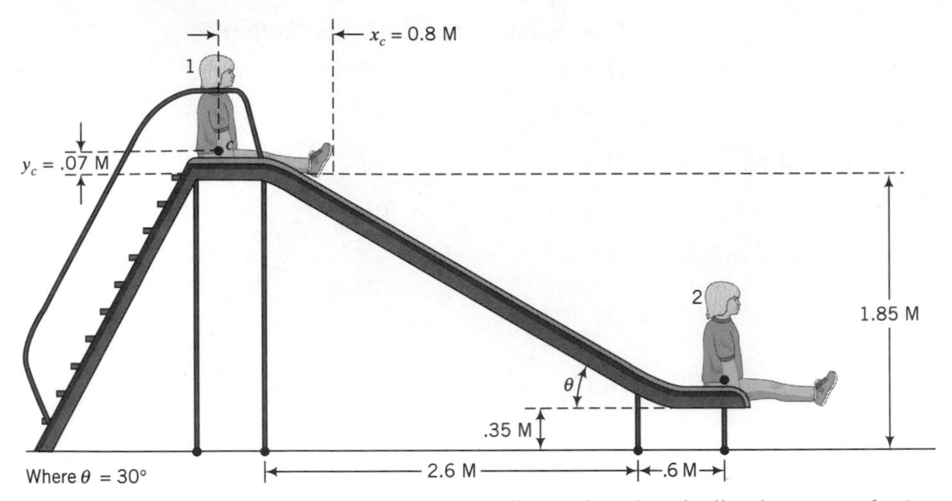

Figure 4.1 Person sliding down a frictionless diagonal surface inclined at an angle θ.

4.1 (where M = length in meters). It is foreseeable that on some days ice may cover the side rails and sliding surface, so that there is zero coefficient of sliding friction ($\mu = 0$). If the largest foreseeable slider has a mass of 30 kg, and his initial velocity at the top of the slide is zero, how fast will the slider be going when he reaches the bottom of the slide?

SOLUTION 4.2

Referring to Figure 4.1 for the relevant dimensions:
The length x_f of the slide is:

$$\cos\theta = \frac{2.6\text{M}}{x_f}$$

$$x_f = \frac{2.6\text{M}}{\cos(30°)} = 3.0\text{M} \tag{i}$$

Using rightward and downward positive sign convention, with the origin at the top of the slide:

$$x_0 = 0 \tag{ii}$$

Also given in the problem:

$$v_0 = 0 \tag{iii}$$

From equation 4.1(a)(v):

$$a_0 = g \cdot \sin\theta \tag{iv}$$

Using equation (6) to solve for the final velocity:

$$v_f^2 = v_0^2 + 2a_0(x_f - x_0) \tag{v}$$

Substituting (i), (ii), (iii), and (iv) into (v):

$$v_f^2 = 2 \cdot g \cdot \sin\theta \cdot x_f \tag{vi}$$

Substituting actual values, and solving:

$$v_f = \sqrt{(2)(9.81)(0.5)(3.0)}$$
$$v_f = 5.42 \text{ M/s}$$

We now examine the human body in a dynamic state in which the kinematic equations are applied to biaxial (two-dimensional) motion analysis. By first understanding one-dimensional linear motion analysis (previously discussed), the student can then proceed to analyze motion in two dimensions. This requires that displacement, velocity, and acceleration be analyzed as vectors in two-dimensional space. Two-dimensional linear motion of a body can be analyzed in rectangular coordinate space (an xy plane) as follows. Analyze the motion in the x direction and y direction separately, and then, combine the results by converting the parameters involved from rectangular to polar form.

For an object experiencing linear motion in a two-dimensional plane, when the acceleration of that object is constant, then a_x and a_y are also constants. Consequently, in the x direction, equations (3) and (4) can be redefined explicitly for the x direction:

$$v_x = v_{x_0} + a_{x_0} t \tag{7}$$

$$x = x_0 + v_{x_0} t + \frac{1}{2} a_{x_0} t^2 \tag{8}$$

Equations (3) and (4) can likewise be redefined for the y direction:

$$v_y = v_{y_0} + a_{y_0} t \tag{9}$$

$$y = y_0 + v_{y_0} t + \frac{1}{2} a_{y_0} t^2 \tag{10}$$

Ballistic motion is a particular form of linear motion in a two-dimensional plane when there is constant acceleration. Ballistic motion can be analyzed by redefining equations (7)–(10) as follows. When the speed and release angle of the body is known, the components of the velocity vector (at time equals 0) in the x and y directions are:

$$v_{x_0} = v_0 \cos\theta \tag{11}$$

$$v_{y_0} = v_0 \sin\theta \tag{12}$$

Assuming negligible air resistance on the body, the acceleration of that body in the x direction is zero.

$$a_{x_0} = 0 \tag{13}$$

In the y-direction, the body is acted upon downward by the gravitational constant (g). If positive *upward* sign convention is used for the y-axis, the acceleration of the body in the y direction is:

$$a_{y_0} = -g \tag{14a}$$

However, it is often convenient to use positive *downward* sign convention, in which case the acceleration of the body in the y-direction is:

$$a_{y_0} = g \tag{14b}$$

Given these boundary conditions, equations (11)–(14a) are substituted into equations (7)–(10). Also, equations (12) and (14b) are substituted into equations (9) and (10).

The following set of equations then apply to ballistic motion:

$$x = x_0 + (v_0 \cos\theta)t \tag{15}$$

$$y = y_0 + (v_0 \sin\theta)t - \frac{1}{2}gt^2 \quad \text{(upward positive)} \tag{16a}$$

$$y = y_0 + (v_0 \sin\theta)t + \frac{1}{2}gt^2 \quad \text{(downward positive)} \tag{16b}$$

$$v_x = v_0 \cos\theta \tag{17}$$

$$v_y = v_0 \sin\theta - gt \quad \text{(upward positive)} \tag{18a}$$

$$v_y = v_0 \sin\theta + gt \quad \text{(downward positive)} \tag{18b}$$

HFE Application: Part of product design is defining an upper limit performance analysis (see Example 4.2). Subsequently, the HFE can use the results of that analysis to define and analyze a system reconfiguration in order to assure an acceptable human-technological system interaction.

EXAMPLE 4.3

Based upon the results of Example 4.2, the slide (of Figure 4.1) is to be installed within a rectangle filled with sand. The slider is approximated as having a center of mass (c) located at the hips and having x_c and y_c coordinates as shown in Figure 4.2. Given the results of Example 4.2, what is the required horizontal distance from the end of the slide to the rectangular border that will assure a worst-case slider lands in the sand with the legs fully extended?

SOLUTION 4.3

Referring to Figure 4.2, where the slider is at the end of the slide, continue with a diagram that shows the subsequent ballistic trajectory of the slider's center of mass:

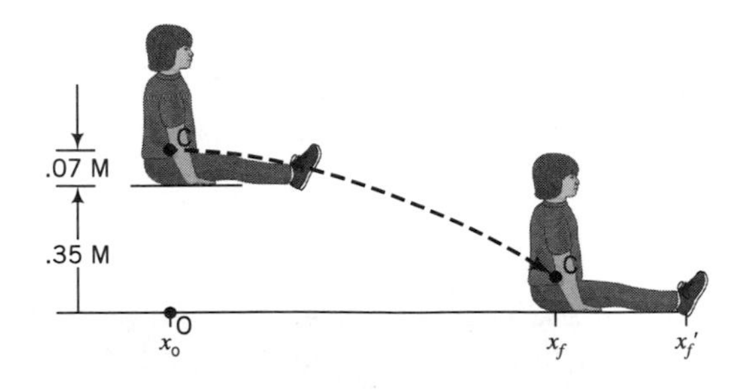

Figure 4.2 Person sliding down a slide installed within a rectangle filled with sand.

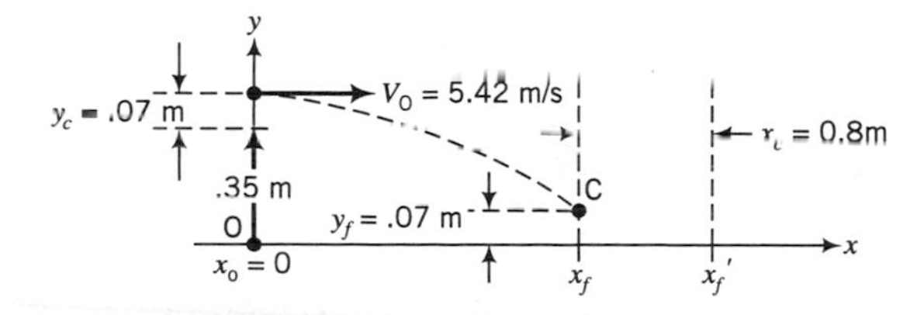

Since the release angle θ is zero degrees, equations (15), (16), (17), and (18a) may be rewritten as follows:

$$x = x_0 + v_0 t \tag{i}$$

$$y = y_0 - \frac{1}{2} g t^2 \tag{ii}$$

$$v_x = v_0 \tag{iii}$$

$$v_y = -\mathbf{g}t \tag{iv}$$

Note that in this example, the equations use rightward and *upward* positive sign convention, the origin (0) is as per Figure 4.2, and initial time (t_0), is zero.
First, solve for the time (Δt) of free-fall in the y-direction:
Rewrite (ii):

$$y_f = y_0 - \frac{1}{2}\mathbf{g}(\Delta t)^2 \tag{v}$$

and rearrange for Δt:

$$\Delta t = \sqrt{\frac{2(y_0 - y_f)}{\mathbf{g}}} \tag{vi}$$

From the diagram (where M = meters in length):

$$y_0 = .35\text{ M} + .07\text{ M} = .42\text{ M}$$
$$y_f = .07\text{ M}$$

Approximating $\mathbf{g} = 9.81\text{ M/s}^2$

$$\Delta t = \sqrt{\frac{0.70}{9.81}} = 0.267\text{ sec}$$

Second, solve for the distance traveled (Δx) in the x-direction:
Rewrite (i):

$$x_f = x_0 + v_0 \Delta t$$

Since $x_0 = 0$ and $v_0 = 5.42$ M/s

$$x_f = (5.42)(.267) = 1.45\text{ M}$$

The problem is completed by noting (from Figures 4.1 and 4.2):

$$x_f' = x_f + x_c$$
$$x_f' = 2.25\text{ M}$$

For this situation, the horizontal length of cord in front of the slide should be even greater than the vertical height of the slide.

b. Angular Kinematics

The kinematic equations for rotational movements consider angular position and displacement, angular velocity, and angular acceleration. Of particular interest to the human factors engineer is rotational motion about a fixed axis, and specifically, circular motion.

When considering circular motion, the velocity and acceleration vectors are defined with respect to two orthogonal directions to the circular path. One direction is normal (n) and the other tangential (t) to the circular path. The velocity vector ($\mathbf{v}$) is always considered as tangent to the path of the body's motion and is designated as the tangential or linear velocity. The magnitude of the linear velocity is the rate of change of the relative position of the body along a segment (s) of the circular path:

$$\mathbf{v} = \frac{ds}{dt} \tag{19}$$

Recall that for circular motion, the acceleration vector has two orthogonal (tangential and normal) components. The tangential acceleration, $\mathbf{a}_t$, is the rate of change of the linear velocity vector:

$$\mathbf{a}_t = \frac{dv}{dt} \tag{20}$$

The normal acceleration, $\mathbf{a}_n$, is the rate of change in the direction of the velocity vector:

$$\mathbf{a}_n = \frac{v^2}{r} \tag{21}$$

Recall that since $s = r \cdot \theta$, and for a circular motion the radius (r) is a constant, then equation (19) can be redefined as:

$$\mathbf{v} = \frac{d(r\theta)}{dt} = r\frac{d\theta}{dt} \tag{22}$$

Since the rate of change of angular displacement is the angular velocity, then:

$$\mathbf{v} = r \cdot \boldsymbol{\omega} \tag{23}$$

Substituting equation (23) into equation (20) allows us to define the tangential acceleration for motion in a circular path in terms of the angular velocity:

$$\mathbf{a}_t = \frac{d}{dt}(r\boldsymbol{\omega}) = r\frac{d\boldsymbol{\omega}}{dt} \tag{24}$$

Recall that the rate of change of the angular velocity is the angular acceleration:

$$\mathbf{a}_t = r \cdot \boldsymbol{\alpha} \tag{25}$$

Substituting equation (23) into equation (21) results in:

$$\mathbf{a}_n = r \cdot \boldsymbol{\omega}^2 \tag{26}$$

The significance of equations (23), (25), and (26) is that they relate the linear parameters ($\mathbf{v}$, $\mathbf{a}_t$, and $\mathbf{a}_n$) to the angular parameters (r, θ, and $\boldsymbol{\alpha}$).

Uniform circular motion is defined as rotational motion about a fixed axis when the linear velocity of the body being rotated is constant. Therefore, with respect to linear rotational parameters:

$$\mathbf{a}_t = 0 \tag{27}$$

$$\mathbf{a}_n = \frac{v^2}{r} = r\omega^2 \tag{28}$$

With respect to angular rotational parameters for uniform circular motion:

$$\boldsymbol{\omega} = \frac{v}{r} = \text{constant} \tag{29}$$

$$\boldsymbol{\alpha} = 0 \tag{30}$$

A set of kinematic equations can be derived for the rotational motion of a body about a fixed axis when there is *constant angular acceleration* α_0. We have previously derived a set of kinematic equations for the movement of a body undergoing one-dimensional linear motion with constant acceleration [equations (3)–(6)]. The derivation of the equations for rotational motion with constant acceleration proceeds in a similar manner and results in:

$$\boldsymbol{\omega} = \omega_0 + \alpha_0 t \tag{31}$$

$$\theta = \theta_0 + \omega_0 t + \frac{1}{2}\alpha_0 t^2 \tag{32}$$

$$\theta = \theta_0 + \frac{1}{2}(\boldsymbol{\omega} + \omega_0)t \tag{33}$$

$$\boldsymbol{\omega}^2 = \omega_0^2 + 2\alpha_0(\theta - \theta_0) \tag{34}$$

Equations (31)–(34) may be used only for rotational motion about a fixed axis with constant angular acceleration. However, such is not always the case and we conclude this section by considering the case of rotational motion about a fixed axis with nonconstant acceleration.

In this case, the angular acceleration is due to the gravitational constant, g.

EXAMPLE 4.4

a. Consider a body whose center of mass is attached to one end of a light but strong cord. The cord is of a fixed length (r) and its other end is attached to a fixed, and frictionless, axis of rotation. At $t = 0$, the extended cord (of length, r) is perfectly horizontal, and the body has an initial velocity of v_0. The body then proceeds to swing downward under the action of the gravity vector, $\mathbf{g}$, so that the cord becomes more vertically oriented. If the angle θ is the inclusive angle between the initial horizontal axis (at 0°) and the current, more vertical orientation, find an expression for the tangential acceleration $\boldsymbol{\alpha}$.
b. Using the result of Example 4.4a, derive an expression for angular velocity ($\boldsymbol{\omega}$), angular displacement (θ), and angular acceleration ($\boldsymbol{\alpha}$) in a form analogous to equation (34).

SOLUTION 4.4(a)

Diagram the system, as it rotates from $\theta = \theta_0 = 0°$ to $\theta = 90°$, and use rightward and downward positive sign convention:

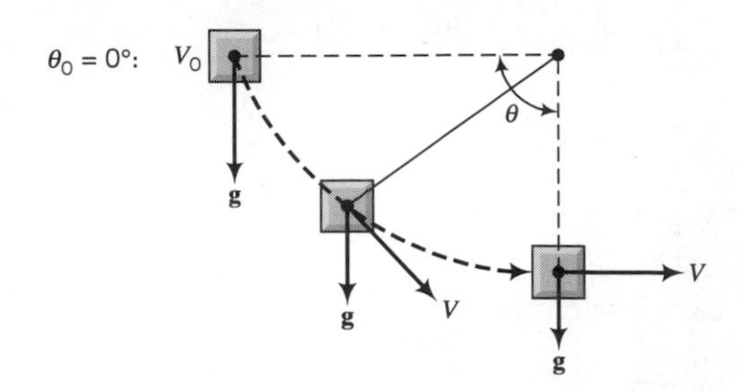

Draw a free-body diagram of the system at some angle, θ:

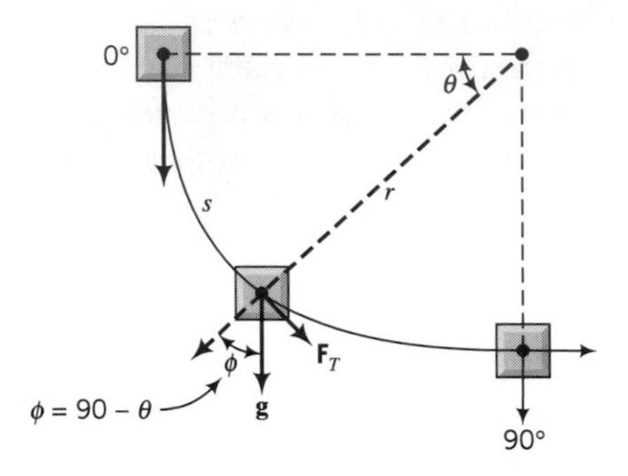

Solve for $\mathbf{a}_t$:

$$\begin{aligned} F_T &= m \cdot g \cdot \sin\phi \\ F_T &= m \cdot g \cdot \sin(90 - \phi) \\ F_T &= m \cdot g \cdot \cos\theta \\ \mathbf{a_T} &= \frac{F_T}{m} = \mathbf{g} \cdot \cos\theta \end{aligned} \tag{i}$$

SOLUTION 4.4(b)

From equation (20):

$$\begin{aligned} \mathbf{a_T} &= \frac{dv}{dt} \\ dv &= \mathbf{a_T} \cdot dt \end{aligned} \tag{i}$$

Substituting equation 4.4(a)(i) into (i):

$$dv = \mathbf{g} \cdot \cos\theta \cdot dt \tag{ii}$$

Recall form equation (22):

$$\mathbf{v} = r \cdot \frac{d\theta}{dt}$$

Rearranging:

$$dt = \frac{r \cdot d\theta}{\mathbf{v}} \tag{iii}$$

Substituting (iii) into (ii):

$$dv = \mathbf{g} \cdot \cos\theta \cdot \frac{r \cdot d\theta}{\mathbf{v}}$$

Separating variables:

$$\mathbf{v} \cdot dv = \mathbf{g} \cdot r \cdot \cos\theta \cdot d\theta$$

Defining the initial condition, $v_0(\theta_0)$, and integrating:

$$\int_{v_0} \mathbf{v} \cdot d\mathbf{v} = \int_{\theta_0} g \cdot r \cdot \cos\theta \cdot d\theta$$

$$\left.\frac{\mathbf{v}^2}{2}\right]_{v_0} = (\mathbf{g} \cdot r) \cdot \sin\theta]_{\theta_0} \qquad \text{(iv)}$$

$$\frac{\mathbf{v}^2}{2} - \frac{v_0^2}{2} = \mathbf{g} \cdot r[\sin\theta - \sin\theta_0]$$

$$\mathbf{v}^2 = v_0^2 + 2\mathbf{g} \cdot r[\sin\theta - \sin\theta_0]$$

Since $v = r \cdot \omega$

$$\omega^2 = \omega_0^2 + \frac{2\mathbf{g}}{r}(\sin\theta - \sin\theta_0) \qquad \text{(v)}$$

Recall:

$$\boldsymbol{\alpha} = \frac{d\boldsymbol{\omega}}{dt}$$

$$\mathbf{g} = \frac{d\mathbf{v}}{dt}$$

Rewriting **g** in angular notation:

$$\mathbf{g} = \frac{d(r \cdot \boldsymbol{\omega})}{dt} = r \cdot \frac{d\boldsymbol{\omega}}{dt} = r \cdot \boldsymbol{\alpha} \qquad \text{(vi)}$$

Substituting (vi) into (v):

$$\boldsymbol{\omega}^2 = \omega_0^2 + 2 \cdot \boldsymbol{\alpha}(\sin\theta - \sin\theta_0) \qquad \text{(vii)}$$

where $\boldsymbol{\alpha} = \mathbf{g}/r$ = constant, which is in the form of equation (34).

Note that this special case reduces to a constant angular acceleration multiplied by sin θ. Equation (vii) is analogous to *sliding* over a frictionless *curved* surface of constant radius of curvature, r.

HFE Application. Adequate design of a human-technological system involves a performance analysis (as shown in Example 4.2). This analysis may identify an unacceptable human operator technological system interaction. The system may then need to be modified or reconfigured as in Example 4.3. This is often a *sequential* process in that once the system modification or reconfiguration has been defined, it is then subjected to analysis. Consequently, system redefinition and analysis can be performed consecutively.

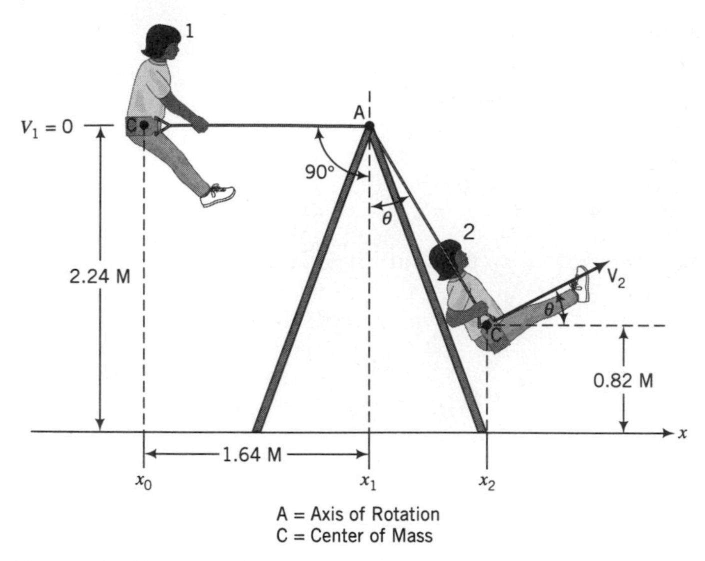

Figure 4.3 Person swinging on a swing surrounded by a rectangular border which will enclose sand under the swing area.

EXAMPLE 4.5

A playground swing is being designed as a human-technological system. The swing is to be surrounded by a rectangular border that will enclose sand under the swing area. An upper limit swinger is approximated as a 50 kg mass (m), and their final swing (before jumping out) is shown in Figure 4.3. The swinger's center of mass is located at the hips, which is 0.82 M above their feet. Upon jumping out, his body is subjected to ballistic motion, and the swinger lands on both feet (which are directly below their center of mass) as shown in Figure 4.4. Find the horizontal distance (as measured beginning directly under the swing axis) to the border directly in front of the swing which must be covered with sand.

SOLUTION 4.5

First, determine the additional parameter values not shown in Figure 4.3.

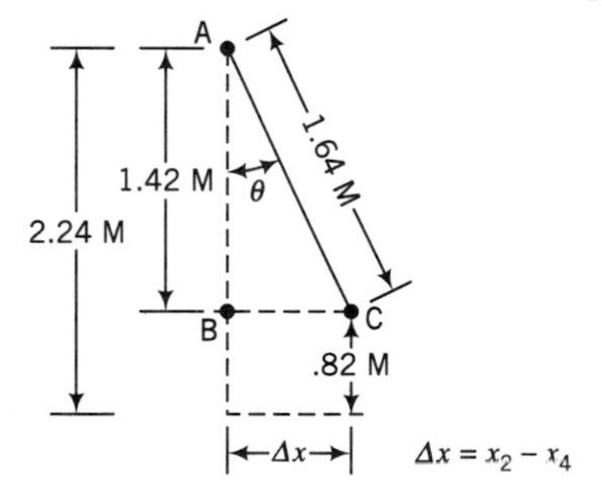

Referring to the above diagram.

$$\overline{AB} = 2.24 - 0.82 = 1.42 \text{ M}$$

$$\theta = \cos^{-1}\left(\frac{1.42}{1.64}\right) = 30.0°$$

$$\Delta x = (1.64)\sin[30°] = 0.82 \text{ M}$$

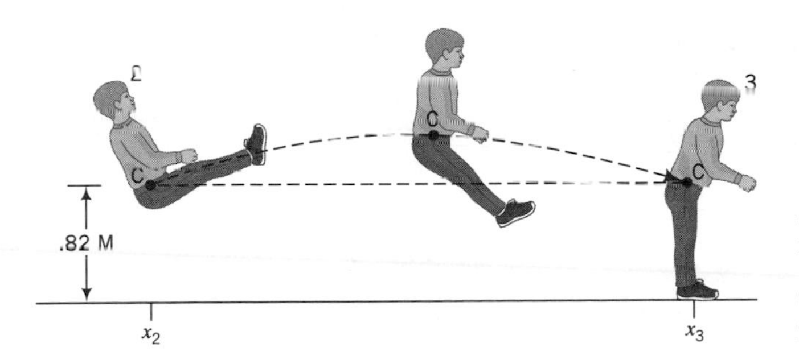

Figure 4.4 Swinger from Figure 4.3 jumps out of swing and lands on both feet.

Set $x_1 = 0$, then

$$x_2 = \Delta x = 0.82 \text{ M}$$

Second, calculate $\mathbf{v}_2$:
Define position 1 of Figure 4.3 at $\theta_1 = 0°$. θ then increases counter clockwise as the person rotates about A.
Since there is negligible air resistance, and rotation about A is frictionless, we can involve the argument of symmetry. From equation 4.4(a)(i), the system accelerates ($\mathbf{a} = \mathbf{g} \cdot \cos\theta$) for $0° < \theta < 90°$. Symmetry then requires that the system will decelerate ($\mathbf{a} = \mathbf{g} \cdot \cos\theta$) for $90° < \theta < 180°$, that is, $\cos\theta$ will be negative. Consequently, equation 4.4(a)(i) can be applied to $0° < \theta < 180°$.
At position 2:

$$\theta_2 = 90° + 30° = 120°$$
$$\mathbf{v}_1 = 0 \text{ (at } \theta_1 = 0)$$
$$r = 1.64 \text{ M}$$

Substituting into equation 4.4(b)(iv):

$$\mathbf{v}^2 = v_0^2 + 2\mathbf{g}r(\sin\theta - \sin\theta_0)$$

and setting $\mathbf{v} = \mathbf{v}_2$, $v_0 = \mathbf{v}_1$, $\theta = \theta_2$, $\theta_0 = \theta_1$:

$$\mathbf{v}^2 = (2)(9.81)(1.64)(\sin[120°])$$
$$\mathbf{v} = \sqrt{27.9} = 5.28 \text{ M/s}$$

Third, analyze the swinger who has now gone ballistic:
From Figure 4.4, note that the vertical height of c above ground level at position 2 is the same as the vertical height of c above the feet when the swinger has landed (position 3).
Diagram the system shown in Figure 4.4 (and use rightward and upward positive sign convention):

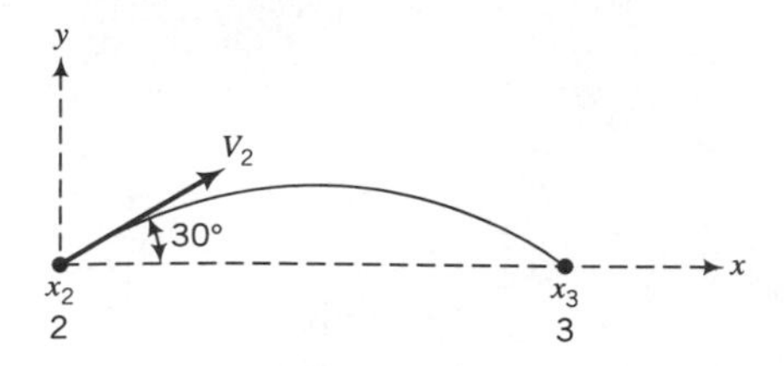

Solve for the time airborne by solving for the time to rise to maximal vertical height and then doubling that time:
Using equation (18a):

$$v_y = \mathbf{v}_2 \sin\theta - \mathbf{g}t$$

At maximum vertical height, $v_y = 0$, and this occurs at $\hat{t}$:

$$0 = (5.28)\sin(30°) - (9.81)\hat{t}$$

$$\hat{t} = \frac{2.64}{9.81} = 0.269 \text{ sec}$$

Total time airborne is:

$$\Delta t = 2 \cdot \hat{t} = 0.538 \text{ s}$$

Fourth, solve for the total horizontal displacement (x_3) that occurs beginning at position 2 (x_2), and ending at position 3 (x_3):
Using equation (15):

$$x_3 = x_2 + (\mathbf{v}_2 \cos\theta)\Delta t$$

$$x_3 = 0.82 + (5.28)(.866)(.538)$$

$$x_3 = 0.82 + 2.46 = 3.28 \text{ M}$$

4.2 HUMAN BODY KINETICS

a. Mass Systems

The previous section has considered human body kinematics and defined some of the relevant equations. Recall that these equations described the motion of a body with respect to its displacement, velocity, and acceleration independent of the forces which caused that motion. In this section we will review human body kinetics. The relevant equations, while based upon the kinematic equations, also incorporate the parameters of force and momentum, which actually cause the motion.

Conveniently, human body kinetics can be divided into linear (translational) kinetics and angular (rotational) kinetics. Translational motion occurs as a consequence as the net force is applied to the body, and rotational motion is caused by the net moment acting upon the body. The field of human body kinetics can be further subdivided into particle kinetics and rigid body kinetics. In this chapter, our purpose will be to obtain some general insight into the motion of a body without specifying the geometric properties of that body. Consequently, we can treat the body as a particle located at its center of mass. This particle will have a mass equal to the total mass of the body. Recall that in Chapter 3 the statics of rigid bodies was used. Similarly, rigid body kinetics is important when the body or body segment is undergoing a rotational motion. In Chapter 7, we shall use rigid body kinetics to describe the general motion of limb segments. In that chapter, human body limb segments will translate and rotate simultaneously in response to both a net force and a net moment acting upon that segment.

Many problems of interest to the human factors engineer can be addressed with the equations of two-dimensional translational motion. The rectangular components for the force and acceleration vectors are then expressed as:

$$\sum F_x = ma_x \tag{35}$$

$$\sum F_y = ma_y \tag{36}$$

One-dimensional or uniaxial (translational) motion occurs along a straight line. For these systems, a single coordinate (x) whose axis is along that straight line can be used for the equations of motion in that direction. Four special cases of translational motion are of particular interest to the human factors engineer: The force is constant, the force is a function of time, the force is a function of velocity, or the force is a function of displacement.

According to Newton's second law, when the force applied to a body is of constant magnitude and applied in a constant direction, that body will move at a constant acceleration in the direction of force application:

$$a_x = \frac{F_x}{m} = \text{constant} \tag{37}$$

A force may be applied in a constant direction, however, the magnitude of the force on the body may vary over time. The resulting acceleration of the body is:

$$a_x(t) = \frac{F_x(t)}{m} \tag{38}$$

Force can also be applied as a function of velocity. Since acceleration is the rate of change of velocity, Newton's second law may be rewritten for this special case as:

$$m \cdot \frac{dv_x}{dt} = F_x(v_x) \tag{39}$$

Equation (39) can be treated mathematically through separation of variables:

$$m \frac{dv_x}{F_x(v_x)} = dt \tag{40}$$

In some cases, the human factors engineer will find it more convenient to express force as a function of displacement. In a manner analogous to equation (39), Newton's second law may be rewritten as:

$$m \frac{dv_x}{dt} = F_x(x) \tag{41}$$

Using the chain rule, the rate of velocity change is:

$$\frac{dv_x}{dt} = \frac{dv_x}{dx}\frac{dx}{dt} = \frac{dv_x}{dx} v_x \tag{42}$$

Substituting equation (42) into equation (41) results in an equation of motion in the x direction:

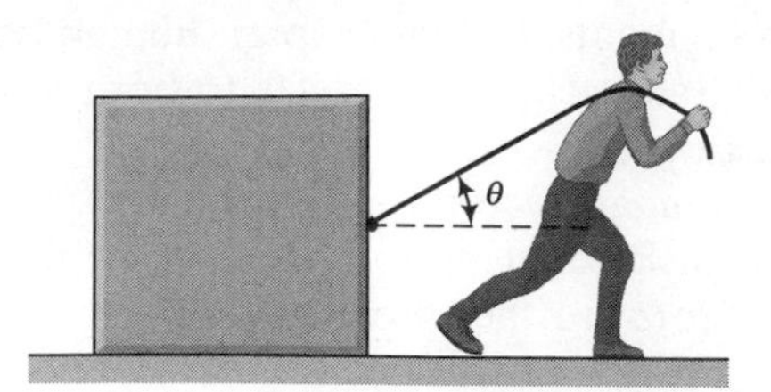

Figure 4.5.a Person pulling a crate along a rough cement horizontal surface.

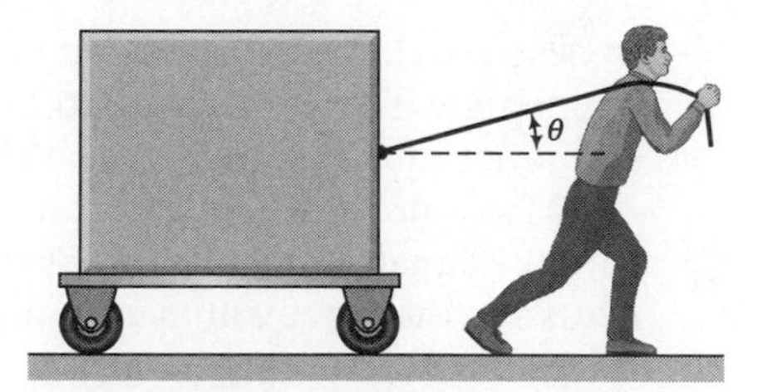

Figure 4.5.b Person pulling a crate placed on a wheeled pallet.

$$mv_x \frac{dv_x}{dx} = F_x(x) \tag{43}$$

Using separation of variables results in a mathematically useful form of this equation:

$$mv_x dv_x = F_x(x)\, dx \tag{44}$$

HFE Application. A HFE will analyze a task in which the human is interacting as part of a system by doing work. Initially, the HFE will analyze the task that the human is currently performing, and then using analytic techniques redefine how the task might be performed more easily and efficiently.

EXAMPLE 4.6

a. Consider a person pulling a crate (with a loaded mass, M = 60 kg) along a rough cement horizontal surface (with a coefficient of sliding friction, $\mu = 0.3$) as shown in Figure 4.5.a. If the person exerts a constant force ($T = 250$ N) at an angle ($\theta = 30°$), find the acceleration of the loaded crate (assuming the entire bottom of the crate remains in continuous contact with the ground).

b. The task of Example 4.6(a) is now modified by placing the crate on a wheeled pallet, as shown in Figure 4.5.b. This elevates the load and reduces the angle (θ) of the rope pull to 15°. If the coefficient of rolling friction is now ($\mu = 0.1$), find the new force (T) that must be exerted on the rope to achieve the same horizontal acceleration obtained in Example 4.6(a).

SOLUTION 4.6(a)

Given:

$$\begin{aligned} \mathrm{m} &= 60\ \mathrm{kg} \\ \theta &= 30° \\ T &= 250\ \mathrm{N} \\ \mu &= 0.3 \end{aligned}$$

Find a:
Draw FBD 1 (of the rigid body) shown in Figure 4.5.a:

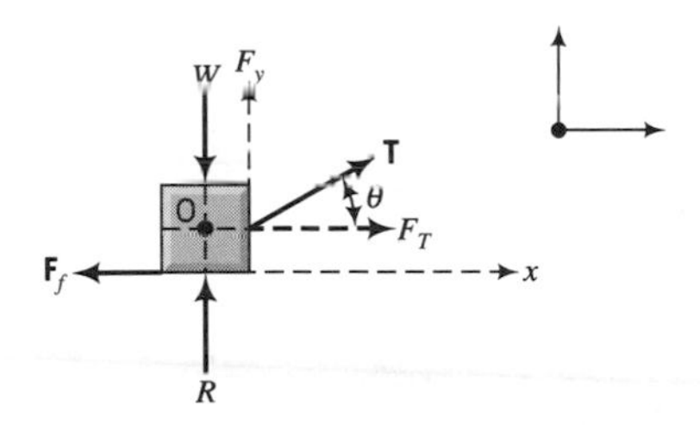

For dynamic equilibrium (*of a particle*):
The FBD 1 reduces to FBD 2:

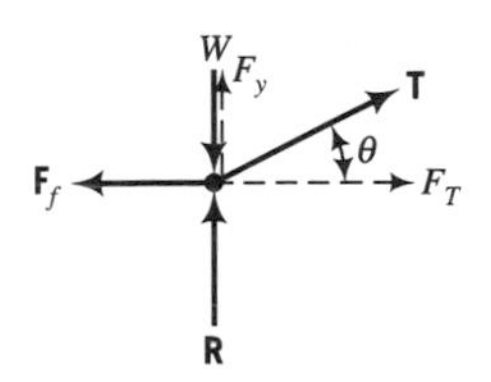

In this problem, it is justifiable to treat an obviously rigid block (FBD 1) as a particle (FBD 2) with a particle representing the entire mass located at the block's center of mass. Since the entire block remains in contact with the floor, it is *not* subjected to rotational motion, and all motion is completely translational, so the governing equations are:

$$\sum F_y = ma_y \quad \text{(i)}$$

$$\sum F_x = ma_x \quad \text{(ii)}$$

Now since the floor is perfectly horizontal, there is no vertical components of displacement (velocity, or acceleration), so that equation (i) reduces to:

$$\sum F_y = 0$$

$$R - W + F_y = 0 \quad \text{(iii)}$$

$$R - m\mathbf{g} + T \cdot \sin\theta = 0$$

$$R = m\mathbf{g} - T \cdot \sin\theta \quad \text{(iv)}$$

Solving $\Sigma F_x = ma_x$

$$\left.\begin{aligned} F_T - F_f &= ma_x \\ T\cos\theta - \mu R &= ma_x \end{aligned}\right\} \quad \text{(v)}$$

Dividing through by m, and substituting equation (iv) into equation (v):

$$a_x = \frac{T \cdot \cos\theta}{m} - \frac{\mu}{m}(m\mathbf{g} - T\sin\theta)$$

Rearranging:

$$a_x = \frac{T}{m}(\cos\theta + \mu\sin\theta) - \mu\mathbf{g} \quad \text{(vi)}$$

Substituting:

$$a_x = \frac{250}{60}(.866 + [.3][.5]) - (.3)(9.81)$$

$$a_x = 4.23 - 2.94 = 1.29\ \text{M/sec}^2$$

SOLUTION 4.6(b)

Given:

$$m = 60 \text{ kg}$$
$$\theta = 15°$$
$$\mu = 0.1$$
$$\alpha_x = 1.29 \text{ M/sec}^2$$

Find T:
Rearrange equation 4.4(a)(vi):

$$T = \frac{m[a_x + \mu g]}{\cos\theta + \mu \cdot \sin\theta}$$

Substituting:

$$T = \frac{(60)[1.29 + (.1)(9.8)]}{[.966 + (.1)(.259)]}$$

$$T = \frac{(60)(2.27)}{(.992)} = 137 \text{ N}$$

The tension exerted by the person on the rope was reduced by almost half when using a wheeled platform. Note that *two* contributing factors occurred. The coefficient of friction was reduced *and* the load was elevated (decreasing the angle, θ).

When a constant, horizontal force F is applied on a body and moves it from position 1 to position 2 over a distance s, the work done on the body is:

$$W_{12} = F \cdot s \tag{45}$$

Recall that the gravitational potential energy PE_g of a body at a vertical position 1 relative to a vertical position 2 is $W \cdot h = mg$

$$PE_g = W \cdot h = m\mathbf{g}h \tag{46}$$

Since W is the weight of the body, and h is the vertical distance between position 1 and position 2, the potential energy of gravity represents the work that the force of gravity would do upon the body in moving it over a vertical distance h.

While potential energy is associated with position (displacement), kinetic energy is associated with motion. For an object with mass (m) that is moving with a speed ($\mathbf{v}$), the equation for kinetic energy is:

$$KE = \frac{1}{2} m \cdot \mathbf{v}^2 \tag{47}$$

For systems of particles, there is a relationship between the kinetic energy and the work done. The net work done on a body (whether these forces are internal or external from the point of view of the body) to displace that body from position 1 to position 2 when added to the kinetic energy at position 1 will equal the kinetic energy at position 2. This relationship is known as the work-energy theorem and can be expressed as:

$$KE_1 + W_{12} = KE_2 \tag{48}$$

The principle of conservation of mechanical energy may be stated as follows:

$$KE_1 + PE_1 = KE_2 + PE_2 \tag{49}$$

This equation states that when a system of particles is moved by the action of forces, then for a conservative system, the sum of the kinetic energy and the potential energy of the system remains constant. The sum $KE + PE$ is the total mechanical energy of the system (denoted by E). It should be noted that with respect to equation (49), when the particles of the system move under the action of internal forces, the potential energy of the system must include the potential energy of those internal forces.

It is important to note that when a mechanical system involves friction, the total mechanical energy does not remain constant but decreases in moving from position 1 to position 2. However, the principle of conservation of energy requires that the mechanical energy of the system is not lost but rather is transformed into heat. Therefore, in the case of friction, the sum of the mechanical energy and of the thermal energy of the system remains constant.

Other forms of energy may also be involved in a human-technological system. For example, a technological power generator converts mechanical energy into electrical energy and a human muscle generator converts chemical energy into mechanical energy. When all of the forms of energy are considered, the energy of any system may be considered as constant and the principle remains valid under all conditions. This application of conservation of energy, when applied to other forms of energy, will be considered in Chapter 6.

HFE Application. The HFE will sometimes consider various methods for lowering the upper performance limit of certain human-technological systems. Often an acceptable system design will involve a modification to an existing system element. Even apparently simple system design modifications require analysis to drive the design process.

EXAMPLE 4.7

a. The slide shown in Figure 4.1 is now modified by coating the sliding surface and band rails with a teflon-like material that retards ice formation. If all other parameters of Example 4.2 remain the same, except the coefficient of sliding friction is now $\mu = 0.2$, use the work-energy method to find the slider's velocity at the bottom of the slide.
b. Repeat Example 4.7(a), this time using the conservation of energy method.

SOLUTION 4.7(a)

Work-energy method:

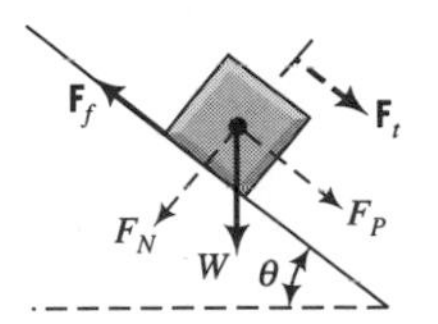

First, solve for F_T (the net force acting on the mass):

$$W = m \cdot \mathbf{g} = (30)(9.81) = 294\ \text{N}$$
$$F_N = W \cdot \cos\theta = (294)(.866)$$
$$F_N = 255\ \text{N}$$
$$F_f = \mu \cdot F_N = (.2)(255)$$
$$F_f = 51\ \text{N}$$
$$F_p = W \cdot \sin\theta = (294)(.5)$$
$$F_p = 147\ \text{N}$$
$$F_T = F_p - F_f = 147 - 51$$
$$F_T = 96\ \text{N}$$

Second, solve for the work (Wk_{12}) performed in moving from the top of the slide (1) to the bottom of the slide (2):
Use rightward and downward positive sign convention with the origin at the top of the slide (1):
Length of the slide surface:

$$\sin\theta = \frac{1.85\ \text{M} - 0.35\ \text{M}}{L}$$
$$L = \frac{1.5\ \text{M}}{\sin(30°)} = 3.0\ \text{M}$$
$$Wk_{12} = F_T \cdot L = (96)(3)$$
$$Wk_{12} = 288\ \text{J}$$

Third, solve the work-energy equation:

$$KE_1 + Wk_{12} = KE_2$$
$$KE_1 = \frac{1}{2}mv_1^2 = 0$$

(Since $v_1 = v_0 = 0$)

$$KE_2 = \frac{1}{2}mv_2^2 = \frac{1}{2}(30)v_f^2$$

(Since $v_2 = v_f$).
Therefore:

$$Wk_{12} = KE_2$$
$$288 = (15)v_f^2$$
$$v_f = \sqrt{19.2} = 4.38\ \text{M/s}$$

SOLUTION 4.7(b)

Conservation of energy method:

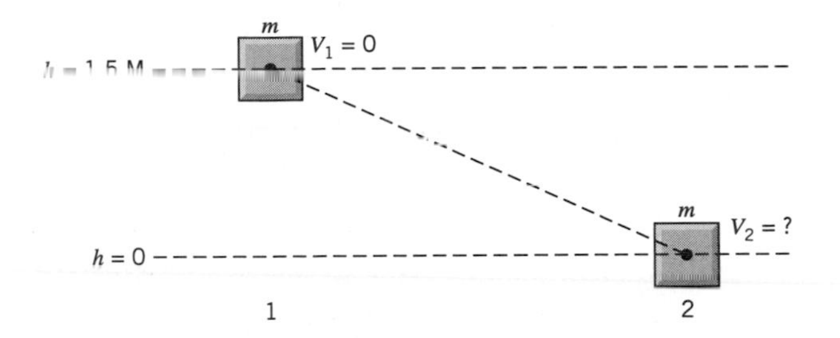

State 1 is at the top of the slide, and state 2 is at the bottom of the slide. As a variation, define rightward and *upward* positive sign convention. Set the zero point of the vertical axis at state 2.
For conservation of mechanical and thermal energy:

$$KE_1 + PE_1 = KE_2 + PE_2 + W_f$$

$$KE_1 = \frac{1}{2} m \cdot v_1^2 = 0 \qquad (\text{since } v_1 = v_0 = 0)$$

$$PE_1 = PE_{g1} = m \cdot \mathbf{g} \cdot h$$

$$PE_1 = (30)(9.81)(1.5) = 441 \text{ J}$$

$$KE_2 = \frac{1}{2} m v_2^2 = 15 v_2^2$$

$$PE_2 = PE_{g2} = 0 \qquad (\text{since } h = 0)$$

W_f = Work done against friction in going from state 1 to state 2

$W_f = F_f L = (51)(3)$ [see Solution 4.7(a) for F_f and L]

$W_f = 153$ J

Substituting into the conservation of energy equation:

$$0 + 441 = 15v_2^2 + 0 + 153$$

$$15v_2^2 = 288$$

$$v_2 = \sqrt{19.2} = 4.38 \text{ M/s}$$

By whichever method (work-energy or conservation of energy), increasing the coefficient of sliding friction (μ) from zero to 0.2, decreased the final slider velocity by almost 20%.

b. Elastic Systems

Recall that when an applied force is the function of displacement, then the work performed is:

$$W_s = \int_{x_1}^{x_2} F_x \cdot dx \tag{50}$$

The ability of a material to deform with the subsequent storage of energy can be characterized as the elastic potential energy. In order to analyze the elastic behavior of materials, it is convenient to model the material as a mechanical spring. In the case of a linear spring, the applied force is a function of the deformation of the spring and is a straight line relationship:

$$F = kx \tag{51}$$

where k is the spring constant.

The work (W_s) done by the force (F) on the spring is calculated by substituting equation (51) into equation (50):

$$W_s = \int_0^x kx \cdot dx = \frac{1}{2}kx^2 \tag{52}$$

The work done on the spring by deforming the spring as described by equation (52) is stored in the spring as potential energy as long as the force is applied. Once removed, however, the spring will release this potential energy as it returns to its undeformed configuration. Consequently, the work done on the spring (W_s) is also equal to the elastic potential energy of the spring (PE_s) and may be expressed as:

$$PE_s = \frac{1}{2}kx^2 \tag{53}$$

A mass in series with a linear spring is subject to both gravitational effects and spring (elastic) effects. The total potential energy (PE) of a body (mass) in series with a linear spring is:

$$PE = PE_g + PE_s \tag{54}$$

where PE_g is from equation (46) and PE_s is from equation (53).

HFE Application. Many examples of human-technological systems have elastic elements as part of the technological system that deform when undergoing human interaction. Adequate design for functional performance requires analysis of this interaction.

EXAMPLE 4.8

a. A company is developing a fireman's safety net to catch people jumping from burning buildings. It is anticipated that for an operational upper limit, a person (with a mass of 90 kg) may jump from a height of 30 M, and the net be held by short firemen so that it is only 1 M above the ground (as shown in Figure 4.6). The company currently uses an elastic net material

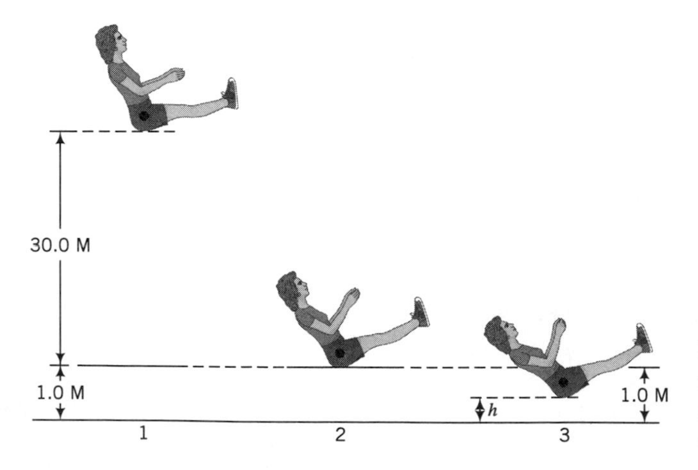

Figure 4.6 Person jumping from a burning building to a fireman's safety net located beneath them.

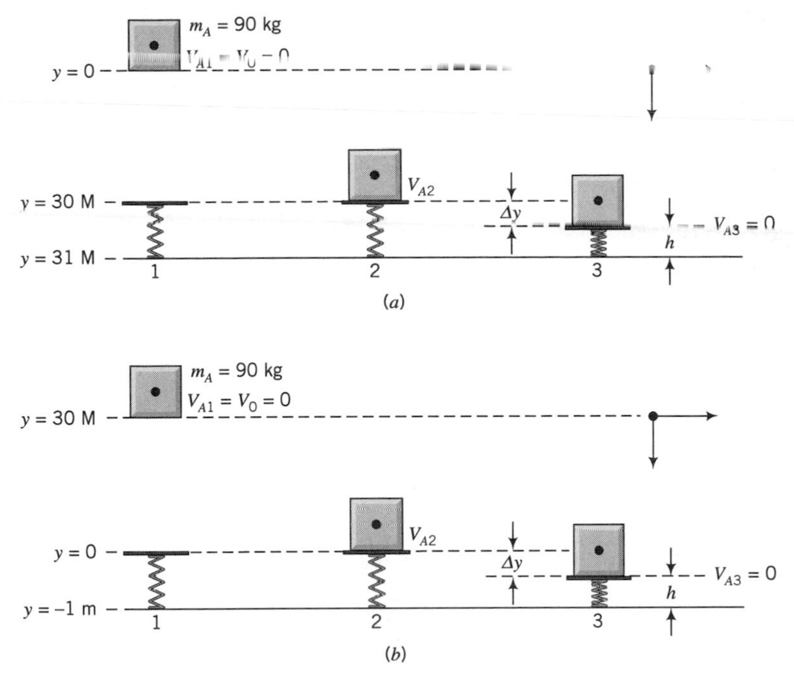

Figure 4.7.a Free-body diagram of a person jumping onto a fireman's safety net from Example 4.8(a). **b** Free-body diagram of a person jumping onto a fireman's safety net from Example 4.8(b).

with a spring constant (K = 200 N/cm), and wants you to determine if this net material will stop a jumper's fall before the jumper (now caught by the net) contacts the ground. For this example, refer to Figure 4.7.a and use the kinematic equations for the free-fall (assume no air resistance) and the work-energy method for the elastic deformation of the net.

b. Repeat Example 4.6(a), above, but now refer to Figure 4.7.b and use the conservation of energy method for both the free-fall (assume negligible air resistance) and the elastic deformation of the net.

SOLUTION 4.8(a)

For the system shown in Figure 4.6, the analytical model is shown in Figure 4.7.a:

a. Kinematic equations $(1 \rightarrow 2)$

$$y = \cancelto{0}{y_0} + \cancelto{0}{y_0 t} + \frac{1}{2} g t^2$$

$$t_{1\rightarrow 2} = \sqrt{\frac{2 \cdot y}{\mathbf{g}}} = \sqrt{\frac{(2)(30)}{9.81}} = 2.47 \text{ sec}$$

$$v = \cancelto{0}{v_0} + \mathbf{g}t$$

$$v_{f2} = (9.81)(2.47) = 24.2 \text{ M/s}$$

b. Work-energy method $(2 \rightarrow 3)$

First, determine the work in going from State 2 to State 3:

FBD:

$W = m \cdot \mathbf{g}$

$\mathbf{F} = K \cdot y$

Net force (F_T) on the person:

$$F_T = m \cdot \mathbf{g} - ky$$

Incremental work:

$$dW_{2\to3} = F_T \cdot dy = m \cdot \mathbf{g} \cdot dy - ky \cdot dy$$

Work from state 2 to state 3:

$$W_{2\to3} = \int_{y_2}^{y_3} F_T \cdot dy = \int_{y_2}^{y_3} m \cdot \mathbf{g} \cdot dy - \int_{y_2}^{y_3} ky \cdot dy$$

Set $y_2 = 0$, and recall downward positive sign convention, so $y_3 = \Delta y$:

$$W_{23} = m \cdot \mathbf{g} \cdot \Delta y - \frac{1}{2}k(\Delta y)^2$$

Substituting:

$$W_{23} = (90)(9.81)\Delta y - (10{,}000)(\Delta y)^2$$

Second, determine the change in kinetic energy between state 2 and state 3:

$$\Delta KE = KE_3 - KE_2$$

$$\Delta KE = \frac{1}{2}m(v_{A3})^2 - \frac{1}{2}m(v_{A2})^2$$

Substituting (recall $v_{A3} = 0$):

$$\Delta KE = -(45)(24.2)^2 = -26{,}354 \text{ J}$$

Third, solve the work-energy equation:

$$W_{23} = \Delta KE$$

$$883\Delta y - 10{,}000(\Delta y)^2 = -26354$$

$$(\Delta y)^2 - \left(\frac{883}{10{,}000}\right)\Delta y - \left(\frac{26{,}354}{10{,}000}\right) = 0$$

$$(\Delta y)^2 - .0883\Delta y - 2.635 = 0$$

Solving by the quadratic formula:

$$\Delta y = \frac{.0883 \pm \sqrt{(.0883)^2 - (4)(-2.635)}}{(2)(1)}$$

$$\Delta y = \frac{.0883 \pm \sqrt{10.54}}{2}$$

$$\Delta y = 1.67 \text{ M}$$

Since $\Delta y > 1$ m, the safety net manufacturer will need to reconsider using its current elastic net material based upon this worst case analysis.

SOLUTION 4.8(b)

For the system shown in Figure 4.6, the analytical model is shown in Figure 4.7b.

a. Conservation of energy (1 → 2):

$$\cancelto{0}{KE_1} + PE_{g1} + \cancelto{0}{PE_{s1}} = KE_2 + \cancelto{0}{PE_{g2}} + \cancelto{0}{PE_{s2}} + \cancelto{0}{W_{f12}}$$

$$PE_{g1} = KE_2$$

$$m_A \cdot \mathbf{g} \cdot y = \frac{1}{2} m_A (v_{A2})^2$$

$$(90)(9.81)(30) = (0.5)(90)(v_{A2})^2$$

$$26487 = 45(v_{A2})^2$$

$$v_{A2} = \sqrt{\frac{26487}{45}} = 24.2 \text{ M/s}$$

b. Conservation of energy (2 → 3):

$$KE_2 + \cancelto{0}{PE_{g2}} + \cancelto{0}{PE_{s2}} = \cancelto{0}{KE_3} + PE_{g3} + PE_{s3} + \cancelto{0}{W_{f23}}$$

$$KE_2 = PE_{g3} + PE_{s3}$$

$$\frac{1}{2} m_A (v_{A2})^2 = m_A \cdot \mathbf{g} \cdot (-\Delta y) + \frac{1}{2} \cdot k \cdot (-\Delta y)^2$$

$$(45)(24.2)^2 = -(90)(9.81)\Delta y + (10{,}000)(-\Delta y)^2$$

$$26354 = -883\Delta y + 10{,}000(\Delta y)^2$$

$$(\Delta y)^2 - .0883\Delta y - 1.635 = 0$$

Proceeding per Solution 4.6(a).

$$\Delta y = 1.69 \text{ M}$$

4.3 HUMAN BODY IMPACT AND COLLISION

a. Human Body Impact

An understanding of the effects of impact on the human body begins with the application of the equations of linear momentum and impulse. The momentum ($\mathbf{p}$) of an object is its mass times its velocity:

$$\mathbf{p} = m\mathbf{v} \tag{55}$$

When the mass of an object is constant, Newton's second law can be expressed in terms of the momentum of that object:

$$F = \frac{dp}{dt} \tag{56}$$

Separating the variables of equation (56), and integrating both sides results in:

$$\int_{t_1}^{t_2} F \cdot dt = \Delta p \tag{57}$$

Equation (57) represents the impulse-momentum theorem. The integral function on the left is the linear impulse generated by force (F) on the body over a specific time interval. The delta function on the right is the change in momentum of the body during the time interval.

When the impulsive force is constant, then equation (57) reduces to:

$$F \cdot \Delta t = \Delta p \tag{58}$$

Equation (58) is useful in that the impulse of a time varying force may be determined by approximating that force as an average force over the time interval, Δt.

HFE Application. The human body is quite vulnerable to injury from impact forces. The HFE will develop and evaluate technological subsystems that minimize the effect of impact on the human subsystem. The result is to improve the human-technological system performance.

EXAMPLE 4.9

An anthropomorphic test device (ATD or simply test dummy) stands erect and is shown in profile in Figure 4.8.

a. The following experiment is conducted: a baseball (with a mass of 0.25 kg) is fired from a ball gun at the ATD's chest. The trajectory of the ball is straight horizontal, and the muzzle velocity of the ball gun is 30 M/s. There is no air resistance, and the muzzle of the barrel is 1.5 M above the level horizontal ground. When the ball strikes the chest, strain gauges in the ATD's chest record a contact time of .05 s. The hardball then rebounds in the reverse but still horizontal direction. The ball then takes a ballistic trajectory and contacts the ground 6 M (horizontally) from the point of chest impact (Figure 4.8). Calculate the impact force sustained by the unprotected chest of the ATD.
b. The manufacturer of human safety equipment now wishes to evaluate its new design for a baseball chest protector (BCP). The BCP is placed securely over the chest of the ATD and the experiment of Example 4.7(a) is repeated. Only two parameters change. The chest contact time is now recorded as 0.15 s and the ball contacts the ground 2 M horizontally from the point of chest impact. Calculate the impact force sustained by the chest of the ATD when using the BCP.

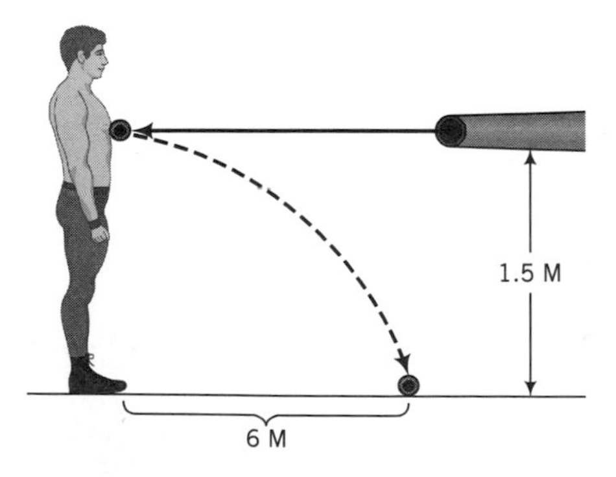

Figure 4.8 A baseball is fired horizontally from a ball gun at the anthropomorphic test device.

SOLUTION 4.9(a)

With no BCP:
Given:

$$h = 1.5\text{ M}$$
$$v_m = 30\text{ M/s}$$
$$\Delta t = .05\text{ s}$$
$$d = 6.0\text{ M}$$

Find *F*:

Referring to Figure 4.8, draw a ballistic diagram of the system:

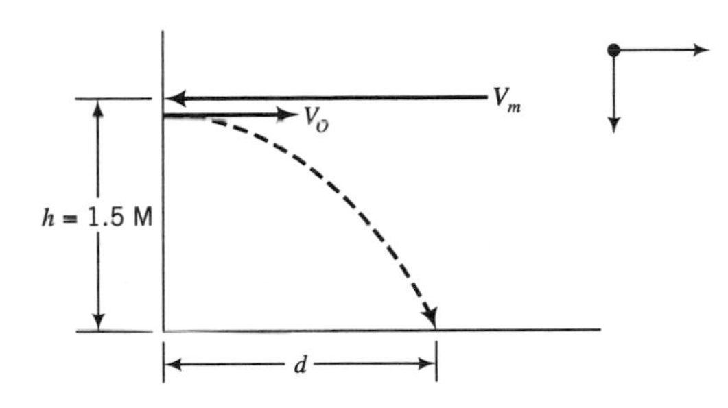

We proceed in three sequential steps:
First, solve for the time (t) that the ball is in free-fall.

$$y = y_0 + v_0 t + \frac{1}{2}\mathbf{g}t^2$$

Set $y_0 = 0$, $v_{0y} = 0$, and $y = h$.

$$h = \frac{1}{2}\mathbf{g}t^2$$

$$t = \sqrt{\frac{2h}{\mathbf{g}}} = \sqrt{\frac{3.0}{9.81}} = .553\text{ sec}$$

Second, proceed to solve for v_{0x}:
Assuming negligible air resistance and;

$$x = v_0 \cdot t$$

Set $x = d$, $v_0 = v_{0x}$

$$d = v_{0x} \cdot t$$

$$v_{0x} = \frac{d}{t} = \frac{6.0}{.553} = 10.8\text{ M/s (rightward)}$$

Third, calculate F:
Assume constant force over time interval, Δt; and define

$$v_0 = v_{0x} = -30\text{ M/s} \quad \text{(leftward)}$$

So:

$$F \cdot \Delta t = m \cdot v_{0x} - m \cdot v_m$$

$$F = \frac{(.25)[10.8 - (-30)]}{(.05)} = 204 \text{ N (rightward)}$$

Thus, the force exerted by the chest on the ball (F) = 204 N.

SOLUTION 4.9(b)

With the BCP:
Given:

$$h = 1.5 \text{ M}$$
$$v_m = 30 \text{ M/s}$$
$$\Delta t = .15 \text{ s}$$
$$d = 2.0 \text{ M}$$

Find F:
Referring again to the same ballistic diagram used in Solution 4.9(a), we repeat the same three step process.

First, solve for t:

$$y = y_0 + v_0 t + \frac{1}{2}\mathbf{g}t^2$$

As per Solution 4.9(a):

$$t = \sqrt{\frac{2h}{\mathbf{g}}} = .553 \text{ s}$$

Second, solve for v_{0x}:
Assuming negligible air resistance,

$$x = v_0 t$$

As per Solution 4.9(a):

$$v_{0x} = \frac{d}{t} = \frac{2.0}{0.553} = 3.6 \text{ M/s (rightward)}$$

Third, calculate F:
As per Solution 4.9(a):

$$F \cdot \Delta t = m \cdot v_{0x} - m \cdot v_m$$

$$F = \frac{(.25)[3.6 - (-30)]}{(.15)} = 56 \text{ N (rightward)}$$

Force (F) exerted by the chest on the ball-BCP combination is only 56 N!

---

b. Human Body Collision

In general, a body may experience various types of collisions. Such collisions may be characterized in the following two broad categories. First is the orientation of the impact velocities of the two bodies involved. Second is the nature of the deformation that each body experiences at impact. We shall now review these.

When two bodies experience a head-on collision, they undergo *direct* central impact. The line of impact is an imaginary straight line which passes through the centers of mass of each of the colliding bodies. Orthogonal to this line is the line of contact which represents a tangent line through the contacting surfaces of the bodies. In direct central impact, the collision is uniaxial in that the velocity vectors of each of the bodies are entirely directed along the line of impact.

In the case of *oblique* central impact of two bodies, the collision is biaxial. In this case, the velocity vectors of each body in a plane are defined by the line of contact as the ordinate and the line of impact as the abscissa. The origin of the two axes is at the point of contact.

The second broad category of collisions is based upon the type of deformations occurring during the time course of the collision. It is convenient to consider three subcategories. A *perfectly elastic* collision (an elastic collision) is one in which the total momentum is conserved and also the total kinetic energy is conserved. In a *perfectly inelastic* collision (or plastic collision), the conservation of the total momentum only is conserved. In a *perfectly inelastic* collision, the two colliding bodies remain together after the collision and move as a single unit.

The third subcategory is represented by an *elasto-plastic collision*. In this situation, the material nature of the body will undergo both an elastic and plastic deformation. Recall that the elastic deformation is recoverable upon release of the deforming forces. However, the plastic deformations are not recoverable, and will remain after removal of the deforming forces.

Let us now review the governing mathematical equations for impact and collisions. There are two governing equations for one-dimensional collisions, and four governing equations for two-dimensional collisions.

Direct central impact is by definition a one-dimensional collision since it occurs along a straight line. Consequently, the equation for conservation of linear momentum can be written in scalar form:

$$m_A v_A + m_B v_B = m_A v'_A + m_B v'_B \tag{59}$$

This equation simply states that the total momentum of the system before the collision must equal the total momentum of the system after the collision.

The second governing equation for uniaxial collisions relates the relative velocity of separation to the relative velocity of approach. In scalar form:

$$v'_B - v'_A = e(v_A - v_B) \tag{60}$$

This equation states that the relative velocity of any two particles after impact can be obtained by multiplying the relative velocity of the particles before impact by the coefficient of restitution.

For a perfectly elastic collision (or elastic collision), the coefficient of restitution (e) equals one. For a perfectly inelastic collision (or plastic collision), the coefficient of restitution (e) equals zero. For an elastoplastic collision, the coefficient of restitution is a positive number greater than zero but less than one.

An oblique central impact is, by definition, a biaxial collision. For such a two-dimensional collision, the conservation of the linear momentum equation must be applied in two coordinate directions and the nature of the deformation must be known. In solving oblique central impact problems, the x-axis is chosen along the line of impact. The y-axis is chosen along the common tangent to the surfaces in contact. We make the approximation that the collision of two bodies represented as particles will be both smooth and frictionless. Consequently, the impulsive forces acting on the particles during impact will be internal forces directed along the line of impact (x-axis). The four governing equations are then as follows:

1. With respect to particle A, the y component of the momentum is conserved:

$$m_A v_{AY} = m_A v'_{AY} \tag{61}$$

2. With respect to particle B, the y component of the momentum is conserved:

$$m_B v_{BY} = m_B v'_{BY} \tag{62}$$

3. With respect to both particles A and B, the x component of the total momentum of the system is conserved:

$$m_A v_{AX} + m_B v_{BX} = m_A v'_{AX} + m_B v'_{BX} \tag{63}$$

4. With respect to both particles A and B, the x component of the relative velocity after impact is equal to the x component of the relative velocity before impact multiplied by the coefficient of restitution.

$$v'_{BX} - v'_{AX} = e(v_{AX} - v_{BX}) \tag{64}$$

HFE Application. The HFE must consider a variety of human-technological systems in which the human subsystem is the elastic element and will deform when interacting with the mass elements of the technological subsystem. System design requires analysis of this interaction and design modifications as appropriate.

EXAMPLE 4.10

The design of a steel safety helmet is being evaluated in the human factors engineering laboratory. As shown in Figure 4.9.a, the helmet is rounded in shape and has a mass of 1 kg. The helmet rests on top of a cadaver's skull (with a spring

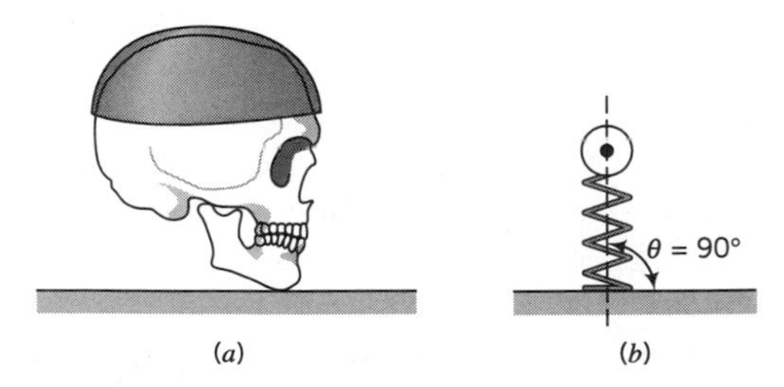

Figure 4.9.a A rounded steel safety helmet.
b Simplified physical model of Figure 4.9.a.

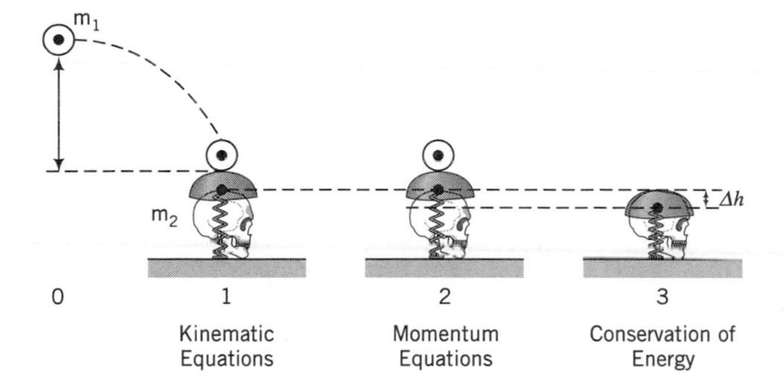

Figure 4.10 Free-body diagram of the various states of the solid steel ball being projected off a ledge onto the safety helmet from Figure 4.9.a.

constant K = 25,000 N/cm), and the base of the skull is affixed to a rigid test bench. Figure 4.9.b shows a simplified physical model of the system in which the center of mass of the 1 kg helmet is modeled as a sphere resting on top of a vertically oriented linear spring. The other end of the spring is affixed to a rigid, massive base.

The experiment begins when a 2 kg solid steel ball is projected off a ledge at an initial velocity (V_0 = 3.6 M/s). The ledge is elevated vertically 2 M above the top of the steel safety helmet, and at a horizontal distance, Δx. As shown in Figure 4.10 (between state 0 and state 1), the subsequent ballistic trajectory of the 2 kg steel ball is such that it strikes directly on top of the steel safety helmet (so that the centers-of-mass of the steel ball and safety helmet are aligned vertically with the axis of the skull (linear spring) and orthogonal to the horizontal ground plane).

a. Using kinematic equations, find the velocity of the 2 kg steel ball just before helmet contact, the angle of that contact with respect to the horizontal, and the distance, Δx.
b. Using the momentum equations for the impact and collision (state 2 of Figure 4.10), find the velocity of the 2 kg steel ball immediately after impact, the angle of that velocity with respect to the horizontal, the velocity of the 1 kg steel safety helmet immediately after impact, and the angle of that velocity with respect to the horizontal. You may assume that the helmet is stationary just prior to impact.
c. Using the conservation of energy method (state 3 of Figure 4.10), find the maximal displacement (depression) of the skull (linear spring) just prior to rebound.

SOLUTION 4.10(a)

Referring to Figure 4.10, draw a ballistic diagram for the system between 0 and 1:

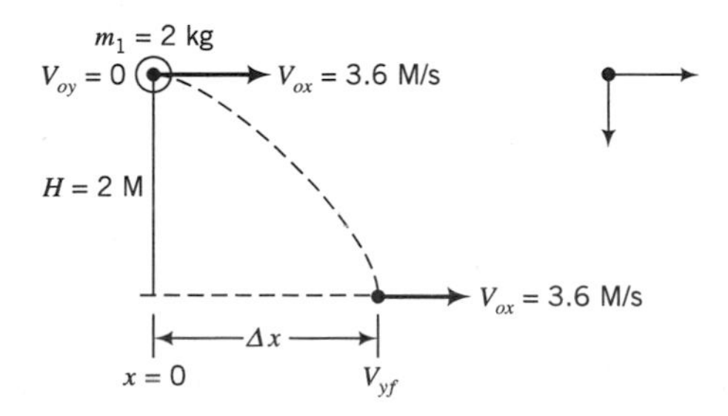

First, solve for t:

$$y = \cancelto{0}{y_0} + \cancelto{0}{v_{0y}} t + \frac{1}{2}\mathbf{g}t^2$$

$$y = H = \frac{1}{2}\mathbf{g}t^2$$

$$t = \sqrt{\frac{2 \cdot H}{\mathbf{g}}} = \sqrt{\frac{(2)(2)}{9.81}} = 0.639 \text{ s}$$

Second, solve for Δx:

$$\Delta x = v_{0x} \cdot t$$
$$\Delta x = (3.6)(.639)$$
$$\Delta x = 2.30 \text{ M}$$

Third, solve for v_{yf}:

$$v_{yf} = \cancelto{0}{v_{y0}} + \mathbf{g}t$$
$$v_{yf} = (9.81)(0.639) = 6.27 \text{ M/s}$$

Fourth, solve for θ_f and v_R:

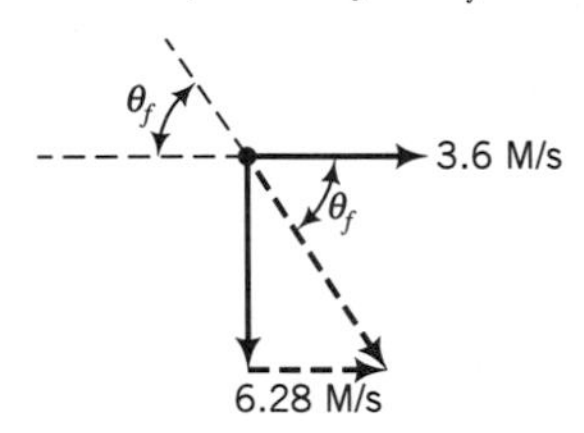

$$\theta_f = \tan^{-1}\left(\frac{6.27}{3.6}\right) = 60.1°$$
$$v_R = \sqrt{(6.27)^2 + (3.6)^2} = 7.23 \text{ M/s}$$

SOLUTION 4.10(b)

Referring again to Figure 4.10, draw the free-body diagram for the system at state 2:

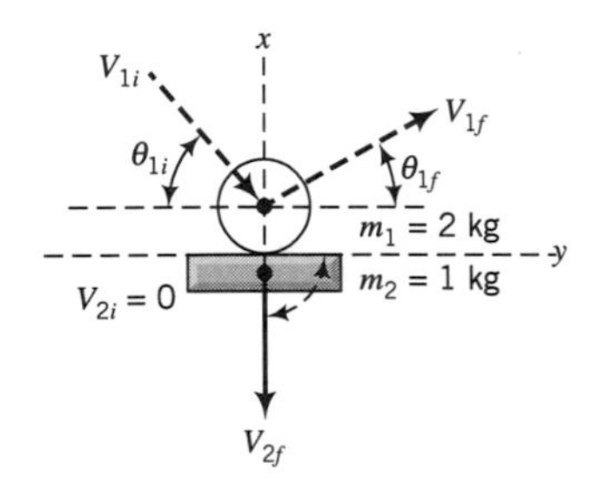

The known parameters are:

$$v_{1i} = v_R = 7.23 \text{ M/s}$$
$$\theta_{1i} = \theta_f = 60.1°$$
$$e = 0.9$$

Solve for v_{2f}; v_{1f}; θ_{1f}:

$$v_{2ix} = 0$$
$$v_{2iy} = 0$$
$$(v_{1fy} = 3.6\,\text{M/s} \qquad v_{2fy} = 0)$$
$$m_1 v_{1ix} + m_2 v_{2ix} = m_1 v_{1fx} + m_2 v_{2fx}$$
$$(2)(6.27) + (1)(0) = (2)v_{1fx} + (1)v_{2fx}$$
$$2v_{1fx} + v_{2fx} = 12.6 \qquad \text{(i)}$$
$$v_{2fx} - v_{1fx} = e[v_{1ix} - v_{2ix}]$$
$$= (.9)(6.27 - 0)$$
$$v_{2fx} - v_{1fx} = 5.65 \qquad \text{(ii)}$$

Solving equations (i) and (ii) simultaneously (multiplying equation (ii) by −2):

$$2v_{1fx} + v_{2fx} = 12.6$$
$$-2v_{1fx} + 2v_{2fx} = 11.3$$
$$3v_{2fx} = 23.9$$
$$v_{2fx} = 7.97\,\text{M/s}$$
$$v_{1fx} = 2.32\,\text{M/s}$$

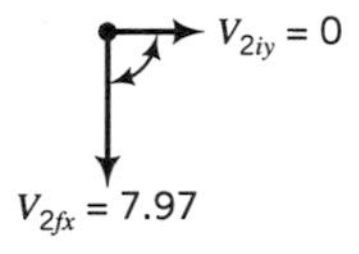

$$\theta_{2i} = \theta_{2f} = 90^\circ$$

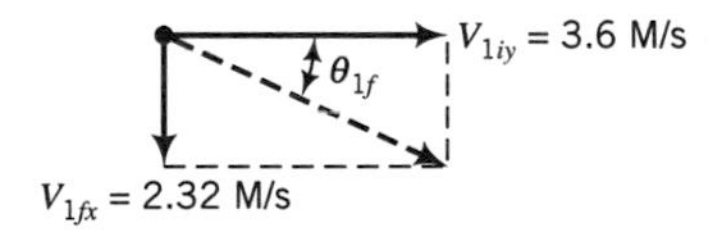

$$\theta_{1f} = \tan^{-1}\left(\frac{2.32}{3.60}\right) = 32.8^\circ$$

SOLUTION 4.10(c)

Referring again to Figure 4.10, draw the analytic diagram for the system between 2 and 3:
Redefine 2 → 1, and 3 → 2:

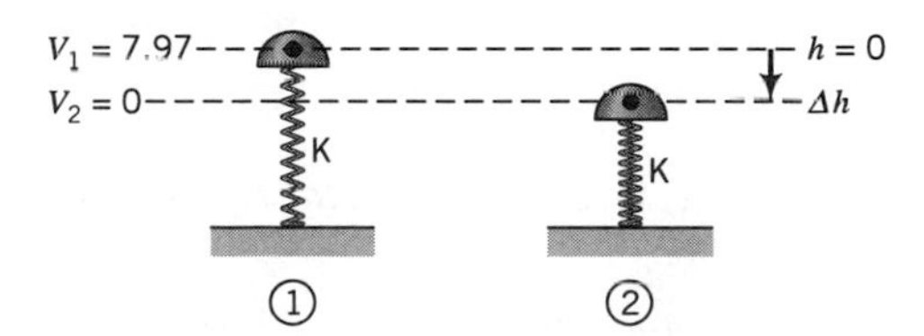

Given:

$$m = 1 \text{ kg}$$

$$K = 25{,}000 \text{ N/cm}$$

$$\cancelto{0}{PE_{g1}} + \cancelto{0}{PE_{s1}} + KE_1 = PE_{g2} + PE_{s2} + \cancelto{0}{KE_2} + \cancelto{0}{W_{f12}}$$

$$KE_1 = \frac{1}{2}mv_1^2 = (.5)(1)(7.97)^2$$

[Note $v_1 = v_{2fx}$ from 4.10(b)]

$$KE_1 = 31.8 \text{ J}$$

$$PE_{g2} = m \cdot \mathbf{g} \cdot h = (1)(9.81)(-\Delta h)$$

$$PE_{s2} = \frac{1}{2}K(\Delta h)^2$$

(Note: $k = 25{,}000$ N/cm, so that $K = 2{,}500{,}000$ N/M)

$$PE_{s2} = (0.5)(2{,}500{,}000)(\Delta h)^2$$

$$KE_1 = PE_{g2} + PE_{s2}$$

$$31.8 = -9.98\Delta h + 1{,}250{,}000(\Delta h)^2$$

$$(\Delta h)^2 - .000008 - .000025 = 0$$

Solving by using the quadratic equation rule:

$$\Delta h = \frac{.000008 \pm \sqrt{(.000008)^2 - (4)(1)(-.000025)}}{(2)(1)}$$

$$\Delta h = \frac{.000008 \pm \sqrt{.0001}}{(2)}$$

$$\Delta h \approx \frac{.01}{2} = .0005 \text{ M}$$

$$\Delta h \approx 0.5 \text{ cm (or 5 mm).}$$

FURTHER INFORMATION

D.B. Chaffin, G.B.J. Andersson and B.J. Martin: *Occupational Biomechanics.* 3rd Edition. John Wiley. New York. 1999.

C.E. Ewing, et al. (eds.): *Impact Injury of the Head and Spine.* Charles C. Thomas. Springfield, IL. 1983.

J.L. Meriam and L.G. Kraige: *Engineering Mechanics: Dynamics.* 4th Edition. John Wiley. New York. 1997.

N. Ozkaya and M. Nordin: *Fundamentals of Biomechanics.* 2nd Edition. Springer-Verlag. New York. 1999.

C.A. Phillips and J.S. Petrofsky (eds.): *Mechanics of Skeletal and Cardiac Muscle.* Charles C. Thomas. Springfield, IL. 1983.

PROBLEMS

4.1. A ski instructor is evaluating the "SKEE-EEZ" elastic trainer cord as a possible aid in instructing novice skiers. The skier (mass = 75 kg) holds onto the elastic trainer cord (tied to a tree at the top of a hill) and start skiing down the hill, as shown in Diagram 4.1. The slack runs out of the elastic cord when it is 20 M long (up to this point its spring constant, $K = 0$). The elastic cord then starts stretching with a spring constant, $K = 0.1$ Newtons per centimeter. If the coefficient of sliding friction down the entire course of the hill is $\mu = 0.2$, how far along the slope of the hill does the skier travel before coming to a stop?

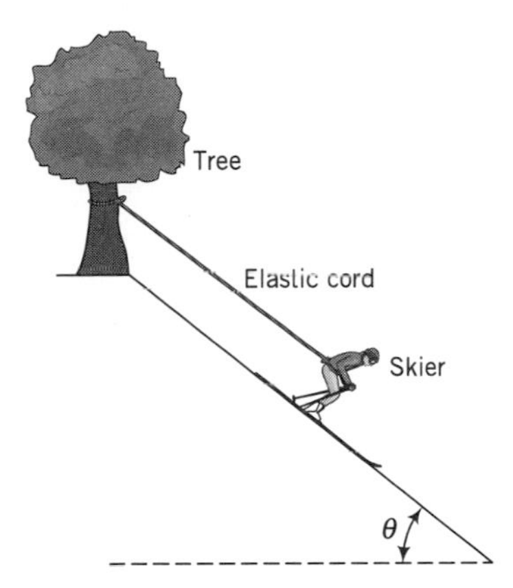

Diagram 4.1 $\theta = 35°$.

4.2. A pair of sledders are using two newly designed "ultralight" sleds. Sledder A (mass = 40 kg) is sledding up a hill, as shown in Diagram 4.2 (inclined at an angle of 20° from the horizontal) and has slowed down to a velocity of 5 M/s when the sledder collides with sledder B. Sledder B (mass = 80 kg) was sledding downhill and had sped up to 20 M/s at the time of collision. The two sledders entangle (a perfectly inelastic collision, $e = 0$) and start rolling down the hill. What is their initial velocity (V_{AB})? If the coefficient of rolling friction is $\mu = 0.15$, over what distance do the entangled sledders roll, before coming to a stop? Because of the new "ultralight" design, you may neglect the mass of the sleds.

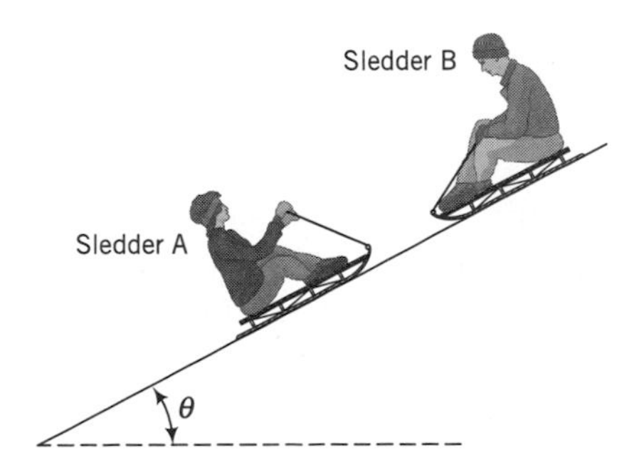

Diagram 4.2

4.3. As shown in Diagram 4.3, Car A (mass of 2000 kg) is stopped half-way downhill on a rain slicked road. Car B (mass of 2000 kg) is simultaneously coming straight uphill on the same rain slicked road at a velocity of 20 M/s and headed toward Car A.

Halfway up the hill, the two cars have a head-on perfectly elastic collision (where $e = 1$) and the collision is direct-central impact. Since both cars are equipped with the latest version of "Gator-Gripper" tread tires, the coefficient of sliding friction (μ) is 0.3. How far apart (along the road) are the two cars when each finally comes to rest?

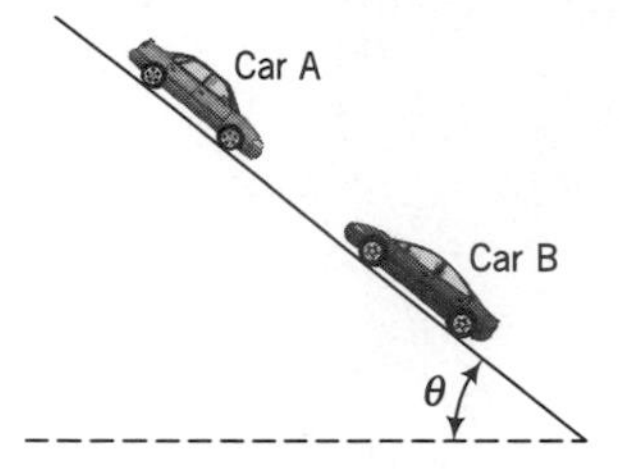

Diagram 4.3

4.4. Two fire and rescue workers, attached by a safety spring ($K = 50$ N/M) were climbing an ice-covered hill in order to rescue a stranded dog. The second climber (with a total mass of 70 kg distributed as 50 kg body mass and 20 kg of backpack and other equipment) slipped at the top (where the safety spring was already prestretched to its zero-load length). The ice that the person was sliding down (Diagram 4.4) had a slope of 60° and a coefficient of friction of 0.2. Assuming the climber had no initial velocity:

1. How far down the slope (from the top) will the person travel before being stopped?
2. What amount of mass must the climber then instantly unload in order to rebound back up the slope and stop at the top?

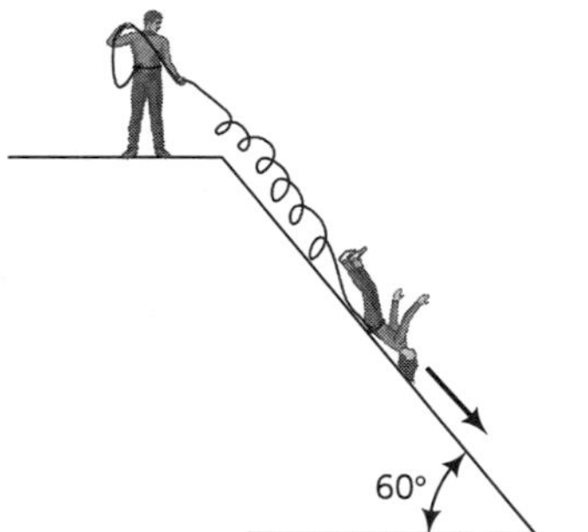

Diagram 4.4

4.5. A military volunteer is using "ELASTO-ROPE" which has been recently designed for rapid descent and deployment of special-forces soldiers. The volunteer is laying on their side at the edge of a vertical cliff (slope = 90°) with one end of the elastic safety cord attached to the person's belt and the other end attached to a boulder 5 M from the cliff edge. Assume the zero-load length of the safety cord is 5 M and $K = 50$ N/M beyond the zero-load length. The volunteer's mass is 75 kg, and the person then rolls over and falls off the cliff (see Diagram 4.5). Assuming that the initial velocity is zero, how far vertically down will the person have fallen at the moment when the maximum vertical velocity occurs?

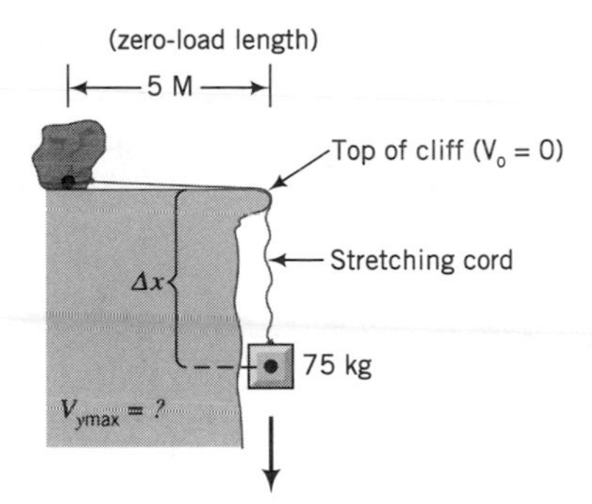

Diagram 4.5

4.6. A human factors engineer is now testing a civilian version of "ELASTO-ROPE" for the manufacturer. The civilian version (called "ESCAPO-ROPE") has been redesigned so that the zero-load length is now zero (and not the 5 M of Problem 4.5). To simulate a jump from a burning 10th-floor apartment, the test engineer jumps off a 30 M high fire tower.

Just prior to the jump the engineer secures one end of the "ESCAPO-ROPE" to the tower (at the person's shoulder height) and ties the other end attached to a vest at a point between the shoulders.

The person's mass is one hundred kilograms, and the elastic safety rope spring constant (K) is one Newton per centimeter. What is the vertical distance that this person will fall (while holding onto the elastic cord) before his vertical velocity is reduced to zero?

4.7. Two cars are moving toward each other on a level (left to right) one way street, with a coefficient of sliding friction, $\mu = 0.15$. Car A on the left weight 2200 kg and is moving rightward at 22 M/sec. Car B on the right weights 1000 kg andis moving rightward at 10 M/sec (see Diagram 4.6). The two cars then collide.

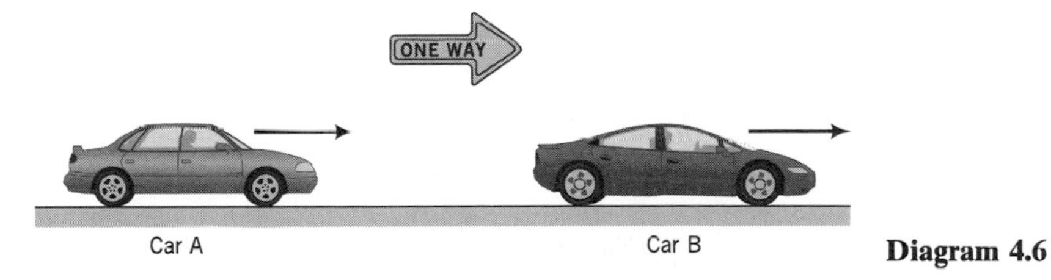

Diagram 4.6

Because both cars are equipped with "Bumper Guard" (an energy absorbing collision bumper), the result is an elastoplastic collision ($e = 0.5$).

1. Immediately after the collision, what is the new velocity of car A and what is the new velocity of car B?
2. When car A and car B finally come to a stop, what will the distance be between the two cars?

4.8. In subduing an evil-doer, a policeman swings his "SOCKO" (an infinitely rigid) night stick (mass of 1 kg) and it strikes the top of the villian's skull at a velocity of 10 M/s. If the mass of the skull is 5 kg and it has a spring constant of 25,000 N per centimeter, how much is the skull indented? You may assume that the neck and shoulders of the culprit do not move in response to the blow.

4.9. Two forest rangers, attached by a safety rope (of 5 M fully extended length), were climbing an ice-covered mountain slope in order to reach a fire look-out station. The second forest ranger, with a mass of 79.2 kg, slipped when she was 1 M behind the first ranger. The ice she was sliding down has a slope of 45° and a coefficient of friction of 0.2. Assuming she was at zero initial velocity, how fast was she going when she used up her remaining 4 M of rope?

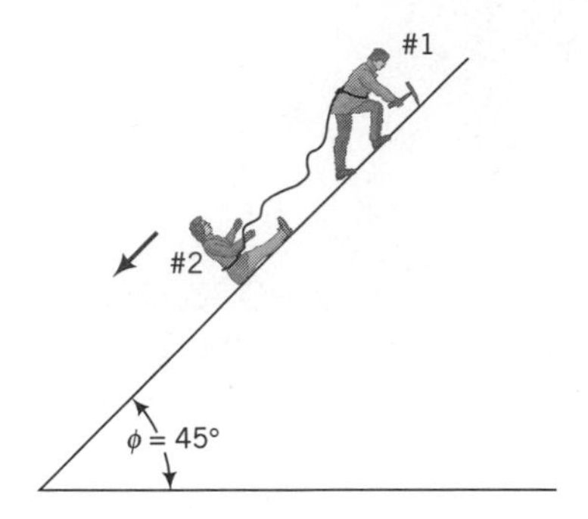

Diagram 4.7

4.10. If the first forest ranger in Problem 4.9 saw his friend slip, how much time did he have to dig in his "SPEED-O" (fast-action) pick axe and prepare for the shock before the slack in the rope ran out?

4.11. A man (m = 80 kg) on a cliff 10 M high jumps into a lake below (see Diagram 4.8).

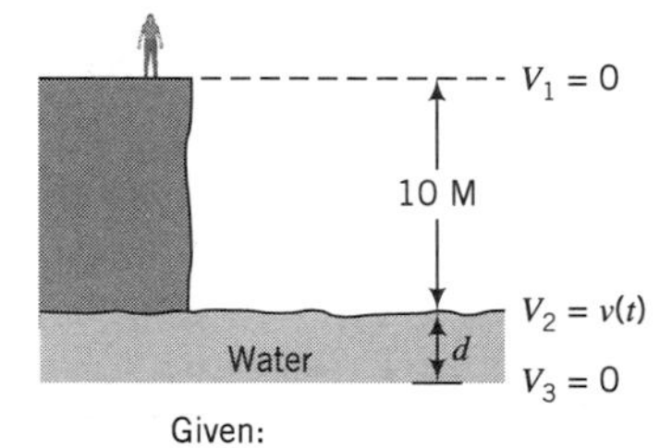

Diagram 4.8

1. Determine the force of impact on the surface of the water, first as a *pancake impact* (Δt = 0.2 s) then as a *normal dive* (Δt = 1 s). Δt is the time period of uniform deceleration from V_2 (impact velocity) to V_3 (the final zero velocity).
2. Determine the depth (d) of the dive for the *pancake impact* and also for a *normal dive.*

4.12. Two boxers are fighting. Boxer A throws a jab to Boxer B's head with a speed of 7 M/s. Boxer A is using a pair of the new "PUNCH-OUT" leather-foam composite gloves so that impact with Boxer B's head has a coefficient of restitution (e) of 0.8. Approximating the glove-head collision as direct-central impact, how far does Boxer B's head snap backwards after receiving the jab from Boxer A?

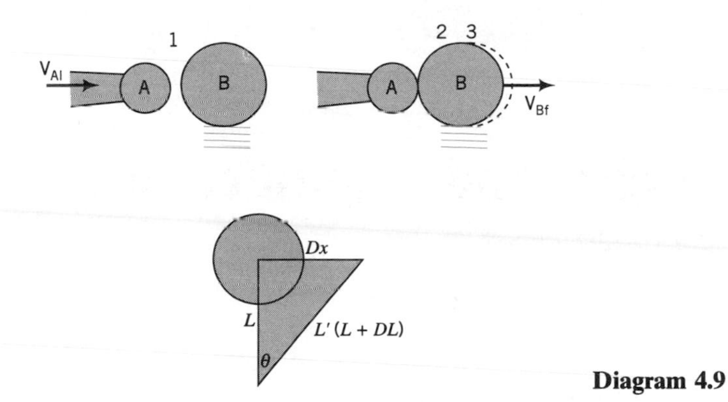

Diagram 4.9

You may approximate the movement of Boxer B's head as directly rearward. The shoulders do not move so that the neck (of spring constant, $K = 30{,}000$ N/m) is stretched as indicated.

In solving this problem, you may use the following information:

Mass of Boxer A = 100 kg; mass of hand, glove and forearm = 2.65 kg

Mass of Boxer B = 102 kg; mass of head and neck = 8.26 kg.

Velocity of Boxer B's head (immediately pre-impact) is zero.

CHAPTER 5

Bioelectricity and Bioelectronics

Bioelectricity is a subdivision of biophysics, as previously noted in Table 3.1. In this chapter, we will denote the relationship between bioelectricity and biophysics by using the term *electrobiophysics*. Bioelectricity, by definition, focuses upon electrical phenomena inherent in biological systems. Bioelectricity is the result of electrical gradients within organisms that represents a spatial separation of charged particles. Section 5.1.a will examine the biophysical basis for these electrical gradients. As a consequence of this separation of charge, currents will flow whenever an electrical circuit is completed. Section 5.1.b will examine the voltage-current relationship that occurs in the nerve cell membrane.

There are many manifestations of bioelectricity within the human organism, depending upon the specific tissue type and anatomical part of the body system generating the electricity. Two important examples are the electromyogram (EMG) of skeletal muscle and the electrocardiogram (ECG) of cardiac muscle. Section 5.2.a will examine the electrobiophysics of muscle fibers (electromyogram).

Bioelectronics is a subdivision of electronic engineering (and biomedical engineering) which addresses the acquisition and processing of biologically generated signals using various electronic modalities. The three elements of a bioelectronic system are signal transduction (various types of transducers), signal conditioning (amplification, filtering, etc.), and signal output (storage, retrieval, and/or display).

Bioelectronics is a very broad field since the resulting instrumentation systems depend (in part) upon the type of signal acquired, the necessary signal conditioning required, and the desired signal output. The biologically generated signal obtained may be a bioelectric signal. Important examples are the bioelectronics of the electromyogram (described in Section 5.2.b) and the bioelectronics of the electrocardiogram (as per Section 5.3.a). The biologically generated signal acquired may also be a physical variable other than electrical. An important example for the human factors engineer is the bioelectronics of air flow measurement (the electropneumogram) as described in Section 5.3.b.

Bioelectricity and bioelectronics are collectively a very broad field of study. Since the material in this chapter is so selective, a specific rationale is necessary. First, three areas of bioelectricity are addressed in which the electrobiophysical foundations are fairly well understood: the action potential of a nerve membrane (nerve cells), the motor unit action potential (skeletal muscle fibers), and the ventricular myocardial cell action potential (heart muscle fibers). Bioelectronics is a rapidly changing engineering field in which new circuit configurations and component devices emerge regularly. However, electrobiophysics represents the fundamental scientific basis for the bioelectrically generated signals to which the bioelectronic systems are subsequently applied. To rationally specify, identify, and select the appropriate bioelectronic system for a specific human-technological system evaluation, the human factors engineer should understand the electrobiophysical principles relevant to the specific bioelectrical signal of interest.

Second, those elements of bioelectricity and bioelectronics that focus upon noninvasive applications are presented. The various biologically generated signals are transduced with either skin-surface electrodes or airflow-monitoring facemask. Minimally invasive applications [e.g., skin-needle electrodes, tympanic membrane (ear) thermometers] are not considered. A noninvasive approach for the measurement of physiological signal activity should be the initial design goal of the human factors engineer. When evaluating a specific human-technological system interaction, a noninvasive data acquisition system will usually be the most comfortable and therefore acceptable alternative for the user. If properly designed and applied, it will expose the user to the lowest risk (among available alternatives). Indeed, the human factors engineer may subsequently consider minimally invasive or other more invasive approaches, but only after a noninvasive approach has been thoroughly evaluated and ruled out for the appropriate reasons.

Third, the material presented in this chapter collectively supports the quantitative analysis of neuromuscular and cardiopulmonary system activity. Data regarding the activity of these systems are acquired with the bioelectronic systems described in this chapter. Such data can then be analyzed by means of sets of mathematical equations (as described in Chapter 8). As a result, the material in this chapter can be applied to the quantitative analysis of physical workload, both static work and dynamic work (also described in Chapter 8).

5.1 ELECTROBIOPHYSICS OF NERVE CELLS

a. Electrical Gradients

In biological systems, the cell membrane separates a solution outside the cell (the extracellular fluid) from a solution inside the cell (the intracellular fluid). A solution is a system of particles (solute) mixed within a body of water (solvent). When a particle carries an electrical charge, it is said to be *ionized*. A positively charged ion is termed a *cation*, and a negatively charged ion is termed an *anion*.

The cell membrane itself may be freely permeable to particles, or it may be completely impermeable to particles. Most cell membranes are semipermeable. Such membranes are freely permeable to some (usually small) particles, but completely impermeable to other (usually larger) particles. These impermeable particles are often referred to as "trapped particles" since they cannot cross the cell membrane. Let us now consider a system having a semipermeable membrane.

Imagine that we have a rectangular fish tank with a semipermeable membrane dividing two equal halves: a left half (Side A) and a right half (Side B). Each side is then filled with a solution consisting of water (solvent) and ionized particles (solute). Solutions A and B both contain the cation potassium (K^+), which is univalent (valence = +1). Both solutions also contain the anion chloride (Cl^-), which is also univalent (valence = −1). Both potassium and chloride are present in some initial amount (at time $t = 0$) on each side of the membrane as Solutions A and B. Finally, there is a trapped protein anion (X^-), which is univalent (valence = −1). This trapped anion is present only on the left side (in Solution A).

For the two ionized solutions (A and B) filling compartments of constant volume and separated by a semipermeable membrane, we can state the Donnan membrane equation, which describes the *final* state (at $t = \infty$) when the system has reached *equilibrium* (between Solutions A and B). Two parts of the Donnan membrane equation must be satisfied *simultaneously:* electrical neutrality and product of diffusable ions.

i. Electrical Neutrality

At equilibrium, the solution on each side of the membrane will be electrically neutral: The *sum* total of charges on the cations will equal the *sum* total of charges on the anions for each solution *separately*. For Solutions A and B, this statement is expressed mathematically as electrical neutrality:

$$\sum_{i=1}^{n} [z_i Q_{Ci}]_A + \sum_{j=1}^{m} [z_j Q_{Aj}]_A = 0 \qquad (1)$$

$$\sum_{i=1}^{k} [z_i Q_{Ci}]_B + \sum_{j=1}^{h} [z_j Q_{Aj}]_B = 0 \qquad (2)$$

where:

z_i = valence of the *i*th ion
z_j = valence of the *j*th ion
Q_{C_i} = amount of the *i*th cation (in millimoles)
Q_{A_j} = amount of the *j*th anion (in millimoles)

It is very important to note that Q_C and Q_A in equations (1) and (2) refer to *all* ions in solution, both the *diffusable* (membrane permeable) ions *as well as* the *nondiffusable* (trapped) ions.

ii. Product of Diffusable Ions

At equilibrium, the *product* of the sum total diffusable cation charge and the sum total diffusable anion charge (on one side of the membrane) will be equal to the *product* of the sum total diffusable cation charge and the sum total diffusable anion charge (on the other side of the membrane). Mathematically, this may be stated as follows: The product of diffusable charge in Solution A will equal the product of diffusable charge in Solution B:

$$\sum_{i=1}^{n} [z_i Q'_{Ci}]_A \times \sum_{j=1}^{m} [z_j Q'_{Aj}]_A = \sum_{i=1}^{k} [z_i Q'_{Ci}]_B \times \sum_{j=1}^{h} [z_j Q'_{Aj}]_B \qquad (3)$$

Note that Q'_C in equation (3) refers to the amount (in millimoles) of *diffusable* (membrane permeable) cations and *excludes* the nondiffusable (trapped) cations. Likewise, Q'_A in equation (3) represents the millimolar amount of *diffusable* (membrane permeable) anions and *excludes* the "trapped" (nondiffusable) anions.

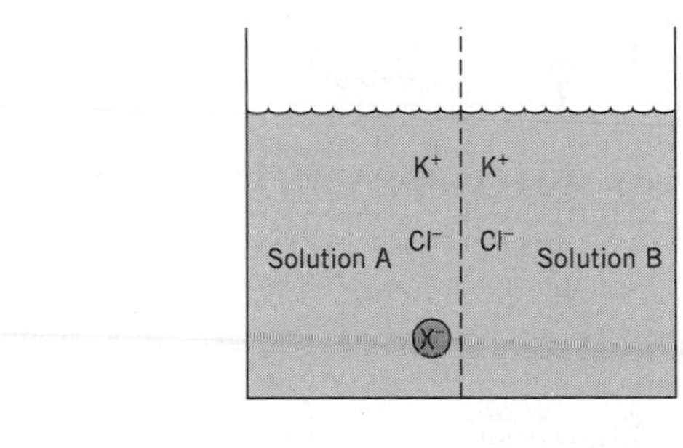

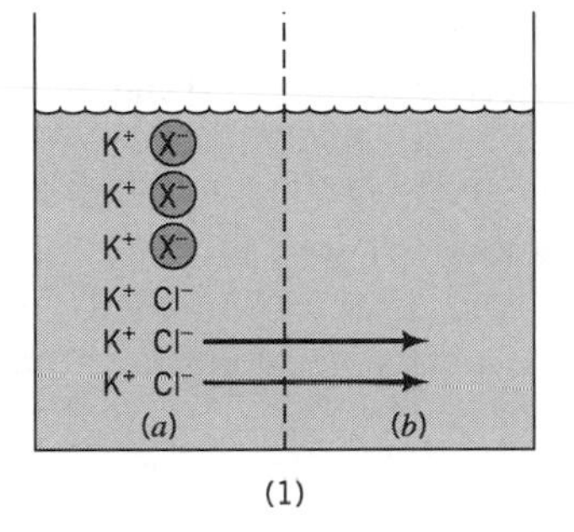

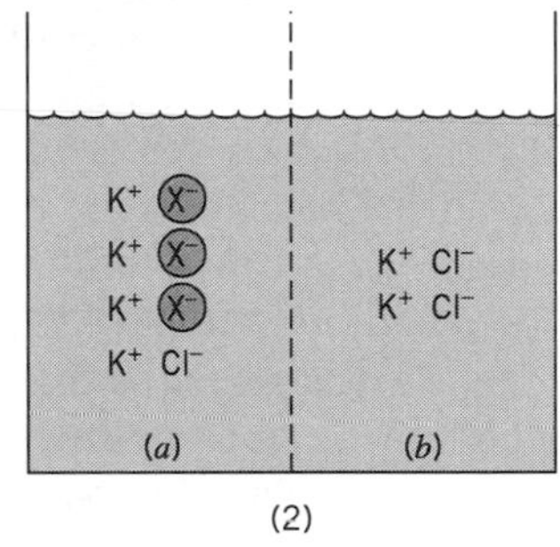

Figure 5.1.a Fish-tank with a semipermeable membrane. **b** State 1 ($t = 0$) and state 2 (subsequent time when Donnan equilibrium is reached) of the system depicted in Figure 5.1.a.

In the remainder of Section 5.1, all ions will be *univalent* (either +1 or −1), and all amounts will be specified in millimoles (and the volume of distribution will usually not be specified).

Consider the system depicted in Figure 5.1.a. At an initial time ($t = 0$) there will be a certain amount of permeable ions (K^+, Cl^-) in each Solution (A and B). There will also be a certain amount of trapped anion (X^-) in Solution A only. Over time, the permeable ions (K^+, Cl^-) will diffuse until a Donnan equilibrium is attained as follows:

$$\left.\begin{aligned} [K^+]_a &= [Cl^-]_a + [X^-]_a \\ [K^+]_b &= [Cl^-]_b \end{aligned}\right\} \quad \text{electrical neutrality}$$

and

$$[K^+]_a \times [Cl^-]_a = [K^+]_b \times [Cl^-]_b\} \quad \text{products of diffusable ions}$$

For example, consider Figure 5.1.b in which State 1 occurs at $t = 0$. All cations and anions are initially in Solution A. No cations or anions are in Solution B at $t = 0$. There are three nondiffusable (trapped) anions (X^-) in Solution A. The other cation and anion are freely diffusable across the membrane. State 2 of Figure 5.1.b occurs at some subsequent time at which Donnan equilibrium has been obtained. The permeable cations and anions have now crossed the membrane from Solution A to Solution B. However, the trapped anions (X^-) remain in Solution A for all time subsequent to $t = 0$.

The equilibrium state must satisfy the Donnan membrane equation (*both* Part I and Part II). The condition of electrical neutrality and the condition for product of diffusable ions are both *simultaneously* satisfied [as per equations (1), (2), and (3)]. For this example (Figure 5.1), we state this mathematically as follows:

$$\left.\begin{aligned} 4(K^+)_a &= 3(X^-)_a + 1(Cl^-)_a \\ 2(K^+)_b &= 2(Cl^-)_b \end{aligned}\right\} \quad \text{electrical neutrality}$$

$$[4]_a[-1]_a = [2]_b[-2]_b \quad \} \quad \text{product of diffusable ions}$$

The Donnan membrane equation predicts a spatial separation of ions so that we may now consider the concept of a transmembrane voltage. Figure 5.2 represents State 2 of Figure 5.1.b with the following modifications. The fish-tank semipermeable membrane is now a *cell* semipermeable membrane. Compartment I is now on the left side of the cell membrane (formerly Side A) and has a reference volume of 1 L. Solution A is now replaced with the *intracellular fluid (ICF)*. Compartment O is now on the right side of the membrane (formerly Side B) and has a reference volume of 1 L. Solution B is now replaced with the *extracellular fluid (ECF)*.

Let us determine the transmembrane voltage of a single ion species, the cation potassium (K^+). Figure 5.2 indicates only the potassium ions at their equilibrium distribution as represented by State 2 of Figure 5.1.b. The transmembrane voltage for potassium ions (V_{mK}) is the potential difference across the membrane that would be measured by a high-impedance voltmeter. The convention is that the *positive* electrode of the voltmeter is placed outside the cell (in Compartment O), and the *negative* electrode of the voltmeter is placed inside the cell (in Compartment I). This is demonstrated in Figure 5.2.

The amount of an ith-ion species (Q_i) in millimoles per compartment volume (Vol) may now be expressed as a concentration of that ith-ion species (C_i), as follows:

$$C_{iO} = \frac{Q_{iO}}{Vol_O} \tag{4}$$

$$C_{iI} = \frac{Q_{iI}}{Vol_I} \tag{5}$$

where:

subscript O denotes the ECF
subscript I denotes the ICF

In order for an ion to be subject to Donnan equilibrium (as defined by the Donnan membrane equation), a thermodynamic parameter, the *electrochemical potential* (μ) must be the same on both sides of the membrane. This electrochemical potential is a function of both the chemical activity of the ion and the effect of the electric field present. This condition of equilibrium may be stated as follows:

$$\mu_{iI} = \mu_{iO} \tag{6}$$

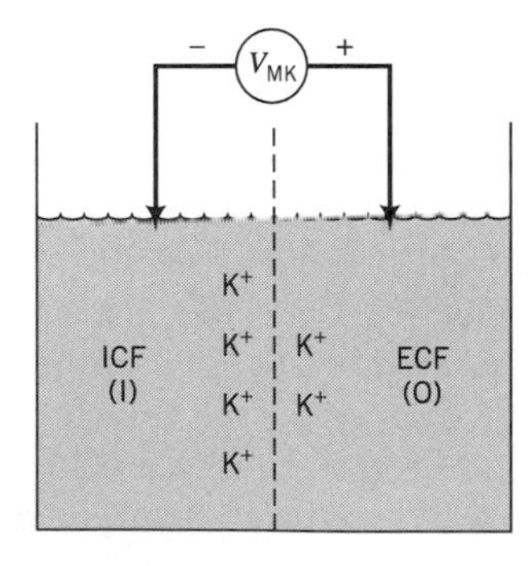

Figure 5.2 State 2 of Figure 5.1.b except that it is now a cell semipermeable membrane.

where the left-hand term is the electrochemical potential of the *i*th ion species on the inside of the membrane (the ICF) and the right-hand term is the electrochemical potential of the *same* *i*th ion species on the outside of the membrane (the ECF).

Defining the electrochemical potential mathematically, and after some algebraic manipulation and simplifying approximations, the following expression for the transmembrane potential for a single *i*th-ion species (V_{mi}) may be derived:

$$V_{mi} = -\frac{RT}{z_i F} \ln\left(\frac{C_{iI}}{C_{iO}}\right) \tag{7}$$

where:

z_i = the valence of ion i
R = the universal gas constant
T = temperature (absolute)
F = the Faraday constant

Substituting the following:

$$R = 8.2 \frac{\text{J}}{\text{mol} \cdot \text{K}}$$

$$T = 310°\text{K } (37°\text{C})$$

$$F = 96{,}500 \frac{\text{C}}{\text{mol}}$$

results in:

$$V_{mi} = -\frac{.0263}{z_i} \ln\left[\frac{C_{iI}}{C_{iO}}\right] \tag{8a}$$

For V_{mi} in millivolts:

$$V_{mi} = -\frac{26.3}{z_i} \ln\left(\frac{C_{iL}}{C_{iO}}\right) \tag{8b}$$

Equation (7) is a definition of the Nernst equilibrium potential, and it is referred to as the Nernst equation. The Nernst equilibrium potential is like a miniature "battery" inside the cell membrane that represents (for a specific ion) a specific separation of charge on each side of the cell membrane. In this particular case, the concentration of the cation (potassium) on the inside of the membrane is twice that of its concentration on the outside of the membrane. Substituting this ratio into equation (7) and setting $z_i = +1$, indicates that V_K (the transmembrane potential for the ion species potassium) will be *negative*. Recall that by convention, the positive electrode of the voltmeter is on the outside of the membrane. With respect to the potassium ion, the positive electrode of the voltmeter will be *less* positive than the negative electrode of the voltmeter (on the inside of the membrane). The Nernst equilibrium potential equation therefore tells us that a negative potential across a cell membrane is not necessarily due to an excess of a negative ion on the *outside* of the membrane, but rather may be due to an excess of a positive ion on the *inside* of the membrane.

Equation (7) is the traditional form of the Nernst equation, and it may be used to calculate a transmembrane voltage for a given distribution of a single specific ion species. Recall that this calculation is based upon the approximation that the

Table 5.1.a V_m for Individual Ions

Ion (i)	C_I $\left[\frac{\text{mMol}}{\text{L}}\right]$	C_O $\left[\frac{\text{mMol}}{\text{L}}\right]$	V_m (mV) [Equation 8(b)]
Cations:			
Na^+	12	145	+66
K^+	155	4	−97
Anions:			
Cl^-	4	120	−90
A^-	155	0	—

Table 5.1.b V_m for All Ions Collectively

Ion	V_I	V_O	V_m
Na^+, K^+, Cl^-, A^-	−90	0	−90

ion is distributed according to its electrochemical equilibrium. The Nernst potential may be calculated *separately* for the ions K^+, Na^+, and Cl^- (V_K, V_{Na}, and V_{Cl}) as indicated in Table 5.1.a. The *actual measured* transmembrane potential for *all* ion species bathing a cell membrane (V_m) is given in Table 5.1.b. Note that the equilibrium potential for K^+ (V_K) is close to V_m. The chloride ion is distributed exactly according to the transmembrane electrochemical equilibrium condition ($V_{Cl} = V_m$). Finally, the equilibrium potential for the Na^+ ion (V_{Na}) is far from the transmembrane equilibrium value (V_m). This finding is but one piece of the evidence in favor of the hypothesis that there is an active process known as the sodium pump, that extrudes this ion from the interior of the cell against both a concentration gradient and an electrical gradient.

In order to account for the collective effect of *all* the individual ions, we must also consider the mobility of each individual ion. By introducing their permeability constants (P), we can derive the Goldman equation from the Nernst equation:

$$V_m = -\frac{RT}{zF}\ln\left[\frac{P_1C_{1I} + P_2C_{2I} + \cdots P_nC_{nI}}{P_1C_{1O} + P_2C_{2O} + \cdots P_nC_{nO}}\right] \tag{9}$$

We can now derive a modified form of the Goldman equation to account for the collective effects of the major cations and anions of Table 5.2.a. Note that the protein anion (A^-) is a trapped anion, to which the membrane is impermeable (so that $P_A = 0$). Since we are now dealing with only the sodium, potassium, and chloride ions, we have a system in which all three ions are univalent. However, by setting $z = +1$ in equation (9), we must remember that the chloride ion has a valence of -1. The rules of logarithmic manipulation require that for this ion, we invert C_I and C_O within the argument of the logarithm. The modified Goldman equation at this point is:

$$V_m = -\frac{RT}{(1)F}\ln\left[\frac{P_K \cdot C_{KI} + P_{Na} \cdot C_{NaI} + P_{Cl} \cdot C_{ClO}}{P_K \cdot C_{KO} + P_{Na} \cdot C_{NaO} + P_{Cl} \cdot C_{ClI}}\right] \tag{10}$$

Taking the ratio of permeabilities with respect to the chloride ion and substituting for the other constants results in the "modified Goldman equation":

$$V_m = -26.3 \ln \left[\frac{\overline{\alpha} C_{KI} + \overline{\beta} C_{NaI} + C_{ClO}}{\overline{\alpha} C_{KO} + \overline{\beta} C_{NaO} + C_{ClI}} \right] \tag{11}$$

where

$$\overline{\alpha} = \frac{P_K}{P_{Cl}}$$

$$\overline{\beta} = \frac{P_{Na}}{P_{Cl}}$$

Since the permeability of the *resting* membrane to potassium ion is close to the permeability of the membrane to chloride ion, we may approximate $\overline{\alpha} = 1.00$. Since the membrane is relatively impermeable to sodium ion (due to the action of the sodium pump) with respect to the permeability of the membrane to chloride ion, we may approximate $\overline{\beta} = 0.01$ for the *resting* membrane. Substituting these values of $\overline{\alpha}$ and $\overline{\beta}$ and the individual ion data of Table 5.2.a, equation (11) predicts a transmembrane voltage (V_m) of -88.7 mV. This value is very close to the V_m of -90 mV for all ions collectively as per Table 5.2.b.

b. Membrane Models

Let us now consider a complete nerve cell. Recall that a *nerve cell* is synonymous with a *neuron* and the two terms are used interchangeably in this text. A nerve cell (in part) consists of a cell body, which contains the nucleus, and a long tubular extension of that cell body, referred to as the *axon.* The cell body itself has constituents of a cell membrane, which surround the *intracellular fluid* and separates it from the *extracellular fluid.* The cell body with its nucleus is responsible for maintaining the viability of the cell. The axon is a long cylindrical (tubular) extension of the cell membrane (of the cell body) so that this axonal membrane is bathed on its inside surface by the intracellular fluid. The axonal membrane is surrounded on its outside surface by the extracellular fluid. The purpose of the axon is to transmit a nerve cell action potential from its origin (at the cell body) to its destination (the dendrites at the end of the axon). Consequently, the axon of a nerve cell behaves somewhat analogously to a transmission line cable.

Figure 5.3.a illustrates an experimental arrangement that will allow us to evaluate the electrical properties of the axonal membrane. A rectangular current pulse is generated by a stimulator (which may be considered as being located nearer the cell body). The negative electrode (cathode) of the stimulator is positioned intracellularly and the positive electrode (anode) of the stimulator is located in the extracellular fluid. Some distance farther along the axon (nearer the dendrites), a recording amplifier is positioned and its input is a cathodic electrode located on the inside of the membrane (in the intracellular fluid). The other recording (anodal) electrode is positioned outside the axon (in the extracellular fluid) and is referenced to ground potential.

Figure 5.3.b indicates the voltage recordings, as displayed on the cathode ray oscilloscope (CRO), in response to a series of 2-ms-duration rectangular current pulses. Each rectangular current pulse is of increasing current amplitude (a', b', c', and d'). In response to each of these 2-ms current pulses, the transmembrane voltage potential (V_m) is indicated at the location of the recording amplifier (a, b, c, and d of Figure 5.3.b). With respect to voltage tracings a, b, and c, the membrane is said to demonstrate passive circuit characteristics. This is because the time course

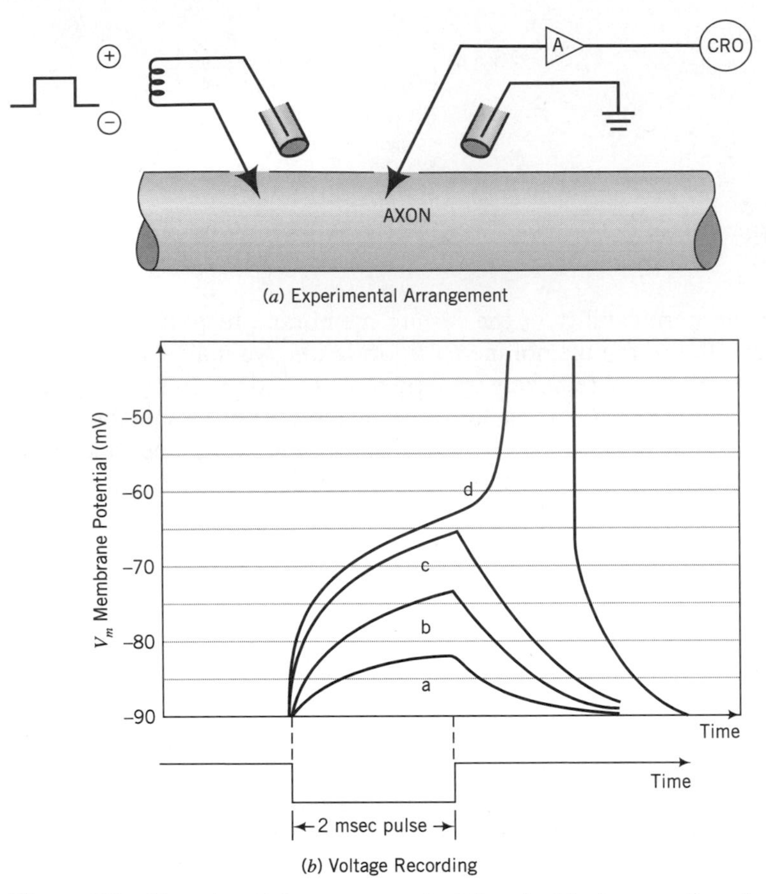

Figure 5.3.a Experimental arrangement of the electrical properties of the axonal membrane. **b** Voltage recordings in response to a series of two-ms duration rectangular current pulses.

of the membrane voltage resembles the charging and discharging of a capacitor across a resistor. At any time, V_m may be *either* a negative or positive voltage.

Figure 5.4.a is an electrical schematic (analog) for the axonal membrane at the location of the recording amplifier. This figure represents the unit membrane. V_r of Figure 5.4.a is the *resting* membrane potential (i.e., the Nernst equilibrium poten-

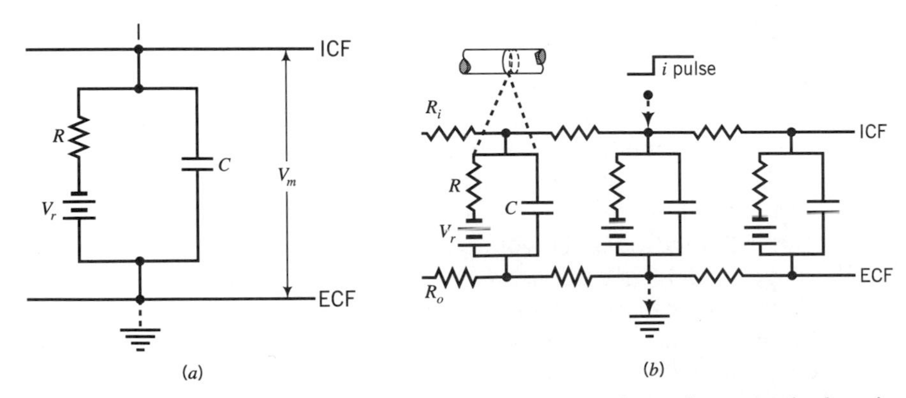

Figure 5.4.a Electrical schematic (analog) for the axonal membrane at the location of the recording amplifier. **b** Equivalent electrical model of a series of unit membranes.

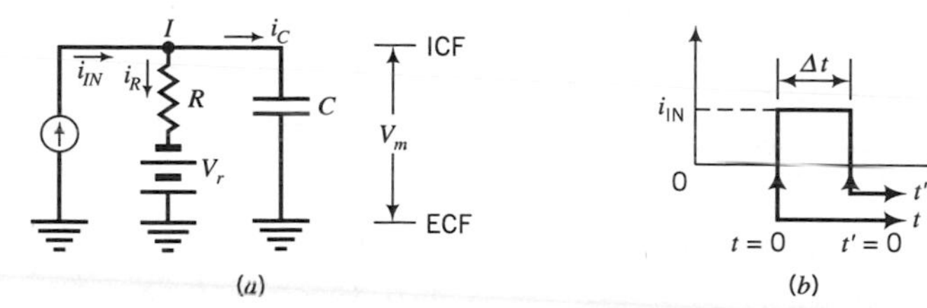

Figure 5.5.a Electrical circuit completed when a rectangular current pulse is injected into the node of the unit membrane at the membrane-intracellular fluid interface. **b** Graph showing the application of the current pulse for Example 5.0.

tial for all cations and ions collectively, when the membrane is not electrically stimulated). It should be noted that V_r is *always* a negative constant voltage (Table 5.1.b). R is the transmembrane resistance for all cations and anions collectively which cross (flow through) the membrane. Other resistive components shown in Figure 5.4.a are: R_0 (resistance of extracellular fluid) and R_I (resistance of the intracellular fluid). C is the transmembrane capacitance which accounts for the dielectric properties of the membrane.

At adjacent locations along the entire length of the axon, these unit membranes are repeated so that the equivalent electrical model is a series of unit membranes (Figure 5.4.b). This figure demonstrates that the passive electrical components of a nerve cell axon resemble a distributed transmission line model. For the purpose of this chapter, only the passive electrical characteristics of a single unit membrane will be considered.

Figure 5.5.a represents the electrical circuit completed when a rectangular current pulse is injected into the node of the unit membrane at the membrane–intracellular fluid interface. Note that the membrane–extracellular fluid interface is at ground potential and that the transmembrane potential (V_m) is measured between the ICF and the ECF. In effect, V_m is the voltage between node I and ground (of Figure 5.5.a).

EXAMPLE 5.0

a. For the circuit of Figure 5.5.a, which is the electrical analog of a unit membrane, derive the governing differential equation for a current pulse injected into node I;

b. For the unit step current pulse of Figure 5.5.b, solve for the transmembrane voltage (V_m) *during* the period of current application $0 \leq t \leq \Delta t$;

c. Solve for the transmembrane voltage (V_m) *following* the application of the current pulse shown in Figure 5.5.b.

SOLUTION 5.0(a)

For the circuit of Figure 5.5.a we write Norton's theorem (with assumed current directions) for node *I*:

$$i_{IN} - i_R - i_C = 0; \tag{i}$$

Now define a differential voltage (V') as follows:

$$V' = V_m - V_r \tag{ii}$$

Recall that V_r itself is always a negative constant voltage so that $-V_r$ is always a *positive* constant voltage.

Define the resistive current and the capacitive current in terms of V':

$$i_R = \frac{V_m - V_r}{R} = \frac{V'}{R} \tag{iii}$$

$$i_c = C \cdot \frac{dV_m}{dt} = C \cdot \frac{dV'}{dt} \tag{iv}$$

Substituting equations (iii) and (iv) into equation (i) and rearranging:

$$C \cdot \frac{dV'}{dt} + \frac{V'}{R} = i_{IN}\,(0 \leq t \leq \Delta t) \tag{v}$$

Proceeding to develop the governing differential equation from equation (v):

$$\frac{dV'}{dt} + \frac{V'}{RC} = \frac{i_{IN}}{C} \tag{vi}$$

The system time constant (T) is:

$$T = RC \tag{vii}$$

The input voltage (V_{IN}) is defined as:

$$V_{IN} := R \cdot i_{IN} \tag{viii}$$

Substituting equations (vii) and (viii) into equation (vi) results in:

$$\frac{dV'}{dt} + \frac{V'}{T} = \frac{V_{IN}}{T} \tag{ix}$$

SOLUTION 5.0.b

Solve equation 5.0(a)(ix) by first defining the initial condition (see Figure 5.5.b):

$$\text{At } t = 0, V_m = V_r$$

So,

$$V'(0) = V_m - V_r = 0$$

Now recall equation (A2) of the appendix:

$$\dot{y} + by = c \tag{A2}$$

The solution of equation (A2) is equation (A22) of the appendix:

$$y = \frac{c}{b}(1 - e^{-bt}) + y_0 e^{-bt} \tag{A22}$$

Equation (A2) is analogous to equation 5.0(a)(ix) when:

$$y = V'(t)$$
$$y_0 = V'(0) = 0$$
$$b = \frac{1}{T}$$
$$c = \frac{V_{IN}}{T}$$

Substituting the above into equation (A22) results in the solution of equation 5.0(a)(ix):

$$V'(t) = V_{IN}(1 - e^{-\frac{t}{T}}) \tag{i}$$

Substituting equation 5.0(a)(ii) into equation (i):

$$V_m(t) - V_r = V_{IN}(1 - e^{-\frac{t}{T}}) \tag{ii}$$
$$V_m(t) = V_r + V_{IN}(1 - e^{-\frac{t}{T}}) \tag{iii}$$

SOLUTION 5.0.c

For $t' \geq 0$ (see Figure 5.5.b), we may rewrite equation (v):

$$C \cdot \frac{dV'}{dt} + \frac{V'}{R} = 0 \tag{i}$$

The governing differential equation now becomes:

$$\frac{dV'}{dt} + \frac{V'}{T} = 0 \tag{ii}$$

Solve equation 5.0(c)(ii) by first defining the initial condition, $t'(0)$, per Figure 5.5.b:

$$\text{At } t' = 0,\ V_m = V_r + V_{IN}(1 - e^{-\frac{\Delta t}{T}})$$

So,

$$V'(0) = V_m - V_r = V_{IN}(1 - e^{-\frac{\Delta t}{T}})$$

Now recall equation (A2) of the appendix:

$$\dot{y} + by = c \tag{A2}$$

The solution of equation (A2) is equation (A22) of the appendix:

$$y = \frac{c}{b}(1 - e^{-bt}) + y_0 e^{-bt} \tag{A22}$$

Equation (A2) is analogous to equation 5.0(c)(ii) when:

$$y = V'(t')$$
$$y_0 = V_{IN}(1 - e^{-\frac{\Delta t}{T}})$$
$$b = \frac{1}{T}$$
$$c = 0$$

Substituting the above into equation (A22) results in the solution of equation 5.0(c)(ii):

$$V'(t') = V_{IN}(1 - e^{-\frac{\Delta t}{T}})[e^{-\frac{t'}{T}}] \quad \text{(iii)}$$

Substituting equation 5.0(a)(ii) into equation (iii):

$$V_m(t') - V_r = [V_{IN}(1 - e^{-\frac{\Delta t}{T}})][e^{\frac{-t'}{T}}] \quad \text{(iv)}$$

$$V_m(t') = V_r + [V_{IN}(1 - e^{-\frac{\Delta t}{T}})][e^{-\frac{t'}{T}}] \quad \text{(v)}$$

The following example demonstrates an actual application of the unit membrane model.

EXAMPLE 5.1

From the voltage recordings in Figure 5.3.b, an HFE prepares the following:

Table of Data

Recording	V_m ($t = 0$)	V_m ($t = 2$ ms)	V_m ($t = 4$ ms)
A	−90	−83	−89
B	−90	−75	−88
C	−90	−68	−87

If the transmembrane capacitance (C) is 1 nF, find $i_{IN}(a')$, $i_{IN}(b')$, and $i_{IN}(c')$ which are the amplitudes of the 2 millisecond square wave current pulses that produced recordings a, b, and c, respectively.

SOLUTION 5.1

First determine the membrane RC time constant from the voltage recording interval between $t = 2$ ms and $t = 4$ ms (zero current pulse):

V_{IN} is not provided, but Δt and V_m (Δt) are, so that substituting for $t = \Delta t$, in equation 5.0(b)(ii) results in:

$$V_m(\Delta t) - V_r = V_{IN}[1 - e^{-\frac{\Delta t}{T}}] \quad \text{(i)}$$

Now substituting equation (i) into equation 5.0(c)(iv) yields:

$$V_m(t') - V_r = [V_m(\Delta t) - V_r][e^{-\frac{t'}{T}}] \quad \text{(ii)}$$

By definition:

$$V_m(t = 0) = V_r \quad \text{(iii)}$$

Substituting equation (iii) into equation (ii):

$$V_m(t') - V_m(0) = [V_m(\Delta t) - V_m(0)][e^{-\frac{t'}{T}}] \quad \text{(iv)}$$

For recording (a), at $t' = 2$ ms, substitute from the table of data into equation (iv):

$$1 = 7 \cdot e^{-\frac{2}{T}}$$

where T (ms)

$$T = \frac{-2}{\ln\left(\frac{1}{7}\right)} = 1.03$$

Similarly, for recording (b) at $t = 2$ ms:

$$2 = 15 \cdot e^{-\frac{2}{T}}$$

So:

$$T = \frac{-2}{\ln\left(\frac{2}{15}\right)} = 0.99 \text{ ms}$$

And in a like manner, for recording (c) at $t' = 2$ ms:

$$3 = 23 \cdot e^{-2/T}$$

So:

$$T = \frac{-3}{\ln\left(\frac{3}{23}\right)} = 0.98 \text{ ms}$$

Taking a geometric average:

$$T = 1.00 \text{ ms}$$

Second, calculate R_m (through which I_m flows):

Given that $C_m = 1.0$ nF, T (s) = RC

So:

$$.001 = R(1)(10^{-9})$$

$$R = \frac{(1)(10^{-3})}{(1)(10^{-9})} = 1 \text{ M}\Omega$$

Third, calculate I_{IN} from the voltage recording interval between $t = 0$ ms, and $t = 2$ ms:

Substituting equation 5.1(iii) into equation 5.0(xi):

$$V_m(t) - V(0) = V_{IN}(1 - e^{-\frac{t}{T}}) \qquad \text{(v)}$$

where V_{IN} (millivolts);

For recording (a) at $t = 2$ ms, substitute from table of data into equation (v):

$$(-83) - (-90) = V_{IN}(1 - e^{-2/1})$$

$$V_{IN} = \frac{7}{(1 - e^{-2})} = 6.14 \text{ mV}$$

Since $V_{IN} = I_{IN}R$ from equation 5.0(a)(viii), then:

$$i_{IN}(a') = \frac{6.14 \text{ mV}}{1.0 \text{ m}\Omega} = 6.14 \text{ nA}$$

Similarly, for recording (b) at $t = 2$ ms:

$$15 = V_{IN}(1 - e^{-\frac{2}{1}})$$

$$V_{IN} = \frac{15}{(1 - e^{-2})} = 17.3 \text{ mV}$$

So that:

$$i_{IN}(b') = \frac{17.3 \text{ mV}}{1.0 \text{ m}\Omega} = 17.3 \text{ nA}$$

Finally, in a similar manner for recording (c) at $t = 2$ ms:

$$23 = V_{IN}(1 - e^{-\frac{2}{1}})$$

$$V_{IN} = \frac{23}{(1 - e^{-2})} = 26.6 \text{ mV}$$

So that:

$$i_{IN}(c') = \frac{26.6 \text{ mV}}{1.0 \text{ m}\Omega} = 26.6 \text{ nA}$$

Let us now consider the *active* electrical properties of a nerve axon. When the stimulus current is of sufficient amplitude (d') the electrical membrane characteristics become nonlinear (voltage recording d of Figure 5.3.b). A self-generating electrical pulse is created, and its voltage-time activity is termed the "action potential." It is interesting to note that the action potential is also a self-propagating electrical pulse in that it will physically move along the course of the axon not only at a constant velocity, but also at a constant amplitude. In this respect, the transmission cable properties (of the passive membrane) are not satisfied. Rather, the action potential is more like the flame moving along a lighted fuse. Let us now analyze this action potential further so that we may develop an appropriate electrical equivalent model.

Voltage recording (d) of Figure 5.3.b demonstrates an initial period (referred to as "hypopolarization") in which there is a progressive rise of the transmembrane potential from its resting value (of about −80 mV) toward a threshold voltage (approximately −55 mV). When the transmembrane voltage exceeds this −55 mV "switching" voltage, an action potential pulse is generated and the membrane electrical characteristics become nonlinear. Below the threshold ("switching") voltage the membrane is passive, as previously described. Stimuli a', b', and c' represent increasing amounts of hypopolarization of the membrane [voltage recordings a, b, and c of Figure 5.3.b]. Once the threshold voltage is exceeded, an action potential pulse is generated, and there is an initial rapid rise in the transmembrane voltage (referred to as the "depolarization" phase). At the peak of the action potential (which is the conclusion of the "depolarization" phase), a reversal of the transmembrane potential has occurred so that the transmembrane voltage is actually positive (by approximately 20 mV) as shown in Figure 5.6.a. The end result of depolarization is a potential difference between the stimulated unit membrane and the adjacent resting unit membranes along the axon. Membrane charge is redistributed in adjacent unit membranes, which then undergo hypopolarization as previously described. Once threshold voltage is exceeded, depolarization occurs. A new action potential

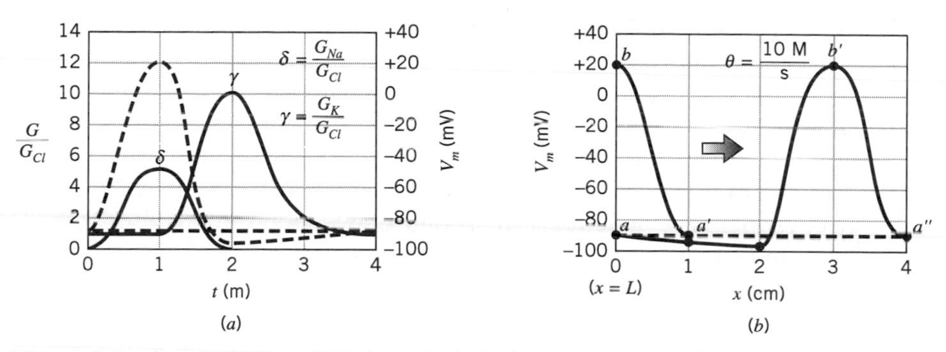

Figure 5.6.a Conductance ratios (α and γ) during time course of an action potential. **b** The spatial distribution during the time course of an action potential.

then appears at the adjacent unit membrane that was not initially directly stimulated. This results in the self-propagating nature of the action potential spatially along the nerve cell axon.

The depolarization phase occurs within about 1 ms and ends at the peak of the action potential "spike." Repolarization then occurs. Over the next 1 ms period, the transmembrane voltage returns to baseline. This is followed by hyperpolarization (increasing the transmembrane voltage to an even more negative value than the resting membrane voltage), which continues until the membrane voltage returns to the original resting membrane voltage value in about 4 ms. The voltage-time events between the action potential peak and final return to resting membrane voltage (events between 1 ms and 4 ms) collectively represent the membrane "repolarization" phase.

Figure 5.6.a shows that the depolarization phase is predominantly due to an increase in the conductance of sodium ions across the membrane (G_{Na}). Repolarization is initially due to a decrease in the conductance of the membrane to sodium ions *and* a concurrent increase in the conductance of the membrane to potassium ions (G_K). The final phase of repolarization is due predominantly to a subsequent decrease of the membrane to the conductance of potassium ions.

In order to describe this behavior schematically, the electrical equivalent model for a unit membrane of Figure 5.4.a must be modified as per the electrical equivalent model of Figure 5.7.a. This revised electrical equivalent model identifies three ionic

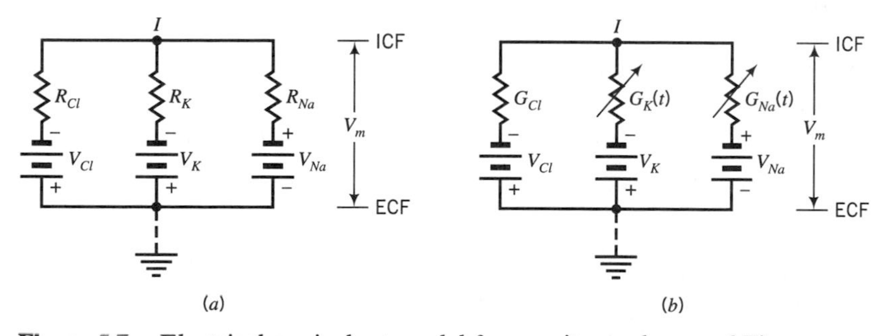

Figure 5.7.a Electrical equivalent model for a unit membrane of Figure 5.4.a. **b** Revised equivalent circuit of Figure 5.7.a to describe an action potential.

currents that flow through the membrane: a sodium current, a potassium current, and a chloride current. However, all three currents flow into and out of a single node of the unit membrane as was the case of the electrical equivalent model of Figure 5.4.a.

In order to reasonably describe an action potential, however, the revised equivalent circuit of Figure 5.7.a must be further modified, as shown in Figure 5.7.b. This final equivalent circuit demonstrates some of the important features of the Hodgkin-Huxley model. Hodgkin and Huxley developed a model to describe the voltage-time characteristics of an action potential in terms of current-time variations of the three ionic currents. In Figure 5.7.b, resistances have been replaced by conductances since that is more consistent with electrobiophysical convention. Note that potassium conductance (G_K) and sodium conductance (G_{Na}) are now time-varying functions. This is consistent with the sodium conductance time variations and potassium conductance time variations shown in Figure 5.6.a. As a side note, the student should appreciate that in the actual Hodgkin-Huxley model, G_K and G_{Na} are complex functions of both time *and* the instantaneous transmembrane voltage (V_m). In the electrical equivalent model of Figure 5.7.b, the chloride conductance is constant. In the Hodgkin-Huxley model this term refers to the conductance of *all other* ions (excluding sodium and potassium) including chloride. The Nernst equilibrium potential for potassium ion, sodium ion, and chloride ion are modeled as constant-voltage batteries. Finally, it should be mentioned that the transmembrane capacitance (C) has been deleted from the electrical equivalent models of Figure 5.7.a and 5.7.b. This electrical parameter is represented in the Hodgkin-Huxley equations, but is not necessary for the mathematical analysis which will now follow.

For the equivalent electrical model of Figure 5.7.a, define the chloride current (I_{Cl}), the potassium current (I_K), and the sodium current (I_{Na}):

$$I_{Cl} = \frac{V_m - V_{Cl}}{R_{Cl}} \tag{12}$$

$$I_K = \frac{V_m - V_K}{R_K} \tag{13}$$

$$I_{Na} = \frac{V_m - V_{Na}}{R_{Na}} \tag{14}$$

V_m is the transmembrane voltage (between node I and ground), measured with a *very* high-impedance voltmeter, so that effectively zero current flows into or out of the measuring device. Recall that V_m may be *either* a positive voltage or a negative voltage at any point in time. From Table 5.1.a, it is apparent that V_{Cl} and V_K are *always* negative constant voltages, and V_{Na} is *always* a positive constant voltage. Therefore, $-V_{Cl}$ and $-V_K$ in equations (12) and (13) are *positive* constant voltages. It also follows that $-V_{Na}$ in equation (14) is then a negative constant voltage.

Invoking Norton's theorem (in its general form) for current node I of Figure 5.7.a:

$$I_{Cl} + I_K + I_{Na} = 0 \tag{15}$$

Substituting equations (12), (13), and (14) into equation (15):

$$\frac{V_m - V_{Cl}}{R_{Cl}} + \frac{V_m - V_K}{R_K} + \frac{V_m - V_{Na}}{R_{Na}} = 0 \tag{16}$$

Separating the variable, V_m:

$$\frac{V_m}{R_{Cl}} + \frac{V_m}{R_K} + \frac{V_m}{R_{Na}} = \frac{V_{Cl}}{R_{Cl}} + \frac{V_K}{R_K} + \frac{V_{Na}}{R_{Na}} \tag{17}$$

Simplifying:

$$V_m = \frac{\dfrac{V_{Cl}}{R_{Cl}} + \dfrac{V_K}{R_K} + \dfrac{V_{Na}}{R_{Na}}}{\dfrac{1}{R_{Cl}} + \dfrac{1}{R_K} + \dfrac{1}{R_{Na}}} \tag{18a}$$

Equation (18a) can be applied to the electrical equivalent model of Figure 5.7.a in terms of conductance by recalling that:

$$G = \frac{1}{R}$$

Substituting this relationship into equation (18a):

$$V_m = \frac{V_{Cl} + \left(\dfrac{G_K}{G_{Cl}}\right) V_K + \left(\dfrac{G_{Na}}{G_{Cl}}\right) V_{Na}}{1 + \dfrac{G_K}{G_{Cl}} + \dfrac{G_{Na}}{G_{Cl}}} \tag{18b}$$

If $G_{Cl} \neq 0$, then equation (18b) can be applied to the electrical equivalent model of Figure 5.7.b in terms of time-varying conductance by noting that:

$$G_K = G_K(t)$$

and

$$G_{Na} = G_{Na}(t)$$

Substituting these two relationships into equation (18b):

$$V_m(t) = \frac{V_{Cl} + \left[\dfrac{G_K(t)}{G_{Cl}}\right] V_K + \left[\dfrac{G_{Na}(t)}{G_{Cl}}\right] V_{Na}}{1 + \dfrac{G_K(t)}{G_{Cl}} + \dfrac{G_{Na}(t)}{G_{Cl}}} \tag{19}$$

If we define γ and δ as conductance ratios (that are valid if $G_{Cl} \neq 0$):

$$\gamma(t) = \frac{G_K(t)}{G_{Cl}} \tag{20}$$

$$\delta(t) = \frac{G_{Na}(t)}{G_{Cl}} \tag{21}$$

Then by substituting equations (20) and (21) into equation (19):

$$V_m(t) = \frac{V_{Cl} + \gamma(t) V_K + \delta(t) V_{Na}}{1 + \gamma(t) + \delta(t)} \tag{22}$$

EXAMPLE 5.2

For the equivalent electrical model of Figure 5.7.b, the Nernst equilibrium voltages have been previously determined as follows:

$$V_{Cl} = -90 \text{ mV}$$
$$V_K = -97 \text{ mV}$$
$$V_{Na} = +66 \text{ mV}$$

From Figure 5.6.a, the following table is developed for the time variation of γ and δ:

t	γ	δ
0	0.8	0.01
1	0.8	5.0
2	10.0	0.01
3	2.0	0.01
4	0.8	0.01

Find the value of the transmembrane voltage (V_m) at $t_0 = 0$, $t_1 = 1$, $t_2 = 2$, $t_3 = 3$, and $t_4 = 4$ ms.

SOLUTION 5.2

Substitute into equation (22):

$$V_{Cl} = -90 \text{ mV}$$
$$V_K = -97 \text{ mV}$$
$$V_{Na} = +66 \text{ mV}$$

At $t_0 = 0$: $\gamma = 0.8$, $\delta = .01$:

$$V_m(t_0) = \frac{(-90) + (0.8)(-97) + (.01)(66)}{1 + 0.8 + .01}$$

$$V_m(t_0) = \frac{-90 - 77.6 + .66}{1.81} = -92.2 \text{ mV}$$

At $t_1 = 1$: $\gamma = 0.8$, $\delta = 5$:

$$V_m(t_1) = \frac{(-90) + (0.8)(-97) + (5)(66)}{1 + 0.8 + 5}$$

$$V_m(t_1) = 23.9 \text{ mV}$$

At $t_2 = 2$: $\gamma = 10$, $\delta = .01$:

$$V_m(t_2) = \frac{(-90) + (10)(-97) + (.01)(66)}{1 + 10 + .01}$$

$$V_m(t_2) = -96.2 \text{ mV}$$

At $t_3 = 3$: $\gamma = 2$, $\delta = .01$:

$$V_m(t_3) = \frac{(-90) + (2)(-97) + (.01)(66)}{1 + 2 + .01}$$

$$V_m(t_3) = -94.1 \text{ mV}$$

At $t_4 = 4$: $\gamma = 0.8$, $\delta = .01$:

$$V_m(t_4) = -92.2 \text{ mV}$$

5.2 MUSCLE ELECTROBIOPHYSICS AND THE ELECTROMYOGRAM

a. Electrobiophysics of Muscle Fibers

The fundamental basis of the electromyogram is that muscle fibers conduct electrical potentials in a manner that is rather similar to the way in which nerve cells conduct action potentials. Figure 5.8 shows the anatomical arrangement of the basic neuromuscular unit, which is referred to as the *motor unit.* The cell body of the alphamotoneuron resides within the tissue of the spinal cord. Its long axon is just one of a large number of neurons that compose the spinal nerve root. This long axon subsequently branches and merges with other nerve axons as it progresses through various divisions of the peripheral nervous system. The axon of the alphamotoneuron terminates in a number of branches (previously referred to as dendrites) with each dendrite innervating a single muscle fiber. Recall that *muscle fiber* is synonymous with *muscle cell*, and the two terms are used interchangeably in this text. The exact ratio of the number of individual muscle fibers innervated by a single alphamotoneuron varies, but this anatomical unit of one neuron to three or more muscle fibers represents the "motor unit." As described in Section 5.1.b, the nerve action potential is generated in its cell body and conducted as a self-generating (self-propagating) wave along the axon. This nerve action potential is then transmitted to each of the individual muscle fibers via the terminal dendrites. The specific electrical potential that is then generated by the muscle fibers and conducted along their length is referred to as the *motor unit action potential* (MUAP). The electromyogram is the algebraic sum of all the MUAPs that are generated by the collective muscle fibers at any point in time. Consequently, the electrobiophysics of the MUAP will first be considered prior to describing the surface electromyogram itself.

An action potential varies *spatially* along the membrane (Figure 5.6.b) in addition to changing with *time*. Consider the time recording in Figure 5.6.a, and assume that it occurs at spatial position, $x = 0$ of Figure 5.6.b. This figure indicates that the action potential is propagated at a constant velocity (θ) of ten meters per second (or 1 cm/ms). At $t = 0$, V_m at position $x = 0$ is -90 mV (point a of Figure 5.6.b). At $t = 1$ ms, V_m at position $x = 0$ is $+20$ mV (point b of Figure 5.6.b). Note that in the spatial dimension (x) the leading edge of the depolarization wave has moved a distance of 1 cm (to spatial position $x = 1$ cm), which is point a' of Figure 5.6.b.

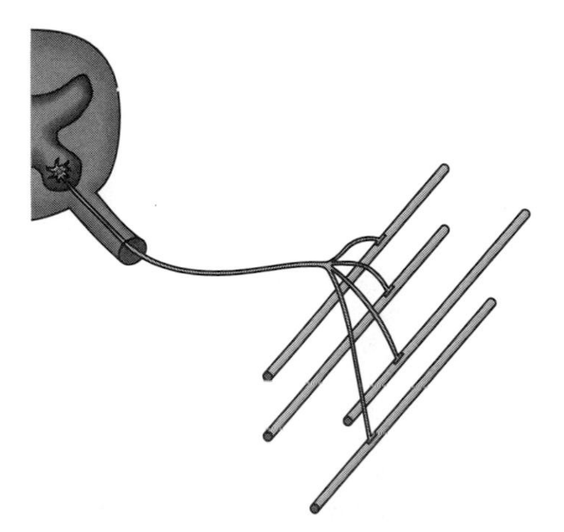

Figure 5.8 Diagram of a motor unit.

This process continues until at $t = 4$ ms, V_m at spatial position $x = 0$ has returned to -90 mV (point a of Figure 5.6.b). Note that at a constant velocity of propagation the leading edge of the depolarization wave has now moved to spatial position, $x = 4$ centimeters (point a″ of Figure 5.6.b). The maximal height (at the end of the depolarization wave) has now moved to a spatial position $x = 3$ centimeters (point b′, of Figure 5.6.b). The repolarization wave sequentially follows the depolarization wave and is distributed spatially between $x = 3.0$ cm and $x = 0$ cm.

Consequently, an action potential is propagated spatially along a membrane as an advancing wave of depolarization (which is associated with a peak positivity) followed by a trailing wave of repolarization (with a peak negativity). This system may be represented by an electrical dipole as follows. The source current (i_{IN}) is located at a point in the region of the depolarization wave (peak δ of Figure 5.6.a). The sink current (i_{OUT}) is located at a point in the region of the repolarization wave (peak γ of Figure 5.6.a). These two current points are then separated by a physical distance (a) along the membrane surface.

Referring to Figure 5.9, the cylindrical axon of the nerve cell has been replaced with a cylindrical muscle cell (fiber). The propagating wave that was the action potential for a nerve cell (Figure 5.6), is now replaced with a propagating wave on a muscle fiber (Figure 5.9), and is denoted as the myopotential. Consider the electrical activity that is recorded at a surface electrode point. This myopotential is spatially propagated along the course of the muscle fiber. Figure 5.9 represents this self-propagating action potential as an electrical dipole. With respect to a *surface electrode point*, the source current is at a distance, r_1, and the sink current is at a distance, r_2. Making the approximation that the conducting medium is isotropic (i.e., the same electrical properties in all three orthogonal directions), the relationship between the electrical potential (Φ) at the electrode point and the dipole current (i) is:

$$\Phi = \frac{i}{4\pi\sigma}\left(\frac{1}{r}\right) \tag{23}$$

where

Φ = electrical potential at the skin surface (millivolts) when
i = dipole current at muscle fiber (milliamps);
σ = conductivity of the medium (megaohms per centimeter); and
r = electrode-to-dipole current distance (centimeters)

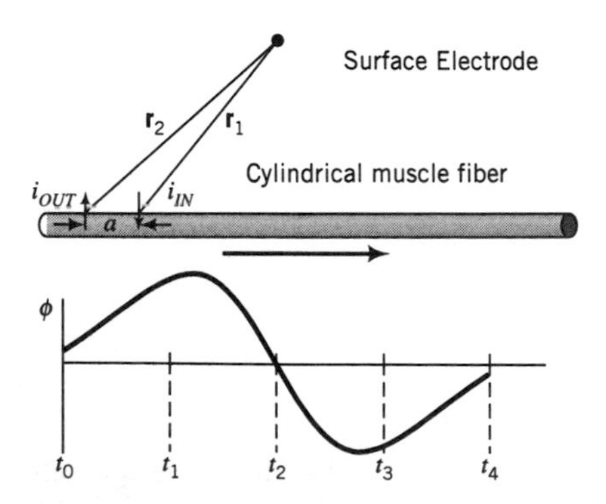

Figure 5.9 Propagating wave on a cylindrical muscle cell (fiber).

Since there is both a source current and a sink current, and by approximating i_{IN} (at r_1) to be of equal magnitude to i_{OUT} (at r_2), the net electrical potential recorded at the point electrode is:

$$\Phi = \frac{i}{4\pi\sigma}\left(\frac{1}{r_1} - \frac{1}{r_2}\right) \tag{24}$$

where

r_1 = electrode-to-depolarization current (centimeters)
r_2 = electrode-to-repolarization current (centimeters)

As the action potential propagates along the muscle fiber, its time history will be a function of r_1 and r_2. As the action potential approaches the electrode (at t_1) r_1 will be less than r_2, so that Φ is positive (Figure 5.9). As the action potential passes directly under the electrode (at t_2) r_1 will equal r_2, and the potential at the point electrode will be zero. Finally, when the action potential is moving away from the point electrode (at t_3) r_1 will be greater than r_2, and Φ has a negative value [equation (24)]. The net result is that a biphasic wave will be recorded when a *single point electrode* is used.

Two active electrodes (located at the skin surface) are utilized in order to record the surface EMG over a particular muscle. Each active electrode is input into a biopotential amplifier so that the recorded output is the difference in electrical potentials between these two active electrodes. Each active electrode is separated from the other by a specific spatial distance (Δx). Since the action potential of the motor fiber (i.e., the myopotential) is propagated at a constant velocity (θ), the relationship between action potential propagation distance and time is:

$$dx = \theta\, dt \tag{25}$$

where θ = velocity of propagation of the motor fiber action potential (meters per second). Integrating equation (25) over the spatial distance separating the two electrodes results in an action potential time shift that occurs at each electrode. Since the action potential waveform at each point electrode is the same, the differential signal Φ', which appears at the output of the biopotential amplifier, will be triphasic, as per Figure 5.10. Φ' is referred to as the MUAP.

The duration of the motor unit action potential is a function of the action potential time shift between the two active electrodes. For a given pair of active electrodes at a specific distance (Δx), equation (25) predicts that the action potential time shift is inversely proportional to the velocity (θ) of the propagating action

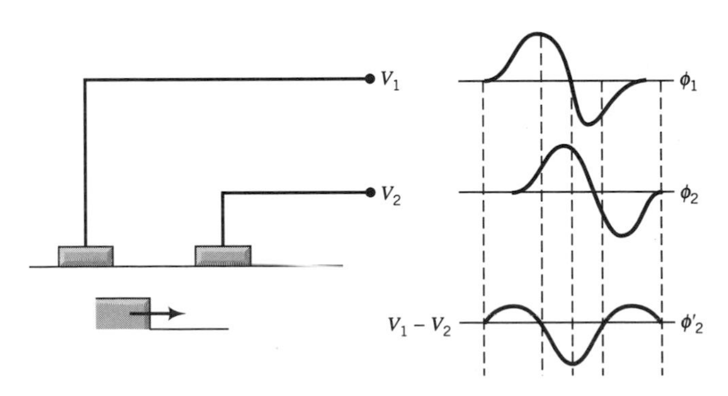

Figure 5.10 Recording of a motor unit action potential (MUAP).

potential. This is quite useful in quantifying muscle fatigue (see Chapter 8). With progressive muscle fatigue, there is a proportional reduction in the average velocity (θ) of the action potential propagation. Consequently, the inter-electrode action potential time shift is increased so that the duration of the MUAP is increased. Since the peak positive and peak negative magnitudes of the triphasic MUAP remain constant [equation (24)], the area under each phase (with respect to resting potential) will increase. Consequently, the average amplitude of the full wave rectified EMG will appear to increase as the muscle experiences progressive fatigue. Since the duration of the MUAP is increased, the center frequency of the full wave rectified EMG (see Section 5.2.b) will progressively decrease with increasing muscle fatigue.

For a pair of active recording electrodes, located upon the surface of a contracting muscle, the electromyogram will be recorded as the algebraic sum of all of the motor unit action potentials that are propagated along all of the muscle fibers at any specific point in time ($\hat{t}$). Each MUAP (at any specific point in time) is inversely proportional to its distance from the pair of active recording electrodes. Consequently, larger MUAPs will be recorded closer to the active point electrode site than those located farther away from the electrode site [see equation (23)]. The electromyogram may be defined mathematically as:

$$EMG(\hat{t}) = \sum_{i=1}^{n} \Phi_i'(\hat{t}) \tag{26}$$

where $\hat{t}$ is a specific point in time and at a specific point in space.

b. Bioelectronics and the Electromyogram

A bioelectronic system for the processing of a surface electromyogram is shown in Figure 5.11. Such bioelectronic systems are developed to obtain specific quantitative parameters for the analysis of the surface electromyogram. The specific configuration of Figure 5.11 provides output parameters (A_{rms}, P_{ave}, and f_c) that are used for the quantitative analysis of neuromuscular activity during physical workload (see Chapter 8). Other bioelectronic system configurations have been used extensively in the past and include the linear envelope of the electromyogram and various definitions of the integrated EMG (IEMG) as output parameters.

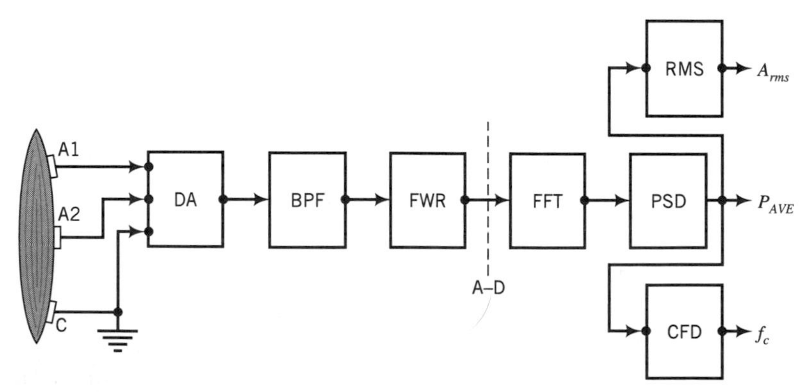

Figure 5.11 A bioelectronic system for the processing of a surface electromyogram.

The system operation is as follows (see Figure 5.11). On the skin surface overlaying the muscle of interest, three electrodes are placed (usually of silver-chloride composition). One active electrode (A1) is placed near either the proximal or distal end of the muscle. A second active electrode (A2) is placed at the center of the muscle. The third common electrode (C) is placed near the other end of the muscle (with respect to A1). This configuration is important for two reasons. The center of the muscle (sometimes referred to as the "belly" of the muscle) has the largest cross-sectional area, and therefore will generate the largest amplitude EMG. Second, motor nerves generally enter the muscle in its central region so that the subsequent motor unit action potentials are then conducted axially toward the ends of the muscle. Inadvertently placing an active electrode near each end of the muscle (with the common electrode located in the central region) can result in various amounts of EMG signal cancellation. For further information regarding bioelectrode theory and design, the interested student is referred to other texts.

The raw EMG signal is then processed sequentially by a series of signal conditioning units (Figure 5.11). These units in sequence are: a biopotential (or differential) amplifier (DA); a band-pass filter (BPF); a full-wave rectifier (FWR); and a Fourier analyzer producing a fast Fourier transform (FFT). The signal is then processed by three additional units, which results in the three quantitative parameters of interest. The EMG average power (P_{AVE}) is obtained by using a power spectral density converter (PSD). The rms amplitude of the EMG (A_{rms}) is then obtained by using a root mean square converter (RMS). Finally the center frequency of the EMG power spectrum (f_c) is obtained by use of a center frequency detector (CFD). Each of these various signal conditioning stages will be discussed in more detail shortly.

Active circuits represent the basis for a large majority of bioelectronic systems and their constituent units. It will be noticed in Figure 5.11 that the band-pass filter could be composed entirely of passive circuit elements (an RLC circuit) and the full-wave rectifier could be composed of diodes combined with passive elements. However, active circuits (particularly operational amplifiers) have the advantage of high input impedance, low output impedance, and a variable amount of gain. Moreover, the use of integrated circuits (as contrasted to discrete active component circuits) results in a very straightforward approach to the design of specific signal conditioning units.

Consider an operational amplifier that is operating in its linear active range: The output of the operational amplifier is directly proportional to the input. A first approach to the design of linear active circuits is based upon a customary set of approximations that are made regarding ideal operational amplifier behavior (unit A of Figure 5.12.a). These may be summarized as follows:

Approximation 1. The input impedance (R_i) is very high (greater than 10 MΩ), so that essentially no current flows between point A (the inverting input) and point B (the noninverting input).

Approximation 2. The output impedance (R_o) is very low (less than 50 Ω), so that essentially V_o' appears at the output as V_o.

Approximation 3. The open-loop operational amplifier gain (A) is very high (greater than 10,000).

Approximation 4. The open-loop bandwidth is very large (DC to greater than 1 MHz) and the amplifier phase shift is negligible.

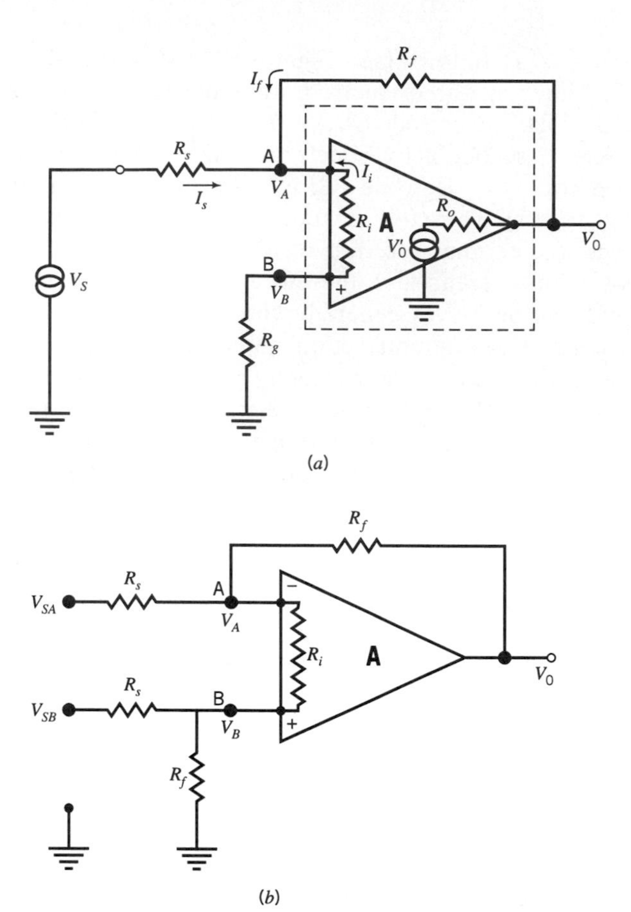

Figure 5.12.a Ideal operational amplifier. **b** Differential amplifier.

Consider the case of an op-amp used as an inverting amplifier as shown in Figure 5.12.a. An equivalent electrical model of the actual operational amplifier itself is shown as unit A. External resistors (R_s and R_f) are configured with respect to the inverting input (node A). A resistor (R_g) is connected between the noninverting input and ground. R_g will minimize the output voltage offset due to input bias current. R_g is necessary because Approximation 1 does not apply to *nonideal* operational amplifiers. Assuming that the signal source itself has a very low output resistance, the design value for R_g is the parallel combination of R_s and R_f:

$$\frac{1}{R_g} = \frac{1}{R_s} + \frac{1}{R_f} \tag{27}$$

so that:

$$R_g = \frac{R_s \cdot R_g}{R_s + R_g} \tag{28}$$

EXAMPLE 5.3

For the inverting amplifier of Figure 5.12.a, find the circuit (closed-loop) voltage gain, $G = V_0/V_s$. You may invoke suitable approximations as necessary.

SOLUTION 5.3

Norton's theorem at Node A:

$$I_s + I_f + I_i = 0 \tag{i}$$

Writing Ohm's law and Norton's theorem (with positive current flow into the node):

$$\frac{V_s - V_A}{R_s} + \frac{V_o - V_A}{R_f} + \frac{V_B - V_A}{R_i} = 0 \tag{ii}$$

Defining the open-loop gain (A) of the operational amplifier itself, *not* the circuit (closed-loop) gain:

$$A = \frac{V_o'}{V_B - V_A} \tag{iiia}$$

V_B is the result of a voltage divider:

$$V_B = \frac{V_A \cdot R_g}{R_i + R_g} \tag{iva}$$

Per Approximation 1, we note that $R_i >> R_g$, so that:

$$V_B \cong 0 \tag{ivb}$$

A fundamental property of an ideal operational amplifier is that:

$$V_B \cong V_A \tag{v}$$

There is a dual basis for this:

1. The input impedance (R_i) is very high:

   By Ohm's law:

   $$I_i = \frac{V_B - V_A}{R_i} \tag{vi}$$

   Since the input impedance (R_i) is very high, Approximation 1 constrains $I_i \cong 0$. So that equation (vi) reduces to:

   $$V_B - V_A \cong 0$$

   so that equation (v) is realized.
2. The open-loop amplifier gain (A) is very high.

   Rearrange equation (iiia):

   $$V_B - V_A = \frac{V_0'}{A} \tag{iiib}$$

   Since V_0' is finite, and A is very high (per Approximation 3), substituting into equation (iiib):

   $$V_B - V_A \cong 0$$

   so that equation (v) is realized.

At this point, we have derived equation (v) as a fundamental property of an ideal operational amplifier. Having solved for V_B above (equation [ivb]), then invoking equation (v) for the inverting amplifier of Figure 5.12.a:

$$V_B \cong V_A \cong 0 \tag{vii}$$

Substituting equation (vii) into equation (ii) yields:

$$\frac{V_s}{R_s} + \frac{V_0}{R_f} = 0 \tag{viii}$$

Since we define the circuit (closed loop) gain (G) as:

$$G = \frac{V_0}{V_s} \tag{ix}$$

rearranging equation (viii) for G results in:

$$G = \frac{V_o}{V_s} = -\frac{R_f}{R_s} \tag{x}$$

Note that in the configuration of Figure 5.12.a, the signal is inverted!

The differential amplifier is the first signal conditioning unit which receives input from the surface electrodes (as seen in Figure 5.11). Differential amplifiers are commonly used when very small voltages (such as the electromyogram) are encountered. Since bioelectric potentials are generally very small, such amplifiers are commonly referred to as biopotential amplifiers. A major advantage of a differential amplifier is that any signals common to both channels (e.g., 60 Hz noise) are suppressed (the common mode rejection).

EXAMPLE 5.4

For the differential amplifier of Figure 5.12.b, find the voltage output (V_0) as a function of the differential voltage input ($V_{SB} - V_{SA}$). You may invoke approximations as appropriate.

SOLUTION 5.4

Write the node A current equation in an analogous manner to equation 5.3(ii):

$$\frac{V_{SA} - V_A}{R_s} + \frac{V_0 - V_A}{R_f} + \frac{V_B - V_A}{R_i} = 0 \tag{i}$$

Invoking Approximation 1, solve for V_B as a simple voltage divider (since essentially no current flows through R_i):

$$V_B = \frac{V_{SB} \cdot R_f}{R_s + R_f} \tag{ii}$$

Invoking equation 5.3(v), V_A effectively equals V_B, so substituting equation 5.3(v) into equation (i):

$$\frac{V_{SA} - V_B}{R_s} + \frac{V_0 - V_B}{R_f} = 0 \tag{iii}$$

Substituting equation (ii) into equation (iii):

$$\frac{V_{SA} - \left(\frac{V_{SB} \cdot R_f}{R_s + R_f}\right)}{R_s} + \frac{V_0 - \left[\frac{V_{SB} \cdot R_f}{R_s + R_f}\right]}{R_f} = 0 \tag{iv}$$

Rearranging and simplifying:

$$\frac{V_0}{R_f} - \frac{V_{SB}}{R_s + R_f} = \frac{V_{SB} \cdot R_f}{R_s(R_s + R_f)} - \frac{V_{SA}}{R_s} \tag{v}$$

Separating variables and simplifying:

$$\frac{V_0}{R_f} = \frac{V_{SB}(R_f + R_s)}{R_s(R_s + R_f)} - \frac{V_{SA}}{R_s} \tag{vi}$$

which results in:

$$V_0 = \frac{(V_{SB} - V_{SA}) \cdot R_f}{R_s} \tag{vii}$$

The band-pass filter is the next sequential unit to process the now amplified electromyogram. The use of an operational amplifier allows the band-pass filter to be an active filter. A main advantage of these active filters is that they do not require inductors. Active filters are generally preferred to passive filters for frequencies, such as the EMG, which are below 1 kHz (Figure 5.13). There are two reasons why a band-pass filter is required at this stage of the bioelectronic system. The low frequency cut-off (f_{c_1}) can be set at approximately 4 Hz, so that effectively any DC component in the signal will be removed. This can also be accomplished with capacitive coupling of the source signal at the input of the differential amplifier. This input capacitor effectively blocks the DC signal from contaminating the output signal. Low frequency electrode movement artifact can also be minimized. The high frequency cut-off (f_{c_2}) can be set between 400 Hz and 500 Hz. At this cutoff frequency approximately 95% of the EMG signal energy is passed. About 5% of the EMG signal energy is between 400 to 500 and 1000 Hz. Selection of this upper band-pass limit allows the removal of high frequency noise while creating only minimal distortion of the true EMG signal.

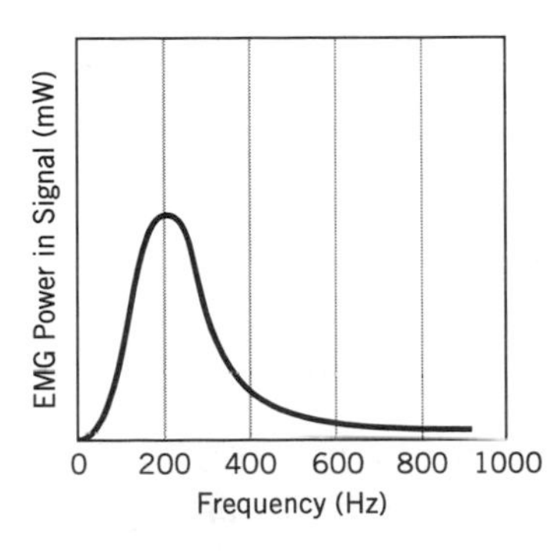

Figure 5.13 A typical graph of power vs. frequency for an active filter.

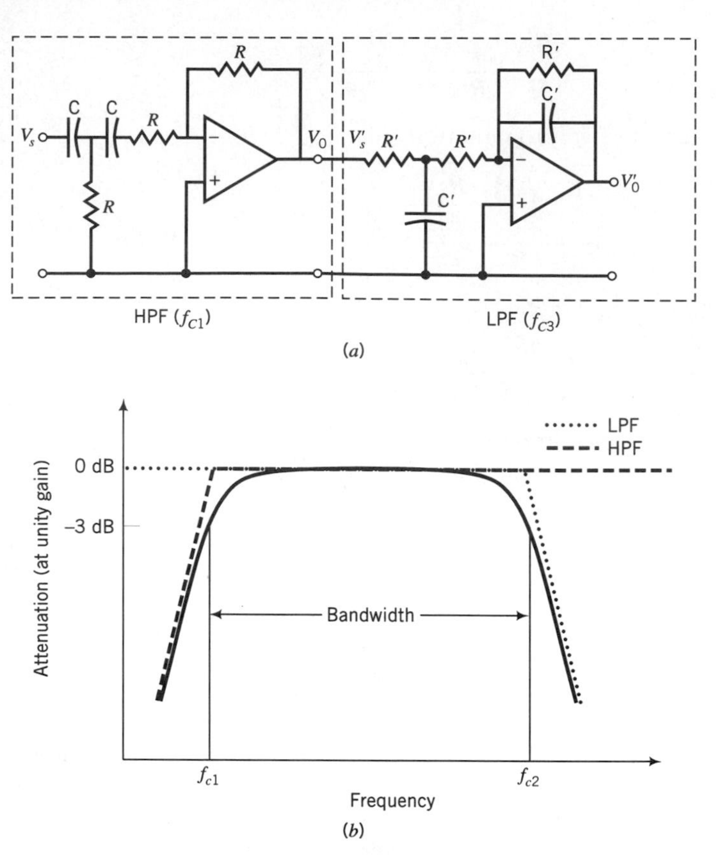

Figure 5.14.a Cascaded band-pass filter. **b** Relatively broad pass band can be attained using the cascading method.

A band-pass filter may be designed by cascading a low-pass filter and a high-pass filter. An example of a cascaded band pass filter using two operational amplifiers in the inverting feedback position is shown in Figure 5.14.a. The advantages of the cascading method are that a relatively broad pass band can be attained, and the low cut-off frequency (f_{c_1}) and high cut-off frequency (f_{c_2}) of the filter can be adjusted separately (Figure 5.14.b).

Analysis of the band pass filter is accomplished by first drawing a block diagram of a single op-amp filter element (either the low-pass filter or the high-pass filter) as shown in Figure 5.15.a. By analogy to the inverting amplifier gain (equation 5.3(ix)), the filter gain is:

$$\frac{V_0(j\omega)}{V_i(j\omega)} = -\frac{Z_f}{Z_{IN}} \tag{29}$$

Note that in active filter design, Node B (at the noninverting input) is connected directly to ground. Invoking equation 5.3(v), which states that Node A be at the same potential as Node B, means that Node A in Figure 5.15.a is also at ground potential. Consequently, the input impedance (Z_{IN}) and the feedback impedance (Z_f) of Figure 5.15.a are determined using the equivalent electrical model of Figure 5.15.b.

The full-wave rectifier is the next sequential unit in the processing of an electromyogram. Active circuit elements (operational amplifiers) allow us to design a

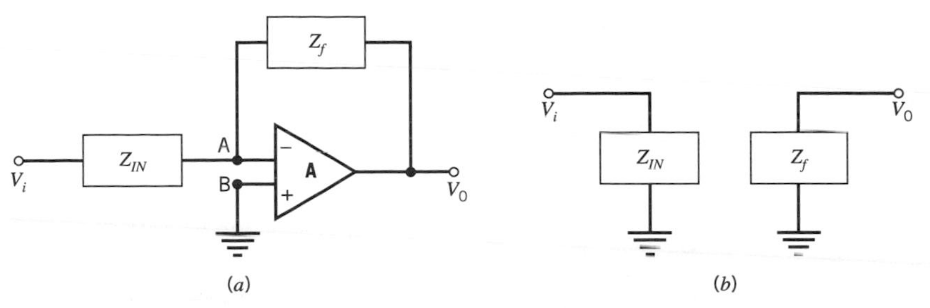

Figure 5.15.a Block diagram of a single op-amp filter element. **b** The equivalent electrical model.

precision full-wave rectifier. Silicon diodes must be forward-biased to approximately 0.7 V before they will effectively conduct, and semiconductor diodes alone introduce too much noise for small signal rectification. However, when silicon diodes are used in the feedback loop of an operational amplifier, the resultant circuit will precisely rectify even very small source signals. In order to understand the operation of a precision full-wave rectifier, consider Figure 5.16.a, which is divided into a precision half-wave rectifier (HWR) and a summing amplifier (SA). Consider first the HWR. The output of the HWR is taken from a connection in the amplifier feedback loop between R2 and D2. (This is *not* the little robot in the Star Wars movie.) When the source signal is positive the feedback current will flow through D1, so that the output voltage of the rectifier (although not of the amplifier) will be zero. When the source signal is negative, the feedback current will flow through D2 and R2 so that the output voltage, which is the voltage drop across R2, is present at the output of the rectifier. The high gain of the operational amplifier is important because even a very small negative source input can adequately forward bias D2. Consequently, the HWR of Figure 5.16.a will precisely rectify even very small input signals. Precision full-wave rectification is obtained by following the HWR with a summing amplifier (Figure 5.16.a). Note that the input resistors of the summing amplifier are selected with R3 at twice the value of R4. The voltage output from the HWR will generate a current flow through R4, which is twice the current flow of that of R3. The net result is that twice the output of the HWR ($2V_R$ of Figure 5.16.b) will be added to the original source signal (V_S of Figure 5.16.b). These two

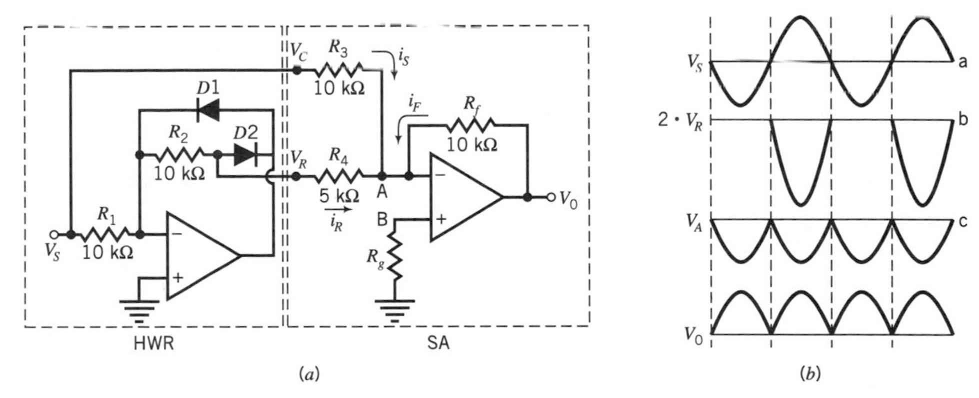

Figure 5.16.a Precision full-wave rectifier, divided into a precision half-wave rectifier (HWR) and a summing amplifier (SA). **b** Output of the precision full-wave rectifier of Figure 5.16.a.

waveforms sum at the inverting input to the operational amplifier (V_A of Figure 5.16.b). They are subsequently inverted so that the output waveform of the precision full-wave rectifier is the inverse of the input waveform to the summing amplifier and will appear as positive full-wave rectification (V_0 of Figure 5.16.b).

EXAMPLE 5.5

For the summing amplifier section (of the precision full-wave rectifier) of Figure 5.16.a:

a. Find V_0 as a function of V_S and V_R;
b. Find the value of the offset balance resistor, R_g.

SOLUTION 5.5.a

Norton's Theorem at Node A:

$$i_S + i_R + i_F = 0 \tag{i}$$

Per Ohm's law at Node A:

$$\frac{V_S - V_A}{R_3} + \frac{V_R - V_A}{R_4} + \frac{V_0 - V_A}{R_f} = 0 \tag{ii}$$

Applying equation 5.3(vii) in which $V_A \cong 0$, into equation (ii):

$$\frac{V_S}{R_3} + \frac{V_R}{R_4} + \frac{V_0}{R_f} = 0 \tag{iii}$$

Rearranging for V_0 and multiplying by R_f:

$$V_0 = -\left(\frac{R_f}{R_3}\right) V_S - \left(\frac{R_f}{R_4}\right) V_R \tag{iv}$$

which represents an active summing amplifier.

SOLUTION 5.5.b

The DC offset-bias current flows through the inverting input into the resistive network. We then make the approximation that the same DC bias current flows through the non-inverting input. If it flows to a ground, there is an offset voltage at the output, so we set R_g equal to the equivalent resistance that appears at the inverting node. There will then be no voltage differential at the input.

Therefore, the R_g for this circuit is:

$$\frac{1}{R_g} = \frac{1}{R_3} + \frac{1}{R_4} + \frac{1}{R_f} \tag{i}$$

so that:

$$R_g = \frac{R_3 R_4 R_f}{R_3 R_4 + R_3 R_f + R_4 R_f} \tag{ii}$$

At this point in Figure 5.11 linear analog processing is complete and digital processing ensues. The resultant analog signal output from the FWR undergoes A-to-D conversion (vertical dashed line of Figure 5.11) so that a time series of discrete data is generated and processed (to the left side of the vertical dashed line of Figure 5.11).

The FFT stage then performs a Fourier analysis on the discrete data, resulting in an output:

$$y(t_i) = \sum_{k=1}^{n} a_k \cos(k\omega_0 t_i) + \sum_{k=1}^{n} b_k \sin(k\omega_0 t_i) \tag{30}$$

The data are sampled over some fundamental time interval (T), at specific increments of time (Δt), so that:

$$\omega_0 = \frac{2\pi}{T} \tag{31}$$

$$t_i = \sum_{i=1}^{n} i \cdot \Delta t \tag{32}$$

Recall that for the EMG signal, $y(t)$ of equation (30) has amplitude coefficients (a and b) in millivolts. The power dissipated in a 1-Ω resistor is the voltage squared divided by 1 Ω, and equation (30) can be used to calculate signal power. Using the convention of a 1-Ω resistor, power (in mV) is directly proportional to millivolts squared. P_{AVE} is the average power in the EMG signal and can be shown to be:

$$y^2(t) = \frac{1}{2} \sum_{k=1}^{n} (a_k^2 + b_k^2) \tag{33}$$

where the term $(1/2)(a_k^2 + b_k^2)$ is the average of the square of the term $(a_k \cos[k\omega_0 t] + b_k \sin[k\omega_0 t])$.

The power effectiveness of the EMG signal is the root-mean-square (rms) value:

$$A_{rms} = \left[\frac{1}{2} \sum_{k=1}^{n} (a_k^2 + b_k^2)\right]^{\frac{1}{2}} \tag{34}$$

Recall that we are using the convention of a 1-Ω resistor, so that A_{RMS} is an rms voltage amplitude (in millivolts).

The center frequency (f_c) is based upon the mth Fourier component. f_c is then defined as the frequency that divides the signal power into equal upper and lower halves:

$$\sum_{k=1}^{m} (a_k^2 + b_k^2) = \sum_{k=m}^{n} (a_k^2 + b_k^2) \tag{35}$$

At the mth harmonic of the fundamental frequency when the left-hand term equals the right-hand term of equation (35), then:

$$f_c = m\omega_0 \tag{36}$$

where ω_0 is per equation (31).

At this point, all digital processing (indicated in Figure 5.11) has been completed and the quantitative parameters A_{RMS}, P_{AVE}, and f_c appear at the output.

5.3 THE ELECTROCARDIOGRAM AND ELECTROPNEUMOGRAM

a. Electrocardiogram (ECG)

The electrocardiogram represents the electrical activity of cardiac muscle that is recorded with surface electrodes. The basis for the electrocardiogram is ultimately the action potentials generated by the myocardial fibers (cardiac muscle cells). In this context, there is a similarity with the electromyogram, which is ultimately based upon the action potentials generated by skeletal muscle fibers (skeletal muscle cells). Referring to Figure 5.17.a the surface electrocardiogram waveform is indicated and consists of: a higher frequency, short duration QRS-wave (of approximately 1 mV amplitude) and a lower frequency, long duration T-wave (of approximately 0.3 mV amplitude). Figure 5.17.b shows a myocardial fiber (cardiac cell) action potential (AP). This AP is recorded from an intracellular electrode, in a manner similar to that in which a nerve cell action potential is recorded.

It is immediately apparent that this myocardial cell action potential has a very long duration (400 ms) compared to the action potential of a nerve cell (with an overall duration of about 4 ms). Because of this long duration, a cardiac cell action potential is divided into five phases (as indicated numerically by 0, 1, 2, 3, and 4 of Figure 5.17.b). Note that the resting membrane potential (phase 4) is at −90 mV for the cardiac cell (which is similar to the nerve cell of Section 5.1.a). The depolarization wave is represented by phase 0. There is an initial (very rapid) rise to a peak positive transmembrane voltage of +20 mV. This depolarization wave is

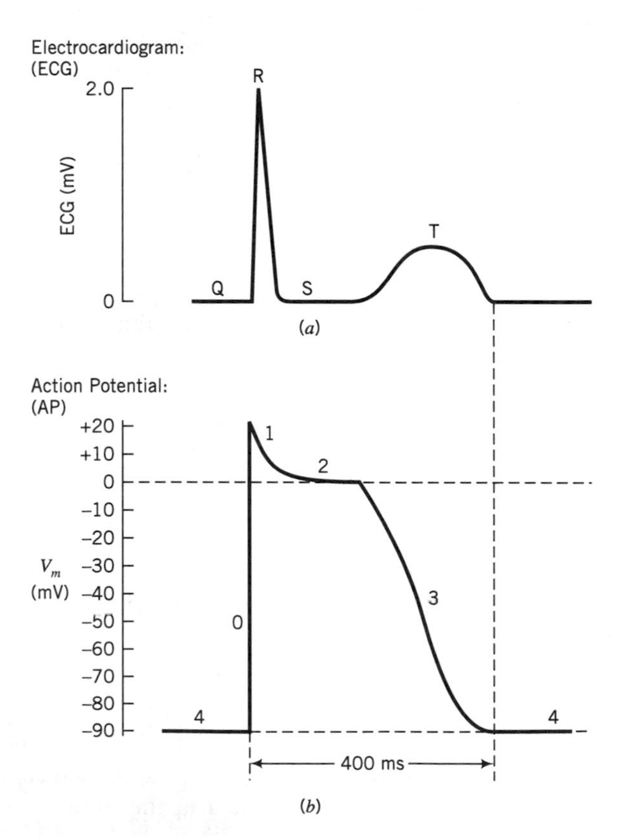

Figure 5.17.a Surface electrocardiogram. **b** Ventricular myocardial cell action potential.

due to a rapid increase in the conductance of the membrane to sodium (as was characteristic for the nerve cell membrane of Section 5.1.b). The repolarization wave of the cardiac cell action potential occurs in three phases. Phase 1 is the initial phase of repolarization in which the transmembrane potential decreases to near 0 mV. Phase 2 is a plateau phase of repolarization (at approximately 0 mV) in which there is a slow influx of calcium ion. Phase 3 is the final repolarization phase in which the transmembrane potential slowly decreases from 0 mV back to the resting membrane potential of −90 mV. Phase 3 is due to an increase in the membrane conductance for potassium, so that there is a potassium efflux, as was characteristic of a nerve membrane.

The electrobiophysical basis for the mathematical relationship of the myocardial cell action potential to the surface electrocardiogram is much more complex than the mathematical relationship previously discussed for a skeletal muscle cell action potential and the surface electromyogram. First, the myocardial cell architecture is less well organized than skeletal muscle cell architecture. Each myocardial cell is approximately 40 μm wide by 100 μm long giving the cell more of a "boxcar" shape as compared to a long cylindrical cell (fiber) of skeletal muscle. Anatomically these myocardial cells are joined (end-on-end) in long series which branch back and forth between adjacent series of myocardial cells. The result is roughly analogous to a bird's-eye view of a very large railroad switching yard in which thousands of boxcars intertwine as they move along the interconnecting railroad tracks. Second, at the tissue-organ level, the myocardium is not axisymmetric (as is the case for many of the skeletal muscles). Rather, myocardial muscle surrounds the four chambers of the heart with highly varying degrees of thickness, and at interacting angles somewhat analogous to the windings of toroidal motor coil. Third, the electrical depolarization wave (phase 0) propagates from the inside of the heart wall (the endocardium) to the outside of the heart wall (the epicardium). After the isoelectric plateau (phase 2), the final (phase 3) repolarization wave propagates in the reverse direction from the outside of the heart wall (the epicardium) back to the inside of the heart wall (the endocardium).

All of these factors make it much more complex to mathematically describe the electrobiophysics of how the cardiac cell action potential is related to the surface electrocardiogram. The interested reader is referred to other texts for the mathematical treatment of this subject. However, some temporal relationships are useful at a qualitative level to the human factors engineer.

In Figure 5.18, a three electrode configuration is shown for the recording of a surface electrocardiogram. The three surface-type electrodes are of silver-chloride

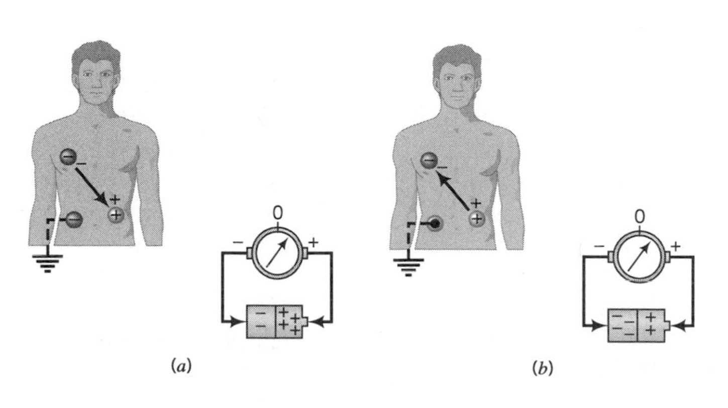

Figure 5.18.a A positive depolarization wave moves toward the positive surface recording electrode.
b A negative repolarization wave moves away from the positive electrode.

compositions and are configured on the front of the person's chest. The positive electrode is placed on the left-lower chest wall of the subject. The negative electrode is positioned at the right-upper chest wall of the subject. These two electrodes are the active electrodes, and are positioned in what is termed a *Lead II configuration*. A third neutral (ground) electrode is located at the right-lower chest wall.

The depolarization wave (phase 0) can be viewed as a time varying and spatially varying electrical dipole which is measured between the two widely separated active electrodes. The heart is located diagonally between the two electrodes so that the body tissues become a volume conductor. This pair of active electrodes records a scalar version of the ECG at a fixed spatial position defined by the Lead II configuration. The ECG itself is actually a time-varying cardiac vector in which the electrical dipole has both variable angle (spatial orientation) and magnitude. The depolarization wave is a positive wave (the transmembrane voltage goes from -90 mV to $+20$ mV) and that wave of depolarization moves from the endocardium (inside surface layer of the heartwall muscle) to the epicardium (outside surface layer of the heart). A positive surface ECG potential is recorded because the positive depolarization wave moves toward the positive surface recording electrode (Figure 5.18.a). The R-wave is upright and positive because the positive electrode is relatively more positive than the negative electrode (Figure 5.17.a).

Sequentially in time, there is an isoelectric period (phase 2 of repolarization) in which a plateau phase of the myocardial action potential occurs (Figure 5.17.b). During this isoelectric period, the two surface electrodes will record zero potential difference so that the surface electrocardiogram will return to its baseline value which occurs during the interval between the *R*-wave and the *T*-wave (Figure 5.17.a).

The final phase of myocardial repolarization (phase 3) is a negative-wave (the transmembrane potential goes from approximately 0 mV to -90 mV), and the repolarization wave moves in a reverse direction (from the epicardium toward the endocardium). The same positive electrode that recorded a positive depolarization wave now records a positive repolarization wave (*T*-wave of Figure 5.17.a). This occurs because the positive electrode will now detect a negative potential (wave front) that is moving progressively in time and space away from that positive electrode (Figure 5.18.b). The *T*-wave is upright and positive because the negative electrode is relatively more negative than the positive electrode.

An electrocardiograph is the bioelectronic system that is used to record, process, and display the surface electrocardiogram. These systems consist of a number of fundamental units: switching circuits for different configurations of external surface electrodes, electrical isolation stages for the subject's safety, amplifiers, filters and various output display devices. However, the essential characteristics of the electrocardiogram are determined by two sequential stages (top half of Fig. 5.19.a). A differential (biopotential) amplifier (DA) is necessary to amplify the 1-mV peak potentials that are recorded by the skin surface electrodes. The subsequent band-pass filter (BPF) specifies the range of frequencies which give the ECG its unique electrical signature. Commonly, the frequency response should be flat (0 dB attenuation) between 4 and 40 Hz, and the lower and upper cut-off frequencies (3 dB attenuation) at or below 1 Hz and around 150 Hz, respectively.

EXAMPLE 5.6

For the high-pass filter section (of the cascaded band-pass filter) in Figure 5.14.a, find the filter gain, $[V_0(j\omega)]/[V_s(j\omega)]$.

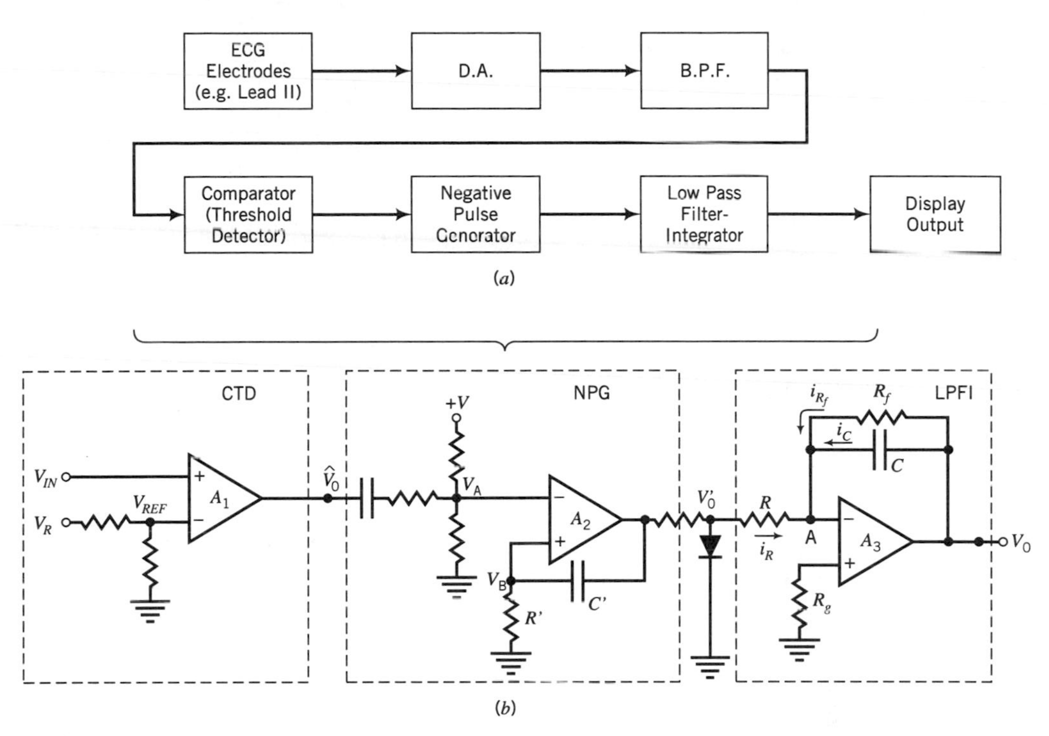

Figure 5.19.a Stages of the cardiotachometer. **b** CTD, NPG and LPFI of the cardiotachometer.

SOLUTION 5.6

Thevenize the input:

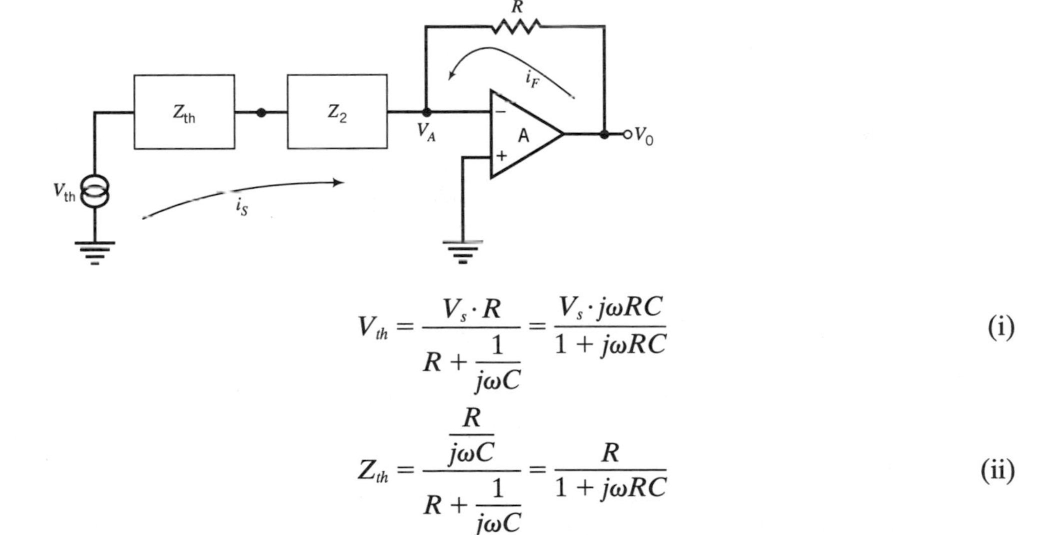

$$V_{th} = \frac{V_s \cdot R}{R + \dfrac{1}{j\omega C}} = \frac{V_s \cdot j\omega RC}{1 + j\omega RC} \tag{i}$$

$$Z_{th} = \frac{\dfrac{R}{j\omega C}}{R + \dfrac{1}{j\omega C}} = \frac{R}{1 + j\omega RC} \tag{ii}$$

$$Z_2 = R + \frac{1}{j\omega C} = \frac{R(1 + j\omega RC)}{j\omega RC} \tag{iii}$$

Recalling that $V_A \cong 0$:

$$i_S + i_F = 0 \tag{iv}$$

Ohm's law for an AC circuit:

$$\frac{V_{th}}{Z_{th} + Z_2} + \frac{V_0}{R_f} = 0 \tag{v}$$

Substituting equations (i), (ii), and (iii) into equation (iv):

$$\frac{\left(\dfrac{V_s \cdot j\omega RC}{1 + j\omega RC}\right)}{\left(\dfrac{R}{1 + j\omega RC}\right) + \dfrac{R(1 + j\omega RC)}{j\omega RC}} + \frac{V_0}{R_f} = 0 \tag{vi}$$

Rearranging and simplifying:

$$\frac{V_0}{R_f} = -\left[\frac{V_s \cdot j\omega RC}{R + \dfrac{R(1 + j\omega RC)^2}{j\omega RC}}\right] \tag{vii}$$

Transposing and factoring:

$$\frac{V_0}{V_s} = -\left[\frac{R_f \cdot j\omega RC}{R\left[1 + \dfrac{(1 + j\omega RC)^2}{j\omega RC}\right]}\right] \tag{viii}$$

Factoring and rearranging:

$$\frac{V_0}{V_s} = -\left[\frac{R_f}{R}\right]\left[\frac{1}{\dfrac{1}{j\omega RC} + \dfrac{(1 + j\omega RC)^2}{(j\omega RC)^2}}\right] \tag{ix}$$

The circuit (filter) gain is then:

$$\frac{V_0}{V_S} = -\left[\frac{R_f}{R}\right]\left[\frac{1}{\dfrac{1}{j\omega RC} + \left[1 + \dfrac{1}{j\omega RC}\right]^2}\right] \tag{x}$$

The physiological application of the electrocardiogram is to quantify cardiac activity. Referring to Figure 5.20, various aspects of cardiac activity are presented. The amplitudes of these activities are indicated on the right side of Figure 5.20. All events proceed with respect to time and the surface ECG is useful for the timing of cardiac events (vertical lines of Figure 5.20). The most common application of the surface ECG is to indicate heart rate. The time interval between the peak of an R-wave and the next successive R-wave peak is termed the *R-R interval.* By approximating the R-R interval as relatively constant over a 60-s interval, the heart rate may be calculated in beats per minute as follows:

$$\bar{f}_{HR} = \frac{60\left(\dfrac{\text{s}}{\text{min}}\right)}{R - R(\text{s})} \tag{37}$$

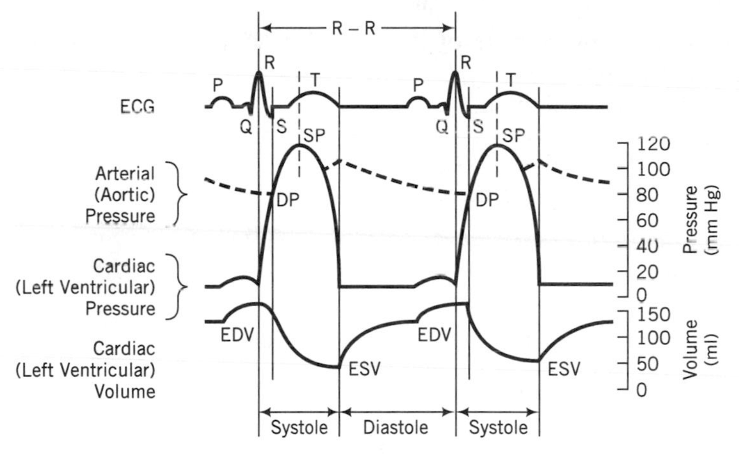

Figure 5.20 Various aspects of cardiac activity.

Since in reality the *R-R* interval can and does vary significantly over time, the heart rate is commonly measured using a bioelectronic system known as a cardiotachometer as per Figure 5.19.a. A surface electrode configuration is selected, such as lead II described above. This surface ECG signal is then processed by a biopotential differential amplifier (DA) and a band pass filter (BPF) as described above for the top half of Fig. 5.19.a. A comparator threshold detection circuit (CTD) is the next stage and is set at an amplitude sufficiently above isoelectric baseline so that only the upper part of the tall *R*-wave is detected (rises above threshold level) as per CTD in Figure 5.19.b. All electrical activity below the threshold value (such as the *T*-wave) remains undetected. The resultant *R*-wave exceeding the threshold results in a positive trigger input (V_0) to the negative pulse generator (NPG). Referring to the NPG block of Figure 5.19.b, this stage is a monostable pulse generator initially biased (no trigger input) so that $V_A < V_B$ and the output (V_0') is positive. A positive trigger input ($\hat{V}_0$) arrives at node A so that $V_A > V_B$ and the output (V_0') swings negative. The feedback capacitor (C') then charges through resistor R' (at a time constant $\tau' = R'C'$) until $V_A < V_B$. The output (V_0') then returns to its original positive value. This negative pulse output is then connected to a low pass filter-integrator (LPFI), which is the final stage in the cardiotachometer. Referring to the LPFI block of Figure 5.19.b, R and C act as an integrator of the rectangular pulse over its pulse duration. The combination R and C define the integration time constant for the low-pass filter/integrator (LPFI of Figure 5.19.b). C will discharge through R_f during the interpulse interval so that the amplitude at the output of LPFI (V_0) will be a time-varying voltage that is proportional to the frequency of the rectangular pulses.

EXAMPLE 5.7

Refer to the low-pass filter-integrator section (the LPFI block of the cardiotachometer system) as shown in Figure 5.19.b. A series of negative square wave pulses has already been integrated by the low-pass filter-integrator section, so that at $t = 0$, the filter-integrator output (V_0) is $\overline{V}_0$. At $t = 0$, a square-wave pulse at the output from the negative pulse generator (V_0' of the NPG block of Figure 5.19.b) occurs and is of magnitude, $-V_s$. The pulse duration is Δt, and at $t = \Delta t$ ($t' = 0$), the square-wave pulse magnitude of the NPG block returns to zero and remains at zero until the next pulse.

For the events just described:

a. Find V_0 during the time interval, Δt;
b. Find V_0 during the prior time, $t' = t - \Delta t$.

SOLUTION 5.7.a

For the time interval, Δt (at the LPFI block of Figure 5.19.b):

Norton's theorem at Node A:

$$i_R + i_{R_f} + i_C = 0$$

Then apply Ohm's law (when $V_0' = -V_S$):

$$\frac{-V_s}{R} + \frac{V_0}{R_f} + C \cdot \frac{dV_0}{dt} = 0$$

Rearranging:

$$\frac{dV_0}{dt} + \frac{V_0}{R_f C} = \frac{V_s}{RC} \qquad \text{(i)}$$

Solve equation 5.7(a)(i) by first defining the initial condition at $t = 0$:

Define $V_0(t = 0) = \overline{V}_0$

Now recall equation (A2) of the appendix:

$$\dot{y} + by = c \qquad \text{(A2)}$$

The solution of equation (A2) is equation (A22) of the appendix:

$$y = \frac{c}{b}(1 - e^{-bt}) + y_0 e^{-bt} \qquad \text{(A22)}$$

Equation (A2) is analogous to equation 5.7(a)(i) when:

$$y = V_0(t)$$
$$y_0 = \overline{V}_0$$
$$b = \frac{1}{R_f C}$$
$$c = \frac{V_S}{RC}$$

Substituting the above into equation (A22) results in the solution of equation 5.7(a)(i):

$$V_0(t) = V_s \left[\frac{R_f}{R}\right] [1 - e^{-\frac{t}{R_f C}}] + \overline{V}_0 e^{-\frac{t}{R_f C}} \qquad \text{(ii)}$$

SOLUTION 5.7.b

For prior time, $t' = t - \Delta t$:

At $t_0' = 0$, when $V_0' = 0$:

Define: $V_0(t_0')$ by substituting $t = \Delta t$ into equation 5.7(a)(ii):

$$V_0(t_0') = V_s \left[\frac{R_f}{R}\right](1 - e^{-\frac{\Delta t}{R_f C}}) + \overline{V}_0 e^{-\frac{\Delta t}{R_f C}} \qquad \text{(i)}$$

Now at t' onward, since $V_0' = 0$, then at Node A:

$$i_{R_f} + i_C = 0$$

The capacitive current discharges through the feedback resistor (as a resistive current).

$$\frac{V_0}{R_f} + C \cdot \frac{dV_0}{dt} = 0$$

And rearranging:

$$\frac{dV_0}{dt} + \frac{V_0}{R_f C} = 0 \qquad \text{(ii)}$$

Now recall equation (A2) of the appendix:

$$\dot{y} + by = c \qquad \text{(A2)}$$

The solution of equation (A2) is equation (A22) of the appendix:

$$y = \frac{c}{b}(1 - e^{-bt}) + y_0 e^{-bt} \qquad \text{(A22)}$$

Equation (A2) is analogous to equation 5.7(b)(ii) when:

$$y = V_0(t')$$

$$y_0 = V_0(t_0')$$

$$b = \frac{1}{R_f C}$$

$$c = 0$$

Substituting the above into equation (A22) results in the solution of equation 5.7(b)(ii):

$$V_0(t') = V_0(t_0') e^{-\frac{t'}{R_f C}} \qquad \text{(iii)}$$

where $V_0(t_0')$ is defined by equation (i).

Before proceeding to discuss another physiological application of the ECG, and prior to the derivation of the Fick equation (see Section 5.3.b), it is important to understand some basic cardiopulmonary anatomy. Figure 5.21 indicates that the cardiopulmonary circuit may be functionally characterized as a series circuit. There are four functional subdivisions to this in-line flow system. First, there is a "right heart," which receives venous (deoxygenated) blood flowing from the venous circuit (vena cavae) into the right atrium, and then right ventricle (RV). Second, there is the pulmonary system (the right and left lung), which oxygenates *all* the venous blood from the right heart. Third, there is the "left heart," which receives *all* the arterial (oxygenated) blood from the lungs. This blood first flows through the left atrium and then through the left ventricle (LV). The major "pumping" chamber of the heart is the left ventricle. Even though flow is closed and uniform through the circuit, the LV generates pumping pressures normally five times (or more) higher than the RV. It is therefore not surprising that the LV muscle mass is about 80% of the total mass of the entire heart (all four chambers collectively). Fourth, there is the arterial blood flow, generated by the LV, which initially flows from the left heart into the aorta. The systolic blood pressure (SP) and diastolic blood pressure (DP) of the upper arm brachial artery (a peripheral artery) are very close to the SP and DP of the aorta (a central artery). The interested student is referred to as a number of standard texts if more detail is desired.

With this basic understanding of cardiopulmonary anatomy, we can now proceed to a second physiological application of the electrocardiogram. The ECG is a useful representation of the electrical events that initiate the mechanical pumping action of cardiac muscle. This phenomenon is termed *electromechanical coupling* so that the electrocardiogram has a temporal relationship to the mechanical performance of the heart (see Figure 5.20). With respect to the left ventricle, the *R-R* interval is functionally divided into two phases: systole and diastole. At about the peak of the first *R*-wave, the left ventricle is at end-diastolic pressure (EDP) and at end-diastolic volume (EDV). During the remainder of the *R*-wave, there is a rapid rise in ventricular pressure, and at the conclusion of the *R*-wave, the ventricular pressure has reached the arterial diastolic pressure (DP) in the region of the aorta, which is the outflow tract for the systemic arterial circulation.

At the beginning of the *T*-wave, the arterial systolic pressure (SP) is reached, and this is coincident with the peak left ventricular pressure. Meanwhile, the left

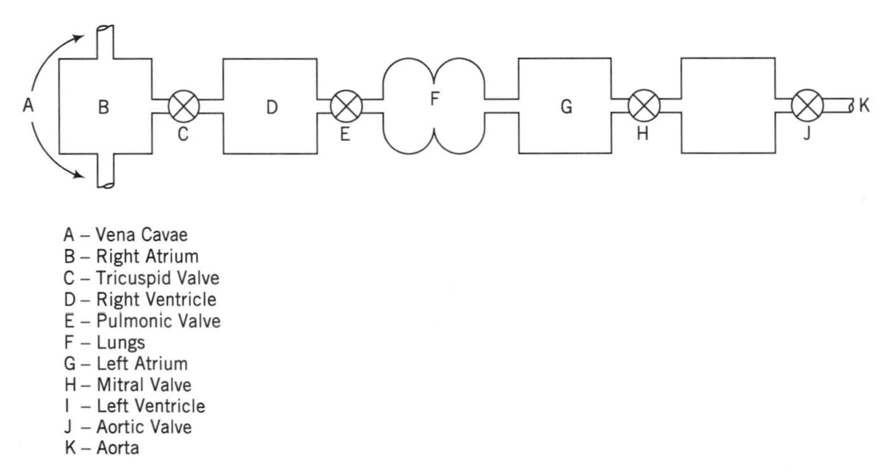

Figure 5.21 Cardiopulmonary circuit functionally characterized as a series circuit.

ventricular volume has remained relatively constant until arterial diastolic pressure is exceeded after which cardiac volume decreases during systole as blood flows from the left ventricle into the aorta and thence to the systemic circulation. During the time course of the T-wave, left ventricular pressure falls rapidly to a very low level, while arterial pressure falls from SP to a point approximately midway between SP and DP. Meanwhile, during the time course of the T-wave, the ventricular volume continues to decrease to its minimum end-systolic volume (ESV), which also occurs near the end of the T-wave and during the rapid fall of the left ventricular pressure. At this point ventricular systole has concluded.

Ventricular diastole occurs during a time interval beginning at the end of the T-wave and continuing until the peak of the subsequent R-wave. During this time interval, arterial pressure slowly falls from a point approximately between SP and DP until it returns to the arterial DP (as described earlier). During this diastolic interval, ventricular pressure remains relatively low although there is a small progressive increase until the ventricular EDP is reached at the peak of the subsequent R-wave. Finally, the left ventricular volume progressively increases (during filling with a new volume of blood) from an initial ESV (at the beginning of diastole) to its final EDV (at the end of diastole).

It is very clear from Figure 5.20 that there is an intimate temporal relationship between the electrical activity (ECG) of the cardiac muscle and the mechanical activity of that muscle. Furthermore, it is apparent that the mechanical activity of cardiac muscle may be characterized by time varying pressure changes and associated time varying volume changes. Within this context, the heart (and its cardiac muscle) may be viewed as an electrically activated pulsatile mechanical pump. The physiological events just described represent the necessary starting point for a set of equations that will be used to quantify performance of the cardiopulmonary system (see Chapter 8).

b. Electropneumogram (EPG)

The electropneumogram (EPG) can be defined as electrical activity produced by the movement of air into and out of the lungs. The EPG is useful to the HFE for two reasons. First, it represents an important example of the application of transducers as applied to human physiology. The "electrical activity" is *not* indigenously generated by the human body but is the result of a transducer (the thermal-convection velocity probe), which converts a physical variable (pulmonary air flow) into an electrical signal (current flow) by means of a power-dissipation thermistor. Second, the EPG provides physiological data that (by applying suitable approximations) represent a quantitative approach to the analysis of the cardiopulmonary system (see Chapter 8). This section will first examine the bioelectrical and bioelectronic fundamentals of the EPG. The physiological application of the EPG as represented by the Fick equation will then be described.

i. The Bioelectronic System

The bioelectronic system for measuring, processing and outputting the EPG is shown in Figure 5.22.a. There are five elements of this bioelectronic system beginning with the thermal-convection velocity sensor (T of Figure 5.22.a). A heated platinum wire (R_w) is positioned within an airflow tube (of cross-sectional area, A), which in turn is connected to a face half-mask worn by the human operator. As a transducer,

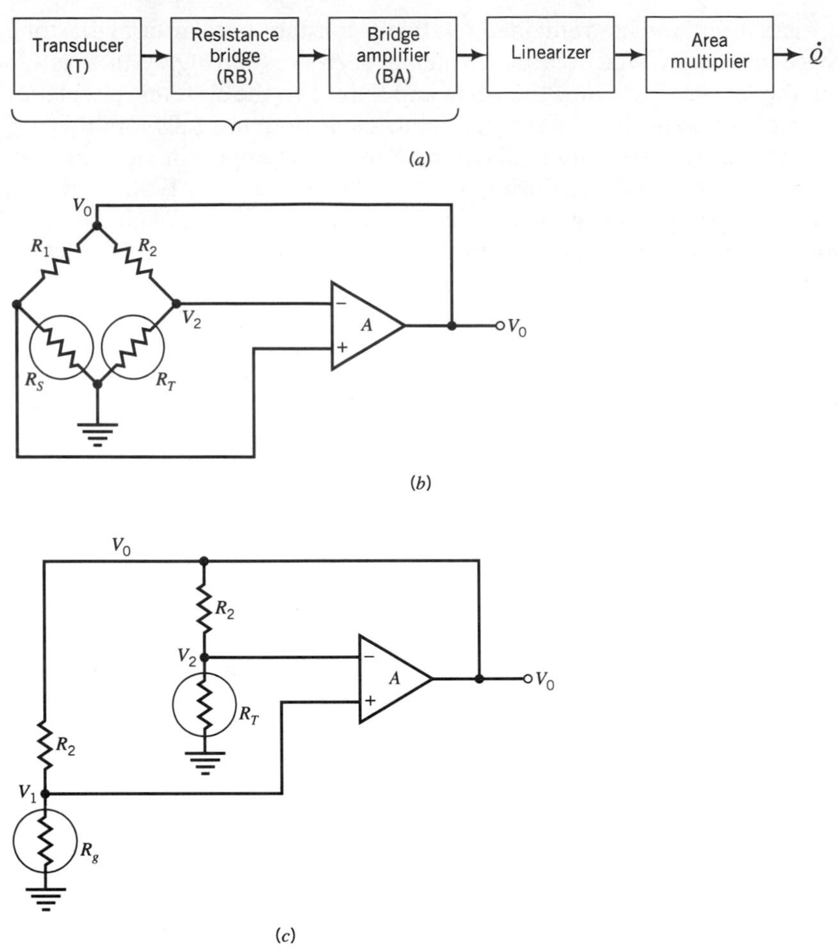

Figure 5.22.a Thermal-velocity flow meter. **b** Thermal-velocity sensor circuit. **c** Equivalent circuit of Figure 5.22.b.

R_w is heated to a temperature differential (ΔT) above air temperature. This is accomplished by a current source, so that the power (W) that is dissipated by the platinum wire is a function of the current passing through R_w:

$$W = i_w^2 \cdot R_w \tag{38}$$

The airflow velocity (V) has the following empirical relationship to the variables W and ΔT:

$$\frac{W}{\Delta T} = a + b \cdot \ln V \tag{39}$$

where a is a zero-velocity offset constant, and b is constant of proportionality.

In operation, R_w is cooled by convective heat loss as airflow moves through the breathing tube and across the heated platinum wire. This convective heat loss is a function of the velocity of airflow in contact with the heated platinum wire and this velocity can vary over the cross-sectional area of the breathing tube. As a result, R_w is heated over its entire length (which represents the diameter of the breathing tube itself) so that the total convective heat loss may be averaged over the entire length of the wire. The thermal-convection velocity sensor (T) output is then interconnected with a resistance bridge circuit (RB), which in turn is interconnected with a bridge amplifier (BA). The arrangement of these three units (T, RB, and BA) are depicted as an electrical schematic in Figure 5.22.b and represents a *constant-temperature circuit*, which may be described as follows.

In order for the circuit of Figure 5.22.b to operate, the circuit must be initially unbalanced so that R_1 does not equal R_2 (see Example 5.8). The unbalanced resistance bridge creates a differential voltage ($V_1 - V_2$) and the amplified output (V_0) is then fed back to the resistance bridge. Initially, the heated platinum wire ($R_S \equiv R_W$) might be 5°C higher than the air temperature (due to self-heating in a quiescent air stream). When the velocity of airflow increases across R_s, this thermal-convection velocity sensor will cool, and its resistance will increase. V_1 will then increase due to the voltage divider effect of R_1 and R_s) and the differential input to the bridge amplifier ($V_1 - V_2$) will increase so that the bridge amplifier output (V_0) will increase. The power dissipation of the resistance bridge will increase (as a function of V_0^2) so that R_s will further heat itself, thus reversing the original cooling. The resistance of R_s will now decrease so that V_1 decreases returning the bridge to its original balance. The overall result of this constant-current sensor circuit is that R_s will remain fairly constant so that its operating temperature remains fairly constant. Consequently, Figure 5.22.b is referred to as a *constant-temperature sensor circuit.*

R_T is a temperature-compensating thermistor, which is needed to keep the bridge circuit in balance when the airflow temperature varies. Since heated air is exhaled from the lungs, and ambient air temperature is inhaled into the lungs, there is a continual variation of airflow temperature during the operation of the heated platinum wire (R_S). The resistance-temperature coefficient of R_T should be much smaller than that of R_S in order to assure that its rise and fall in temperature is very small with changes in the convective airflow. This will assure that R_T will act as a sensor of the surrounding (ambient) air temperature and *not* of the surrounding airflow velocity. By positioning R_T in the resistance bridge as shown in Figure 5.22.b, any resistance changes due to temperature will equally affect R_S and R_T so that the differential voltage ($V_1 - V_2$) at the bridge amplifier input will remain constant with the surrounding (ambient) air temperature.

EXAMPLE 5.8

For the thermal-velocity sensor circuit (of the thermal velocity flow meter) of Figure 5.22.b, find V_0 as a function of the differential input, $V_1 - V_2$.

SOLUTION 5.8

Figure 5.22.c represents the equivalent electrical circuit of Figure 5.22.b. Since by Approximation 1, negligible current is drawn at the inverting and noninverting inputs of the op-amp (A) of Figure 5.22.c, then:

$$\frac{V_0 - V_1}{R_1} = \frac{V_1}{R_S} \qquad \text{(i)}$$

$$\frac{V_0 - V_2}{R_2} = \frac{V_2}{R_T} \qquad \text{(ii)}$$

Rearranging equations (i) and (ii):

$$\frac{V_0}{R_1} = \frac{V_1}{R_1} + \frac{V_1}{R_S} \quad \text{(iii)}$$

$$\frac{V_0}{R_2} = \frac{V_2}{R_2} + \frac{V_2}{R_T} \quad \text{(iv)}$$

Differencing equations 5.8(iii) and 5.8(iv):

$$V_0\left[\frac{1}{R_1} - \frac{1}{R_2}\right] = V_1\left[\frac{1}{R_1} + \frac{1}{R_S}\right] - V_2\left[\frac{1}{R_2} + \frac{1}{R_T}\right] \quad \text{(v)}$$

Rearranging:

$$V_0 = \frac{V_1\left[\dfrac{R_1 + R_S}{R_1 R_S}\right] - V_2\left[\dfrac{R_2 + R_T}{R_2 R_T}\right]}{\left[\dfrac{R_2 - R_1}{R_1 R_2}\right]} \quad \text{(vi)}$$

so that for $R_1 \neq R_2$:

$$V_0 = V_1\left[\frac{R_2}{R_S}\right]\left[\frac{R_1 + R_S}{R_2 - R_1}\right] - V_2\left[\frac{R_1}{R_T}\right]\left[\frac{R_2 + R_T}{R_2 - R_1}\right] \quad \text{(vii)}$$

The output (V_0) of the constant-temperature sensor circuit must subsequently be processed by a linearizer circuit (Figure 5.22.a). This linearizer circuit will perform a series of mathematical operations on V_0 in order to solve equation (39) for the airflow velocity (V). The derivation of the linearizing equation begins by setting R_w of equation (38) to R_S, so that:

$$W_{R_S} = (i_{R_S})^2 \cdot R_S \quad (40)$$

From the schematic of Figure 5.22.b; and op-amp theory:

$$i_{R_S} = i_{R_1} = \frac{V_0}{R_1 + R_s} \quad (41)$$

Substituting equation (41) into equation (40):

$$W_{R_S} = \frac{(V_0)^2 \cdot R_s}{(R_1 + R_s)^2} \quad (42)$$

Substituting equation (42) into equation (39) by setting $W = W_{Rs}$:

$$(V_0)^2\left[\frac{R_s}{(R_1 + R_s)^2 \cdot \Delta T}\right] = a + b \cdot \ln(V) \quad (43)$$

Since a constant-temperature sensor circuit is used:

$$\alpha \equiv \frac{R_s}{(R_1 + R_s)^2 \cdot \Delta T} \cong \text{constant} \quad (44)$$

Substituting equation (44) into equation (43):

$$\alpha \cdot V_0^2 = a + b \cdot \ln(V) \tag{45}$$

Rearranging:

$$\frac{\alpha \cdot (V_0)^2 - a}{b} = \ln(V) \tag{46}$$

so that:

$$V = e^{\frac{1}{b}(\alpha V_0^2 - a)} \tag{47}$$

Consequently, the sequential series of mathematical operations which the linearizer must perform are represented by equation (47). The final stage to receive the output from the linearizer is the area multiplier and converts air velocity into actual airflow (Figure 5.22.a). Mathematically, this may be represented as follows.

Since the heated-platinum wire extends across the diameter of the breathing tube, the flow area is:

$$A = \frac{(L_{R_S})^2 \cdot \pi}{4} \tag{48}$$

Now since:

$$\dot{Q} = V \cdot A \tag{49}$$

substituting equations (47) and (48) into equation (49) results in:

$$\dot{Q} = \left(\frac{\pi}{4}\right)(L_{R_S})^2 \cdot \ln^{-1}\left[\frac{\alpha \cdot (V_0)^2 - a}{b}\right] \tag{50}$$

so that $\dot{Q}$ represents the final output of the thermal-convection airflow meter of Figure 5.22.

ii. The Physiological Application

A physiological application of the electropneumogram is that pulmonary airflow ($\dot{Q}$) can be related to the oxygen consumption of the human body ($\dot{V}_{O_2}$). $\dot{V}_{O_2}$ can then be related to the blood flow output of the heart, the cardiac output ($\dot{V}_{CO}$). The mathematical basis for the relationship of these important parameters is described by the Fick principle and will now be considered.

Referring to Figures 5.21 and 5.23, the Fick principle for the pulmonary system may be stated as follows: the rate of oxygen entering the lung compartment (L)

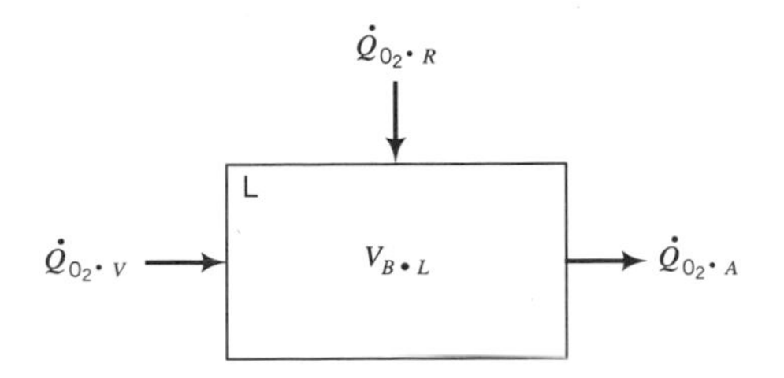

Figure 5.23 Graphical representation of the Fick principle for the pulmonary system.

via the inflowing blood from the right heart ($\dot{Q}_{O_2V}$) plus the rate of oxygen entering the blood in the lung compartment (L) from the respiratory airflow ($\dot{Q}_{O_2R}$) plus the rate of oxygen leaving the lung compartment (L) via the outflowing blood to the left heart ($\dot{Q}_{O_2A}$) must equal zero. It is apparent from this description that the Fick principle is applied to analyze compartments (such as the pulmonary compartment) when the system is in steady state. Also proper attention must be paid to sign convention with respect to quantities entering into or exiting from the compartment. The Fick principle can be applied to the calculation of the arterial (aortic) blood flow out of the heart, the cardiac output ($\dot{V}_{CO}$), as follows.

Let C_{O_2} represent the concentration of oxygen dissolved in a volume of blood:

$$C_{O_2} = \frac{Q_{O_2}}{V_B} \tag{51a}$$

(milliliters of oxygen per 100 milliliters of blood).

Rearranging equation (51a):

$$Q_{O_2} = C_{O_2} \cdot V_B \tag{51b}$$

Making the approximation that C_{O_2} is constant in the venous blood flowing through the right side of the heart ($C_{O_2 \cdot V}$), then:

$$\dot{Q}_{O_2 \cdot V} = C_{O_2 \cdot V} \cdot \dot{V}_{B \cdot V} \tag{52}$$

where $\dot{V}_{B \cdot V}$ is the venous blood flow *into* the lungs from the right side of the heart (Figures 5.21 and 5.23).

Making the approximation that the oxygen concentration in the arterial blood ($C_{O_2 \cdot A}$) leaving the lungs and flowing through the left side of the heart is constant, then:

$$\dot{Q}_{O_2 \cdot A} = C_{O_2 \cdot A} \cdot \dot{V}_{B \cdot A} \tag{53}$$

where $\dot{V}_{B \cdot A}$ is the arterial blood flow *from* the lungs into the left side of the heart (Figures 5.21 and 5.23).

Making the approximation that the volume of blood in the lung compartment ($V_{B \cdot L}$) is constant, then the rate of oxygen entering the blood in the lung compartment from the respiratory airflow ($\dot{Q}_{O_2 \cdot R}$):

$$\dot{Q}_{O_2 \cdot R} = \dot{C}_{O_2 \cdot L} \cdot V_{B \cdot L} \tag{54}$$

We can now state Fick's principle mathematically for the system in Figure 5.23:

Assign a positive sign to quantities entering into or in the lung compartment and a negative sign to quantities leaving the lung compartment:

$$-\dot{Q}_{O_2 \cdot A} + \dot{Q}_{O_2 \cdot R} + \dot{Q}_{O_2 \cdot V} = 0 \tag{55}$$

Substituting equations (52), (53), and (54) into equation (55):

$$-C_{O_2 \cdot A}\dot{V}_{B \cdot A} + \dot{C}_{O_2 L} \cdot V_{BL} + C_{O_2 \cdot V}\dot{V}_{B \cdot V} = 0 \tag{56}$$

In order for $V_{B \cdot L}$ to be constant, we make the approximation:

$$\dot{V}_{B \cdot A} = \dot{V}_{B \cdot V} = \dot{V}_B \tag{57}$$

where $\dot{V}_B$ is the pulmonary blood flow.

Also, the conventional symbol for defining the oxygen consumption of the human body is:

$$\dot{V}_{O_2} := \dot{Q}_{O_2 \cdot R} = \dot{C}_{O_2 \cdot L} \cdot V_{B \cdot L} \tag{58}$$

This is a symbolic change only!
Substituting equations (57) and (58) into equation (56):

$$-C_{O_2 A} \cdot \dot{V}_B + C_{O_2 V} \cdot \dot{V}_B + \dot{V}_{O_2} = 0 \tag{59}$$

Solving for the pulmonary blood flow:

$$\dot{V}_B = \frac{\dot{V}_{O_2}}{(C_{O_2 \cdot A} - C_{O_2 \cdot V})} \tag{60}$$

It is apparent that the cardiopulmonary system is an *in-line* pulsatile mechanical pump, so that arterial blood flow ($\dot{V}_{B \cdot A}$) from the lungs to the left heart must equal aortic blood flow ($\dot{V}_{CO}$) from the left heart (Figure 5.21):

$$\dot{V}_{CO} = \dot{V}_{B \cdot A} \tag{61}$$

Substituting equation (61) into equation (57):

$$\dot{V}_{CO} = \dot{V}_{B \cdot V} = \dot{V}_B \tag{62}$$

Solve for the cardiac output by substituting equation (62) into equation (60):

$$\dot{V}_{CO} = \frac{\dot{V}_{O_2}}{(C_{O_2 \cdot A} - C_{O_2 \cdot V})} \tag{63}$$

Equation (63) is referred to as the *Fick equation*. This equation allows the human factors engineer to translate the pulmonary airflow ($\dot{Q}$) entering an individual's lungs with respect to the cardiac blood flow ($\dot{V}_{CO}$) entering the systemic circulation. The EPG combined with the Fick equation are central elements in the quantitative analysis of cardiopulmonary function and human operator workload (see Chapter 8).

FURTHER INFORMATION

L.A. Geddes: *Medical Device Accidents*. CRC Press. Boca Raton, FL. 1998.

L.A. Geddes and L.E. Baker: *Principles of Applied Biomedical Instrumentation*. 3rd Edition. John Wiley. New York. 1989.

R.K. Hobbie: *Intermediate Physics for Medicine and Biology*. 3rd Edition. Springer-Verlag. New York. 1997.

C. A. Phillips: *Engineering Biophysics*. ClassNote Publications (Wright State University). Dayton, OH. 1997.

R. Plonsey and R.C. Barr: *Bioelectricity: A Quantitative Approach*. Plenum Press. New York. 1988.

PROBLEMS

5.1 For Cl^- in the ICF equal to 5 μMol/mL, and V_I in the ICF equal to -95 mV, what is Cl^- in the ECF when V_0 in the ECF equals -10 mV? You may assume that $RT/F = 26.3$ mV.

ICF(I)	ECF(O)
$Cl^- = 5$ mM	$Cl^- = ?$
$V_I = -95$ mV	$V_O = -10$ mV

Diagram 5.1

5.2 In an experimental cell culture the following concentrations have been measured:

Inside cells: $Na^+ = 10$ mM/L and $Cl^- = 4$ mMol/L

Outside cells: $Na^+ = 150$ mM/L and $Cl^- = 120$ mMol/L

For an $\alpha = P_{Na}/P_{Cl} = 0.01$, what is the potential inside the cells? You may assume $RT/F = 26.3$ mV.

ICF(I)	ECF(O)
$Na^+ = 10\,\frac{mM}{L}$	$Na^+ = 150\,\frac{mM}{L}$
$Cl^- = 4\,\frac{mM}{L}$	$Cl^- = 120\,\frac{mM}{L}$
$V^I = ?$	

Diagram 5.2

5.3 Two sides of a fish tank, each of 1-L water volume, are separated by a semipermeable membrane that is permeable to all ions *except* phosphate and sulfate. At time = 0 in the left-hand (A) compartment: zinc, Zn (+2 valence), chloride, Cl (−1 valence) and phosphate, PO_4 (−3 valence) are present in amounts of 200 mMol, 50 mMol and 100 mMol, respectively.

Also at $t = 0$ in the right-hand (B) compartment: Iron, Fe (+3 valence), chloride, Cl (−1 valence) and sulfate, SO_3 (−2 valence) are present in amounts of 100 mMol, 100 mMol and 125 mMol, respectively. The system then proceeds over time to go to equilibrium. Using the appropriate Donnan equation:

1. What is the final amount of chloride (Cl) on side A at equilibrium?
2. What is the final amount of chloride (Cl) on side B at equilibrium?

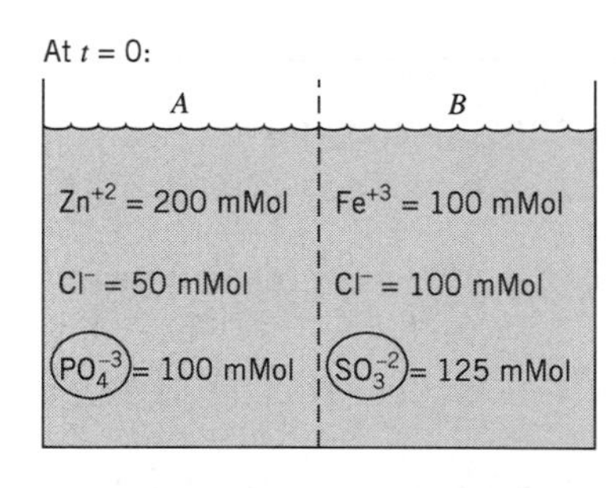

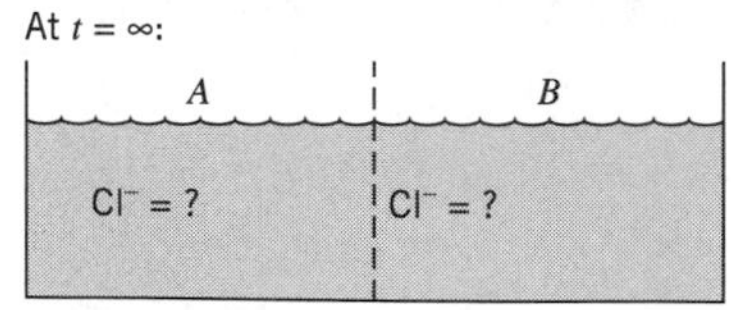

Diagram 5.3

5.4 On the inside of an experimental reconstituted cell, the calcium ion (Ca^{+2}) concentration is 100 mMol/L and the magnesium ion (Mg^{+2}) concentration is 200 mMol/L.

On the outside of the cell, the voltage is -15 mV (relative to the earth grounded table surface), the calcium concentration is 20 mMol/L, and the magnesium concentration is 2 mMol/L. α is defined as the ratio of magnesium permeability to calcium permeability ($\alpha = P_{Mg}/P_{Ca}$), and is known to be 16 for this type of membrane. Assume $RT/F = 26.3$ mV. What is the potential of the inside of the cell relative to the earth grounded table surface?

I	O
$Ca^{+2} = 100 \frac{\text{mMol}}{\text{L}}$	$Ca^{+2} = 20 \frac{\text{mMol}}{\text{L}}$
$Mg^{+2} = 200 \frac{\text{mMol}}{\text{L}}$	$Mg^{+2} = 2 \frac{\text{mMol}}{\text{L}}$
$V_I = ?$	$V_O = -15$ mV

Diagram 5.4

5.5 A passive nerve membrane has a resting membrane voltage (V_r) of -90 millivolts. A subthreshold current (at time equal zero) is then injected (as a unit step) on the inside of the membrane at an axial distance (x) equal to zero. After a steady state is reached (the transient response has died out), the differential voltage (V') at an axial distance ($x =$ zero centimeters) is 10 mV. Given the following transmembrane values:

$$C = 1.2 \text{ mF/cm}$$
$$G_{Na} = 0.05 \text{ m}\cdot\text{mho/cm}$$
$$G_K = 2.0 \text{ m}\cdot\text{mho/cm}$$
$$G_{Cl} = 1.2 \text{ m}\cdot\text{mho/cm}$$

and given the following values of the Nernst potentials:

$$V_{Na} = 80 \text{ mV}$$
$$V_K = -90 \text{ mV}$$
$$V_{Cl} = -70 \text{ mV}$$

Using any and all appropriate equations, what are the values and signs (plus or minus) for the following transmembrane currents at an axial distance (x) equal to zero: I_c, I_{Na}, I_K, I_{Cl}?

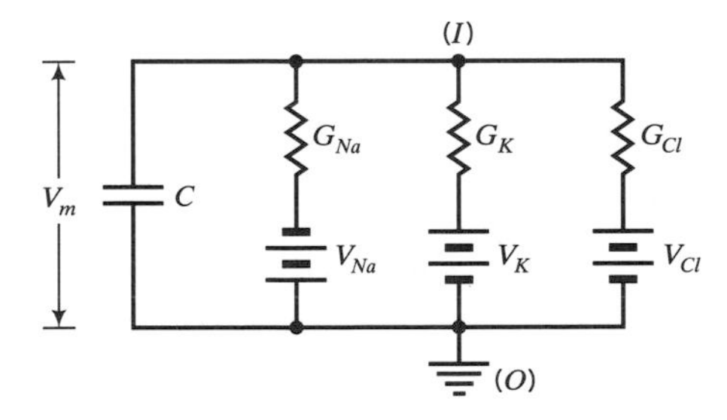

Diagram 5.5

5.6. A freely permeable membrane has both +2 valence ions [Calcium (Ca) and Iron (Fe)] and a -2 valence ion [carbonate (CO_3)] on each side of the membrane. In the intracellular fluid (ICF), the concentrations of calcium, iron, and carbonate are 5 mMol/L, 2mMol/L, and 5 mMol/L, respectively. In the extracellular fluid (ECF), the concentrations of calcium, iron and carbonate are 20 mMol/L, 8 mMol/L, and 1 mMol/L, respectively. It also happens that the extracellular voltage (V_0) is 20 mV and that the constant (RT/F) is 26.3 mV. In addition, the ratio of the calcium permeability to the carbonate permeability ($\alpha = P_{Ca}/P_{CO_3}$) is 0.05, and the ratio of the iron permeability to the carbonate permeability ($\beta = P_{Fe}/P_{CO_3}$) is 0.125 for this problem. Using an appropriate form of the modified Nernst equation, what is the value of the intracellular voltage (V_I)?

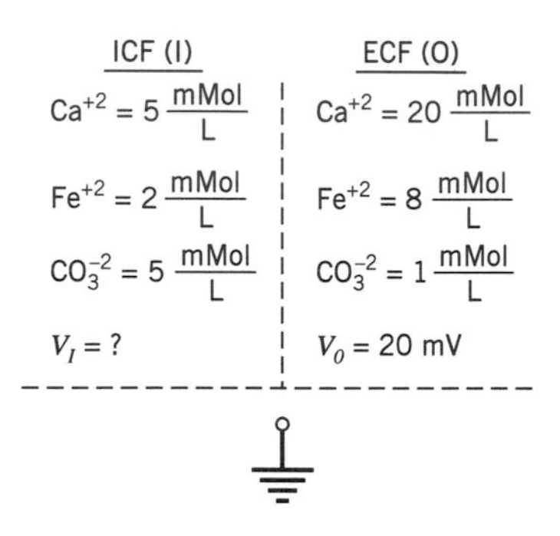

Diagram 5.6

5.7. Given the following cell characteristics: K_i = 155 mMol/L, Na_i = 12 mMol/L, K_0 = 4 mMol/L, and Na_0 = 145 mMol/L. Experimental results indicate a membrane potential of −88 mV under normal conditions (i.e. RT/F = 26.3 mV). Find the ratio of permeabilities (α), where ($\alpha = P_{Na}/P_K$). All ions are +1 valence.

5.8 For the low-pass filter section (of the cascade amplifier band-pass filter) of Figure 5.14a, find the filter gain, $[V_0'(j\omega)]/[V_s'(j\omega)]$.

5.9 For the low-pass filter section of Problem 5.8, plot the unity gain attenuation (A) in decibels *and* the phase angle (ϕ) in degrees as a function of the frequency ratio (f/f_{c_2}) where f_{c_2} is the low-pass cutoff frequency.

5.10 The single op-amp differential amplifier of Figure 5.12.b is adequate for applications with low-resistance signal sources (such as a strain gage bridge circuit, as per Figure 5.22.b). A 3 op-amp differential amplifier (referred to as an *instrumentation amplifier*) is often used to amplify biopotentials (such as the ECG), which can represent a high-resistance signal source.

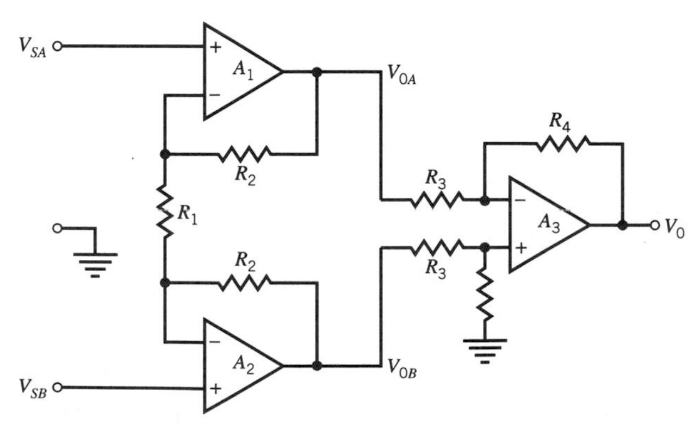

Diagram 5.7

For the biopotential amplifier shown in this diagram, calculate the differential voltage gain, $V_0/(V_{SB} - V_{SA})$.

5.11. For each of the op-amp circuits shown below:

a. Find the voltage output (V_0) as a function of the voltage input (V_I).
b. Define the function of the op-amp circuit.

Circuit A:

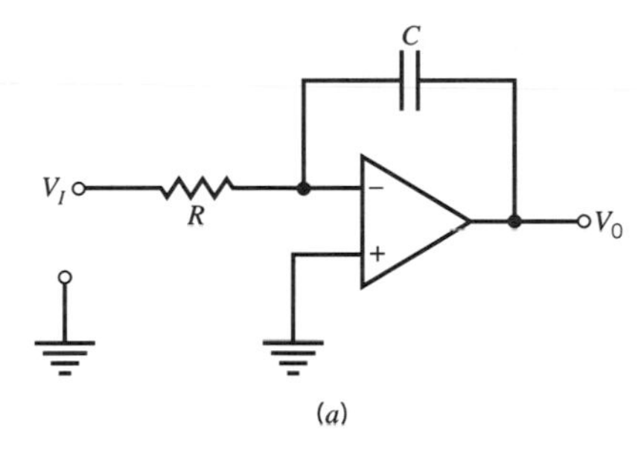

Diagram 5.8.a

Circuit B:

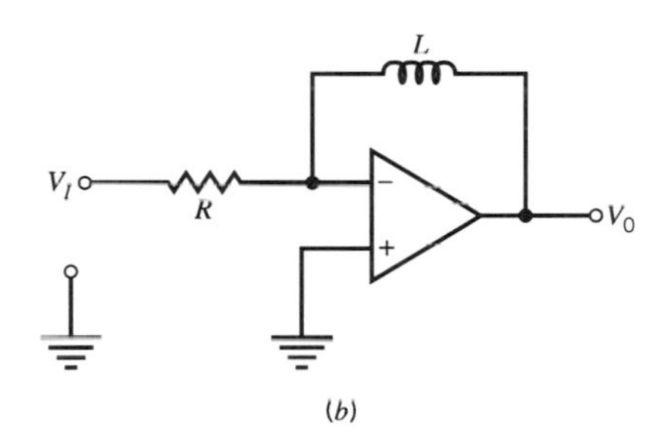

Diagram 5.8.b

5.12. For each of the op-amp circuits shown below:

a. Find the input resistance (R_{IN}).
b. Find the circuit (closed-loop) voltage gain, V_0/V_I.

Circuit A:

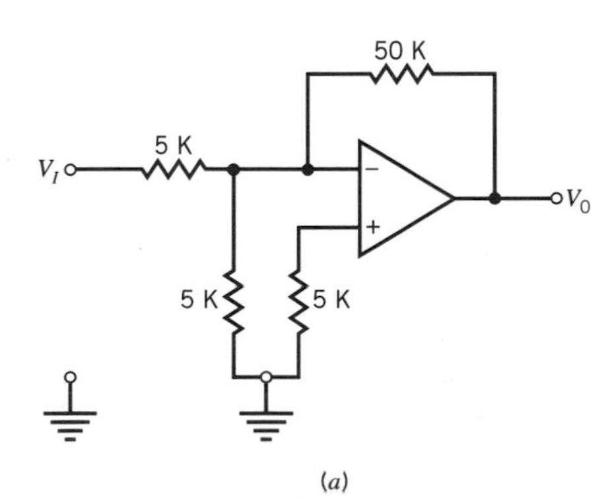

Diagram 5.9.a

Circuit B:

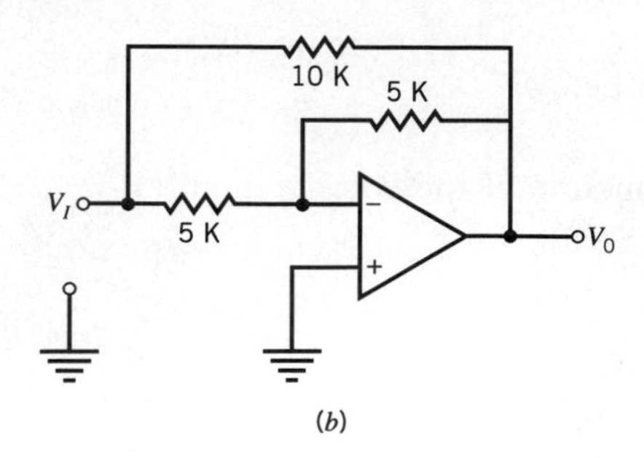

Diagram 5.9.b

CHAPTER 6

Biothermodynamics and Bioenergetics

Biothermodynamics is the study of energy, heat, and work in living systems. Biothermodynamics includes the materials involved and the processes used in the generation and transformation of energy, heat, and work. As noted previously in Table 3.1, biothermodynamics is a subdiscipline of both biophysics and biomedical engineering. Science, and biophysics in particular, pursue fundamental information with respect to cause and effect in living systems. Biomedical engineering (and in particular human factors engineering) then applies this information for the benefit of the individual person. Bioenergetics is the study of energy transformation in living systems and may include (in part) biothermodynamic processes, as well as biomechanics and physical biochemistry (chemical thermodynamics).

This chapter has been divided into three sections that represent those areas of biothermodynamics and bioenergetics, which are of specific interest to the human factors engineer. The first section addresses biothermal fundamentals and is a review of both thermoregulatory physiology and also human system bioenergetics. The second section specifically considers human operator heat transfer. This section includes both a review of heat transfer fundamentals and also the applications of heat transfer with respect to the human operator. The third section addresses human operator thermoregulation. The human operator is first considered in a passive state (in which there is no physical workload). The active human operator is then considered with respect to physical workload (both static work and dynamic work).

6.1 BIOTHERMAL FUNDAMENTALS

a. Thermoregulatory Physiology

The normal human body temperature (T_B) is between 36.1°C and 37.2°C as measured by oral thermometry. The mean T_B is actually about 36.5°C (oral) although a nominal value of 37°C is universally accepted. The rectal temperature of the human body is 0.6°C higher than the oral temperature and represents the core body temperature (T_c) compared to the mean body temperature (T_B).

The human body temperature is close to the lethal temperature (45°C), but this high temperature promotes cellular enzymatic activity and therefore enables the biochemical reactions of the body to take place more efficiently. Human beings regulate their body temperatures and this homeothermy can be maintained by a nude person over an external ambient temperature range of 13°C to 65°C. This ability of individuals to regulate their core body temperature is based upon a balance between heat production mechanisms and heat loss mechanisms as shown in Figure 6.1.

Heat loss from the human body generally results because the ambient (*external*) temperature (T_a) is usually lower than the core body (*internal*) temperature (T_c). Under normal circumstances, heat is continuously lost from the body as it follows the thermal gradient from the highest to the lowest heat energy. In a resting nude individual, approximately 60% of body heat is lost by radiation. This heat loss occurs in the form of electromagnetic waves in the infrared wavelength spectrum.

Other mechanisms by which heat loss from the body may occur are conduction, convection, and evaporation (Table 6.1). Conduction is heat loss from the body to cooler materials (solids) by means of molecular motion. Convection is heat carried away from the surface of the body by the movement of the surrounding fluid (water or air) by means of bulk hydrodynamic motion. It should be noted that the body may also gain heat by any or all of the above mechanisms (radiation, conduction, or convection). This occurs when the thermal gradient is reversed and the body is the cooler object (at a lower temperature than the surrounding environment).

Evaporation is the final mechanism by which heat loss from a person can occur. For each gram of water that evaporates, 0.58 kcal of heat energy is lost from the body. Recall that human nutritional Calories (indicated with a capital C) are equivalent to the conventional kilocalories of heat energy. Insensible water loss may occur from both the skin surface and the respiratory tract. When the ambient (external environment) is at a higher temperature than the body core temperature, evaporation is the only method by which the individual may lose heat. However, this requires a low relative humidity.

Regulation of the deep body (core) temperature is accomplished by a thermostatic controlled center located in the hypothalamus (at the base of the brain). There is a set point for temperature control (T_{SP}) so that T_c is regulated by a negative feedback mechanism, which will be discussed in more detail in Section 6.3. The human being has both warm and cold receptors (located throughout the body) that monitor regional body temperatures and transmit this information to the temperature control center (the hypothalamus). At the skin surface, these warm and cold receptors monitor the ambient (environmental) temperature. Warm and cold receptors located in the hypothalamus directly monitor the core blood temperature.

Behavioral thermoregulation may occur when an individual perceives an uncomfortable temperature. In contrast to physiological thermoregulation, which is an example of automatic (involuntary) control, behavioral thermoregulation is under voluntary control and is apparent from the observed behavior of the individual.

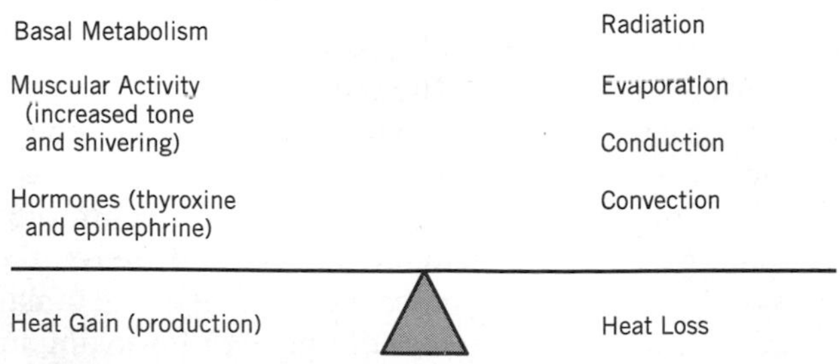

Figure 6.1 Regulation of core body temperature.

Table 6.1 Relative Significance of Heat Loss Terms (eq. [67])

External Heat Loss Mechanism	Fraction $\dot{Q}'$ (as a fraction of $\dot{M}_0$)
Conduction	0.10
Convection	0.20
Radiation	0.60
Evaporation	0.10

Physiological thermoregulation begins with stimulation of either the cold receptors or the warm receptors. When the cold receptors are stimulated, nerve impulses are conducted to the posterior half of the hypothalamus, which then initiate reflex (automatic) adaptive mechanisms in order to decrease heat loss from the body. Such adaptive mechanisms include the following: (1) increased sympathetic stimulation of the peripheral blood vessels resulting in vasoconstriction of arteries underneath the skin; (2) increased neural stimulation of skeletal muscle resulting in an increase in muscular tone (tension) and may also include shivering; and (3) the release of hormones into the circulating blood in order to increase a person's metabolic rate.

When warm receptors are stimulated, nerve impulses are transmitted to the anterior half of the hypothalamus, which results in reflex (automatic) adaptive mechanisms by which to increase heat loss from the body. Such adaptive mechanisms include: (1) a decrease in the metabolic rate (usually achieved by cessation of any external work); (2) an increase in colonergic sympathetic nerve stimulation, which results in peripheral vasodilation so that blood flow under the skin surface is increased; and (3) perspiration at the skin surface resulting in evaporative heat loss (at the expense of losing some body water and body salt).

Two types of altered physiological thermoregulation are fever and heat stroke. Fever is caused by toxic substances (pyrogens), which increase the set point of the hypothalamus (T_{SP}) above its normal value (37°C). Consequently the individual will perceive that they are cold and the automatic adaptive mechanisms will be initiated to increase body heat by decreasing heat loss from the individual. During this phase the individual will experience chills and cover themselves with blankets. When the fever breaks, the set point of the hypothalamus returns to its normal level (37°C). However, the body temperature is now significantly higher so that the individual perceives that they are hot and reflex adaptive mechanisms are initiated to lose heat from the body. During this phase of fever, the individual will perspire freely and completely uncover themselves.

Heat stroke results in very high body temperatures, which can reach a lethal level. Heat stroke is initiated by prolonged and strenuous activity in both a hot and humid climate. Radiative, conductive, and convective heat loss mechanisms are not effective due to the thermal gradient. Furthermore, evaporative heat loss is not effective due to the relatively high humidity. The individual will experience a very large water and electrolyte loss as profuse perspiration fails to evaporate but rather drips off the skin surface. Consequently, there is a decreased blood volume, which results in inadequate venous return and a low cardiac output. Blood pressure falls and the individual collapses. As the core body temperature (T_c) rises, there is a progressive widening between the hypothalamic T_{SP} and T_c. At some point thermoregulatory (negative feedback) fails and the human temperature regulation system goes open loop. When this occurs, core body temperature rises precipitously and can result in coma and convulsions. Death may be imminent. Early recognition of impending heat stroke by the individual themselves is the best treatment. The human operator should be continuously aware of the environmental risk factors (high physical workload in a hot and humid climate) and be prepared to initiate behavioral thermoregulatory mechanisms (e.g., move into the shade, cool the body externally with fanning and cold moist towels, and drink liquids). With the onset of heat stroke, a true medical emergency exists and medical attention should be obtained immediately.

b. Human System Bioenergetics

Bioenergetics deals with the generation and transformation of energy in the human body. In this section we will first examine the various forms of mechanical and thermal energy. We will then consider the first law of thermodynamics with respect to conservation of energy and also internal energy. Finally we shall develop the basic equation for metabolic energy transfer.

The student is already familiar with various forms of mechanical energy as discussed in Chapter 4. Recall that potential energy (PE) resulted when a mass was moved in a gravitational field or a spring was compressed/extended by a certain distance. In effect, PE may be thought of as the energy level as a consequence of position. Recall that kinetic energy (KE) was also defined and may be viewed as the energy resulting from relative motion. Newton's second law was applied in order to demonstrate that work is required to change the KE content of a given amount of mass.

In addition to these forms of mechanical energy, there is also thermal energy. Matter itself can possess relative energy as a function of its molecular state. In thermodynamics two measures of relative molecular activity are used. One measure of the relative molecular activity of matter is temperature. Another measure of the molecular state of matter is its composition. Molecular state energy (due to the macroscopic composition of matter) is the internal energy (U). Specific energy content can be symbolically noted as $\widehat{PE}$, $\widehat{KE}$, and $\hat{U}$, which, respectively, identify the potential, kinetic, and internal energy per unit mass.

Chapter 4 demonstrated that the energy content of a body could be changed by doing work on that body. It was also demonstrated that one form of energy could be changed to another, as with a person's body on a slide in which there is conversion of the potential energy (at the top of the slide) to kinetic energy (at the bottom of the slide). In thermodynamics it is important to define a system as that portion of all the matter that is under consideration. A closed system is defined as a system in which there is no material transfer across the boundaries. An open system will be one in which material does cross the boundaries. In addition to work, the energy content of a system may also be changed as a consequence of heat (Q). Q is defined as a transfer of energy in response to a temperature difference. Consequently, the term *heat transfer* is interchangeable with *heat*.

The first law of thermodynamics was historically regarded as a statement of the conservation of energy. However, it is also a definition of internal energy in terms of only the thermodynamic state of the system. In its simplest formulation (for a single phase system, consisting of one component, which has closed boundaries and is fixed in space) the first law may be stated as:

$$\Delta\hat{U} = \hat{Q} - \hat{W} \tag{1}$$

Alternatively, the differential form is:

$$\dot{U} = \dot{Q} - \dot{W} \tag{2}$$

in which:

$$\dot{U} = \frac{d}{dt}\left(m_{TOT}\hat{U}\right) \tag{3}$$

$$\dot{Q} = \frac{d}{dt}\left(m_{TOT}\hat{Q}\right) \tag{4}$$

$$\dot{W} = \frac{d}{dt}\left(m_{TOT}\hat{W}\right) \tag{5}$$

where:

m_{TOT} = the total amount of mass in the system

$\dot{Q}$ = the amount of energy transferred as heat *into* the system per unit mass of material in the system

$\dot{W}$ = amount of energy transferred *out* of the system (by virtue of it doing *external* work) per unit mass

The sign convention for equation (2) is demonstrated in Figure 6.2.a.

Note that the internal energy change from some process (as defined by equation (1)) is independent of the thermodynamic path followed by the system. The internal energy change will also be independent of the paths followed by the heat (Q) and work (W) processes which appear on the right-hand side of equation (1). In its differential form, the first law of thermodynamics states that the rate of internal energy change is a function of only the rates at which heat is added and external work is performed.

When the closed system is moved in a potential field, it is necessary to account for changes in the mechanical energy of the system in addition to the thermal energy. Equation (2) is then stated as:

$$\frac{d}{dt} = (PE + KE + U) = \dot{Q} - \dot{W} \tag{6}$$

where:

$$PE = m_{TOT}\widehat{PE} \tag{7}$$

$$KE = m_{TOT}\widehat{KE} \tag{8}$$

Equation (6) relates how the various energy forms of a closed system may be interconverted (left side of equation (6)). This equation also describes how the transfer of energy between the system and its surroundings can affect the system

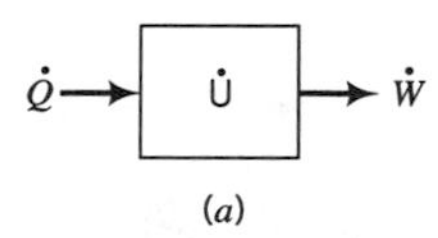

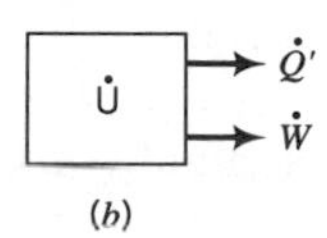

Figure 6.2.a Sign convention for equation (2). **b** $\dot{Q}'$ represents the energy loss from the system to its surrounding in the form of heat energy.

(right side of equation (6)). Using this basic formulation of the first law of thermodynamics, let us proceed to define enthalpy and specific heat. For many thermodynamic processes in closed systems, we may define a special case. Internal energy changes represent the only significant energy changes, and pressure-volume expansion work is the only significant work done by the system (when frictional work can be neglected). In this case, equation (1) may be expressed as:

$$d\hat{U} = \delta\hat{Q} - Pd\hat{V} \tag{9}$$

where:

$\delta\hat{Q}$ = a discrete (not differential) amount of heat transferred into the system per unit mass
$\hat{V}$ = volume of the system per unit mass (specific volume)

Since this simple situation occurs relatively frequently, it is convenient to introduce a new parameter, the enthalpy ($\hat{H}$), which may be defined as:

$$\hat{H} = \hat{U} + P\hat{V} \tag{10}$$

where $\hat{H}$ is a function only of the thermodynamic state.

Equation (10) in its differential form:

$$d\hat{H} = d\hat{U} + Pd\hat{V} + \hat{V}dP \tag{11}$$

Rearranging:

$$d\hat{U} = d\hat{H} - Pd\hat{V} - \hat{V}dP \tag{12}$$

Substituting equation (12) into equation (9):

$$d\hat{H} - Pd\hat{V} - \hat{V}dP = \delta\hat{Q} + Pd\hat{V} \tag{13}$$

Simplifying and rearranging:

$$d\hat{H} = \delta\hat{Q} + \hat{V}dP \tag{14}$$

When systems or processes are at constant pressure (isobaric), equation (14) reduces to:

$$d\hat{H} = \delta\hat{Q} \tag{15}$$

Therefore enthalpy reduces to only the heat generated or absorbed by the system. For systems with constant volume (isochoric), equation (9) reduces to:

$$d\hat{U} = \delta\hat{Q} \tag{16}$$

And substituting (15) into equation (16):

$$d\hat{H} = d\hat{U} \tag{17}$$

So for constant volume systems, the enthalpy change will equal the internal energy change.

The thermal dynamic parameter, heat capacity (C) may now be defined. C relates how the enthalpy (or internal energy) variation is related to temperature. The heat capacity per unit mass ($\hat{C}$) may be defined separately for a constant pressure process and a constant volume process. For a system at constant pressure (isobaric), the heat capacity is:

$$\hat{C}_P = \left(\frac{\partial \hat{H}}{\partial T}\right)_P \tag{18}$$

For a constant volume system, the heat capacity is:

$$\hat{C}_V = \left(\frac{\partial \hat{U}}{\partial T}\right)_V \tag{19}$$

Water (in its liquid phase at 4°C) has a heat capacity of 1.0 kcal/kg°C. The specific heat of liquids is defined as the heat capacity of the particular substance divided by the heat capacitance of water. It will be seen that the average heat capacity of the human body is approximately 0.86 kcal/kg°C. Consequently, the specific heat of the human body is 0.86 (somewhat less than water).

Considering the heat capacity of water, the enthalpy change (amount of heat required) that occurs when the temperature of a given mass of water is changed (under isobaric conditions) may be computed as follows:

Rearranging equation (18), and for an isobaric system:

$$d\hat{H} = \hat{C}_P \, dT \tag{20}$$

The total enthalpy of a system is:

$$H = m_{TOT}\hat{H} \tag{21}$$

Or in its differential form:

$$dH = m_{TOT} \, d\hat{H} \tag{22}$$

Then substituting equation (20) into equation (22):

$$dH = m_{TOT}\hat{C}_P \, dT \tag{23}$$

Since the heat capacity of water is relatively constant over a small range of temperature change:

$$\int_{H_1}^{H_2} dH = m_{TOT}\hat{C}_P \int_{T_1}^{T_2} dT \tag{24}$$

which results in:

$$\Delta H = m_{TOT}\hat{C} \, \Delta T \tag{25}$$

EXAMPLE 6.1

Doing an on-site evaluation, a night security guard's task is being evaluated in which a regulation-clothed guard (body mass [m_{TOT}] of 70 kg) is subjected to winter-night temperature. The guard will stand at a plant entrance gate to check identification of persons passing through the gate. By oral thermometry, it is then noted that the guard's body temperature drops by 1°C over a 1-hr period. Due to the continuous nature of the I.D. checking, the work intervals are 4 hr between meals. A candy-bar snack (containing sugar in the form of glucose, $C_6H_{12}O_6$) is being considered as a quick, efficient energy source:

a. Calculate the amount of body heat lost when the body temperature is lowered 1°C;
b. Find the amount of glucose (in grams) that must be ingested and completely converted to heat energy in order to offset the body heat lost in part (a).

SOLUTION 6.1.a

Recall the average heat capacity for the human body:

$$\hat{C}_P = 0.86 \frac{\text{kcal}}{\text{kg} \cdot °\text{C}}$$

Substituting into equation (25), the enthalpy change that results (or the amount of heat lost) when the temperature of the 70 kg body mass is lowered by 1°C.

$$\Delta H = (70 \text{ kg}) \left(0.86 \frac{\text{kcal}}{\text{kg} \cdot °\text{C}}\right) (-1°\text{C})$$

$$\Delta H = -60 \text{ kcal}$$

SOLUTION 6.1.b

To solve for the grams of glucose necessary to offset this heat loss (enthalpy change), we first must define a *heat of combustion* for glucose. This proceeds as follows:

A molar enthalpy ($\overline{H}$) is defined as the enthalpy per mole of a chemical compound. Recall that one mole (Mol) equals one Avogadro's number (6.02×10^{23}) of the specific chemical molecule (in this case glucose). One mole has a mass denoted as the *gram molecular weight* ($\overline{M}$), so that:

$$\overline{H} = (.001)\overline{M}\hat{H} \tag{i}$$

where: .001 kg/g

$$\overline{M} = \text{g/Mol}$$

Since this example involves glucose, it is necessary to calculate $\overline{M}_G$, which is also called a *gram mole*. The atomic weight of carbon (C) is 12, of hydrogen (H) is 1, and of oxygen (O) is 16. Recall that this translates to an equivalent mass of 12 g/Mol, 1 g/Mol, and 16 g/Mol, respectively. A gram per mole of glucose ($C_6H_{12}O_6$) is then calculated as:

$$\overline{M}_G = (6)\left(12 \frac{\text{g}}{\text{Mol}}\right) + (12)\left(1 \frac{\text{g}}{\text{Mol}}\right) + (6)\left(16 \frac{\text{g}}{\text{Mol}}\right)$$

For chemical compounds (e.g., glucose), the enthalpy change that occurs with the formation of one mole of that compound (at standard temperature and pressure) from its constituent elements (also at standard temperature and pressure) is defined as the standard enthalpy of formation ($\Delta\overline{H}^\circ_F$) for that compound. This represents the heat of formation since it defines the amount of energy (or heat) necessary to form the compound from its original elements: Most heats of formation are negative and indicate that heat is transferred to the environment (involved) during formation. Table 6.2 indicates some approximate values for the standard enthalpy of formation ($\Delta\overline{H}^\circ_F$) of various compounds. Note that standard temperature is 298°K, and standard pressure is 760 torr (mm Hg).

These standard enthalpies of formation (Table 6.2) are used to calculate the change in enthalpy that results during a chemical reaction when both the reactants and products are at standard temperature and pressure. For this example, one mole of glucose ($\overline{M}_G$ = 180 g/Mol) is oxidized to produce CO_2 and H_2O as end-products:

$$C_6H_{12}O_6 + 6O_2 \rightarrow 6CO_2 + 6H_2O \tag{ii}$$

Table 6.2 Standard Enthalpy of Formation for Various Compounds ($\Delta \overline{H}^\circ_F$) in kcals/Mol (Example 6.2)

Gas Phase		Liquid Phase	
H_2O	−57.8	C_6H_6	11.7
SO_2	−71.0	CH_3OH	−57.0
CO_2	−94.0	H_2O	−68.3
CO	−26.4	**Solid Phase**	
CH_4	−18.0		
O_2	0.0	$Ca(OH)_2$	−235.5
C_2H_6	−20.2	$C_6H_{12}O_6$	−301.0
		Al_2O_3	−399.0

The standard heat of reaction is the standard enthalpy change and is the heat required at standard temperature and pressure. This standard enthalpy change of reaction ($\Delta \overline{H}^\circ_R$) is calculated as:

$$\Delta \overline{H}^\circ_R\,(\text{reaction}) = \Delta \overline{H}^\circ_R\,(\text{products}) - \Delta \overline{H}^\circ_R\,(\text{reactants}) \qquad \text{(iii)}$$

From Table 4.2, and equations (ii) and (iii):

$$\Delta \overline{H}^\circ_R\,(\text{reaction}) = (6)(-94.0) + 6(-68.3) - [(-301) + (6)(0)]$$

$$\Delta \overline{H}^\circ_R = -974 + 301$$

$$\Delta \overline{H}^\circ_R = -673\ \text{kcal/Mol of glucose oxidized}$$

In other words, when one gram mole (180 g) of glucose ($C_6H_{12}O_6$) is completely oxidized, 673 kcal of energy will be evolved as heat, when this occurs at standard temperature and pressure. Since this is a special category of chemical combustion with oxygen the standard enthalpy of reaction is termed the *heat of combustion*. This represents the energy (or heat) per mole of glucose only when there is *complete* oxidation of the glucose to carbon dioxide and water.

Finally, we calculate the amount of glucose (m_G) that must be completely oxidized to generate the amount of heat energy (ΔH) lost to the environment [part (a)].

This is done by defining the necessary amount of glucose (m_G in grams) as a mole fraction and equating that to the required enthalpy ratio (also with units of mole fraction):

$$\frac{m_G}{\overline{M}_G} = \frac{\Delta H}{\Delta \overline{H}^\circ_R} \qquad \text{(iv)}$$

Rearranging and substituting in the previously calculated values:

$$m_G = \frac{(-60)(180)}{(-673)} = 16\ \text{g}$$

The human factors engineer is now in a position to formulate some of the basic mathematical relationships regarding energy and energy transfer within the human operator. Some simplifying approximations are necessary in order to do this, and they may be summarized as follows. First, the human body is considered to be a closed system. This requires not only that the mass of the body is constant but that there is no material crossing the boundary of the system. Such an approximation neglects gas exchange (oxygen and carbon dioxide) between the respiratory system and the environment. It can be shown that any additional energy necessary for heating or cooling the respiratory gases is minimal. Another example would be energy losses associated with evaporated perspiration from the body surface, which certainly represents mass transfer across the system boundary. It will be shown subsequently that such energy loss is accounted for by a heat transfer parameter. Second, the time interval over which the human operator is evaluated is limited to the relatively short period between food intake. This approximation is a consequence of the closed system approximation since food represents mass that would cross the system boundary. Food contains caloric energy (which can be calculated by enthalpy considerations per Example 6.1) and represents an additional energy term.

Accepting these simplifying approximations, the energy balance for the human operator (closed system) can be defined directly from equation (6) as follows:

$$\frac{dU}{dt} + \frac{d(KE)}{dt} + \frac{d(PE)}{dt} = \dot{Q} - \dot{W} \tag{26}$$

Recall that $\dot{Q}$ is the heat energy transferred into the system and that $\dot{W}$ is the external work (energy transferred out of the system).

The third simplifying approximation is that the kinetic energy rate change and the potential energy rate change are significantly less than the other terms in equation (20). With respect to biodynamic mechanics (Chapter 4), the human body may undergo significant changes in both kinetic energy and potential energy. These changes occur in particular situations and circumstances, but are not typical of those cases in which the human factors engineer will desire a thermodynamic (and bioenergetic) energy balance for the human body. In nearly every situation for which an energy balance is of interest, the kinetic energy rate change and the potential energy rate change will be significantly less than the other terms in equation (6).

With this third simplifying approximation, equation (6) reduces to:

$$\frac{dU}{dt} = \dot{Q} - \dot{W} \tag{27}$$

With these simplifying approximations stated, we may proceed to develop the general equation for the human operator. Recall that the internal energy of a system may be divided into two parts. There is the internal energy associated with temperature (relative molecular activity) and the internal energy attributed to composition (the enthalpy of chemical bond formation). It is apparent that since certain chemical compounds have higher enthalpies of formation, they will have relatively higher internal energies (at a specific temperature). Example 6.1 demonstrated that a complicated molecule such as glucose will have a significantly higher quantity of internal energy (residing in the chemical bonds) than would an equivalent amount of carbon dioxide.

The above may be stated mathematically as:

$$U = U_T + U_C \tag{28}$$

And in differential form:

$$\frac{dU}{dt} = \frac{dU_T}{dt} + \frac{dU_C}{dt} \tag{29}$$

Let us define the temperature-dependent internal energy change (dU_T) of the system, where T is the average temperature of the system (which can be the mean body temperature). When the composition of the system is static for a given temperature change, then the differential form of equation (28) reduces to:

$$dU = dU_T \tag{30}$$

For an isochoric system, multiplying both sides of equation (17) by m_{TOT} and transposing:

$$m_{TOT}\, d\hat{U} = m_{TOT}\, d\hat{H} \tag{31}$$

which reduces to:

$$dU = dH \tag{32}$$

Substituting equation (23) and equation (30) into equation (32) results in:

$$dU_T = m_{TOT}\hat{C}_P\, dT \tag{33}$$

Since $\hat{C}_P$ is constant over small temperature changes:

$$\frac{dU_T}{dt} = m_{TOT}\hat{C}_P \frac{dT}{dt} \tag{34}$$

which may be restated as:

$$\alpha\dot{T} = \frac{dU_T}{dt} = (m_{TOT}\hat{C}_P)\frac{dT}{dt} \tag{35}$$

where:

$$\alpha = m_{TOT}\hat{C}_P \tag{36}$$

Recall that the heat capacity of the human body (closed-system) is 0.86 kcal/kg°C.

Let us define the composition-dependent internal energy as:

$$dU_C = dM \tag{37}$$

where M represents the conversion of stored chemical energy into other energy forms. Subsequently, this energy will be lost across the boundaries of the system as either heat or work.

Rewriting equation (37) in its rate form results in:

$$\dot{M} = \frac{dM}{dt} = \frac{dU_C}{dt} \tag{38}$$

where $\dot{M}$ represents the metabolic energy conversion rate (more frequently referred to as the metabolism). It should be noted that $\dot{M}$ is a symbolic notation only. This metabolism term provides no specific information regarding how the stored chemical energy that was converted has been used within the system. For example, energy is required for cardiac pumping, muscular contraction, active transport, and other processes. $\dot{M}$ is only a symbolic notation that describes the metabolic energy conversion rate with respect to the overall system exchanges. Mathematically, this may be formulated as follows.

Equating equation (27) with equation (29):

$$\frac{dU_T}{dt} + \frac{dU_C}{dt} = \dot{Q} - \dot{W} \tag{39}$$

Substituting into equation (39) from equations (35) and (38) results in:

$$\alpha\dot{T} + \dot{M} = \dot{Q} - \dot{W} \tag{40}$$

The human operator as a living closed thermodynamic system will generally lose heat to the environment (rather than gain heat). As we shall see, the human operator is generally at a temperature higher than the ambient environmental temperature so that the thermal gradient favors heat transfer out of the system. Consequently, it is convenient to define a collective heat loss term as:

$$\dot{Q}' = -\dot{Q} \tag{41}$$

Following our thermodynamic sign convention, $\dot{Q}'$ will represent a collective energy loss from the system to its surrounding in the form of heat energy (see Figure 6.2.b).

Substituting equation (41) into equation (40):

$$\alpha\dot{T} + \dot{M} = -\dot{Q}' - \dot{W} \tag{42}$$

Inspection of equation (42) and Figure 6.2.b indicates that for the simplifying approximations used, the human operator is a living system that appears to be continuously decreasing its internal energy state. As we shall see in the next section, the right side of equation (42) is almost always negative, so it is only through the ingestion of higher-energy foods that the energy state of the human operator can be sustained.

A particularly relevant special case of equation (42) is that of the resting human ($\dot{W} = 0$) at a constant temperature ($\dot{T} = 0$). Under such conditions, equation (42) reduces to:

$$\dot{M}_0 = -\dot{Q}_0' \tag{43}$$

where $\dot{M}_0$ is defined as the basal metabolism and has a value of approximately -40 kcal/m^2/hr. $\dot{M}_0$ may be obtained by using calorimetry in which the thermodynamic term $\dot{Q}_0'$ is measured directly in various types of closed thermodynamic calorimeter systems. The interested student is referred to the appropriate literature in this field. An alternative approach to the measurement of basal metabolism is by indirect calorimetry in which the oxygen consumption rate of the human operator is sensed and (after certain thermodynamic formulations are applied) the basal metabolism is calculated. Indirect calorimetry involves significantly less complicated and expensive equipment so that it is generally the method of choice for the human factors engineer (see Example 6.2).

For many of the practical situations of interest to the human factors engineer the metabolic rate will be higher than the basal metabolic rate. This may be expressed mathematically as:

$$\dot{M} = \dot{M}_0 + \Delta\dot{M} \tag{44}$$

Substituting equation (44) into equation (42):

$$\alpha\dot{T} + \dot{M}_0 + \Delta\dot{M} = -\dot{Q}' - \dot{W} \tag{45}$$

where ΔM is an extra metabolic energy evolved in addition to the basal metabolic energy (M_0).

EXAMPLE 6.2

At the HFE laboratory, a warehouse worker and a work task are being evaluated with respect to worker clothing and warehouse ambient temperature.

a. For part (a), the worker is appropriately clothed, seated at rest, and breathing into a half-face mask connected by a flow tube to a pulmonary gas analyzer. The lab is at ambient (warehouse) temperature and the worker has consumed a "maintenance" daily balanced ("mixed") diet (composed of 250 g of carbohydrate, 80 g of fat, and 80 g of protein).

   Over an average 1-min period, the following average pulmonary gas values are measured:

   Liters O_2 (inhaled air) = 1.040 L
   Liters O_2 (exhaled air) = 0.788 L
   Liters CO_2 (inhaled air) = 0.002 L
   Liters CO_2 (exhaled air) = 0.216 L

   For part (a), calculate the basal metabolic rate ($\dot{M}_0$) for this individual.

b. For part (b), the worker is now walking on a treadmill (inclined at the angle of a loading dock ramp) and pushing a dolly (also on the treadmill) that holds weighted boxes. The worker is connected to the same pulmonary gas analyzer as in part (a). At some point a steady state is reached (where body temperature, rate, and depth of respirations are all constant).

   Over an average 1-min (steady-state) period, the following average (steady-state) pulmonary gas values are measured:

   Liters O_2 (inhaled air) = 2.496 L
   Liters O_2 (exhaled air) = 1.884 L
   Liters CO_2 (inhaled air) = .005 L
   Liters CO_2 (exhaled air) = .545 L

   Calculate the extra metabolic rate ($\Delta\dot{M}$) for this individual.

SOLUTION 6.2.a

Indirect calorimetry is a method by which the metabolic conversion rates of a human operator are calculated from the rate of oxygen consumption. The basis of this method is to define an energy (kilocalorie) equivalent per liter of oxygen consumed (i.e., the energy equivalent of oxygen).

To illustrate the principle of indirect calorimetry, let us calculate the equivalent (kilocalorie) oxygen content for the oxidation (combustion) of one mole of glucose.

From equation 6.1(b)(ii), recall that for complete combustion of glucose:

$$C_6H_{12}O_6 + 6O_2 \rightarrow 6CO_2 + 6H_2O \qquad \text{(i)}$$

and the heat of combustion (the enthalpy change of reaction) was calculated as:

$$\Delta \overline{H}^{\circ}_R = -673 \text{ kcal/Mol}$$

Equation (i) states that 6 Mol of oxygen (gas) are necessary to completely combust one mole (180 g) of glucose. In that process, 6 Mol of carbon dioxide (gas) and 6 Mol of water (liquid) are produced.

Recall that 1 Mol of a gas has a molar volume ($\overline{V}$) of 22.4 L at a temperature of 273°K and a pressure of 760 mm Hg (torr). This is from Charles' law for an ideal gas, where $\overline{V}$ has units of liters per mole.

For the combustion of glucose, the following parameters are calculated.

(i) Liters of oxygen (ΔO_2) consumed per gram of fuel (glucose) consumed (Δm):

$$\frac{\Delta O_2}{\Delta m} = \frac{n\overline{V}_{O_2}}{\overline{M}_G} \tag{ii}$$

where n is the stoichiometric number for oxygen [O_2] in equation (i). Substituting known values:

$$\frac{\Delta O_2}{\Delta m} = \frac{\left[(6)\left(22.4\,\frac{\text{L}}{\text{Mol}}\right)\right]}{\left[\left(180\,\frac{\text{g}}{\text{Mol}}\right)\right]}$$

$$\frac{\Delta O_2}{\Delta m} = 0.75\ \text{L/g}$$

(ii) Liters of carbon dioxide produced (ΔCO_2) per gram of fuel (glucose) consumed (Δm):

$$\frac{\Delta CO_2}{\Delta m} = \frac{k\overline{V}_{CO_2}}{\overline{M}_G} \tag{iii}$$

where k is the stoichiometric number for carbon dioxide (CO_2) in equation (i). Substituting known values:

$$\frac{\Delta CO_2}{\Delta m} = \frac{\left[(6)\left(22.4\,\frac{\text{L}}{\text{Mol}}\right)\right]}{\left[\left(180\,\frac{\text{g}}{\text{Mol}}\right)\right]}$$

$$\frac{\Delta CO_2}{\Delta m} = 0.75\ \text{L/g}$$

Note that $\Delta O_2/\Delta m$ and $\Delta CO_2/\Delta m$ are defined as absolute values for our purposes.

(iii) Define the respiratory quotient (RQ) as the ratio of carbon dioxide produced (in liters) to the oxygen consumed (in liters):

$$RQ = \frac{k(CO_2)}{n(O_2)} \tag{iv}$$

where:

$k(CO_2)$ = number of moles of carbon dioxide produced (dimensionless)
$n(O_2)$ = number of moles of oxygen consumed (dimensionless)

Substituting results in:

$$RQ = 1.0$$

(iv) Kilocalories of energy (as heat) evolved per gram of fuel (glucose) consumed ($\Delta E/\Delta m$) is:

$$\frac{\Delta E}{\Delta m} = \frac{\Delta\overline{H}^{\circ}_R}{\Delta\overline{M}_G} \tag{v}$$

Substituting known values:

$$\frac{\Delta E}{\Delta m} = \frac{-673 \frac{\text{kcal}}{\text{Mol}}}{-180 \frac{\text{g}}{\text{Mol}}}$$

$$\frac{\Delta E}{\Delta m} = 3.75 \text{ kcal/g}$$

(v) The energy equivalent (in kcals) of 1 L of oxygen (E_{O_2}) is:

$$E_{O_2} = \frac{\frac{\Delta E}{\Delta m}}{\frac{\Delta O_2}{\Delta m}} \tag{vi}$$

Substituting calculated values:

$$E_{O_2} = \frac{3.75 \frac{\text{kcal}}{\text{g}}}{0.75 \frac{L_{O_2}}{\text{g}}}$$

$$E_{O_2} = 5.00 \frac{\text{kcal}}{L_{O_2}}$$

At this point, the energy equivalent of oxygen (E_{O_2}) has been calculated for only one carbohydrate, a simple sugar (glucose). However, equations (ii), (iii), (iv), (v), and (vi) can be applied to any food molecule. When this is done by food molecule category (carbohydrate, fat, and protein), and the individual food molecule values are averaged within each category, the equivalent energy relationships are obtained as depicted in Table 6.3:

Table 6.3 Equivalent Energy Relationships

Parameter	Carbohydrate	Fat	Protein
$\frac{\Delta O_2}{\Delta m}$ [a]	0.75	2.02	0.97
$\frac{\Delta CO_2}{\Delta m}$ [b]	0.75	1.42	0.78
RQ [c]	1.00	0.70	0.80
$\frac{\Delta E}{\Delta m}$ [d]	4.10	9.30	4.10
E_{O_2} [e]	5.45	4.60	4.25

[a] Liters of oxygen consumed per gram of foodstuff
[b] Liters of CO_2 produced per gram of foodstuff
[c] Respiratory quotient ($\Delta CO_2/\Delta O_2$) [dimensionless]
[d] kcals of energy (as heat) produced per gram of foodstuff
[e] kcals of energy (as heat) produced per liter of oxygen consumed

Utilizing the data in Table 6.3, and the data given in Example 6.2(a), the HFE may now proceed to do some preliminary calculations for this resting human operator.

(i) Determine the actual measured respiratory quotient ($\overline{RQ}$) from the measured pulmonary gas data:

$$\overline{RQ} = \frac{|\Delta V_{CO_2}/\Delta t|}{|\Delta V_{O_2}/\Delta t|} \tag{vii}$$

For an average 1-min period (Δt), this reduces to:

$$\overline{RQ} = \frac{|\Delta V_{CO_2}|}{|\Delta V_{O_2}|} \tag{viii}$$

Substituting:

$$|\Delta V_{CO_2}| = |.002 - .216| = .214 \text{ L}$$

$$|\Delta V_{O_2}| = |1.040 - 0.788| = .252 \text{ L}$$

Substituting into equation (viii):

$$\overline{RQ} = \frac{.214}{.252} = 0.849$$

This value is consistent with the "mixed" diet of this worker. Note from Table 6.3, if this individual were "carbohydrate loading," then $\overline{RQ} \rightarrow 1.00$.

(ii) Determine the calculated respiratory coefficient ($\overline{RQ}'$) from the quantity and variety of foodstuffs ingested:
For an average 1-min period, use a variation of equation (viii):

$$\overline{RQ}' = \frac{|\Delta V_{CO_2}|_T}{|\Delta V_{O_2}|_T} \tag{ix}$$

Define the total body oxygen consumption as:

$$|V_{O_2}|_T = |\Delta O_2|_C + |\Delta O_2|_F + |\Delta O_2|_P \tag{x}$$

Alternatively, equation (x) may be expressed as:

$$|\Delta V_{O_2}|_T = m_C \left|\frac{\Delta O_2}{\Delta m}\right|_C + m_F \left|\frac{\Delta O_2}{\Delta m}\right|_F + m_P \left|\frac{\Delta O_2}{\Delta m}\right|_P \tag{xi}$$

From the nutritional data provided (m_C = 250 g, m_F = 80 g, and m_P = 80 g) and the data from Table 6.3:

$$|\Delta V_{O_2}|_T = (250)(0.75) + (80)(2.02) + (80)(0.97)$$

$$|\Delta V_{O_2}|_T = 187 + 162 + 78$$

So that:

$$|\Delta V_{O_2}|_T = 427 \text{ L}$$

Define the total body carbon dioxide produced as:

$$|\Delta V_{CO_2}|_T = |\Delta CO_2|_C + |\Delta CO_2|_F + |\Delta CO_2|_P \qquad \text{(xii)}$$

Alternatively, equation (xii) may be expressed as:

$$|V_{CO_2}|_T = RQ_C|\Delta O_2|_C + RQ_F|\Delta O_2|_F + RQ_P|\Delta O_2|_P \qquad \text{(xiii)}$$

Substituting the values for $|\Delta O_2|_C$, $|\Delta O_2|_F$, and $|\Delta O_2|_P$ obtained from equation (xi) and the RQ values from Table 6.3:

$$|\Delta V_{CO_2}|_T = (1.00)(187) + (0.70)(162) + (0.80)(78)$$

$$|\Delta V_{CO_2}|_T = 187 + 113 + 62$$

$$|\Delta V_{CO_2}|_T = 362 \text{ L}$$

Substituting the calculated results into equation (ix):

$$\overline{RQ}' = \frac{362L}{427L} = 0.848$$

Recall that actual (measured) $\overline{RQ}$ was 0.849. The HFE has now performed a "validity check" that this particular individual has indeed ingested a balanced ("mixed") diet of the *proportions* of carbohydrate, fat, and protein specified in the example. However, $\overline{RQ}'$ gives no direct information on the actual amounts of these three foodstuffs. This may be demonstrated mathematically by substituting equation (xiii) and equation (x) into equation (ix):

$$\overline{RQ}' = \frac{RQ_C|\Delta O_2|_C + RQ_F|\Delta O_2|_F + RQ_P|\Delta O_2|_P}{|\Delta O_2|_C + |\Delta O_2|_F + |\Delta O_2|_P} \qquad \text{(xiv)}$$

(iii) Calculate the total Calories in the "maintenance" diet of this human operator: Note that human nutritional *Calories* are indicated with a capital C and are equivalent to the conventional kilocalories of heat energy:

$$1 \text{ Calorie} \equiv 1 \text{ kilocalorie} \qquad \text{(xv)}$$

Define the total body Calories as:

$$\Delta E_T = m_C\left(\frac{\Delta E}{\Delta m}\right)_C + m_F\left(\frac{\Delta E}{\Delta m}\right)_F + m_P\left(\frac{\Delta E}{\Delta m}\right)_P \qquad \text{(xvi)}$$

From the nutritional data provided (m_C = 250 g, m_F = 80 g, and m_P = 80 g) and $\Delta E/\Delta m$ data from Table 6.3:

$$\Delta E_T = (250)(4.10) + (80)(9.30) + (80)(4.10)$$

$$\Delta E_T = 1025 + 744 + 328$$

$$\Delta E_T = 2097 \text{ kcals} = 2097 \text{ Cal}$$

A "maintenance" diet is 2000 to 2200 Calories daily and will maintain body weight for an "average" person performing "average" daily activities. An adult performing strenuous daily physical activity (such as this warehouse worker) will require a higher caloric intake. Perhaps this particular human operator is trying to lose a few pounds!

(iv) Determine the actual energy equivalent of oxygen for this industrial worker, as follows.

Define the total body energy equivalent of oxygen as:

$$\overline{E}_{O_2} = \frac{\Delta E_T}{|\Delta V_{O_2}|_T} = 4.905\,\frac{\text{kcal}}{\text{L}} \tag{xvii}$$

Substituting the previously calculated values for ΔE_T and $|\Delta VO_2|_T$ results in:

$$\overline{E}_{O_2} = \frac{2097\text{ kcal}}{427.5\text{ L}} = 4.905\,\frac{\text{kcal}}{\text{L}}$$

Interestingly enough, $\overline{E}_{O_2}$ (much like $\overline{RQ}\,'$) is based upon the *proportions* of carbohydrate, fat, and protein, but not the actual amounts. This may be demonstrated mathematically by substituting equations (xvi) and (xi) into equation (xvii):

$$\overline{E}_{O_2} = \frac{m_C\left|\frac{\Delta E}{\Delta m}\right|_C + m_F\left|\frac{\Delta E}{\Delta m}\right|_F + m_P\left|\frac{\Delta E}{\Delta m}\right|_P}{m_C\left|\frac{\Delta O_2}{\Delta m}\right|_C + m_F\left|\frac{\Delta O_2}{\Delta m}\right|_F + m_P\left|\frac{\Delta O_2}{\Delta m}\right|_P} \tag{xviii}$$

This completes the preliminary calculations, so that the HFE now proceeds to solve for $\dot{M}_O$:

For the resting (basal) metabolic state, using the method of indirect calorimetry:

$$\dot{M}_0 = \left|\frac{\Delta V_{O_2}}{\Delta t}\right| (\overline{E}_{O_2}) \tag{xix}$$

where $|\Delta V_{O_2}/\Delta t|$ is the pulmonary oxygen consumption rate measured (in this case) by pulmonary gas analysis.

Substituting measured $|\Delta V_{O_2}/\Delta t|$ and calculated $\overline{E}_{O_2}$:

$$\dot{M}_0 = \left(\frac{.252\text{ L}}{1\text{ min}}\right)\left(4.905\,\frac{\text{kcal}}{\text{L}}\right)$$

$$\dot{M}_0 = 1.236\,\frac{\text{kcal}}{\text{min}}$$

from which we obtain the hourly basal metabolic rate:

$$\dot{M}_0 = \left(1.236\,\frac{\text{kcal}}{\text{min}}\right)\left(60\,\frac{\text{min}}{\text{hr}}\right)$$

$$\dot{M}_0 = 74.2\,\frac{\text{kcal}}{\text{hr}}$$

SOLUTION 6.2.b

First perform the preliminary calculations in a manner similar to that of part (a).

(i) The actual (measured) respiratory quotient ($\overline{RQ}$):
From the pulmonary gas data:

$$|\Delta V_{CO_2}| = |.005 - .545| = .540\text{ L}$$

$$|\Delta V_{O_2}| = |2.496 - 1.884| = .612\text{ L}$$

Substituting into equation 6.2(a)(viii):

$$\overline{RQ} = \frac{.540\ \text{L}}{.612\ \text{L}} = 0.882$$

Compared to the resting basal value ($\overline{RQ} = 0.849$), it appears that *preferentially* more carbohydrate may be burned (where $\overline{RQ} \rightarrow 1.0$) for this acute work effort. Recall that "sugar" (carbohydrate) is a well-known "quick-energy" source.

It may also be that this modestly increased $\overline{RQ}$ is within the "experimental error" of human data acquisitions; that is, this worker drank a can of soda pop prior to the part (b) evaluation.

(ii) The calculated respiratory quotient ($\overline{RQ}'$) will not change with this work task. Recall that $\overline{RQ}'$ is calculated from tabular data and the quantity and variety of food stuffs. Therefore,

$$\overline{RQ}' = 0.848$$

(iii) The total Calories in the maintenance diet for this human operator will also not change with this work effort for the same reasons:

$$\Delta E_T = 2097\ \text{kcals}$$

(iv) The actual energy equivalent of oxygen for this individual (who is now in a working steady state) is a calculated value based upon results obtained in part (a) preceding. Therefore, $\overline{E}_{O_2} = 4.905$ kcal/L.

For the working (extra-energy) metabolic state, when the human operator is in thermal equilibrium ($\dot{T} = 0$), the governing equation of indirect calorimetry is:

$$\dot{M}_0 + \Delta\dot{M} = \left|\frac{\Delta V_{O_2}}{\Delta t}\right| (\overline{E}_{O_2}) \tag{xx}$$

where $|\Delta V_{O_2}/\Delta t|$ has been previously defined from the pulmonary gas data. Substituting the measured $|\Delta V_{O_2}/\Delta t|$ for this working state, and the previously calculated $\overline{E}_{O_2}$ (for the resting state):

$$\dot{M}_0 + \Delta\dot{M} = \left(.612\,\frac{\text{L}}{\text{min}}\right)\left(4.905\,\frac{\text{kcal}}{\text{L}}\right)$$

$$\dot{M}_0 + \Delta\dot{M} = 3.00\,\frac{\text{kcal}}{\text{min}}$$

from which the hourly rate is:

$$\dot{M}_0 + \Delta\dot{M} = \left(3.00\,\frac{\text{kcal}}{\text{min}}\right)\left(60\,\frac{\text{min}}{\text{hr}}\right)$$

$$\dot{M}_0 + \Delta\dot{M} = 180\,\frac{\text{kcal}}{\text{hr}}$$

Recalling that $\dot{M}_0 = 74.2$ kcals/hr (from part [a]), then:

$$\Delta\dot{M} = 106\,\frac{\text{kcal}}{\text{hr}}$$

6.2 HUMAN OPERATOR HEAT TRANSFER

a. Heat Transfer Fundamentals

In Section 6.1.a the student was introduced to the various mechanisms by which heat loss from the body can occur. This section will develop the fundamental mathematical formulations for these mechanisms. Specifically, this section will develop the physical basis for conductive heat transfer, convective heat transfer, evaporative heat transfer, and radiative heat transfer.

Conductive heat transfer, or more simply conduction, represents the transport of energy as a function of molecular collisions within a material that is subjected to a temperature gradient. Conduction has a strictly molecular basis. Consequently, this transport process can be completely described when the properties of the material are known and the temperature field is specified.

Convective heat transfer, or simply convection, represents the transport of energy as a function of the bulk hydrodynamic motion of a fluid medium (which can be either a liquid or a gas). This bulk or hydrodynamic motion is responsible for the heat flow. As we shall see, there can be forced convection (fluid motion secondary to a superimposed pressure gradient) and also free convection (motion caused by density differences within the fluid).

Evaporative heat transfer, or simply evaporation, represents energy transfer as a function of the evaporation of water vapor from the skin surface. This energy transport is a heat loss term that is proportional to the amount of water that evaporates. Evaporative heat transfer differs mainly from the conductive, convective, and radiative transfer in that it is the only mechanism by which heat loss can occur when the ambient (environmental) temperature is greater than the core body temperature.

Radiative heat transfer, or simply radiation, represents energy transport as a function of electromagnetic radiation (in the infrared spectrum). This heat transfer occurs between the body and a surrounding wall so that it requires no intermediate material phase to be located between the radiating (or exchanging) surfaces.

The fundamental parameter of heat transfer equations is that of heat flux (q_x). Heat flux is a heat flow per unit area, which may be defined as follows:

$$q_x = \frac{\dot{Q}}{A} \tag{46}$$

where:

q_x = heat flux, kcal/m^2 min in the x direction
$\dot{Q}$ = heat flow, kcal/min
A = normal area through which the heat is transferred, m^2

Heat flux will be the dependent variable in all four types of heat transfer (conduction, convection, evaporation, and radiation). We now proceed to address each of these heat transfer mechanisms individually.

i. Conductive Heat Transfer

The fundamental relationship for heat flow as a function of the temperature gradient in a material is represented by Fourier's law which can be expressed in its differential form for one-dimensional heat flow:

$$q_x = -k\left(\frac{dT}{dx}\right) \tag{47}$$

where:

q_x = heat flux, kcal/m^2 min in the x direction
k = thermal conductivity, kcal/m min °C
dT = differential of temperature, °C
dx = thickness differential, M

Recall that the negative sign is employed to indicate that the heat flux will be positive (in the x direction) when there is a negative temperature gradient (in the same x direction). This is a consequence of the second law of thermodynamics in that energy flows downhill.

With respect to steady-state conductive heat transfer, the temperature variation is not a variation of time but only of displacement (x). In this situation, and in the absence of any internal heat generation, heat flow will be constant. This requires that the heat input through a given section of material will equal the output heat for rectangular geometry. The following condition will then result:

$$q_x = \text{constant} \tag{48}$$

For steady-state conductive heat transfer situations, when the thermal conductivity coefficient is approximated as constant, the following relationships apply to rectangular geometry.

For $T = T_0$ at $x = x_0$:

$$q_0 = -k\left(\frac{dT}{dx}\right)_{x=x_0} \tag{49}$$

$$T - T_0 = \frac{-q_0}{k}(x - x_0) \tag{50}$$

Equation (50) can be rearranged to solve for the heat flow for one boundary at $T_0(x_0)$ when the other boundary is at $T_L(x_L)$:

$$q_0 = -k(T_L - T_0)/(x_L - x_0) \tag{51}$$

When there is a constant value of thermal conductivity (k) and the thickness between the two boundaries is Δx, equation (51) may be rearranged to express the steady-state conductive heat transfer rate ($\dot{Q}_{COND}$):

$$\dot{Q}_{COND} = q_0 = -\left(\frac{k}{\Delta x}\right)(T_L - T_0) \tag{52}$$

Equation (52) is an example of a rate equation that expresses energy transport through a material. Equation (52) is analogous to another rate equation which is Ohm's law for the conduction of electricity. In this analogy, thermal conductance would be the thermal conductivity per unit length (analogous to electrical conductance). The temperature difference would be the potential (analogous to the voltage difference of electricity). Finally, thermal heat flow would be the dependent variable (analogous to electrical current).

An important special case of thermal conductivity is that of conductive heat

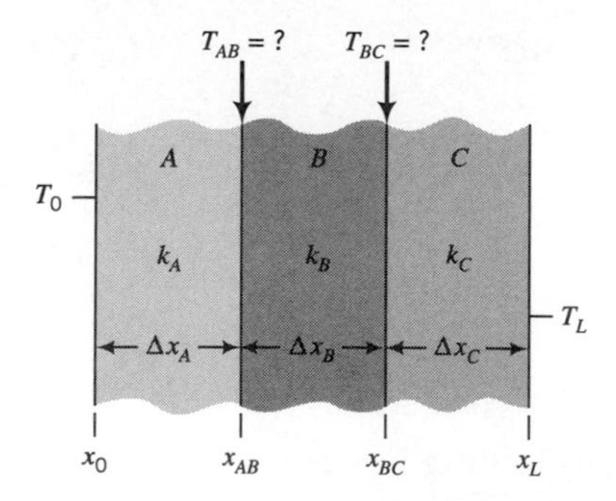

Figure 6.3 Conductive heat transfer through a conductive wall.

transfer through a composite wall material. Figure 6.3 demonstrates this heat transfer problem which will be important in subsequent heat transfer applications (Section 6.2.b). The composite wall system is illustrated in Figure 6.3 in which the temperatures at the inside boundary T_0 and the outside boundary T_L are known, but the temperature at the interface between the A and B layers (T_{AB}), of different materials, and the temperature at the interface between the B and C layers (T_{BC}), also of different materials, are not known. When the relative dimensions of the three walls (Δx_A, Δx_B, and Δx_C) are known and also the thermal conductivities (k_A, k_B, and k_C) of the layers are also known, the human factors engineer may compute the heat flow through the composite wall material. As a first step, equation (50) is applied to each layer individually:

$$T_{AB} - T_0 = \frac{-q_0}{k_A}(x_{AB} - x_0) \tag{53}$$

$$T_{BC} - T_{AB} = \frac{-q_0}{k_B}(x_{BC} - x_{AB}) \tag{54}$$

$$T_L - T_{BC} = \frac{-q_0}{k_C}(x_L - x_{BC}) \tag{55}$$

Multiplying by negative one and using the layer thickness (Ax) indicated in Figure 6.3:

$$T_0 - T_{AB} = q_0\left(\frac{\Delta x_A}{k_A}\right) \tag{56}$$

$$T_{BC} - T_{AB} = q_0\left(\frac{\Delta x_B}{k_B}\right) \tag{57}$$

$$T_{BC} - T_L = q_0\left(\frac{\Delta x_C}{k_C}\right) \tag{58}$$

By superposition theorem:

$$T_0 - T_L = (T_0 - T_{AB}) + (T_{AB} - T_{BC}) + (T_{BC} - T_L) \tag{59}$$

Substituting equations (56), (57), and (58) into equation (59), then rearranging and recalling that $\dot{Q}_{COND} = q_0$ (equation (52)):

$$\dot{Q}_{COND} = q_0 = \frac{T_0 - T_L}{\left(\frac{\Delta x_A}{k_A}\right) + \left(\frac{\Delta x_B}{k_B}\right) + \left(\frac{\Delta x_C}{k_C}\right)} \tag{60}$$

It will be noted that the denominator of equation (60) is the sum of three parameters each of which is the reciprocal of thermal conductance. By analogy to electrical current flow, each of these parameters would be thermal resistance and the total composite wall thermal resistance would summate (analogous to series resistance in an electrical circuit).

ii. Convective Heat Transfer

Recall that the mechanism of convective heat transfer is a function of the bulk hydrodynamic motion of a surrounding fluid (liquid or gas) near the surface from which heat transfer occurs. The fundamental differential equation that describes such convective heat transfer is extremely complex and not well suited for practical applications. Therefore, the human factors engineer should use heat transfer coefficients for convective heat loss problems. This approach recognizes that heat flux away from a surface into a fluid medium could theoretically be computed by Fourier's law:

$$q_x = \left(\frac{k}{\delta}\right)(T_s - T_a) \qquad (61)$$

where:

T_s = surface temperature, °C
T_a = ambient (surrounding fluid) temperature, °C
δ = the film thickness, M

The difficulty is that the film thickness (δ) is generally not known and is rather difficult to estimate for most practical situations. Consequently, an alternative expression of equation (61) is the following:

$$q_x = h(T_s - T_a) \qquad (62)$$

where h is the convective heat transfer coefficient. This parameter is distinctly easier to apply in most practical situations than would be the film thickness (δ). It is apparent that h will not only be a function of the intrinsic properties of the fluid but also of the fluid flow profile. This characteristic feature of the convective heat transfer coefficient will be addressed in Section 6.2.b.

iii. Radiative Heat Transfer

Radiative heat transfer is a phenomenon in which electromagnetic radiation from a material at a high temperature results in a net amount of thermal energy being transferred to the surrounding surfaces, which are at a lower temperature. This electromagnetic radiation results from excited molecules in a material at a high temperature returning to lower energy states. This radiant energy may be represented by photons, which are particles that transmit the thermal energy, but of themselves have neither mass nor charge.

Before proceeding to the radiative heat transfer equation, some basic concepts should be appreciated. When a certain amount of energy contacts a solid surface, it will be either reflected or absorbed by that surface. The absorptivity of a surface is the fraction of the impinging energy that is absorbed. Monochromatic absorptivity represents that fraction of the contacting energy for a given frequency that is absorbed.

A *black body* may be defined as a body that represents a perfect absorber for all frequencies. In effect, no incident energy at any frequency is reflected. With a *gray body,* on the other hand, the body exhibits a uniform absorptivity at all frequencies, but some amount of uniform reflection occurs at all frequencies. With respect to engineering applications, it will be seen that a gray body is very useful.

As noted earlier, a certain amount of thermal energy will be emitted from the surface of a solid material in the form of electromagnetic radiation. Consequently, we may define the emissivity (e) of a substance as that fraction of the energy actually emitted by the substance (at a given temperature) with respect to the energy that would be emitted by a black body (at the same temperature). It should be noted that the monochromatic emissivity (e_λ) would be the fraction of the emitted energy from the substance for a given frequency with respect to the fraction of emitted energy by a black body at the same frequency.

Figure 6.4 illustrates the energy emitted from heated bodies (black, gray, and real) as a function of wavelength. This figure also indicates the variation of emitted energy from heated black bodies as a function of temperature. When the areas under the black-body curves were experimentally or analytically evaluated, the law of Stefan Boltzmann has been shown to apply.

$$q_b^e = \sigma T^4 \tag{63}$$

where:

q_b^e = emitted energy flux from a black body (cal/min M^2)
σ = Stefan-Boltzmann constant (4.88×10^{-8} kcal/M^2 °K^4)
T = absolute temperature (°K)

A good approximation for nonblack bodies is:

$$q_e = e\sigma T^4 \tag{64}$$

where:

q^e = emitted energy flux from real (nonblack) body (cal/min M^2)
e = average emissivity for a real material (0.87 for skin at 37°C)

iv. Evaporative Heat Transfer

The energy loss due to the evaporation of water vapor from the skin may be evaluated as a heat loss term. This is accomplished as follows. In a manner analogous

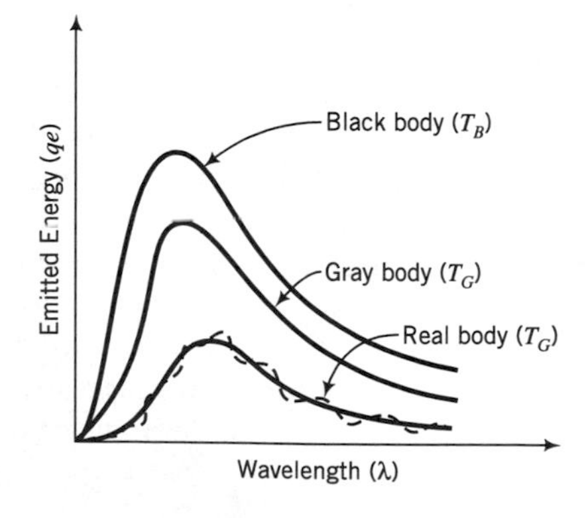

Figure 6.4 Energy emitted from heated bodies as a function of wavelength.

to convective heat transfer, we define a heat transfer coefficient (h) for this situation and replace the temperature driving force by a partial pressure driving force:

$$q_x = h(p_s - p_a) \tag{65}$$

where:

h = heat transfer coefficient for evaporative transfer, kcal/M^2 hr mm Hg
p_s = partial pressure of water at the temperature of the skin, mm Hg
p_a = partial pressure of water vapor in the ambient air, mm Hg

b. Heat Transfer Applications

In this section, we apply the fundamental heat transfer relationships of Section 6.2.a to a living system, the human operator. As a general rule, this section will describe the human operator as a steady state system ($\dot{T} = 0$) when no external work is being performed ($\dot{W} = 0$). Invoking these approximations, equation (45) reduces to:

$$\dot{M}_0 + \Delta\dot{M} = -\dot{Q}' \tag{66}$$

In Section 6.2.b.i, internal heat transfer from the body core (at a temperature of T_c) through the muscle layer (at a temperature T_m) to the skin surface (at a temperature of T_s) will be described. The analytical model used will assume a generic (non-specific) heat transfer mechanism from the skin (at a temperature of T_s) to the ambient (external) environment (at a temperature of T_A).

In Sections 6.2.b.ii, 6.2.b.iii, 6.2.b.iv, and 6.2.b.v, external heat transfer from the surface of the skin (at a temperature of T_s) to the surrounding environment will be considered. Each section will deal with the external heat transfer mechanism of convection, radiation, combined convection/radiation, and evaporation, respectively. When external conduction from the surface of the skin (at a temperature of T_s) to a contacting external material (at a temperature of T_L) is also included, then the overall external heat loss term ($\dot{Q}'$) may be defined as follows:

$$\dot{Q}' = \dot{Q}_{COND} + \dot{Q}_{CONV} + \dot{Q}_{RAD} + \dot{Q}_{EVAP} \tag{67}$$

Table 6.1 provides information regarding the relative significance of each heat loss term on the right side of equation (67) when applied to an average, seated, clothed adult human operator who is exposed to an ambient temperature of 21°C. The human factors engineer will appreciate that the relative significance of any particular heat loss mechanism will be a *combinatorial* function of the human operator environment, activity, clothing, and behavior.

i. Conduction

Within the living system, internal conduction is the significant mechanism by which heat is transferred through the various regions of the human body. Internal conduction may be thought of as passive heat transfer in a manner analogous to heat transfer through a composite wall.

The other significant mechanism for internal heat transfer is that of forced convection in which blood flow carries heat through the various regions of the body via a complex system of concurrent and countercurrent heat exchanges. Forced convection may be considered as an active heat transfer process since regional blood flow and the opening and closing of concurrent and countercurrent heat exchanger shunts are under autonomic nervous control.

With respect to internal heat transfer of the human operator, this section will only consider conduction. The rationale for this is that the steady state approximation ($\dot{T} = 0$) is rather straightforward when applied to a passive material system. Since forced convection is a dynamic system under regulatory control, this mechanism of internal heat transfer plays a significant role with respect to the human operator thermoregulatory system. Consequently, we will defer further discussion of this system until Section 6.3 where we will model human operator thermoregulation.

Consequently, the human factors engineer will appreciate that the subsequent models for internal heat transfer of the human operator will be so formulated to account for conduction as the mechanism by which the entire heat load of the body will be internally transferred. This approach is a generally useful approximation for many practical applications. However, the model developed in this section may indeed require modification to account for forced convection as the other significant mechanism for internal heat transfer. Such a modification would obviously depend on the particular engineering application. Accepting this qualification, we shall proceed to describe internal conductive heat transfer.

This subsection will first consider an internal conduction model for heat transfer between the core region of the body to the surface region of the body. Subsequently, the model shall be modified to account for external heat transfer by means of external conduction. Finally, this subsection will conclude by modifying our model to account for external heat transfer via generic (nonspecific mechanistic) heat loss pathways.

A simplified model can be applied to describe the steady state temperature distribution through the core region, muscle region, and skin region of the human body. Figure 6.5 illustrates this simplified model. The core region temperature (T_c) is the mean operating temperature of the internal organs. In this model, T_c is the mean core temperature which is located spatially at the midpoint of the core region. The muscle temperature (T_m) is the operating temperature for the muscle layer of

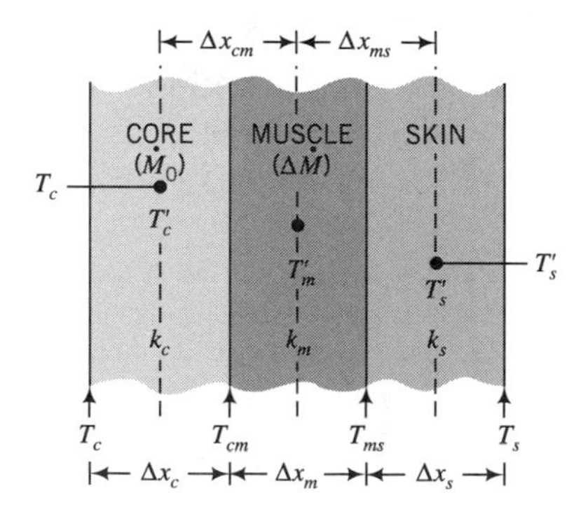

where:

$$\Delta x_{cm} = \frac{\Delta x_c}{2} + \frac{\Delta x_m}{2}$$

$$\Delta x_{ms} = \frac{\Delta x_m}{2} + \frac{\Delta x_s}{2}$$

Figure 6.5 Steady state temperature distribution through the core region, muscle region and skin region of the human body.

the human body. Note that muscle is a shell tissue and can be either resting or actively working. T_m is the mean muscle temperature located spatially at the midpoint of the muscle region. The skin region temperature (T_s) is the operating temperature for the surface region of the body consisting of a subcutaneous fat layer (adjacent to the muscle layer), the dermal layer (overlying the subcutaneous fat), and finally the epidermal layer (which includes the skin itself). T_s is the mean skin region temperature which is spatially located at the midpoint of the skin region.

Mean core temperature (T_c) in the human operator is approximately 37°C and is most accurately approximated with rectal thermometry or tympanic membrane thermometry. The mean body temperature (T_b) is an approximated average temperature of the various regional temperatures throughout the human body. An example of T_b that approximates the core temperature and skin temperature is:

$$T_b = 0.6T_c + 0.4T_s \tag{68}$$

When muscle is actively working, body temperature may experience a significant rise. Therefore, an example of T_b that accounts for the muscle region temperature in addition to core temperature and skin temperature is:

$$T_b = 0.15T_s + 0.30T_m + 0.55T_c \tag{69}$$

Oral thermometry is useful in approximating the mean body temperature (as opposed to core body temperature).

Figure 6.5 represents the three composite regions of the body in cross section and employs a rectangular coordinate system so that the heat conduction pathway is along the x-axis. This model represents internal conduction as a pair of adjacent composite walls. The first composite wall is of thickness Δx_{cm} and is a composite region between the midpoint of the core region and the midpoint of the muscle region. The purpose of the first composite wall is to transfer the basal metabolic heat flux ($\dot{M}_0$) from the core region to the muscle region. Using electrical analogy theory for a composite wall (see equation [60]) results in:

$$\dot{Q}_{COND} = \frac{T_c' - T_m}{\dfrac{\Delta x_c}{2k_c} + \dfrac{\Delta x_m}{2k_m}} \tag{70}$$

With respect to metabolic heat transfer, $\dot{M}_0$ and $\Delta\dot{M}$ are *negative* heat rates. This is consistent with equation (66):

$$\dot{M}_0 + \Delta\dot{M} = -\dot{Q}' \tag{66}$$

Since only $\dot{M}_0$ is the heat rate for the core region ($\Delta\dot{M} = 0$ for the core region), recall equation (43):

$$\dot{M}_0 = -\dot{Q}' \tag{43}$$

Since *all* heat transfer is due *entirely* to conduction, equation (67) reduces to:

$$\dot{Q}' = \dot{Q}_{COND} \tag{71}$$

Substituting equation (71) into equation (70), and then substituting into equation (43) and transposing:

$$-\dot{M}_0 = \frac{T'_c - T_m}{\dfrac{\Delta x_c}{2k_c} + \dfrac{\Delta x_m}{2k_m}} \tag{72}$$

A good approximation of equation (72) is as follows:

$$-\dot{M}_0 \cong \frac{k_{cm}}{\Delta x_{cm}}(T'_c - T_m) \tag{73}$$

where:

$$k_{cm} = \frac{k_c \cdot k_m}{k_c + k_m} \tag{74}$$

and:

$$\Delta x_{cm} = \frac{\Delta x_c + \Delta x_m}{2} \tag{75}$$

The second composite wall is adjacent to the first composite wall and extends from the midpoint of the muscle region to the midpoint of the skin region over total linear displacement of Δx_{cm} (as shown in Figure 6.5). The purpose of this second composite wall is to conduct $\dot{M}_0$ and any additional metabolic energy flux generated in the muscle region ($\Delta\dot{M}$) to the skin region. In a manner similar to the first composite wall, the following relationship may be written for the second composite wall:

$$-\dot{M}_0 - \Delta\dot{M} = \frac{T_m - T'_s}{\dfrac{\Delta x_m}{2k_m} + \dfrac{\Delta x_s}{2k_s}} \tag{76}$$

A good approximation of equation (76) for the second composite wall is:

$$-\dot{M}_0 - \Delta\dot{M} \cong \frac{\Delta k_{ms}}{\Delta x_{ms}}(T_m - T'_s) \tag{77}$$

where:

$$k_{ms} = \frac{k_m k_s}{k_m + k_s} \tag{78}$$

and:

$$\Delta x_{ms} = \frac{\Delta x_m + \Delta x_s}{2} \tag{79}$$

EXAMPLE 6.3

Muscle is a shell tissue and its temperature is affected by two variables: (1) heat transfer by internal conduction between the central body core and the surface (skin) layer; and (2) extra metabolic energy rate ($\Delta\dot{M}$) generated by muscular contraction. Referring to Figure 6.5, develop and solve the relevant equations necessary to answer the following situations. You may approximate the internal conduction system as being in thermal equilibrium ($\dot{T}_c = 0$) at all times.

a. Muscle temperature (T_m) for a resting (inactive) person at normal skin layer temperature:

Given:

$$\hat{\dot{M}} = -37 \frac{\text{kcal}}{\text{M}^2 \cdot \text{hr}}$$

$$\Delta\hat{\dot{M}} = 0 \text{ (inactive muscle)}$$

$$k = k_c = k_m = k_s = 0.425 \frac{\text{kcal}}{\text{M} \cdot \text{hr} \cdot {}^\circ\text{C}}$$

$$T_c' = 37.0^\circ\text{C}$$

$$T_{cm} = 35.25^\circ\text{C}$$

$$T_{ms} = 33.5^\circ\text{C}$$

$$T_s' = 33.05^\circ\text{C}$$

Find T_m, Δx_c, Δx_m, and Δx_s.

Muscle temperature (T_m) for a shivering person at a subnormal skin layer temperature:

Given:

$$\hat{\dot{M}}_0 = -37 \frac{\text{kcal}}{\text{M}^2 \cdot \text{hr}}$$

$$k = k_c = k_m = k_s = 0.425 \frac{\text{kcal}}{\text{M} \cdot \text{hr} \cdot {}^\circ\text{C}}$$

$$T_c' = 37.0^\circ\text{C}$$

$$T_s' = 29.0^\circ\text{C}$$

From part [a]:

$$\Delta x_c = .040 \text{ M}$$

$$\Delta x_m = .020 \text{ M}$$

$$\Delta x_s = .010 \text{ M}$$

Find T_m, $\Delta\hat{\dot{M}}$, T_{cm}, and T_{ms}.

c. Muscle temperature (T_m) for a shivering person at a subnormal skin layer temperature and a falling body temperature.

Given:

$$\hat{\dot{M}}_0 = -37 \frac{\text{kcal}}{\text{M}^2 \cdot \text{hr}}$$

$$k = k_c = k_m = k_s = 0.425 \frac{\text{kcal}}{\text{M} \cdot \text{hr} \cdot {}^\circ\text{C}}$$

$$T_s' = 20.0^\circ\text{C}$$

$$m_B = 60 \text{ kg}$$

$$\dot{T}_c = -2 \frac{^\circ\text{C}}{\text{hr}} \text{ (constant rate)}$$

From part [a]:

$$\Delta x_c = .040 \text{ M}$$
$$\Delta x_m = .020 \text{ M}$$
$$\Delta x_s = .010 \text{ M}$$

From part [b]:

$$\Delta \hat{\dot{M}} = -69 \frac{\text{kcals}}{\text{M}^2 \cdot \text{hr}}$$

Find T_m, T_{cm}, T_{ms}, and T_c'.

SOLUTION 6.3.a

To solve for four unknowns, we must write four independent equations. Refer to Figure 6.5.

(i) Conduction through outer-half of the central core layer [by analogy to equation (73) but with reversal of temperature gradient]:

$$-\hat{\dot{M}}_0 = -\frac{2k_c}{\Delta k_c}(T_{cm} - T_c') \quad \text{(i)}$$

(ii) Conduction through the entire muscle layer [by analogy to equation (77) but with reversal of temperature gradient]:

$$-\hat{\dot{M}}_0 - \Delta \dot{M} = -\frac{k_m}{\Delta x_m}(T_{ms} - T_{cm}) \quad \text{(ii)}$$

(iii) Conduction through the inner one-half of the surface (skin) layer [by analogy to equation (77) but with reversal of temperature gradient]:

$$-\hat{\dot{M}}_0 - \Delta \hat{\dot{M}} = -\frac{2k_s}{\Delta x_s}(T_s' - T_{ms}) \quad \text{(iii)}$$

(iv) Definition of muscle temperature (Figure 6.5):

$$T_m = \frac{T_{cm} + T_{ms}}{2} \quad \text{(iv)}$$

Proceeding with these four independent equations:
Solve equation (iv) by direct substitution of known values:

$$T_m = \frac{35.25 + 33.50}{2} = 34.4°\text{C} \quad \text{(v)}$$

Rearrange equation (i) for Δx_c:

$$\Delta x_c = \frac{2k}{\hat{\dot{M}}_0}(T_{cm} - T_c')$$

Solve equation (v) by direct substitution of known values:

$$\Delta x_c = \frac{(.850)}{(-37)}(-1.75) = .040 \text{ M}$$

Rearrange equation (ii) for Δx_m:

Since $\Delta\hat{M} = 0$, then:

$$\Delta x_m = \frac{k}{\hat{M}_0}(T_{ms} - T_{cm}) \tag{vi}$$

Solve equation (vi) by direct substitution of known values:

$$\Delta x_m = \frac{(.425)}{(-37)}(-1.75) = .020\ \text{M}$$

Rearrange equation (iii) for Δx_s:
Since $\Delta\hat{M} = 0$, then:

$$\Delta x_s = \frac{2k}{\hat{M}_0}(T_s' - T_{ms}) \tag{vii}$$

Solve equation (vii) by direct substitution of known values:

$$\Delta x_s = \frac{(.850)}{(-37)}(-.45) = .010\ \text{M}$$

SOLUTION 6.3.b

To solve for four unknowns, again write the same four independent equations as per part [a]:
Rearrange equation 6.3(a)(i) for T_{cm}.
Separate variables:

$$T_{cm}\left(\frac{2k}{\Delta x_c}\right) = \hat{M}_0 + \left(\frac{2k}{\Delta x_c}\right)T_c'$$

so that:

$$T_{cm} = T_c' + \left(\frac{\Delta x_c}{2k}\right)\hat{M}_0$$

Solve equation (ii) by direct substitution of known values:

$$T_{cm} = 37.0 + \left(\frac{.040}{.850}\right)(-37)$$

$$T_{cm} = 37.0 - (1.74) = 35.3°\text{C}$$

Equate equation 6.3(a)(ii) with equation 6.3(a)(iii) as follows:

$$\left(\frac{k}{\Delta x_m}\right)(T_{ms} - T_{cm}) = \left(\frac{2k}{\Delta x_s}\right)(T_s' - T_{ms}) \tag{iii}$$

Separate variables:

$$\left(\frac{k}{\Delta x_m}\right)T_{ms} + \left(\frac{2k}{\Delta x_s}\right)T_{ms} = \left(\frac{2k}{\Delta x_s}\right)T_s' + \left(\frac{k}{\Delta x_m}\right)T_{cm} \tag{iv}$$

so that:

$$T_{ms} = \left[\frac{1}{\dfrac{1}{\Delta x_m} + \dfrac{2}{\Delta x_s}}\right]\left[\left(\frac{2}{\Delta x_s}\right) T_s' + \left(\frac{1}{\Delta x_m}\right) T_{cm}\right] \tag{v}$$

Solving equation (v) by direct substitution of known values:

$$T_{ms} = \left[\frac{1}{250}\right][(200)(29) + (50)(35.3)]$$

so that:

$$T_{ms} = \frac{7565}{250} = 30.3°\text{C}$$

Solve equation 6.3(a)(iv) by direct substitution of known values:

$$T_m = \frac{35.3 + 30.3}{2} = 32.8°\text{C}$$

Rearrange equation 6.3(a)(ii) for $\Delta\hat{\dot{M}}$:

$$\Delta\hat{\dot{M}} = \left(\frac{k}{\Delta x_m}\right)(T_{ms} - T_{cm}) - \hat{\dot{M}}_0 \tag{vi}$$

Solve equation (vi) by direct substitution of known values:

$$\Delta\hat{\dot{M}} = (21.25)(-5.0) - (-37)$$

so that:

$$\Delta\hat{\dot{M}} = -69 \frac{\text{kcal}}{\text{M}^2 \cdot \text{hr}}$$

SOLUTION 6.3.c

Rewrite equation (45) for the human operator biothermal system when no *external* work is done:

$$\alpha\dot{T} + \dot{M}_0 + \Delta\dot{M} = -\dot{Q}' \tag{i}$$

For heat transfer by internal conduction, equation (i) is normalized per square meter of body surface area (A_B):

$$\hat{\alpha}\dot{T} + \hat{\dot{M}}_0 + \Delta\hat{\dot{M}} = -\dot{Q}_{COND} \tag{ii}$$

where:

$$\hat{\alpha} = \frac{\alpha}{A_B} \tag{iii}$$

The relationship between A_B and body mass (m_B) is:

$$A_B = 0.1(m_B)^{0.67} \tag{iv}$$

Setting $m_B = m_{TOT}$ of equation (36), and then substituting equation (iv) and equation (36) into equation (iii):

$$\hat{\alpha} = \frac{m_B \hat{C}_P}{0.1(m_B)^{0.67}} \tag{v}$$

which simplifies to:

$$\hat{\alpha} = 10(m_B)^{0.33}\hat{C}_P \tag{vi}$$

with units of kcal/M² °C.
For this example, recall that:

$$\hat{C}_P = 0.86 \frac{\text{kcal}}{°\text{C}\cdot\text{kg}}$$

and since $m_B = 60$ kg, then substituting these values into equation (vi):

$$\hat{\alpha} = (10)(3.86)(.86) = 33.2 \frac{\text{kcal}}{\text{M}^2 \cdot °\text{C}} \tag{vii}$$

To solve for four unknowns, we must write four independent equations. Refer to Figure 6.5.

(i) Conduction through the outer one-half of the central core layer:

$$-\hat{\alpha}\dot{T} - \hat{\dot{M}}_0 = -\frac{2k}{\Delta x_c}(T_{cm} - T'_c) \tag{viii}$$

(ii) Conduction through the *outer one-half* of the muscle layer:

$$-\hat{\alpha}\dot{T} - \hat{\dot{M}}_0 - \Delta\hat{\dot{M}} = -\frac{2k}{\Delta x_m}(T_{ms} - T_m) \tag{ix}$$

(iii) Conduction through the inner one-half of the surface (skin) layer:

$$-\hat{\alpha}\dot{T} - \hat{\dot{M}}_0 - \Delta\hat{\dot{M}} = -\frac{2k}{\Delta x_s}(T'_s - T_{ms}) \tag{x}$$

Definition of the *core-to-muscle interface temperature* (T_{cm}) by rearranging equation 6.3(a)(iv):

$$T_{cm} = 2 \cdot T_m - T_{ms} \tag{xi}$$

Rearrange equation (x) for T_{ms}:

$$T_{ms} = T'_s - \left(\frac{\Delta x_s}{2k}\right)(\hat{\alpha}\dot{T} + \hat{\dot{M}}_0 + \Delta\hat{\dot{M}}) \tag{xii}$$

Solve equation (xii) by direct substitution of known values:

$$T_{ms} = 20 - \frac{(.01)(-66.4 - 37.0 - 69.0)}{(.850)}$$

so that:

$$T_{ms} = 20 + 2.0 = 22.0°\text{C}$$

Rearrange equation (ix) for T_m:

$$T_m = T_{ms} - \frac{\Delta x_m}{2k}(\hat{\alpha}\dot{T} + \hat{\dot{M}}_0 + \Delta\hat{\dot{M}}) \tag{xiii}$$

Solve equation (xiii) by direct substitution of known values:

$$T_m = 22.0 - \left[\frac{(.02)}{(.85)}\right](-172.4)$$

so that:

$$T_m = 22.0 + 4.1 = 26.1°\text{C}$$

Solve for T_{cm} by direct substitution of known values into equation (xi):

$$T_{cm} = (2)(26.1) - (22.0) = 30.2°\text{C}$$

Rearrange equation (viii) for T_c':

$$T_c' = T_{cm} - \left(\frac{\Delta x_c}{2k}\right)(\hat{\alpha}\dot{T} + \hat{\dot{M}}_0) \qquad \text{(xiv)}$$

Solve equation (xiv) by direct substitution of known values:

$$T_c' = 30.2 - \left(\frac{.04}{.85}\right)(-103.4)$$

so that:

$$T_c' = 30.2 + 4.9 = 35.1°\text{C}$$

The model presented in Figure 6.5 makes the approximation that the core temperature (T_c') is constant over the inner half of the core region.

The actual central core temperature (T_c) may be calculated as:

$$T_c = T_c' - \frac{\Delta x_C}{2k_C}(\dot{M}_0 + \Delta\dot{M}) \qquad (80)$$

This model makes the additional approximation that the skin region temperature (T_s') is constant over the outer half of the skin region. The actual skin surface temperature (T_s) may be calculated as:

$$T_s = T_s' + \frac{\Delta x_S}{2k_S}(\dot{M}_0 + \Delta\dot{M}) \qquad (81)$$

Equation (81) is necessary when the heat transfer mechanism is convection, radiation, and/or evaporation.

The internal conductive heat transfer model of Figure 6.5 can be modified to account for external heat transport due to external conduction. With external conduction, a cooler (or hotter) material is in physical contact with the body surface. Conductive heat flow then follows the thermal gradient between the external conductive material and the human body. The internal conduction model, as modified for external conduction, is shown in Figure 6.6.a. The essence of the model is a third composite wall that operates (when the external material is cooler than the body) to conduct $\dot{M}_0$ and $\Delta\dot{M}$ (which are *negative* heat rates) away from the human body. This third composite wall consists of a region extending from the midpoint of the skin region to the outside wall of the external conducting material. Per Figure 6.6.a, note that the outside wall may be defined functionally as that point at which the wall material reaches a constant temperature (T_L) with respect to displacement along the x-axis.

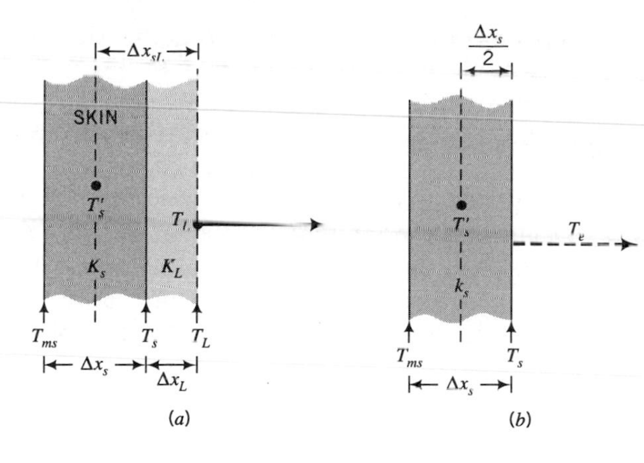

Figure 6.6.a Internal conduction model as modified for external conduction. **b** Generic external heat transfer model.

In a manner similar to the first composite wall, the governing equation for external conduction is:

$$-\dot{M}_0 - \Delta\dot{M} = \frac{T_s' - T_L}{\dfrac{\Delta x_S}{2k_S} + \dfrac{\Delta x_L}{2k_L}} \tag{82}$$

A good approximation for external conductive heat transfer through a third composite wall may be written as:

$$-\dot{M}_0 - \Delta\dot{M} = \frac{k_{SL}}{\Delta x_{SL}}(T_s' - T_L) \tag{83}$$

where:

$$k_{SL} = \frac{k_S k_L}{k_S + k_L} \tag{84}$$

and:

$$\Delta x_{SL} = \frac{\Delta x_S + 2\Delta x_L}{2} \tag{85}$$

The internal conduction model of Figure 6.5 may be modified to account for generic (nonspecific mechanistic) heat energy transfer. This generic external heat transfer model is shown in Figure 6.6.b. The purpose of this model modification is to equate the *internal* heat transfer rate ($\dot{M}_0$ and $\Delta\dot{M}$), which accounts for heat transfer to the skin region, to the *external* heat rate, which accounts for heat transfer into the environment [at some external temperature (T_e)]. The generic external heat transfer model may be mathematically expressed as:

$$-\dot{M}_0 - \Delta\dot{M} = h(T_s - T_e) \tag{86}$$

where:

Mechanism	H	T_s	T_e
Convection	h_C	T_s	T_a
Radiation	h_r	T_s	T_w
Evaporation	h_v	$P_v(T_s)$	$P_v(T_a)$

It is apparent that the model is nonspecific with respect to the mechanism of heat transfer and that the actual heat transfer mechanism may be by convection, radiation, evaporation, or any combination of these. Each heat transfer mechanism, of course, has its own appropriate heat transfer coefficient as indicated in equation (86). An important feature of this generic external heat transfer model is that the actual skin surface temperature (T_s) is used in equation (86). Consequently, if only T_s' is known from prior application of the internal conduction model (see Figure 6.5), then T_s must be calculated from equation (81) before applying the generic external heat transfer model of equation (86). Specific formulations of equation (86) for particular engineering applications will be considered in the following subsections.

ii. Convection

The human factors engineer will appreciate that the application of heat transfer coefficients is extremely useful in the calculation of heat transfer rates from the human operator. The convective heat transfer coefficient (h_C) is a function of physical posture (standing or seated), the clothing worn (unclothed or clothed), the motion of the convective fluid (stationary or dynamic), and the nature of the convective fluid (air or water). As a general approximation, the convective coefficients may be expressed as kcal/M^2 min °C.

We can apply the convective heat transfer equation to the human operator by rewriting equation (46) specifically for the convective heat flux:

$$q_x = \frac{\dot{Q}_{CONV}}{A_C} \tag{87}$$

where:

$\dot{Q}_{CONV}$ = heat flow by convection (kcal/hr)
A_C = area of the convecting surface (M^2)

Substituting equation (87) into equation (62), and rearranging:

$$\dot{Q}_{CONV} = h_C A_C (T_s - T_a) \tag{88}$$

Table 6.4 Convective Heat Transfer Coefficients (Unclothed Person)

Heat Transfer/Posture	Symbol	Value[a]
Free Convection (Standing)	h_{CF}	2.3
Free Convection (Seated)	h_{CF}	2.0
Forced Convection (Standing)	h_{CV}	$7.5\left(\frac{V}{V_0}\right)^{0.6}$
Forced Convection (Seated)	h_{CV}	$6.4\left(\frac{V}{V_0}\right)^{0.6}$
Convection-to-Water (Still Water)	h_{CW}	37.5
Convection-to-Water (Stirred Water)	h_{CW}	50.0

[a] kcals/hr · M^2 · °C
V_0 = 1 M/s

Table 6.5 Convective Heat Transfer Coefficients (Clothed Person)

Heat Transfer/Posture	Symbol	Value[a]
Free Convection	h_{CF}	1.2
Forced Convection ($T > 0°C$)	h_{CV}	$3.0\left(\frac{V}{V_0}\right)^{0.6}$
Forced Convection ($T \leq 0°C$)	h_{CV}	$1.2\left(\frac{V}{V_0}\right)^{0.16}$

[a] kcals/hr · M² · °C
V_0 = 1 M/s

Equation (88) is the general form of the steady-state convective heat transfer equation for a human operator. For most applications of practical interest to the human factors engineer, the general convective heat transfer equation must be modified to account for important physical variables (i.e., posture, clothing, and the nature and motion of the convective fluid). These modifications result in a set of specific convective heat transfer equations, as follows.

a. Human operator heat transfer by *free* convection: heat transfer to still (zero velocity) ambient air (at temperature, T_a).

$$\dot{Q}_{CF} = h_{CF}A_C(T_s - T_a) \tag{89}$$

where h_{CF} is the free convection heat transfer coefficient and is a function of posture and clothing (see Tables 6.4 and 6.5).

b. Human operator heat transfer by *forced* convection: heat transfer to moving ambient air (at some velocity V, and at temperature, T_a).

$$\dot{Q}_{CV} = h_{CV}A_C(T_s - T_a) \tag{90}$$

where h_{CV} is the forced convection heat transfer coefficient and a function of posture, clothing, air velocity, and temperature (per Tables 6.4 and 6.5).

c. Human operator heat transfer by convection-to-water: heat transfer to still or stirred water (at temperature, T_{H_2O}).

$$\dot{Q}_{CW} = h_{CW}A_C(T_s - T_{H_2O}) \tag{91}$$

where h_{CW} is the convection-to-water heat transfer coefficient and is a function of water motion (see Table 6.4).

EXAMPLE 6.4

A human factors engineer is asked to evaluate the interaction of subzero temperatures and windy conditions on construction workers for the Alaskan pipeline.

The engineer elects to construct a wind-chill chart to alert workers to the hazard of cold temperatures combined with high winds. The wind-chill temperature (T_{wc}) is the equivalent temperature at zero wind-velocity that results from the actual wind velocity and temperature combination being experienced by the person.

The approximation is made that convective heat transfer predominates, so that radiative and evaporative heat transfer may be neglected. The worker will be clothed and actively performing physical work.

a. Derive the governing equation for a wind-chill analysis.
b. Prepare a chart of equivalent wind-chill temperature (T_{wc}) for the combination of four actual wind velocities (10, 20, 30, and 40 mph) and four actual air temperatures (0, −10, −20, and −30°C).

SOLUTION 6.4(a)

The mathematical definition of a wind-chill temperature proceeds as follows. The heat transfer due to forced convection (which occurs at some wind-velocity actually experienced) is equivalent to a calculated heat transfer due to free convection (zero-wind velocity) at some wind-chill temperature (T_{wc}). As a general mathematical statement of the above:

$$\dot{Q}_{CV} = \dot{Q}'_{CF} \tag{i}$$

Recall from equations (89) and (90):

$$\dot{Q}_{CF} = h_{CF}A_C(T_s - T_a) \tag{89}$$
$$\dot{Q}_{CV} = h_{CV}A_C(T_s - T_a) \tag{90}$$

Per the definition of a wind-chill temperature (T_{wc}), we define an equivalent (wind-chill) free convective heat transfer ($\dot{Q}'_{CF}$) by modifying equation (89) as follows:

$$\dot{Q}'_{CF} = h_{CF}A_C(T_s - T_{wc}) \tag{ii}$$

Substituting equations (90) and (ii) into equation (i) and simplifying:

$$h_{CV}(T_s - T_a) = h_{CF}(T_s - T_{wc}) \tag{iii}$$

Recall that:

T_a = actual environmental (ambient) temperature (°C)
T_{wc} = equivalent wind-chill temperature (°C)

Next approximate the skin surface temperature (T_s) as 35°C, and use the appropriate thermal conductivity coefficients for a *clothed* person (from Table 6.5):

$$1.2\left(\frac{V}{V_0}\right)^{0.16}(35 - T_a) = 1.2(35 - T_{wc}) \tag{iv}$$

Finally, simplify and rearrange equation (iv) for T_{wc}:

$$T_{wc} = 35 - \left(\frac{V}{V_0}\right)^{0.16}(35 - T_a) \tag{v}$$

This is the governing equation for calculation of the wind-chill chart. Note that it is independent of an individual's basal metabolism rate ($\dot{M}_0$) and activity level ($\Delta\dot{M}$). However, it does require that the skin surface temperature be approximated and that all heat transfer is convective.

SOLUTION 6.4(b)

Substituting the independent variables (T_a and V) into equation 6.4(a)(v) results in calculation of the dependent variable, T_{wc}. The following table of data gives the final form of the wind-chill chart:

Wind-Chill Temperature (T_{wc})

	Wind Velocity (mph)*			
T_a(°C)	10	20	30	40
0°C	–9.5	–14.7	–18.0	–20.5
–10°C	–22.2	–28.9	–33.2	–36.4
–20°C	–34.9	–43.1	–48.4	–52.3
–30°C	–47.6	–57.3	–63.5	–68.1

$*0.448 \dfrac{\text{M/s}}{\text{mph}}$

Note: Frostbite is imminent at a wind-chill temperature (T_{wc}) of –40°C.

iii. Radiation

The fundamental relationships for radiative heat transfer (see Section 6.2.a) may be applied to the human operator with some empirical modifications. For application to the living human, equation (64) may be expressed in a slightly modified form:

$$q^e = e_s\sigma\lfloor T_s^4 - T_w^4\rfloor \tag{92}$$

where:

T_s = average skin temperature
T_w = surrounding wall temperature

To apply the radiative heat transfer equation to the human operator, we rewrite equation (46) with respect to radiative heat flux:

$$q^e = \frac{\dot{Q}_{rad}}{A_r} \tag{93}$$

where:

$\dot{Q}_{rad}$ = heat flow by radiation (kcal/hr)
A_r = effective surface area for radiation. This quantity is often estimated as some fraction of the total body surface area. One popular value of the fraction is 0.77.

Substituting equation (93) into equation (92), and rearranging:

$$\dot{Q}_{rad} = e_s\sigma A_r(T_s^4 - T_w^4) \tag{94}$$

where:

e_s = average emissivity of the skin
σ = Stefan-Boltzmann constant

An alternative approach is to factor equation (94):

$$\dot{Q}_{rad} = e_s\sigma A_r\lfloor T_s^2 + T_w^2\rfloor\lfloor T_s^2 - T_w^2\rfloor \tag{95}$$

which may be factored further:

$$\dot{Q}_{rad} = e_s\sigma A_r\lfloor T_s^2 + T_w^2\rfloor[T_s + T_w][T_s - T_w] \tag{96}$$

Then define:

$$h_r = e_s\sigma\lfloor T_s^2 + T_w^2\rfloor[T_s + T_w] \tag{97}$$

And substituting equation (97) into equation (96); resulting in:

$$\dot{Q}_{rad} = h_r A_r (T_s - T_w) \tag{98}$$

A representative range for h_r for unclothed humans has been determined to be from 3.5 to 5 kcal/m^2 hr °C. A specific value may be calculated as follows:

Recall that:

$$e_s = 0.87$$

$$\sigma = 4.88 \times 10^{-8} \text{ kcal/hr M}^2\ °\text{K}^4$$

When the skin surface temperature (T_s) is 308°K (35°C) and the wall temperature (T_w) is 288°K (15°C), then equation (97) predicts:

$$h_r = 4.5 \text{ kcal/hr M}^2\ °\text{C}$$

Note that °C in the units for h_r replaces °K when the units of T_s and T_w in equation (95) are in °C. This is because h_r multiplies a differential temperature in equation (95) and the *differential* scale factor of °C to °K is unity.

Finally, it should be appreciated that the radiative heat transfer coefficient (h_r) is a function of posture (standing or seated) and clothing (unclothed or clothed). Some representative values are given in Table 6.6.

iv. Convection and Radiation

Having considered human operator heat transfer for convection and radiation separately, there are some issues for which it is useful to consider these heat transfer mechanisms collectively.

The first issue is that of the human operator body surface area which is the "effective" area across which heat transfer occurs. It is obvious that the effective heat transfer surface area is some fraction of the *total* body surface area of the human operator (A_B). This effective area for convective heat transfer is expressed as:

$$A_C = f_C A_B \tag{99}$$

The effective surface area for radiative heat transfer is expressed as:

$$A_r = f_r A_B \tag{100}$$

TABLE 6.6 Radiative Heat Transfer Coefficients

Heat Transfer/Posture	Symbol	Value[a]
Unclothed Person		
Radiation (Standing)	h_r	4.7
Radiation (Seated)	h_r	4.5
Clothed Person		
Radiation	h_r	2.0

[a]kcals/hr · M^2 · °C

where f_C and f_r are the coefficients for the convective and radiative surface areas and are a function of posture (per Table 6.7).

Most practical human factors applications will involve clothed individuals. Therefore, a second issue is to more systemically consider the approaches to quantifying external heat transfer with respect to the clothed individual.

The first approach is to recognize that the equations developed for the unclothed individuals may be generalized to those situations involving a clothed human operator. The specific equation modification requires that the surface skin temperature (T_s) from which convection or radiation occurs should now be replaced with the clothing surface temperature (T_{CS}). In the case of radiative heat transfer, e_S is now the emissivity of the clothing surface. In the particular situation of evaporative heat transfer (see below), A_v would be replaced with the area of the clothing through which the perspiration is evaporated. By employing such straightforward modifications, the various heat transfer equations are still valid.

A second approach to the clothed human operator is to recognize that the various layers of clothing modify the overall heat transfer coefficient by summing the added resistances involved. The general approach is rather similar to that of heat conduction through a composite wall (as previously discussed). The clothing offers a heat transfer resistance between the skin and the environment, so that effectively, the clothing surface temperature (T_{SC}) can be significantly lower than the surface skin temperature (T_s). Effectively, surface skin temperature can remain in a comfortable region despite relatively cold environmental temperature. The general mathematical formulation may be presented as follows:

$$\dot{Q}_{SCE} = h_{TOT} A_{TOT} (T_s - T_e) \tag{101}$$

where:

$\dot{Q}_{SCE}$ = effective heat flow from the human skin surface through the clothing to the external environment, kcals/hr

h_{TOT} = overall heat transfer coefficient from skin surface to environment, kcal/M^2 hr °C

A_{TOT} = effective body surface area (M^2)

T_s = skin surface temperature, °C

T_e = external environment temperature, °C

The overall heat transfer coefficient, h_{TOT}, from the skin surface to the environment may be calculated as:

$$\frac{1}{h_{TOT}} = \frac{1}{h_{SC}} + \frac{1}{(h_{CC} + h_{CR})} \tag{102}$$

where:

h_{SC} = heat transfer coefficient, skin to clothing exterior

h_{CC} = convective heat transfer coefficient, clothing to surroundings

h_{CR} = radiative heat transfer coefficient, clothing to surroundings

Table 6.7 *f*-Coefficients (Dimensionless) for Convective and Radiative Surface Areas

Mechanism	Symbol	Standing	Sitting
Convection	f_C	0.90	0.80
Radiation	f_r	0.75	0.65

Note that this method involves calculating the reciprocal of a total conductance as a function of the sum of the reciprocal of each individual conductance. Since the reciprocal of a conductance is resistance (in electrical analog theory), this is analogous to calculating the total heat flow resistance by summing the individual (clothing layer) resistances.

Equation (102) may be simplified by defining a heat transfer coefficient from the exterior clothing surface to the environment (h_{CE}) as follows:

$$h_{CE} = h_{CC} + h_{CR} \tag{103}$$

Substituting equation (103) into equation (102) results in:

$$\frac{1}{h_{TOT}} = \frac{1}{h_{SC}} + \frac{1}{h_{CE}} \tag{104}$$

Representative values for the reciprocal heat transfer coefficients of equation (104) are presented in Table 6.8 with respect to various clothing categories.

A major limitation of equation (101) is that it does not account for the effects of wind velocity. As formulated, this expression is valid only for still air. A second major limitation is that the environmental temperature (T_e) does not distinguish between the ambient (external) temperature (T_a) and the temperature of the surrounding radiative bodies (T_w). Application of equation (101) requires that T_a be reasonably close to T_w. For many practical engineering applications, the above limitations are overly restrictive, so a third approach is necessary. A useful engineering approximation of equation (101) that accounts for the combined heat loss rate from convection and radiation for the clothed human operator is:

$$\dot{Q}' = \left[3.0\left(\frac{V}{V_0}\right)^{0.6} A_C(35 - T_a)\right] + 2.0A_r(35 - T_w) \tag{105}$$

where:

$V_0 = 1$ M/s
$V =$ wind approach velocity (≥ 0.15 M/s)

Finally, a usual case of the human operator will involve one in which there are both uncovered skin areas (head, neck, arms, etc.) and also clothed areas (chest, hips, feet, etc.). For this situation, heat transfer for both areas should be calculated separately (using the appropriate equations) and then summed together. This is

Table 6.8 Reciprocal Heat Transfer Coefficients[a] for Various Clothing Categories

Clothing Categories	$\frac{1}{h_{SC}}$	$\frac{1}{h_{CE}}$	$\frac{1}{h_{TOT}}$
Light working ensemble	0.18	0.42	0.60
U.S. Army fatigues, man's	0.18	0.52	0.70
Typical American business suit	0.18	0.52–0.82	0.70–1.00
Light outdoor sportswear	0.18	0.72	0.90
Heavy traditional European business suit	0.18	1.02	1.20
U.S. Army standard cold-wet uniform	0.18	1.32–1.82	1.50–2.00
Heavy wool pile ensemble (polar weather suit)	0.18	3.22	3.40

[a]hr · M² °C/kcal

necessary because the surface temperatures will vary depending upon whether it is the exposed skin surface (T_s) or the exterior clothing surface (T_{CS}).

EXAMPLE 6.5

A human factors engineer has been requested to provide input on the environmental design of a business office. Light physical activity will be performed (typing, filing, writing, telephoning) by a group of office workers. The mean individual body mass for an office staff member is 65 kg. A standard deviation of 7.5 kg is common for this office population (mixed men and women).

a. For these clothed workers, the HFE is requested to define a "neutral" environment with respect to human operator heat transfer. It may be approximated that essentially the entire human operator heat transfer is by free convection (still air) and radiation. Heat transfer by conduction and evaporation is negligible.
b. Repeat the analysis performed in part (a), with the exception that forced convection (air moving at velocity, v) replaces free convection (still air).

SOLUTION 6.5(a)

The first step is to define the applicable system equations. Begin by recalling the general equation for the human operator biothermal system:

$$\alpha \dot{T} + \dot{M}_0 + \Delta \dot{M} = -\dot{Q}' - \dot{W} \tag{45}$$

By definition, in a thermally "neutral" environment, the heat input (generation) exactly equals the heat output (transfer to the environment), so that the human operator is in thermal steady state (and, therefore, $\dot{T} = 0$). So equation (45) reduces to:

$$\dot{M}_0 + \Delta \dot{M} = -\dot{Q}' - \dot{W} \tag{i}$$

$\dot{W}$ is the external physical work rate of the human operator, and it is the product of external force applied and external displacement rate. For typical office "work rate," two types of activity predominate:

1. Static work rate (which involves isometric muscle force) such as holding a telephone receiver, gripping a pen when writing, or quietly standing. There is minimal external displacement, so that the external physical work rate is essentially zero.
2. Dynamic work rate (which involves significant rate of external displacement) such as typing, filing folders, or telephone dialing. There is minimal external load (which requires minimal force application), so that the external physical work rate is essentially zero.

Consequently, the external physical work rate for the predominate "office activities" is approximately zero. So equation (i) reduces to:

$$\dot{M}_0 + \Delta \dot{M} = -\dot{Q} \tag{66}$$

A useful first estimate of the basal metabolic rate ($\dot{M}_0$) for a human operator (based upon individual body mass) is:

$$\dot{M}_0 = k \cdot m_B \tag{ii}$$

where:

$$k = -1 \frac{\text{kcal}}{\text{kg} \cdot \text{hr}}$$

m_B = body mass (kg)

A useful approximation of the extra metabolic rate ($\Delta\dot{M}$) is a fraction of the basal metabolic rate (dependent on the level of activity):

$$\Delta\dot{M} = c \cdot \dot{M}_0 \tag{iii}$$

where c = fraction of $\dot{M}_0$ (dependent upon level of activity)
Since heat transfer from the human operator to the environment is by simultaneous free convection and radiation, $\dot{Q}'$ is defined by equation (89) and equation (98):

$$\dot{Q}' = h_{CF}A_C(T_s - T_a) + h_r A_r(T_s - T_w) \tag{iv}$$

Substituting equation (ii), equation (iii), and equation (iv) into equation (66):

$$km_B + c\dot{M}_0 = -h_{CF}A_C(T_s - T_a) - h_r A_r(T_s - T_w)$$

Substituting equation (ii) into the above and factoring results in the applicable equation for this system:

$$(1 + c)k \cdot m_B = -h_{CF}A_C(T_s - T_a) - h_r A_r(T_s - T_w) \tag{v}$$

The second step is to define all equation constants so that equation (v) reduces to three unknowns (m_B, T_a, and T_w).
Recall that k is a proportionality constant:

$$k = -1 \frac{\text{kcal}}{\text{kg} \cdot \text{hr}}$$

Office work is in the general category of light activity, so that from Table 6.9:

$$c = 0.50$$

For clothed individuals, the thermal conductivity coefficients are obtained from Tables 6.5 and 6.6:

$$h_{CF} = 1.2 \frac{\text{kcal}}{\text{hr} \cdot \text{M}^2 \cdot {}^\circ\text{C}}$$

$$h_r = 2.0 \frac{\text{kcal}}{\text{hr} \cdot \text{M}^2 \cdot {}^\circ\text{C}}$$

Table 6.9 Fraction of Basal Metabolic Rate (c) for Various Activity Levels

Activity Level	c^a
Bed rest (hospital patient)	0.10
Sedentary activity, typing	0.30
Light activity, tailor or nurse	0.50
Moderate activity, carpenter, painter	0.75
Severe activity, lumberman, stone mason	1.00

[a]dimensionless

The body surface area (A_B) may be approximated from the body mass using the following equation:

$$A_B = 0.1(m_B)^{0.67} \quad \text{(vi)}$$

The effective surface areas for convection and radiation, respectively, are calculated from equations (99) and (100), using the proportionality coefficients of Table 6.7:

$$A_C = .090A_B \quad \text{(vii)}$$
$$A_r = .075A_B \quad \text{(viii)}$$

Substituting equation (vi) into equation (vii) and equation (viii):

$$A_C = 0.90(m_B)^{0.67} \quad \text{(ix)}$$
$$A_r = 0.75(m_B)^{0.67} \quad \text{(x)}$$

The average skin surface temperature of an individual may be approximated as:

$$T_s = 35°\text{C}$$

Substituting c, h_{CF}, and h_r, and equations (ix) and (x) into equation (v) results in:

$$-1.5m_B = -.108(m_B)^{0.67}(35 - T_a) - .150(m_B)^{0.67}(35 - T_w) \quad \text{(xi)}$$

Rearranging and simplifying:

$$(m_B)^{0.33} = .072(35 - T_a) + .100(35 - T_w) \quad \text{(xii)}$$

The applicable system equation has now been specified with m_B, T_a, and T_w as the three unknowns.

The third step is to solve equation (xii) for the ambient (air) temperature [dependent variable] with respect to surrounding surface (object) temperature (T_w) and body mass (m_B) as the independent variables. Rearrange equation (xii) accordingly:

$$(.072)T_a = (.072)(35) + (.100)(35) - (.100)T_w - (m_B)^{0.33} \quad \text{(xiii)}$$

so that:

$$T_a = 83.6 - 1.4T_w - 13.9(m_B)^{0.33} \quad \text{(xiv)}$$

As a generalization, a comfortable ambient (air) temperature for the clothed working individual is about 20°C. Making this approximation, develop a table of data for $T_a = f(T_w, m_B)$, where m_b is the mean individual body mass, ±1 standard deviation. A representative table of data for equation (xiv) might then be as follows:

T_w(°C)	T_a(°C) (57.5 kg)	T_a(°C) (65.0 kg)	T_a(°C) (72.5 kg)
0°C	30.7	28.5	26.5
5°C	23.7	21.5	19.5
10°C	16.7	14.5	12.5
15°C	9.7	7.5	5.5

SOLUTION 6.5(b)

Since heat transfer from a clothed human operator to the environment is simultaneous *forced* convection and radiation, $\dot{Q}'$ is defined by equation (105) as restated here:

$$\dot{Q}' = \left[3.0\left(\frac{V}{V_0}\right)^{0.6} A_C(35 - T_a)\right] + 2.0A_r(35 - T_w) \tag{105}$$

where $V_0 = 1$ M/s.
Recalling from the left-hand side of equation (v) that:

$$\dot{M}_0 + \Delta\dot{M} = (1 + c)k \cdot m_B \tag{i}$$

Substituting equation (i) and equation (105) into equation (66):

$$(1 + c)k \cdot m_B = -\left[3.0\left(\frac{V}{V_0}\right)^{0.6} A_C(35 - T_a)\right] - 2.0A_r(35 - T_w) \tag{ii}$$

Substituting into equation (ii) the values for k and c given in part [a] and simplifying:

$$-1.5m_B = -\left[3.0\left(\frac{V}{V_0}\right)^{0.6} A_C(35 - T_a)\right] - 2.0A_r(35 - T_w) \tag{iii}$$

Substituting equation 6.5(a)(ix) and equation 6.5(a)(x) into equation (iii) and simplifying:

$$-1.5m_B = -\left[.27\left(\frac{V}{V_0}\right)^{0.6} (m_B)^{0.67}(35 - T_a)\right] - .15(m_B)^{0.67}(35 - T_w) \tag{iv}$$

Rearranging and simplifying:

$$(m_B)^{0.33} = \left[.18\left(\frac{V}{V_0}\right)^{0.6} (35 - T_a)\right] + .10(35 - T_w) \tag{v}$$

where $V_0 = 1$ M/s.

The applicable system has now been specified with m_B, V, T_a and T_w as the four unknowns. At a specified ambient air velocity, solve equation (v) for the ambient (air) temperature (dependent variable) with respect to the surrounding surface (object) temperature (T_w) and body mass (m_B) as the independent variables.

Rearrange equation (v) for $T_a = f(V, T_w, \text{and } m_B)$:

$$.18\left[\frac{V}{V_0}\right]^{0.6} T_a = \left[.18\left(\frac{V}{V_0}\right)^{0.6} (35)\right] + .10(35) - .10T_w - (m_B)^{0.33} \tag{vi}$$

So that:

$$T_a = 35 + \left(\frac{V}{V_0}\right)^{-0.6} [19.44 - 0.56T_w - 5.56(m_B)^{0.33}] \tag{vii}$$

Approximating a comfortable ambient (air) temperature around 20°C, develop a table of data for a set of specified ambient air velocities (V/V_0), at each of which $T_a = f(T_w, m_B)$ where m_B is the mean individual body mass, plus and minus one standard deviation.

A representative table of data for equation (vii) might then be prepared for the following specific velocities (V/V_0):

When $(V/V_0) = 1$, equation (vii) reduces to:

$$T_a = 54.44 - 0.56T_w - 5.56(m_B)^{0.33} \tag{viii}$$

When $(V/V_0) = 2$, equation (vii) reduces to:

$$T_a = 47.83 - 0.37T_w - 3.67(m_B)^{0.33} \quad \text{(ix)}$$

Solution of equation (viii) and equation (ix) then results in the following representative table of data. For $V/V_0 = 1$:

T_w(°C)	T_a(°C) (57.5 kg)	T_a(°C) (65.0 kg)	T_a(°C) (72.5 kg)
0	33.2	32.4	31.6
10	27.7	26.8	26.0
20	22.1	21.2	20.4
30	16.5	15.6	14.8
40	10.9	10.0	9.2

For $V/V_0 = 2$:

T_w(°C)	T_a(°C) (57.5 kg)	T_a(°C) (65.0 kg)	T_a(°C) (72.5 kg)
0	33.9	33.3	32.7
10	30.2	29.6	29.0
20	26.5	25.9	25.3
30	22.8	22.2	21.6
40	19.1	18.5	17.9

v. *Evaporation*

We can apply the evaporative heat transfer equation to the human operator by rewriting equation (46) with respect to evaporative heat flux:

$$q_x = \frac{\dot{Q}_{EVAP}}{A_v} \quad (106)$$

where:

$\dot{Q}_{EVAP}$ = heat flow by evaporation (kcal/hr)
A_V = area from which vaporization is taking place; the equivalent area of skin that is covered with water, m^2

Substituting equation (106) into equation (65), and rearranging:

$$\dot{Q}_{EVAP} = h_v A_v (p_s - p_a) \quad (107)$$

Recall that h_v is the heat transfer coefficient for evaporative heat transfer (kcal/m^2 hr mm Hg). This value varies depending upon whether the evaporative heat transfer is by *free* convection or *forced* convection (as shown in Table 6.10).

Recall that p_s is the vapor pressure of water at the temperature of the skin included in A_v (mm Hg). This vapor pressure of water at skin temperature (T_S) may be approximated by:

$$p_s = .049T_s^2 - .954T_s + 15.6 \quad (108)$$

Table 6.10 Evaporative Heat Transfer Coefficients

Mechanism	h_v[a]
Free convection [1°C ≤ $(T_s - T_a)$ ≤ 20°C	$3.3(T_s - T_a)^{.26}$
Forced convection ($V \leq .50$ m/s)	$9.7\left(\frac{V}{V_0}\right)^{.25}$
Forced convection ($V > .50$ m/s	$12.5\left(\frac{V}{V_0}\right)^{.63}$

[a]kcal/hr · M² · °C
$V_0 = 1$ M/sec

Also recall that p_a is the partial pressure of water vapor in the ambient air (mm Hg). Note that $p_a = 0$ in "dry air," which is 0% humidity. In reality this is rarely the case. A useful approximation is to recall that the vapor pressure of water at 21°C and 760 torr equals 18.8 mm Hg. Then p_a may be calculated as:

$$p_a(T = 21°\text{C}) = H(.188) \tag{109}$$

where H = percent humidity.

Note that in applying equation (107), the "wetted" skin area (due to perspiration) is approximated as only a small proportion of the total body surface area.

6.3. HUMAN OPERATOR THERMOREGULATION

a. Passive Operator Thermoregulation

In Section 6.2 we evaluated human operator heat transfer under steady-state conditions. Effectively, the body temperature remained constant and experienced no change over time. In this section, we shall specifically examine the ability of the human operator to regulate their body temperature, which requires that we evaluate time varying changes in body temperature. This section will consider the simpler case of the passive human operator before proceeding to the more complex case of the active human operator.

The analysis of human operator thermoregulation begins with rearranging equation (45) so as to define the time rate change of temperature:

$$\alpha\dot{T} = \underbrace{-\dot{M}_0 - \Delta\dot{M}}_{\text{Input Heat Rate}} \underbrace{- \dot{Q}' - \dot{W}}_{\text{Output Heat Rate}} \tag{110}$$

where $\dot{T}$ is the mean core temperature ($\dot{T}_c'$) of the human body which is the regulated variable. Equation (110) is a function of the difference between an input heat rate ($\dot{M}_0$ and $\Delta\dot{M}$) and an output heat rate ($\dot{Q}'$ and $\dot{W}$). Recall that $\dot{M}_0$ is the conversion rate of stored chemical energy (into other forms), which is ultimately lost as heat. As such, it represents a negative rate of change of the internal energy for the resting state ($\dot{M}_0$) so that the negative term in equation (110) is a positive (input) heat rate. Since $\Delta\dot{M}$ is the muscle conversion rate of stored chemical energy, the negative term in equation (110) is also a positive (input) heat rate.

Passive Operator:

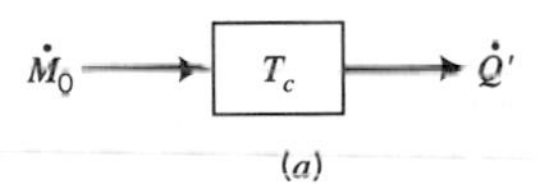

For $T_{sp} > T_c \Rightarrow -(K_p)(T_{sp} - T_c) \Rightarrow$ Positive (Heat Gain)
(cold receptors)

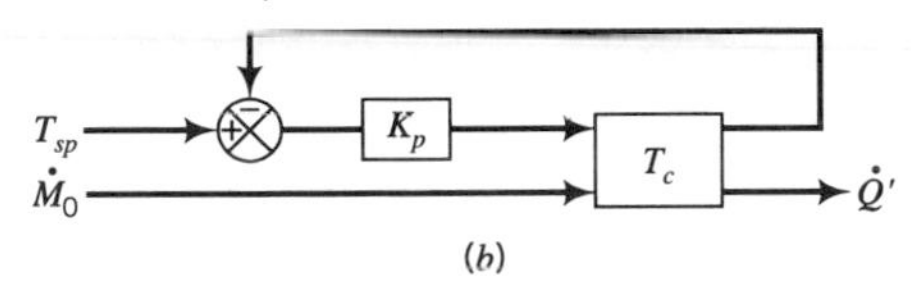

For $T_{sp} < T_c \Rightarrow -(K_p)(T_{sp} - T_c) \Rightarrow$ Negative (Heat Loss)
(warm receptors)

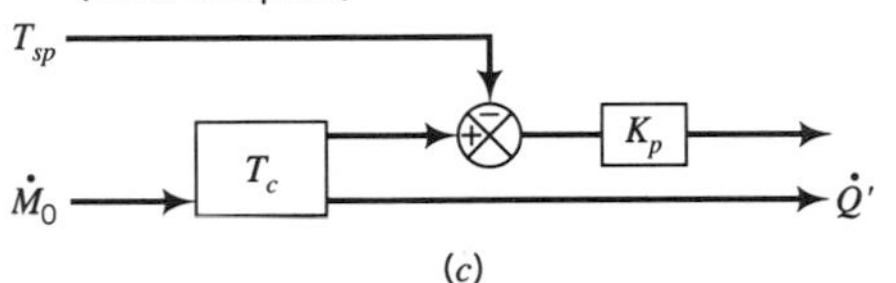

Figure 6.7.a Feed-forward (open-loop) thermoregulatory control system. **b** Human operator thermoregulatory control when mean core temperature is lower than the hypothalamic set point temperature. **c** The thermal regulatory system for the passive human operator when the mean core temperature is higher than the hypothalamic set point temperature.

Recall that $\dot{Q}'$ is the rate of heat energy transferred into the system (and as such carries a positive sign). However, $\dot{Q}'$ carries a negative sign in equation (110) since that parameter represents the rate at which heat energy is lost to the surroundings. Also recall that $\dot{W}$ carries a negative sign in equation (110) when it represents external work done by the system on its surroundings. Consequently, $\dot{Q}'$ and $\dot{W}$ in equation (110) represent negative (output) heat rates.

The passive human operator may be defined as an individual in a resting state. No muscle metabolic energy is being generated, nor is any external work being performed. Mathematically, we may state:

$$\Delta\dot{M} = \dot{W} = 0 \tag{111}$$

Substituting equation (111) into equation (110) results in:

$$\alpha\dot{T} = -\dot{M}_0 - \dot{Q}' \tag{112}$$

This is the feed-forward (open-loop) thermoregulatory control equation for the human operator. This system is shown diagrammatically in Figure 6.7.a where $\dot{M}_0$ is now a positive value.

EXAMPLE 6.6

A variety of conditions affect the ability of a human being to thermoregulate. This variability should be appreciated as the HFE evaluates environmental influences on body temperature.

Thermoregulation is impaired in the very young and the very old. It is also impaired with infirmity (poor hydration and/or nutrition), with debility and disease, and with chemical agents (certain prescription drugs, some over-the-counter drugs, specific street drugs, and alcohol). Simultaneous combinations of these factors result in significant impairment (deregulation) of the thermoregulation system, which may be illustrated as follows.

A 68 kg person has ingested a significant amount of alcohol in order to "cure" a current viral syndrome (cough, congestion, runny nose), which has now lasted five days. Despite these ministrations, the person goes to work at an indoor office job.

For this person, the following parameters exist:

$$\hat{\dot{M}}_0 = -40\frac{\text{kcal}}{\text{hr} \cdot \text{m}^2}$$

$$A_B = 1.80\text{ m}^2$$

$$T_c = 37.0°\text{C}$$

$$T_s = 34.2°\text{C}$$

The ambient (office) temperature is:

$$T_{a \cdot o} = 22.5°\text{C}$$

And the temperature of surrounding (office) objects (e.g., room walls, desks, and chairs) is:

$$T_{w \cdot o} = 21.5°\text{C}$$

The office air is essentially still (zero wind velocity) and the person is clothed for indoor office work. The initial environmental (office) temperature is:

$$T_{a \cdot o} \cong 22.0°\text{C}$$

The effective heat transfer coefficient and effective body surface area is:

$$h_{TOT} = 4.0\frac{\text{kcal}}{\text{hr} \cdot \text{m}^2}$$

$$A_{TOT} = (0.82)A_B$$

At some point in time ($t_0 = 0$), the person leaves the upstairs office and goes into a basement room full of filing cabinets and proceeds to file a number of invoices.

The basement room environmental temperature surrounding object temperature and wind velocity are:

$$T_{a \cdot B} = T_{w \cdot B} = 12°\text{C}$$

$$V = 0$$

Given the above information:

1. Determine the governing differential equation for the core body temperature (T_c) by approximating this individual as an unregulated biothermal system.
2. Can this person maintain their core body temp (T_c) at office environmental temperature?
3. Beginning with the initial time ($t_0 = 0$), what would T_c be at the end of a 2-hr period in the basement?
4. If the worker were to fall asleep and remain in the basement room (undiscovered overnight), what would T_c be after a 12-hr period?

SOLUTION 6.6

1. This is an example of a relatively passive human operator, and we may invoke the governing equation for the unregulated state.

$$\alpha\dot{T}_c = -\dot{M}_0 - \dot{Q}' \tag{112}$$

For a clothed person, we may equate the external heat transfer ($\dot{Q}'$) to the effective heat flow from the human skin surface through the clothing to the external environment ($\dot{Q}_{SCE}$):

$$\dot{Q}' = \dot{Q}_{SCE} \tag{i}$$

Substituting equation (101) into equation (i) results in:

$$\dot{Q}' = h_{TOT}(T_s - T_a)A_{TOT} \tag{ii}$$

Substituting equation (ii) into equation (102):

$$\alpha \dot{T}_c = -\dot{M}_0 - h_{TOT}(T_s - T_a)A_{TOT} \tag{iii}$$

From the given data:

$$\Delta T_{cs} = T_c - T_s = 2.8°\text{C} \tag{iv}$$

Then rearranging equation (iv):

$$T_s = T_c - 2.8 \tag{v}$$

Making the approximation that ΔT_{cs} remains constant over the ambient (environmental) temperature range (12°C–22°C), we may substitute equation (v) into equation (iii):

$$\alpha \dot{T}_c = -\dot{M}_0 - h_{TOT}(T_c - 2.8 - T_a)A_{TOT} \tag{vi}$$

We perturb the system with the environmental (office-to-basement) temperature change:

$$\Delta T_a = T_{a \cdot B} - T_{a \cdot 0} = -10°C$$

So that:

$$T_a(t = 0^-) = T_{a \cdot 0} \tag{vii}$$

$$T_a(t = 0^+) = T_{a \cdot 0} + \Delta T_a \tag{viii}$$

Substituting equation (vii) into equation (vi):

$$\alpha \dot{T}_c = -\dot{M}_0 - h_{TOT}A_{TOT}(T_c - 2.8 - T_{a \cdot 0}) \tag{ix}$$

Substituting equation (viii) into equation (vi):

$$\alpha \dot{T}_c = -\dot{M}_0 - h_{TOT}A_{TOT}(T_c - 2.8 - T_{a \cdot 0} - \Delta T_a) \tag{x}$$

Now we proceed to simplify equations (ix) and (x) in order to find the relationship of $\dot{T}_c$ to T_c at the office environmental temperature ($T_{a \cdot 0}$). We calculate the equation constants:

$$\dot{M}_0 = A_B \hat{\dot{M}}_0$$

$$\dot{M}_0 = (1.80)(-40.0)$$

$$\dot{M}_0 = -72 \frac{\text{kcal}}{\text{hr}} \tag{xi}$$

$$h_{TOT}A_{TOT} = (4.0)(0.82)(1.80)$$

$$h_{TOT}A_{TOT} = 5.90 \frac{\text{kcal}}{°\text{C} \cdot \text{hr}}$$

Recall that:

$$\alpha = m_{TOT}\hat{C}_P \tag{36}$$

and that:

$$\hat{C}_P = 0.86\frac{\text{kcal}}{°\text{C}\cdot\text{kg}}$$

Substituting:

$$\alpha = (68)(.86) = 58.5\frac{\text{kcal}}{°\text{C}}$$

Substituting the above relationship into equation (ix) and equation (x):

$$58.5\dot{T}_c = 72 - 5.90(T_c - 2.8 - T_{a\cdot 0}) \tag{xii}$$
$$58.5\dot{T}_c = 72 - 5.90(T_c - 2.8 - T_{a\cdot 0} - \Delta T_a) \tag{xiii}$$

Separating variables and simplifying results in the two governing differential equations:

$$58.5\dot{T}_c + 5.90T_c = 88.5 + 5.90T_{w\cdot 0} \tag{xiv}$$
$$58.5\dot{T}_c + 5.90T_c = 88.5 + 5.90(T_{a\cdot 0} + \Delta T_a) \tag{xv}$$

2. *Can this person maintain their core body temperature (T_c)?*
Define a steady-state temperature (T_{ss}) as the temperature at which $\dot{T}_0 = 0$. Solve equation (xiv) for T_{ss} when $T_{a\cdot 0} = 22\cdot$C:

$$5.90T_{ss} = 88.5 + (5.90)(22.0)$$

Rearranging:

$$T_{ss} = \frac{218.3}{5.90} = 37.0°\text{C}$$

Therefore when $T_{a\cdot 0} = 22°$C, this person's steady-state temperature, $T_{ss} = T_c = 37.0°$C. Essentially, the basal metabolic heat rate (heat input) exactly equals the combined convective and radiative heat transfer (heat output) to the environment:
3. *In order to calculate T_c after 2 hr in the basement, proceed to solve equation (xv) for $T_c = f(t)$:*
Recalling that $T_{a\cdot 0} = 22°$C and that $\Delta T_a = -10°$C, substitute these values into equation (xv):

$$58.5\dot{T}_c + 5.90T_c = 88.5 + 70.8$$

So that the governing differential equations reduces to:

$$\dot{T}_c + .101T_c = 2.723 \tag{xvi}$$

Solve equation 6.6(xvi) by first noting that:

$$T_c(0) = 37.0°\text{C}$$

Now recall equation (A2) of the Appendix:

$$\dot{y} + by = c \tag{A2}$$

The solution of equation (A2) is equation (A22) of the Appendix:

$$y = \frac{c}{b}(1 - e^{-bt}) + y_0 e^{-bt} \tag{A22}$$

Equation (A2) is analogous to equation 6.6(xvi) when:

$$y = T_c(t)$$
$$y_0 = 37.0$$
$$b = 0.101$$
$$c = 2.723$$

Substituting the above into equation (A22) results in the solution of equation 6.6(xvi):

$$T_c(t) = (37.0)e^{-(.101)t} + \left(\frac{2.723}{.101}\right)(1 - e^{-(.101)t})$$
$$T_c(t) = (37.0)e^{-(.101)t} + 27.0 - (27.0)e^{-(.101)t}$$

so that:

$$T_c(t) = 27.0 + (10.0)e^{-(.101)t} \tag{xvii}$$

Now, solve for T_c [from equation (xvii)] after 2 hr of basement environmental temperature:

$$T_c = 27.0 + (10.0)e^{-.202}$$
$$T_c = 27.0 + 8.2 = 35.2°\text{C}$$

Hypothermia begins at a core temperature of 35.0°C and lower. Our worker is at borderline hypothermia after only 2 hr in the basement!

4. Solve for T_c after 12 hours of basement environmental temperature:

$$T_c = 27.0 + (10.0)e^{-1.21}$$
$$T_c = 27.0 + 3.0 = 30.0°\text{C}$$

Severe hypothermia has now occurred. Immediate emergency medical treatment will be required *if* the worker's biothermal system remains completely unregulated for this period of time.

Feedback (closed-loop) thermoregulatory control of the human operator is accomplished by introducing a passive proportional heat rate into equation (112) as follows:

$$\alpha\dot{T} = -\dot{M}_0 - \underbrace{K_P(T_{SP} - T)}_{\text{Proportional Heat Rate (Passive)}} - \dot{Q}' \tag{113}$$

where:

K_P = passive heat rate constant, kcal/hr m² °C
T_{SP} = hypothalamic set point temperature, °C
T = mean core temperature, °C

It is apparent from inspection of equation (113) that $K_P(T_{SP}-T)$ is a heat rate that is proportional to the difference between the set point temperature and mean core temperature. In the simplest case, where the mean core temperature is equal to the set point temperature, equation (113) will reduce to equation (112) as shown in Figure 6.7.a.

The *passive heat rate constant* (K_P) is intrinsically a negative heat rate. However, the proportional heat rate term may be either positive or negative depending on the sign of ΔT. This is of both physiological significance and thermoregulatory control significance as illustrated in Figures 6.7.b and 6.7.c. Human operator thermoregulatory control is illustrated in Figure 6.7.b when the mean core temperature (T_c) is lower than the hypothalamic set point temperature (T_{SP}). In this situation, the passive proportional heat rate term is negative and this becomes a positive (input) heat rate with respect to equation (113). Figure 6.7.b is consequently a proportional heat rate gain. The obvious effect of this negative feedback would be to increase the mean core temperature (T_c) toward the hypothalamic set point temperature (T_{SP}). The physiological interpretation of this thermoregulatory control is as follows. When the mean core temperature (T_c) is less than the hypothalamic set point temperature (T_{SP}), cold receptors are stimulated and neural impulses to the posterior hypothalamus initiate reflex adaptive mechanisms that result in an overall heat gain by the system. This is accomplished by a combination of increasing heat production by the human body as well as decreasing heat loss. Specifically, the passive proportional heat rate in Figure 6.7.b is a quantitative representation of the following mechanisms:

1. Peripheral vasoconstriction, which significantly decreases the forced convection of heat transfer between the core body region and the skin region. Since the fat layer beneath the skin's surface is a good insulator, skin surface temperature is also lowered effectively reducing the thermal gradient for external heat transfer.
2. There is an increase in muscular tone ($d\dot{M}$), which is manifested as shivering. This represents an extra conversion rate of stored chemical energy in the muscle region that is ultimately dissipated as internal heat.
3. Certain hormones are released (thyroxin and epinephrine), which increase the basal metabolic rate ($\dot{M}_0$) in the core body region. This also results in an extra stored chemical energy conversion rate that is ultimately dissipated as internal heat within the core body region.

The thermal regulatory system for the passive human operator is shown in Figure 6.7.c for the case in which the mean core temperature (T_c) is higher than the hypothalamic set point temperature (T_{SP}). The temperature differential (ΔT) is then a negative value and when multiplied by the negative value of the passive heat rate constant (K_P) results in a positive value for the passive proportional heat rate. When this parameter is positive, equation (113) indicates that it will sum as a negative (output) heat rate. As shown in Figure 6.7.c the resultant thermoregulatory system can be diagramed with $K_P\,(T_c - T_{SP})$ as a proportional heat rate loss. The physiological interpretation of this thermoregulatory system for the passive human operator is as follows. When the mean core temperature (T_c) is greater than the hypothalamic set point temperature (T_{SP}) warm receptors are stimulated. Consequently, nerve impulses to the anterior hypothalamus initiate adaptive reflex mechanisms which result in an overall human operator heat loss. This overall heat loss is a combination of both decreased heat production and also an increased heat dissipation. Specifically, the passive proportional heat rate in Figure 6.7.c represents the following mechanisms:

1. Stimulate peripheral vasodilation in the skin region. This effectively increases forced convection between the core region and the skin region. The result is an increase in the internal heat transfer toward the periphery of the body. A second result of peripheral vasodilation is to increase skin surface temperature in order to enhance a thermal gradient that favors external heat transfer to the environment.
2. Stimulate perspiration at the skin surface for evaporative heat transfer. This may be the only mechanism for body heat loss when the surface skin temperature (T_s) cannot be increased above the environmental temperature (T_e).
3. Decrease any extra metabolic rate so that the human operator experiences only basal metabolic rate ($\dot{M}_0$). This is an indirect effect associated with the sensation of fatigue (from elevated body temperature) combined with dehydration (the loss of water and salt due to evaporative heat transfer), which will result in the human operator becoming as quiescent as possible.

Review of Figures 6.7.a–c indicates that the passive operator thermoregulatory system is a combination of a feed-forward (open-loop) control system and also a feedback (closed-loop) control system. Figures 6.7.b and 6.7.c depict the feedback control system for the human operator superimposed upon the feed-forward control system. This is apparent when the mean core temperature (T_c) is equal to the hypothalamic set point temperature (T_{SP}). When this condition occurs, Figures 6.7.b and 6.7.c reduce to Figure 6.7.a. A characteristic feature of human operator control systems is indeed the combination of feedback control with continuous feed-forward control. This will also be apparent with respect to the human operator neuromuscular control system (see Chapter 9).

EXAMPLE 6.7

One week later, the same office worker as in Example 6.6 is again at work. This time the person is well (no flu syndrome) and sober. You may assume that all other parameters and conditions as per part (a) remain the same, except:

Approximate the individual as a regulated biothermal system with the following characteristics:

$$T_{SP} = 37.0°\text{C}$$

$$\hat{K}_P = -25 \frac{\text{kcal}}{\text{hr} \cdot \text{M}^2 \cdot °\text{C}}$$

1. What is this person's steady-state temperature (T_{ss}) in the office environment?
2. Determine the equation for the core body temperature [$T_c = f(t)$].
3. What is T_c at the end of a 2-hr period in the basement environment?
4. What is T_c at the end of a 12-hr period in the basement environment?

SOLUTION 6.7

For this example of the passive human operator, we invoke the governing equation for a regulated system:

$$\alpha \dot{T}_c = -\dot{M}_0 - K_P(T_{SP} - T_c) - \dot{Q}' \tag{113}$$

For a clothed human operator, by analogy to equation 6.6(iii):

$$\alpha \dot{T}_c = -\dot{M}_0 - K_P(T_{SP} - T_c) - h_{TOT}A_{TOT}(T_s - T_a) \tag{i}$$

Noting (as in Example 6.6) that $\Delta T_{cs} = 2.8°C$ and $\Delta T_a = -10°C$, recall:

$$T_a(t = 0^-) = T_{a \cdot 0} \tag{ii}$$
$$T_a(t = 0^+) = T_{a \cdot 0} + \Delta T_a \tag{iii}$$

Substituting equation (ii) into equation (i):

$$\alpha \dot{T}_c = -\dot{M}_0 - K_P(T_{sp} - T_c) - h_{TOT}A_{TOT}(T_c - 2.8 - T_{a \cdot 0}) \tag{iv}$$

Substituting equation (iii) into equation (i):

$$\alpha \dot{T}_c = -\dot{M}_0 - K_P(T_{sp} - T_c) - h_{TOT}A_{TOT}(T_c - 2.8 - T_{a \cdot 0} - \Delta T_a) \tag{v}$$

Now, simplify equations (iv) and (v) in terms of the equation constants. Recall that:

$$K_P = \hat{K}_P A_B$$
$$K_P = (-25)(1.8) = -45 \frac{\text{kcal}}{\text{hr} \cdot °\text{C}} \tag{vi}$$

Substitute these results and the other equation constants obtained in Example 6.6 into equations (iv) and (v):

$$58.5\dot{T}_c = 72 + 45(37 - T_c) - 5.90(T_c - 2.8 - T_{a \cdot 0}) \tag{vii}$$
$$58.5\dot{T}_c = 72 + 45(37 - T_c) - 5.90(T_c - 2.8 - T_{a \cdot 0} - \Delta T_a) \tag{viii}$$

We may now answer the specific questions:

1. Solve equation (vii) for the steady-state temperature, T_{ss}, when $T_{a \cdot 0} = 22°C$:

$$72 + 45(37 - T_{ss}) - 5.90(T_{ss} - 24.8) = 0$$
$$72 + 1665 - 50.9T_{ss} + 146 = 0$$
$$T_{ss} = \frac{1883}{50.9} = 37.0°C$$

Since $T_{ss} = T_c$ at the office ambient temperature ($T_{a \cdot 0}$) of 22°C, our worker is also in heat balance [as per part (a)]. Note that this is now a regulated biothermal system. The proportional heat rate, $K_P(T_{sp} - T_c)$, is zero because:

$$T_{sp} = T_c = T_{ss} \tag{ix}$$

2. Solve equation (viii) for $\dot{T}_c = f(t)$. Separate variables:

$$58.5\dot{T}_c + 50.9T_c = 72 + 1665 + 16.5 + 5.90T_{a \cdot 0} + 5.90\Delta T_a$$

Recalling that $T_{a \cdot 0} = 22°C$ and $\Delta T_a = -10°C$:

$$58.5\dot{T}_c + 50.9T_c = 1754 + 71$$
$$\dot{T}_c + .870T_c = 31.2 \tag{x}$$

Solve equation 6.7(x) by first noting that:

$$T_c(0) = 37.0°C$$

Now recall equation (A2) of the Appendix:

$$\dot{y} + by = c \qquad \text{(A2)}$$

The solution of equation (A2) is equation (A22) of the Appendix:

$$y = \frac{c}{b}(1 - e^{-bt}) + y_0 e^{-bt} \qquad \text{(A22)}$$

Equation (A2) is analogous to equation 6.7(x) when:

$$y = T_c(t)$$
$$y_0 = 37.0$$
$$b = 0.870$$
$$c = 31.2$$

Substituting these values into equation (A22) results in the solution of equation 6.6(x):

$$T_c(t) = (37.0)e^{-(.870)t} + \left(\frac{31.2}{.870}\right)(1 - e^{-(.870)t})$$

$$T_c(t) = (37.0)e^{-(.870)t} + 35.9 - (35.9)e^{-(.870)t}$$

so that:

$$T_c(t) = 35.9 + (1.1)e^{-(.870)t} \qquad \text{(xi)}$$

It is instructive to compare equation (xi) with equation 6.6(xvii). Note that for this regulated biothermal system the *magnitude* of the transient response is reduced to about one-tenth that of the unregulated system. Furthermore, the *rate constant* of the transient response for the regulated system is increased over eight times that of the unregulated system.

This translates into a regulated biothermal system that exhibits distinctly less temperature variation (in response to environmental perturbations) and distinctly more rapid rates of change, compared to an unregulated biothermal system. These characteristic features are apparent in the remaining elements of this part [b] solution.

3. *Solve for T_c [from equation (xi)] after two hours of basement environmental temperature:*

$$T_c = 35.9 + (1.1)e^{-1.74}$$
$$T_c = 35.9 + 0.2 = 36.1°\text{C}$$

A small magnitude change, but at a rapid rate of change.

Compare to $T_c = 35.2$°C for the unregulated system. After two hours, there is one-half the magnitude of change (from 37.0°C) but this has occurred at a distinctly faster rate of change.

4. *Solve for T_c [from equation (xi)] after 12 hr of basement environmental temperature.*

$$T_c = 35.9 + (1.1)e^{-10.4}$$
$$T_c = 35.9°\text{C}$$

Because of the large rate constant, the regulated system is essentially at a new steady-state temperature (T'_{ss}) after about 3.5 hr (i.e., three time constants). Compare to 30.0°C for the unregulated system.

b. Active Operator Thermoregulation

The active human operator will perform work upon the environment at an external work rate ($\dot{W}$). This will require that an extra metabolic energy rate ($\Delta\dot{M}$) occur in the muscle region. Consequently, the feed-forward (open-loop) thermoregulatory system for the active human operator is a direct application of equation (110):

$$\alpha\dot{T} = -\dot{M}_0 - \Delta\dot{M} - \dot{Q}' - \dot{W} \tag{110}$$

The external work rate ($\dot{W}$) will be some function of the extra metabolic energy conversion rate ($\Delta\dot{M}$) generated in the muscle region. It is useful to characterize this thermodynamic process with respect to its thermal efficiency.

The purpose of an engineering thermodynamic cycle is the conversion of heat into work. This concept of a thermodynamic cycle may be characterized by a thermal efficiency. Conventionally, thermal *work* efficiency ($\overline{\eta}$) is the ratio of the net external work output from a process to the heat added to that process:

$$\overline{\eta} = \frac{\text{external work output}}{\text{added heat input}} \tag{114}$$

In an analogous manner, it is often convenient to define a thermal *power* efficiency (η as follows):

$$\eta = \frac{\text{external work rate output}}{\text{added heat rate input}} \tag{115}$$

With respect to the human operator, recall that $\Delta\dot{M}$ is the conversion rate of stored chemical energy (into other forms) within the muscle region. As stated previously, this decrease in the internal energy of composition is ultimately lost as heat (if no external work is performed). Consequently, $\Delta\dot{M}$ represents the added heat rate input into the system. When the system generates an external work rate output (to the environment), equation (115) may be defined for these human operator parameters:

$$\eta = \frac{\dot{W}}{-\Delta\dot{M}} \tag{116}$$

Recall that $\Delta\dot{M}$ is a negative value, so that the denominator of equation (116) must be multiplied by minus one in order to represent a positive (input) heat rate.

The thermal efficiency of the muscle region, as defined by equation (116), has been variably estimated but is always at or below 25%. The specific thermal efficiency depends greatly upon the type of work performed. For many practical engineering applications, it is useful to characterize external work (output) as predominantly dynamic work or predominantly static work (see Chapter 8). The thermal efficiency for a purely dynamic (aerobic-type) task performed by a physically conditioned individual (such as pedaling a bicycle) approaches 0.25. In this case, $\dot{W}$ represents a force-velocity product (applied force required to rotate the pedals multiplied by the external rotational velocity of the pedals).

$$\dot{W} = F_{app}V_{ext} \tag{117}$$

Referring to equation (117), there are two ways by which η can approach zero. When the applied force (F_{app}) is minimal, then even though external velocity (V_{ext}) is high, $\dot{W}$ will be minimal (per equation [117]), and η will approach zero. A second

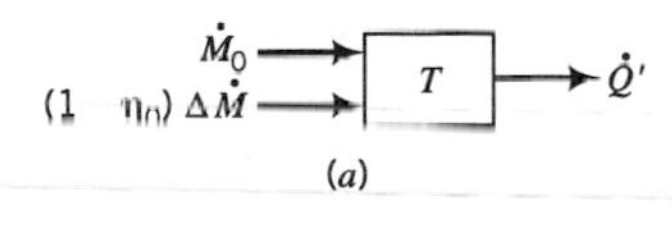

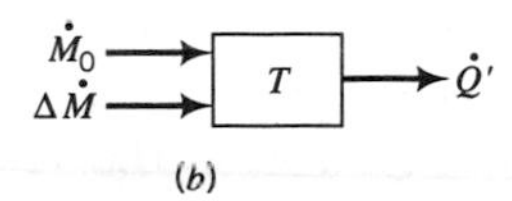

Figure 6.8.a Diagrammatic representation of equation (120). **b** Diagrammatic representation of equation (121).

way that η approaches zero is when the human operator performs purely static (isometric-type) work. An example would be attempting to unscrew a "frozen" bolt with a hand wrench. Significant applied force is generated, but there is minimal external displacement. Consequently, the thermal efficiency for static work (and static work rate) approach zero. For most applications of practical interest to the human factors engineer, η will vary somewhere between 0 and 0.25.

Equation (110) may be modified to account for the thermal efficiency (η) of the active human operator as follows. First, rearrange equation (116):

$$\dot{W} = -\eta\Delta\dot{M} \tag{118}$$

Substituting equation (118) into equation (110):

$$\alpha\dot{T} = -\dot{M}_0 - \Delta\dot{M} - \dot{Q}' + \eta\Delta\dot{M} \tag{119}$$

and after factoring, results in:

$$\alpha\dot{T} = -\dot{M}_0 - (1-\eta)\Delta\dot{M} - \dot{Q}' \tag{120}$$

This equation represents the feed-forward (open-loop) thermal regulatory system for an active human operator when performing external work. It is apparent from equation (120) and Figure 6.8.a that the external work output is accounted for by adjusting the muscle metabolic rate by means of the thermal efficiency. The efficiency of the muscle system will vary not only with respect to the type of external work being done but with the particular muscles involved as well as the general physical conditioning of the human operator. Consequently, equation (120) is a more generally representative model of the human operator than equation (110).

The governing equation for the feed-forward (open-loop) thermal regulatory system for the active human operator when performing only "internal" work (i.e., purely "static" work or purely "velocity" work) may also be defined. In this case $\eta = 0$, so that equation (120) reduces to:

$$\alpha\dot{T} = -\dot{M}_0 - \Delta\dot{M} - \dot{Q}' \tag{121}$$

which is represented diagrammatically and symbolically in Figure 6.8.b.

EXAMPLE 6.8

As noted in Example 6.6, impairment of the ability to thermoregulate can be caused by a variety of factors and combination thereof. Such impairment may result in hypothermia, as described for a passive human operator exposed to a significant decrease in environmental temperature (per Example 6.6). Correspondingly, such impairment may result in hyperthermia, especially in an active human operator who is exposed to a significant increase in environmental temperature. This may be illustrated as follows:

A 62-kg utility worker has just experienced two days of a viral gastrointestinal upset during which the person hasn't eaten very much and also lost about 3 kg of body water. (The worker normally has a 65 kg body mass!) Thanks to the help of certain over-the-counter drugstore remedies (which unfortunately worsen the dehydration and disrupt some thermoregulatory responses), the individual feels better and goes to work, which is at a utility construction site.

For this person, the following parameters exist:

$$\hat{\dot{M}}_0 = -40 \frac{\text{kcal}}{\text{hr} \cdot \text{M}^2}$$

$$A_B = 1.75 \text{ M}^2$$

$$T_c = 37.0°\text{C}$$

$$T_s = 35.0°\text{C}$$

During the morning (A.M.), it is cloudy and the worker is laying an underground utility line (which includes digging the trench). Since this work involved shoveling, it has an isometric component and a dynamic component, so that the overall work efficiency is:

$$\eta = 0.12$$

At a prior time in the HFE laboratory, the extra metabolic rate ($\Delta\dot{M}$) when performing underground cable laying (e.g., shoveling, trenching, cable running, etc.) was determined by indirect calorimetry (see Example 6.2):

$$\Delta\hat{\dot{M}} = -40.4 \frac{\text{kcal}}{\text{hr} \cdot \text{M}^2}$$

During the morning, the work is being done in a well-shaded (woods-like) area. The ambient air temperature is:

$$T_{a\cdot AM} = 20.5°\text{C}$$

The temperature of the surrounding objects (e.g., trees and grass) is:

$$T_{w\cdot AM} = 20.0°\text{C}$$

There is a windy breeze at:

$$V = 1.615 \text{ M/sec}$$

and the person is lightly clothed for outdoor work. Given these conditions, we may approximate a morning environmental temperature:

$$T_{e\cdot AM} \cong T_{a\,AM} \cong T_{w\cdot AM} = 20.25°\text{C}$$

The outside humidity is near 100% so that the person is essentially being cooled by combined forced convection and radiation (without evaporation). Therefore, the effective surface areas (from equation (99), equation (100) and Table 6.7) are:

$$A_c = (0.90)A_B$$

$$A_r = (0.75)A_B$$

1. Does this individual maintain their core body temperature at this activity level within this environment?

It is now early afternoon (P.M.) and the clouds have dispersed; the sun is high and shining. The worker now leaves the woods-like environment and is directly exposed to the hot sun (at an initial time, $t_0 = 0$). Furthermore, the underground cable laying now proceeds between a black asphalt parking lot to one side and a black-top county road to the other.

The ambient air temperature and the surrounding object (black-top surfaces) temperatures are now elevated and equivalent. Therefore, we may approximate an afternoon (P.M.) environmental temperature:

$$T_{e\cdot PM} \cong T_{a\cdot PM} \cong T_{w\cdot PM} = 30.25°\text{C}$$

The wind velocity and relative humidity remain the same and steady as in the morning.

2. When the worker transitions from the shady woods to the open sun environment, what is the equation for the core body temperature [$T_c = f(t)$] by approximating this individual as an unregulated biothermal system?

3. What would T_c be at the end of a two hour period in the open sun environment?

SOLUTION 6.8

This is an example of an active human operator, and we invoke the appropriate equation for the unregulated state:

$$\alpha \dot{T}_c = -\dot{M}_0 - (1 - \eta)\Delta\dot{M} - \dot{Q}' \tag{120}$$

Defining Q' as per equation (105):

$$Q' = \left[(3.0)\left(\frac{V}{V_0}\right)^{0.6} A_C(35 - T_a)\right] + (2.0)A_r(35 - T_w) \tag{105}$$

Recalling that:

$$T_{e\cdot AM} \cong T_{a\cdot AM} \cong T_{w\cdot AM} = 20.25°\text{C}$$

Substituting this relationship into equation (105) and factoring:

$$\dot{Q}' = \left[(3.0)\left(\frac{V}{V_0}\right)^{0.6} A_C + (2.0)A_r\right](35 - T_{e\cdot AM}) \tag{i}$$

where $T_{e\cdot AM}$ = A.M. environmental temperature.

Also recalling that:

$$T_{e\cdot PM} \cong T_{a\cdot PM} \cong T_{w\cdot PM} = 30.25°\text{C}$$

Then substituting this relationship into equation (105) and factoring:

$$\dot{Q}' = \left[(3.0)\left(\frac{V}{V_0}\right)^{0.6} A_C + (2.0)A_r\right](35 - T_{e\cdot PM}) \tag{ii}$$

where $T_{e\cdot PM}$ = P.M. environmental temperature.

From the given data:

$$\Delta T_{cs} = T_c - T_s = 2.0°\text{C}$$

Approximating ΔT_{cs} as constant over the environmental range of 20–30°C:

$$35 = T_c - 2.0$$

Substituting this equation into equations (i) and (ii), and then substituting respectively into equation (120):

$$\alpha \dot{T}_c = -\dot{M}_0 - (1-\eta)\Delta\dot{M} - \left[(3.0)\left(\frac{V}{V_0}\right)^{0.6} A_C + 2.0A_r\right](T_c - 2 - T_{e\cdot AM}) \quad \text{(iii)}$$

We perturb the system with the environmental (woods-to-open sun) temperature change:

$$\Delta T_e = T_{e\cdot PM} - T_{e\cdot AM} \quad \text{(iv)}$$

$$\Delta T_e = 30.25 - 20.25 = 10°\text{C}$$

Restating equation (iv):

$$T_{e\cdot PM} = T_{e\cdot AM} + \Delta T_e \quad \text{(v)}$$

This environmental perturbation in the time domain is:

$$T_e(t = 0^-) = T_{e\cdot AM}$$

$$T_e(t = 0^+) = T_{e\cdot PM} = T_{e\cdot AM} + \Delta T_e \quad \text{(vi)}$$

Substituting equation (vi) into equation (iii) results in:

$$\alpha \dot{T}_c = -\dot{M}_0 - (1-\eta)\Delta\dot{M} - [(3.0)(v)^{0.6}A_C + (2.0)A_r](T_c - 2 - T_{e\cdot AM} - \Delta T_e) \quad \text{(vii)}$$

where $v = V/V_0$.

We may now simplify equations (iii) and (vii) in terms of the equation constants:

Calculate α (from equation [36]):

$$\alpha = m_{TOT}\hat{C}_p \quad (36)$$

$$\alpha = (62)(0.86) = 53.3\,\frac{\text{kcal}}{°\text{C}}$$

Calculate $\dot{M}_0$:

$$\dot{M}_0 = \hat{\dot{M}}_0 A_B$$

$$\dot{M}_0 = (-40.0)(1.75) = -70.0\,\frac{\text{kcal}}{\text{hr}}$$

Calculate $\Delta\dot{M}$:

$$\Delta\dot{M} = \Delta\hat{\dot{M}} \cdot A_B$$

$$\Delta\dot{M} = (-40.0)(1.75) = -70.7\,\frac{\text{kcal}}{\text{hr}}$$

Calculate A_C, A_r, and $(V/V_0)^{0.6}$:

From equation 6.7(i):

$$A_C = (0.9)(1.75) = 1.58\ \text{M}^2$$

From equation 6.7(ii):

$$A_r = (.75)(1.75) = 131\ \text{M}^2$$

And also:

$$\left(\frac{V}{V_0}\right)^{0.6} = \left(\frac{1.615}{1.000}\right)^{0.6} = 1.333$$

Calculate the heat transfer proportionality constants:

$$[(3.0)(v)^{0.6}A_C + (2.0)A_r] = [(3.0)(1.333)(1.58) + (2.0)(1.31)]$$
$$[(3.0)(v)^{0.6}A_C + (2.0)A_r] = [6.32 + 2.62] = 8.94$$

Substitute the above into equation (iii):

$$53.3\dot{T}_c = 70.0 + (.88)(70.7) - (894)(T_c - 2 - T_{e \cdot AM}) \qquad \text{(x)}$$

which simplifies to:

$$53.3\dot{T}_c = 132.2 - (8.94)(T_c - 2 - T_{e \cdot AM}) \qquad \text{(xi)}$$

Substitute the above into equation (vii):

$$53.3\dot{T}_c = 70.0 + (.88)(70.7) - (8.94)(T_c - 2 - T_{e \cdot AM} - \Delta T_e) \qquad \text{(xii)}$$

which simplifies to:

$$53.3\dot{T}_c = 132.2 - (8.94)(T_c - 2 - T_{e \cdot AM} - \Delta T_e) \qquad \text{(xiii)}$$

We now proceed to address the specific questions of this example:

1. *Define the steady-state temperature,* T_{ss}, as that temperature at which $\dot{T}_c = 0$. When $T_{eAM} = 20.25°\text{C}$, using equation (xi):

$$132.2 - (8.94)(T_{ss} - 22.25) = 0$$
$$T_{ss} = \frac{132.2}{8.94} + 22.85 = 37.0°\text{C}$$

Since $T_{ss} = T_c$ at $T_{eAM} = 20.25°\text{C}$, this individual is in thermal equilibrium.

2. *Solve equation (xiii) or* $T_c = f(t)$. Separate variables:

$$53.3\dot{T}_c + 8.94T_c = 132.2 + 17.9 + (8.94)(T_{e \cdot AM} + \Delta T_e)$$

Recalling that $T_{eAM} = 20.25°\text{C}$ and that $\Delta T_e = 10°\text{C}$:

$$\begin{aligned} 53.3\dot{T}_c + 8.94T_c &= 150.1 + 270.4 \\ \dot{T}_c + .168T_c &= 7.89 \end{aligned} \qquad \text{(xiv)}$$

Solve equation 6.8(xiv) by first noting that:

$$T_c(0) = 37.0°\text{C}$$

Now recall equation (A2) of the Appendix:

$$\dot{y} + by = c \qquad \text{(A2)}$$

The solution of equation (A2) is equation (A22) of the Appendix:

$$y = \frac{c}{b}(1 - e^{-bt}) + y_0e^{-bt} \qquad \text{(A22)}$$

Equation (A2) is analogous to equation 6.8(xiv) when:

$$y = T_c(t)$$
$$y_0 = 37.0$$
$$b = 0.168$$
$$c = 7.89$$

Substituting the above into equation (A22) results in the solution of equation 6.8(xiv):

$$T_c(t) = (37.0)e^{-.168t} + \left(\frac{7.89}{.168}\right)(1 - e^{-.168t})$$
$$T_c(t) = (37.0)e^{-.168t} + 47.0 - (47.0)e^{-.168t}$$

So that:

$$T_c(t) = 47.0 - (10.0)e^{-.168t} \qquad \text{(xv)}$$

3. Solve for T_c after 2 hr of the open sun environmental temperature.
From equation (xv):

$$T_c = 47.0 - (10.0)e^{-.336}$$
$$T_c = 47.0 - 7.1 = 39.9°\text{C}$$

Heat stroke begins at a core temperature of 40°C and above. Our worker is at borderline heat stroke after only 2 hr in the open sun environment.

When the active human operator performs external work, the feedback (closed-loop) thermoregulatory system may be characterized by:

$$\alpha\dot{T} = -\dot{M}_0 - \Delta\dot{M} - \underbrace{K_A(T_{SP} - T)}_{\text{Proportional Heat Rate (Active)}} - \dot{Q}' - \dot{W} \qquad (122)$$

where:

K_A = active heat rate constant, kcal/hr M^2°C
T_{SP} = hypothalamic set point temperature, °C
T = mean core temperature, °C

This equation defines a thermoregulatory control system in which an active proportional heat rate parameter, K_A ($T_{SP} - T$), is superimposed upon the feed-forward control system of equation (110). This is analogous to the method applied for the passive human operator thermoregulatory control system (as described by equation (113)). With respect to equation (122), the *active heat rate constant* (K_A) is a negative value parameter as was K_P. The physical and physiological interpretation of K_A is similar to that previously discussed for the K_P parameter (although K_A does not necessarily equal K_P).

The governing equation for the thermoregulatory control system of the active human operator performing external work may also be defined with respect to the thermal efficiency (η) by substituting equation (118) into equation (122):

$$\alpha\dot{T} = -\dot{M}_0 - \Delta\dot{M} - K_A(T_{SP} - T) - \dot{Q}' + \eta\Delta\dot{M} \qquad (123)$$

Rearranging equation (123) results in:

$$\alpha\dot{T} = -\dot{M}_0 - (1 - \eta)\Delta\dot{M} - K_A(T_{SP} - T) - \dot{Q}' \qquad (124)$$

The feedback (closed-loop) thermoregulatory system for the active human operator when performing external work [equation (124)] may be interpreted in the following manner:

When the mean core temperature (T_c) is less than the hypothalamic set point temperature (T_{SP}), equation (124) predicts that the active proportional heat rate will be a positive (input) heat rate as illustrated in Figure 6.9.a. The physical definition and physiological interpretation of the resultant system (Figure 6.9.a) is analogous to that for the passive human operator (Figure 6.7.b). When the mean core temperature (T_c) is greater than the hypothalamic set point temperature (T_{SP}), the active proportional heat rate will be a negative (output) heat transfer as depicted in Figure 6.9.b. The physical definition and physiological interpretation of the active proportional heat rate are analogous to that of a passive human operator (see Figure 6.7.c).

Equation (124) may also be modified to represent the active human operator when performing only "internal" work. Recall that "internal" work is purely "static" work or purely "velocity" work. In this case, $\eta = 0$ so that equation (124) reduces to:

$$\alpha \dot{T} = -\dot{M}_0 - \Delta\dot{M} - K_A(T_{SP} - T) - \dot{Q}' \tag{125}$$

The feedback (closed-loop) thermoregulatory system for the active human operator when performing internal work [equation (125)] may be interpreted as follows: When the mean core temperature (T_c) is less than the hypothalamic set point temperature (T_{SP}), the active proportional heat rate will be a positive (input) heat transfer rate as indicated in Figure 6.10.a. This thermoregulatory system for internal work is analogous to that of Figure 6.9.a for external work.

When the mean core temperature (T_c) is greater than the hypothalamic set point temperature (T_{SP}), the active proportional heat rate of equation (125) will be a negative (output) heat transfer rate as illustrated in Figure 6.10.b. This thermoregulatory system for internal work is analogous to that of Figure 6.9.b for external work. For those situations in which the mean core temperature (T_c) is equal to the hypothalamic set point temperature (T_{SP}) when the active human operator is performing external work, equation (124) and Figures 6.9.a and b reduce to equation (120) and Figure 6.8.a. When the mean core temperature is equal to the hypothalamic set point temperature for the active human operator performing internal work,

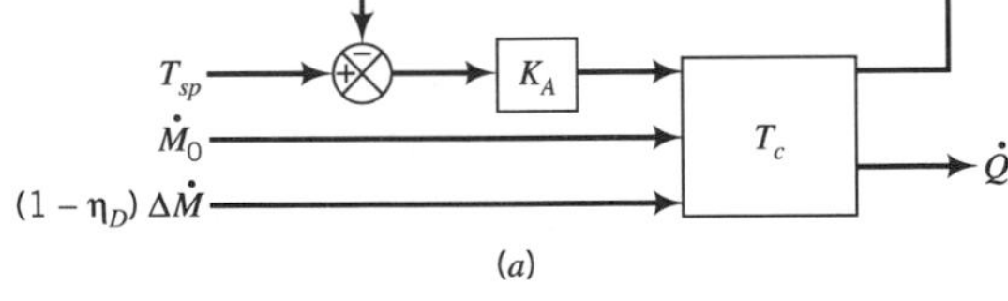

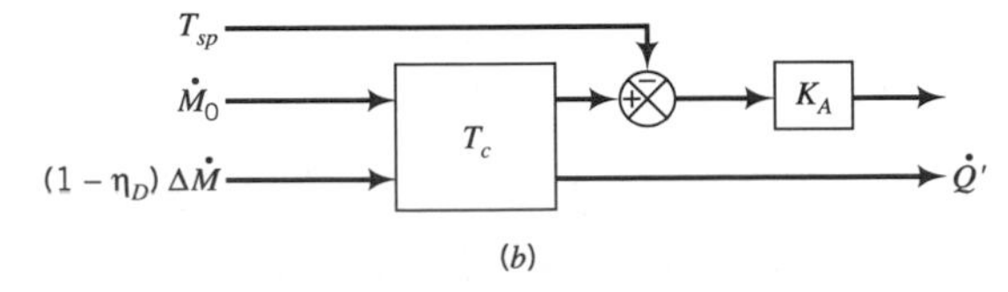

Figure 6.9 For external work: **a** Positive (heat gain): **b** Negative (heat loss).

For $T_{sp} > T_c \Rightarrow -(K_A)(T_{sp} - T_c) \Rightarrow$ Positive (Heat Gain): INTERNAL WORK

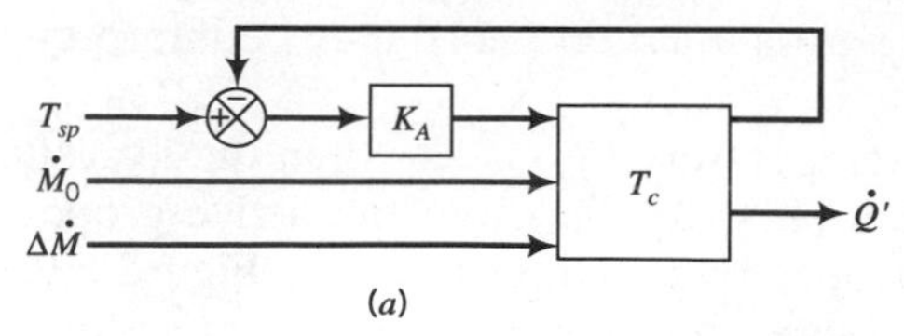

(a)

For $T_{sp} < T_c \Rightarrow -(K_A)(T_{sp} - T_c) \Rightarrow$ Negative (Heat Loss)
(warm receptors)

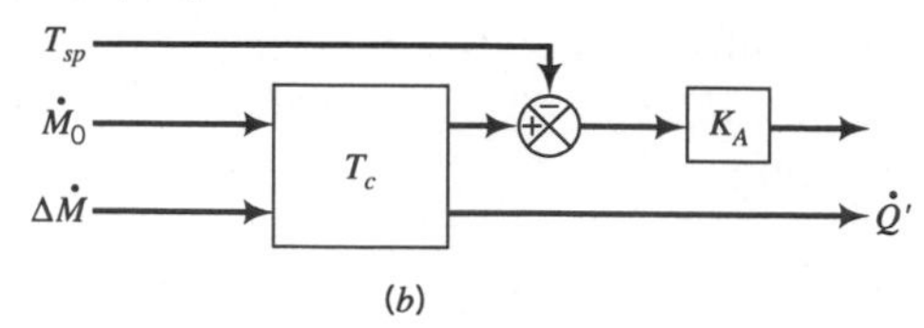

(b)

Figure 6.10 For internal work: **a** Positive (heat gain): **b** Negative (heat loss).

equation (125) and Figures 6.10.a and 6.10.b reduce to equation (121) and Figure 6.8.b, respectively.

EXAMPLE 6.9

A week later, the same utility worker as in Example 6.8 is again at work. This time the worker is healthy (no viral gastroenteritis), well-nourished, well-hydrated, and taking no medications. You may assume that all other parameters and conditions as in Example 6.8 remain the same, except the following.

Approximate this individual as a regulated biothermal system with the following characteristics:

$$T_{SP} = 37°\text{C}$$

$$\hat{K}_A = -25 \frac{\text{kcal}}{\text{hr} \cdot \text{M}^2 \cdot °\text{C}}$$

1. What is this individual's steady-state temperature (T_{ss}) in the shady woods environment?
2. What is the equation for the core body temperature [$T_c = f(t)$]:
3. What is T_c at the end of a 2-hr period in the open sun environment?

SOLUTION 6.9

For this example of an active human operator, we invoke the governing equation for a regulated system:

$$\alpha\dot{T}_c = -\dot{M}_0 - (1 - \eta)\Delta\dot{M} - K_A(T_{SP} - T_c) - \dot{Q}' \qquad (124)$$

For a clothed human operator, when $T_e = T_{e \cdot AM}$, $\dot{Q}'$ is analogous to equation 6.8(i). When $T_e = T_{e \cdot PM}$, $\dot{Q}'$ is analogous to equation 6.8(ii):
From the given data:

$$\Delta T_{cs} = T_c - T_s = 2.0°\text{C} \qquad (\text{i})$$

Approximating ΔT_{cs} as constant over the environmental range of 20°C to 30°C:

$$35 = T_c - 2.0 \qquad (\text{ii})$$

Substituting equation (ii) into equations 6.8(i) and 6.8(ii) and then substituting these results respectively into equation (124):

$$\alpha \dot{T}_c = -\dot{M}_0 - (1-\eta)\Delta\dot{M} - K_A(T_{SP} - T_c) \quad \text{(iii)}$$
$$- [(3.0)(v)^{0.6}A_C + (2.0)A_r](T_s - 2 - T_{eAM}) \quad \text{(iv)}$$

In a manner analogous to Example 6.8, the environmental perturbation (woods to open-sun) in the time domain is:

$$T_e(t = 0^-) = T_{eAM} \quad \text{(v)}$$
$$T_e(t = 0^+) = T_{e\cdot PM} = T_{eAM} + \Delta T_e \quad \text{(vi)}$$
$$\alpha \dot{T}_c = -\dot{M}_0 - (1-\eta)\Delta\dot{M} - K_A(T_{SP} - T_c)$$
$$- [(3.0)(v)^{0.6}A_C + (2.0)A_r](T_s - 2 - T_{eAM} - \Delta T_e) \quad \text{(vii)}$$

where:

$$v = \frac{V}{V_0}$$

Now proceed to simplify equation (iv) and equation (vii) in terms of the equation constants:
Recall that:

$$K_A = \hat{K}_A A_B$$
$$K_A = (-25)(1.75) = -43.8 \frac{\text{kcal}}{\text{hr} \cdot {}^\circ\text{C}}$$

Substitute the above and the other equation constants obtained in Example 6.8 into equation (iv) and equation (vii):

$$53.3\dot{T}_c = 132.2 + 43.8(37 - T_c) - 8.94(T_c - 2 - T_{eAM}) \quad \text{(viii)}$$
$$53.3\dot{T}_c = 132.2 + 43.8(37 - T_c) - 8.94(T_c - 2 - T_{eAM} - \Delta T_e) \quad \text{(ix)}$$

The specific questions presented in this example are now addressed as follows:

1. *Solve equation (viii) for the steady-state temperature T_{ss}, when T_{eAM} = 20.25°C:*

$$132.2 + 43.8(37 - T_{ss}) - 8.94(T_{ss} - 22.25) = 0$$
$$132.2 + 1621 - 52.7T_{ss} + 199 = 0$$
$$T_{ss} = \frac{1952}{52.7} = 37.0^\circ\text{C}$$

Since $T_{ss} = T_c$ at the shady woods environmental temperature (T_{eAM}) of 20.25°C, our worker is in heat balance, as was the case for Example 6.8. Recall from equation 6.7(ix), that for a regulated system (such as this person), the proportional heat rate, $K_A(T_{SP} - T_c)$ is zero since:

$$T_{SP} = T_c = T_{ss}$$

2. *Solve equation (ix) for $T_c = f(t)$. Begin by separating the variables:*

$$53.3\dot{T}_c + 52.7T_c = 132.2 + 1621 + 17.8 + 8.94(T_{e\cdot AM} + \Delta T_e)$$

Recalling that T_{eAM} = 20.25°C and ΔT_e = 10°C:

$$53.3\dot{T}_c + 52.7T_c = 1771 + 270$$
$$\dot{T}_c + .989T_c = 38.3 \tag{x}$$

Solve equation 6.9(x) by first noting that:

$$T_c(0) = 37.0°\text{C}$$

Now recall equation (A2) of the Appendix:

$$\dot{y} + by = c \tag{A2}$$

The solution of equation (A2) is equation (A22) of the Appendix:

$$y = \frac{c}{b}\left(1 - e^{-bt}\right) + y_0 e^{-bt} \tag{A22}$$

Equation (A2) is analogous to equation 6.9(x) when:

$$y = T_c(t)$$
$$y_0 = 37.0$$
$$b = 0.989$$
$$c = 38.3$$

Substituting the above into equation (A22) results in the solution of equation 6.9(x):

$$T_c(t) = (37.0)e^{-.989t} + \left(\frac{38.3}{.989}\right)(1 - e^{-.989t})$$
$$T_c(t) = (37.0)e^{-.989t} + 39.1 - (39.1)e^{-.989t}$$

so that:

$$T_c(t) = 39.1 - (2.1)e^{-.989t} \tag{xi}$$

It should be noted that *magnitude* of the transient response for the regulated system is reduced to about one-fifth that of the unregulated system. Furthermore, the *rate constant* of the transient response for the regulated system is increased about six times that of the unregulated system. The implications of this will follow.

3. Solve for T_c (per equation [xi]) after two hours of open sun environmental temperature.

$$T_c = 39.1 - (2.1)e^{-1.98}$$
$$T_c = 39.1 - 0.3 = 38.8°\text{C}$$

Compare to a T_c = 39.9°C for an unregulated system. After 2 hr, there is about three-fifths the magnitude change (from 37.0°C), but this has occurred at a distinctly faster rate constant.

It is most instructive to compare the regulated response of the *active* human operator [equation 6.9(xi)] with the regulated response of the *passive* human operator [equation 6.7(xi)]. For the same proportional heat rates [$\hat{K}_p = \hat{K}_A$], the *magnitude* of the transient response (for a passive operator) is about half the response magnitude (for an active operator) *with the η and $\Delta\dot{M}$ given in this example.* This is because the active human operator has a distinctly higher heat load (due to the extra metabolic heat rate of the exercising muscles) which must be dissipated. This *extra* heat load must be handled by the active regulating system *in addition* to basal heat load. The passive regulating system handles *only* basal heat load, hence its control margin is tighter.

These differences between passive operator thermoregulation and active operator thermoregulation are independent of whether there is an abrupt *increase or decrease* in the environmental temperature. For example, the interested student should solve Example 6.6 for a ΔT_a of *plus* 10°C and also solve Example 6.7 for a ΔT_a of *minus* 10°C.

FURTHER INFORMATION

D.O. Cooney: *Biomedical Engineering Principles: An Introduction to Fluid, Heat and Mass Transport Processes.* Marcel Dekker. New York. 1976.

C.A. Phillips: *Biomechanics and Biothermodynamics.* ClassNote Publications (Wright State University). Dayton, OH. 1997.

R.C. Seagrave: *Biomedical Applications of Heat and Mass Transfer.* Iowa State University Press. Ames, IA. 1971.

R.E. Sonntag and G.J. Van Wylen: *Introduction to Thermodynamics,* 3rd Edition. John Wiley. New York. 1991.

W.J. Yang: *Biothermal-Fluid Sciences.* Hemisphere Pub. Corp. New York. 1989.

PROBLEMS

6.1. For internal conduction, the core temperature, $T_c = 36°C$, and the skin temperature, $T_S = 33°C$, and the ambient temperature, $T_a = 29°C$. For the composite layers, the core thickness, $\Delta x_C = 2.2$ cm, the muscle thickness, $\Delta x_m = 1.8$ cm, and the skin thickness, $\Delta x_S = 1.2$ cm. The muscles are actively working, and the thermal conductivity coefficient, $k = 0.38$ kcals/M · hr · °C and the thermal coefficient $h = 9.2$ kcals/M^2 · hr · °C. Given the above, find:

a. The muscle temperature, T_m
b. The basal metabolic flux, $\hat{\dot{M}}_0$
c. The muscle metabolic flux, $\Delta\hat{\dot{M}}$

6.2. For a system in *simultaneous* radiation *and* convection, we define State 1 when the differential skin-to-ambient temperature $(T_S - T_a)$ is $-8.9°C$, and the differential skin-to-wall temperature $(T_S - T_w)$ is 15°C. In State 2, the differential skin-to-ambient temperature $(T_S - T_a)$ is 25.8°C and the differential skin-to-wall temperature $(T_S - T_w)$ is $-5°C$. If $\gamma = \dfrac{h_C A_C}{h_r A_r}$, and $\beta = \dfrac{\hat{\dot{M}}_0}{h_r A_r}$, and if γ and β do not change in going from State 1 to State 2, what is the value of γ? What is the value of β?

6.3. Repeat Example 6.6, keeping all parameters and activities the same, *except* as follows:

At some point in time $(t_0 = 0)$, the person leaves the upstairs office and goes into the basement *furnace* room (which is also used as a filing cabinet annex), and proceeds to file a number of invoices.

The basement room environmental temperature, surrounding object temperature, and wind velocity are:

$$T_{a\cdot B} = T_{w\cdot B} = 32°C$$
$$V = 0$$

Given the above information, for this unregulated biothermal system:

a. Beginning with the initial time ($t_0 = 0$), what would T_c be at the end of a 2-hr period in the basement?

b. If the worker were to fall asleep and remain in the basement room (undiscovered overnight), what would T_c be after a 12-hr period?

6.4. Repeat Example 6.7, keeping all parameters and activities the same *except* as follows:

At some point in time ($t_0 = 0$), the person leaves the upstairs office and goes into the basement *furnace* room (with all parameters and activities the same as Problem 6.3).

Given the above information, for this regulated biothermal system:

a. Determine the equation for core body temperature [$T_c = f(t)$].

b. What is T_c at the end of a 2-hr period in the basement environment?

c. What is T_c at the end of a 12-hr period in the basement environment?

6.5. While vacationing at a health spa, a woman *au naturel* is quietly standing alone in a water tub, submerged up to her neck. Assuming her convective heat loss constant (h) to be 16.5 kcal/M^2 · hr · °C, her submerged surface area to be 1.8 M^2, her heat transfer to be 72 kcal/hr, and the water temperature to be 88.84°F, what is her skin temperature?

6.6. At a nudist colony, a man is quietly standing erect in a room. He is alone and is practicing transcendental meditation. His skin temperature is 35°C, the wall temperature is 26.6°C, and the air temperature is 21.1°C. Given the following information:

$$h_r = 4.0 \text{ kcal/M}^2 \cdot \text{hr} \cdot °C$$
$$\Delta Q = \dot{M}_0 = 71 \text{ kcal/hr}$$
$$A_c = 1.8 \text{ M}^2$$
$$A_r = 1.2 \text{ M}^2$$

Determine the value of h_c, the constant for convective heat loss.

6.7. Assume a man is standing in water, half underwater and the other half of his mass above water. If heat loss is the same for both halves, and given the following information:

$$m_{man} = 79.4 \text{ kg}$$
$$T_s = 33°C$$
$$T_a = 30°C$$
$$T_{H_2O} = 31°C$$
$$f_r = 0.75$$
$$f_c = 0.9$$

What is the wind velocity (v)?

6.8. Repeat Example 6.8, keeping all parameters and activities the same *except* as follows:

It is now early afternoon (P.M.) and the worker leaves the woods-like environment and immediately enters a tunnel (at an initial time, $t_0 = 0$).

The ambient air temperature and the surrounding object (tunnel wall and the walkway surface) temperatures are now reduced and equivalent. Therefore, we may approximate an afternoon (P.M.) environmental temperature.

$$T_{e\cdot PM} \cong T_{a\cdot PM} \cong T_{w\cdot PM} = 10.25°\text{C}$$

The wind velocity, and relative humidity remain the same and steady as in the morning.

a. When the worker transitions from shady woods to the tunnel environment, what is the equation for the core body temperature [$T_c = f(t)$] that is obtained by approximating this individual as an unregulated biothermal system?
b. What would T_c be at the end of a 2-hr period in the tunnel environment?

6.9. Repeat Example 6.9, keeping all parameters and activities the same *except* as follows:
At some point in time ($t_0 = 0$), the worker leaves the shady woods and enters a tunnel (with all parameters and activities the same as Problem 6.8).

a. When the worker transitions from shady woods to the tunnel environment, what is the equation for the core body temperature [$T_c = f(t)$] that is obtained by approximating this individual as a regulated biothermal system? You may approximate $T_{e\cdot PM} \cong T_{a\cdot PM} \cong T_{w\cdot PM} = 10.25°\text{C}$.
b. What would T_c be at the end of a 2-hr period in the tunnel environment?

6.10. A man is seated on a raft in a pool, losing heat from both radiation and forced convection, while a woman (wearing a face mask and breathing through a snorkel tube) is standing underwater and losing heat from water convection.
Given the following information:

$$m_{man} = 79.4 \text{ kg}$$
$$m_{woman} = 54.4 \text{ kg}$$
$$T_s \text{ (man)} = 33°\text{C}$$
$$T_s \text{ (woman)} = 33°\text{C}$$
$$T_a = 30°\text{C}$$
$$T_{H_2O} = 31°\text{C}$$
$$f_{cm} = 0.9$$
$$f_{cw} = 0.8$$

What wind velocity (v) is necessary in order for the heat loss from the man to exactly equal the heat loss from the woman?

6.11. A person is riding a bicycle in still air. The air temperature is 30°C, and the skin temperature is 35°C. The mass of the person (m_b) is 70 kg, and given the following:

$\eta = .25$ (estimate based on max $\eta = .25$ for a bicycle ergometer)
$A_b = 1.8 \text{ M}^2$
$k = -1 \text{ kcal/kg} \cdot \text{hr}$
$c = .75$ (from Table 6.9)

Then how fast is the bicyclist riding if the person's heat loss is 100 kcals/hr?

6.12. A man is walking through a lake on a summer day. The water is at mid-thigh level so that approximately twenty percent of the body surface area is in the water. The man is dressed. A wind is blowing at .5 M/s. The body mass (m_b) of the man is 77 kg. Given the following:

$$T_s = 34°\text{C}$$
$$T_a = 30°\text{C}$$
$$T_{H_2O} = 27°\text{C}$$
$$T_w = 27°\text{C}$$
$\eta = .07$ (estimate based on max $\eta = .25$ for a bicycle ergometer and an $\eta = .05$ for walking while pushing a heavy object)
$$A_c = .09 m_b^{.67}$$
$$A_r = .075 m_b^{.67}$$
$$k = -1 \text{ kcal/kg} \cdot \text{hr}$$
$$c = .75$$

What is the extra metabolic energy rate ($\Delta \dot{M}$) [which is in addition to the basal metabolic rate ($\dot{M}_0$)] when performing this activity?

Part III

Human Factors Engineering: Practice

Part III-A

Ergonomic Engineering

Chapters 7 and 8 that follow represent an area of HFE practice that is referred to as Ergonomic Engineering. The following is a brief introduction to this area.

Ergonomics may be defined as the scientific study of work. It focuses upon human capabilities and limitations with respect to the appropriate design of living and working environments. The goal of ergonomics is to design for "ease and efficiency." In other words, minimize human operator stress and fatigue, and also promote work output and productivity. Many different types of professionals practice ergonomics with respect to their disciplines. These include applied psychologists, exercise physiologists, occupational biomechanists, applied physical anthropometrists and also human focused industrial engineers.

Ergonomic engineering may be defined as the engineering practice of ergonomics. The practice of ergonomic engineering involves the analysis and the design of the ergonomic system configuration. Recall that *analysis* is the study of the characteristics of an existing system. The *design* is the selection and arrangement of the various system elements in order to perform a specific task.

The human factors engineer can utilize two different engineering methods for the design phase. The first method is *design by analysis* in which the elements of an existing or prototype ergonomic system are modified. The second method is *design by synthesis* in which the form of the ergonomic system is directly determined from the system specifications. Throughout this textbook, design by analysis is the approach that is utilized. This method allows the student to reinforce and expand their analytical tools and skills with respect to the practice of engineering.

We now proceed to the individual chapters that represent specific areas of ergonomic engineering. Ergonomic biodynamics (Chapter 7) represents the *combination* of biodynamic mechanics (Chapter 4) *and* physical anthropometry as applied to the analysis and design of ergonomic systems. Quantitative workload analysis (Chapter 8) represents the *combination* of bioelectricity and bioelectronics (Chapter 5) *and* quantitative exercise physiology as applied to the analysis and design of ergonomic systems.

Ergonomic Biodynamics

Ergonomic biodynamics is concerned with the motions of the human body and the forces that cause those motions. In this sense, it relies upon the second law of classical Newtonian mechanics. Ergonomic biodynamics is a subset of biodynamics in which the forces are internally generated about the joints of the various body segments, resulting in external motion of those human body segments. Such external body movement is often associated with force application on some external object so that external work is performed by the human. Hence, the "ergonomic" qualifier of the term "biodynamics." Moreover, this external motion (and any associated external work) is intelligently directed by the human operator. It is purposeful movement intended to accomplish a desired action.

Prior to proceeding with this chapter, the student should have reviewed (or already be familiar with) the material presented earlier in Chapters 3 and 4. Chapter 3 has previously reviewed biostatic mechanics with respect to forces acting upon various body segments. Recall that it was necessary to treat the various body segments as rigid bodies in order to perform the required analysis. Chapter 4 has already reviewed biodynamic mechanics with specific reference to particle mechanics.

Ergonomic biodynamics proceeds from biostatic (rigid-body) mechanics and biodynamic (particle) mechanics by examining the dynamics of rigid bodies. As noted in Chapter 4, a particle in a dynamic state is subjected to translational forces only and experiences translational motion only. In a single plane, these forces and the resultant motions have vertical and horizontal components. When undergoing a dynamic motion, a rigid body will also be subjected to rotational forces and will experience rotational movement. As we shall see in this chapter, these translational and rotational forces (and the subsequent movements) require that the body be in a state of dynamic equilibrium as described by Newton's second law.

This chapter is divided into four sections. The first section (segment kinematics) defines the various body segments that will be considered and also defines the kinematic parameters for a body segment. The second section (segment mass and moment) reviews a rigid body as a system with distributed mass. The third section (segment kinetics) examines the relevant kinetic parameters for a segment and interconnects the various segments by means of a link-segment model. The fourth section (segment energetics) examines muscle mechanical power and the muscle mechanical work that is done upon a body segment.

7.1 SEGMENT KINEMATICS

a. Kinematic System

The motion of numerous body segments interconnected together in three-dimensional space is very complex. This complexity is further confounded by the fact that various body segments have multiple degrees of freedom about their associated joints. In order for the human factors engineer to apply the equations of mechanics to this complex system of motion, the specific motion under study and the constituent body segments must be reduced to a deterministic system. This requires the intelligent application of simplifying approximations, a process referred to as *biomathematical modeling* (Chapter 2). The process begins with the selection of a spatial frame of reference and the definition of a spatial coordinate system.

Three spatial frames of reference are generally used in describing the position of the human body in space. Each of these represent two-dimensional planes through which position and movement can be specified. The frontal plane represents a two-dimensional view of the human body as it would appear (for example) when standing directly in front of a person who is moving directly toward you. The transverse plane would be a two-dimensional view of human body movement as it would be seen (for example) by an observer positioned directly above or directly below the person performing the movement. Finally, the sagittal plane would represent position and movements of the human body in a two-dimensional plane as seen by an observer of a person's motion in side-view. An example would be a view of the human body as seen by an observer through a glass window when the person is moving parallel to the plane of the window.

Depending upon the type of body position or movement one desires to analyze, the human factors engineer may need to view the individual in one, two, or all three of these spatial frames of reference. Perhaps the most commonly used view is that of an individual in profile, or side-view. This sagittal plane is the frame of reference which will be consistently used throughout this chapter. Having selected this spatial frame of reference, we may now define the spatial coordinate systems to be used.

Figure 7.1 represents an individual as viewed in the sagittal plane and identifies

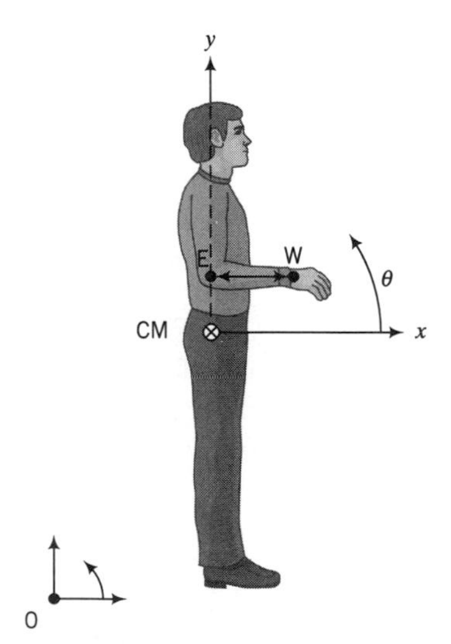

Figure 7.1 Spatial coordinate system.

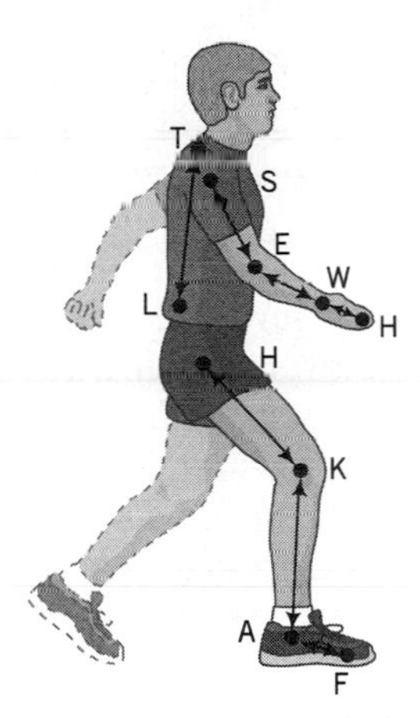

Figure 7.2 Marker locations defining body segments.

the horizontal and vertical coordinates of this plane. In this particular figure, the origin of the coordinate system is at the center of mass of the human body. A positive sign convention is indicated by the arrows. Note that both linear position and angular position are accounted for in this coordinate system. It is important to note that the origin of the spatial coordinate system is arbitrary. It can depend upon the measurement system being used. For example, movement of the entire body recorded and analyzed as a video image might have a quite different coordinate origin than that of an electrogoniometer positioned at one joint to record the relative movements of two interconnected body segments. For the purposes of this chapter, the exact position of the origin of the coordinate system will be variable and unspecified. However, the origin of the rectangular coordinate system will be positioned so that the particular segment (or segments) under consideration will always be within quadrant I, as shown at the lower left of the individual appearing in Figure 7.1. With respect to polar coordinates, however, a particular body segment may appear in quadrants I, II, or IV.

The system of body segments to be analyzed is shown in Figure 7.2. Marker locations represent the ends of individual segments as viewed in right profile. Note that this selection of body segments corresponds to the various rigid bodies that were examined in Chapter 3. The upper extremity link-segment system consists of three segments and two links. Between the shoulder marker (S) and the elbow marker (E) is the upper arm (SE) segment. Between the elbow marker (E) and the wrist marker (W) is the forearm (EW) segment. Between the wrist marker (W) and end of the hand marker (H) is the hand (WH) segment. In this right profile view, the wrist is midway between pronation and supination. When the hand has the fingers fully extended, this would appear in the right side view as if the individual were about to shake hands with someone. When the fingers are fully flexed, this would appear in right side view as if the individual were gripping the handle of a hammer about to strike a nail.

The lower extremity system is a link-segment system consisting of three segments and two links. The hip marker (H) and the knee marker (K) define the thigh (HK) segment. The knee marker (K) and the ankle marker (A) define the foreleg (KA) segment. The ankle marker (A) and the marker located at the ball of the foot (F) represent the foot (AF) segment.

The thoracolumbar spine is identified as a single segment. It is between the thoracic (located at the top of the first thoracic vertebra) marker (T) and the lumbar (located at the base of the fifth lumbar vertebra) marker (T). The thoracolumbar (TL) segment has no link with either the upper extremity system or the lower extremity system.

In reviewing Figures 7.1 and 7.2, it is readily apparent that a number of simplifying approximations have been made. With respect to Figure 7.1, selection of the sagittal frame of reference implies that position and movement in the other two frames of reference (frontal and transverse) can be neglected. This puppet's arm view of the upper extremity may at first appear to be overly simplistic. The model recognizes that human movement occurs in all three planes, but that for the particular task under study, movement in these other two planes is approximated as second order or higher effects. The same reasoning applies also to our puppet's leg view of the lower extremity.

With respect to Figure 7.2, the selection of seven body segments would also appear overly simplistic. This model implies that the head and neck, the shoulders, the pelvis, and the fingers and toes may be neglected as additional segments. The model also implies that the right side of the body is the mirror image of the left side of the body, which neglects anatomical asymmetry (as well as a host of physical disabilities). Therefore, due caution must be exercised when our model is applied to the analysis of human subsystem tasks. Only second order or higher effects should have been deleted from the model for the analysis of the particular task under study.

b. Kinematic Parameters

The complete kinematic description of a body segment when viewed in a single plane (i.e., the sagittal or profile view) requires that nine parameters be used. These parameters describe a particular moment in time and are subject to change during time. These parameters are:

1. The horizontal position (x) of the segment's center of mass.
2. The vertical position (y) of the segment's center of mass.
3. Horizontal linear velocity (V_x) of the segment's center of mass.
4. Vertical linear velocity (V_y) of the segment's center of mass.
5. Horizontal linear acceleration (a_x) of the segment's center of mass.
6. Vertical linear acceleration (a_y) of the segment center of mass.
7. Angle (θ) of the segment in the x-y plane.
8. Angular velocity (ω) of the segments in the x-y plane.
9. Angular acceleration (α) of the segment in the x-y plane.

Referring again to Figure 7.1, consider a simple flexion-extension-flexion task of the forearm (EW) segment. Let the origin of the rectangular coordinate system be at the elbow marker (E) and assume that this marker remains stationary during subsequent movement. The forearm segment is initially aligned colinear with the x-axis (the neutral position) and is initially stationary. Beginning at t_0, the individual flexes the forearm segment raising the wrist marker toward the shoulder (while the elbow marker remains stationary). At t_1, the direction of wrist motion reverses, and the person extends the forearm back to the neutral position (colinear with the x-axis). The person instantaneously hesitates (at t_2) and then continues extending the forearm segment moving the wrist marker further downward. The wrist motion

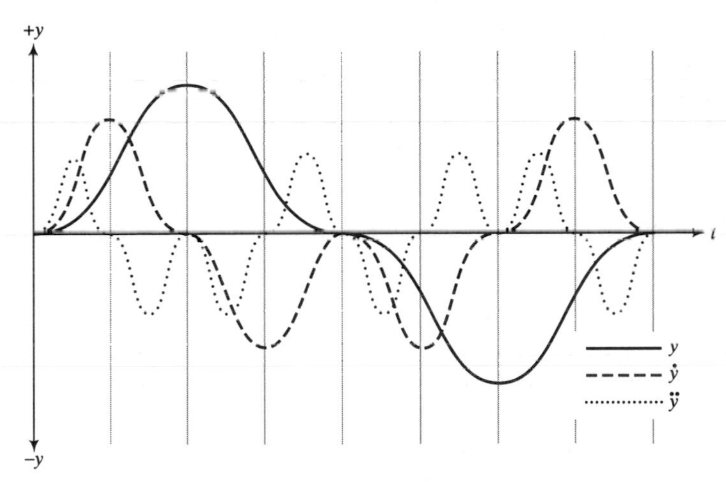

Figure 7.3 Vertical component of the forearm segment center of mass changing with the time.

then continues the same angular distance downward (as it did upward). At t_3, the direction of motion is again reversed, and the wrist is brought upward (by flexing the forearm segment) coming to a rest at its original neutral starting position (at t_4).

Figure 7.3 indicates the vertical component of the forearm segment center of mass as it changes over the time course of performing this task, referred to as the "forearm task." Note that the instantaneous vertical position is biphasic over the time course of the forearm task. Using our specific coordinate system, vertical position of the forearm segment is positive over the first half of the task and negative over the second half of the task. With respect to the vertical velocity of the forearm center of mass, it is biphasic during the first half of the task and also biphasic (but inverted) during the second half of the task. Finally, the vertical acceleration of the forearm segment center of mass is triphasic during the first half of the task and also triphasic during the second half of the task. Note that the triphasic nature of the vertical acceleration is inverted with respect to the two halves of the task.

Table 7.1 presents the associated signs of the individual vertical parameters (displacement, velocity, and acceleration). As per Table 7.1, eight regions during the time course of the forearm task are identified by a unique combination of the vertical parameter signs.

EXAMPLE 7.1

The person (shown in Figure 7.1) is standing at an assembly line facing a conveyor belt upon which a four-cylinder engine is advancing. Beginning with the position

Table 7.1 Regions/Signs of Variable (see Figure 7.3)

Region/Variable	1	2	3	4	5	6	7	8
Displacement	+	+	+	+	−	−	−	−
Velocity	+	+	−	−	−	−	+	+
Acceleration	+	−	−	+	−	+	+	−

indicated in Figure 7.1, the person performs a cyclical upper extremity task, always keeping the E marker stationary. The task is as follows. Initially the person flexes the forearm, raising their wrist and hand, and instantaneously grasps (with thumb and forefinger) a ball bearing (from a ball bearing dispenser). The person extends the forearm back downward and approaches the original position (shown in Figure 7.1) in order to place the ball bearing within a fitted orifice (on upper side of the engine). There is an instantaneous stop in order to seat the ball bearing past a securing O-ring.

The person then opens their hand and continues extending the forearm downward until the hand instantaneously grasps a second ball bearing (from a second ball bearing dispenser). The person then flexes the forearm, raising the wrist and hand upward again. As the person approaches the original starting position (shown in Figure 7.1), they place the ball bearing within a fitted orifice (from the underside of the engine). There is an instantaneous stop in order to seat the second ball bearing within a second securing O-ring.

The engine now proceeds down the conveyor belt and the cycle repeats itself as the next engine approaches the worker. The vertical kinematic parameters for the W marker over the entire cycle (just described) are shown in Figure 7.3.

At some moment in time during this cycle, the six kinematic parameters for the W marker are measured and their associated signs are indicated in the following grid:

a. Indicate with respect to W marker, the six kinematic parameters and their directionality.
b. For the eight time regions that occur during this forearm task (see Figure 7.3 for the vertical kinematic parameters), identify the specific regions in which the six kinematic parameters have the signs given in the grid above.

	x	y	θ
Velocity	−	+	+
Acceleration	+	−	−

SOLUTION 7.1(a)

The required figure is as follows:

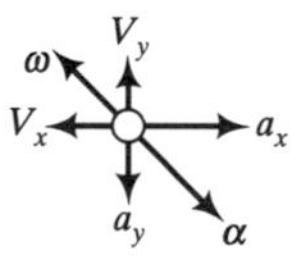

SOLUTION 7.1(b)

The vertical velocity is positive (upward) and the vertical acceleration is negative (downward). This means that the W marker is moving upward but at progressively slower rates (decelerating). Inspection of the eight regions (of Figure 7.3) indicates that region two and region eight both satisfy this combination of vertical kinematic parameters.

With respect to the W marker, the angular velocity is counterclockwise (positive) but the angular acceleration is clockwise (negative). This means that the W marker is rotating in a counterclockwise direction but at a progressively slower rate (decelerating). With respect to the forearm task, this again would occur in either region two or region eight.

With respect to the horizontal kinematic parameters, the horizontal velocity is leftward (negative) and the horizontal acceleration is rightward (positive). This means that the W marker is moving leftward at a progressively slower rate (decelerating). This can only occur in *region two* of the forearm task. In region eight, the horizontal velocity of the W marker would be directed rightward (positive) but at a progressively slower rate (i.e., the W marker would be decelerating so that the horizontal acceleration would be directed leftward [negative]).

In kinesiological measurement systems, at any instant in time, positional coordinate information will be determined for the actual markers that overlie specific joints (see Figure 7.1). The system of interest to the human factors engineer, however, is the segment that directly connects the two adjacent markers (see Figure 7.1). In order to accomplish a segment analysis, it is necessary to determine that point along the segment line at which the center of mass is located and also to determine the angle at which the segment line is oriented. The center of mass location along the segment line requires the use of anthropometric data and will be described in Section 7.2. The determination of the angle of orientation of the segment line will now be considered. Figure 7.4.a indicates two markers located in coordinate space. With respect to vertical orientation, *the ith marker is always inferior, and the jth marker is always superior.*

Figure 7.4.b indicates the specific x and y coordinates in which the ith marker is positioned leftward and jth marker is positioned rightward. As previously mentioned, note that the origin of the rectangular coordinate system is located so that the segment appears in quadrant I. The angle of orientation (θ) of the segment is always defined with respect to the inferior marker. The segment itself may be viewed as rotating counterclockwise around the inferior marker (in this case, the ith marker). The segment angle of rotation (θ) is zero degrees when the segment is parallel to the x-axis.

For an angle of θ between 0 and $\pi/2$ radians, the segment is in quadrant I of a polar coordinate system, so that:

$$\tan\theta_i = \frac{|y_j - y_i|}{|x_j - x_i|} \tag{1}$$

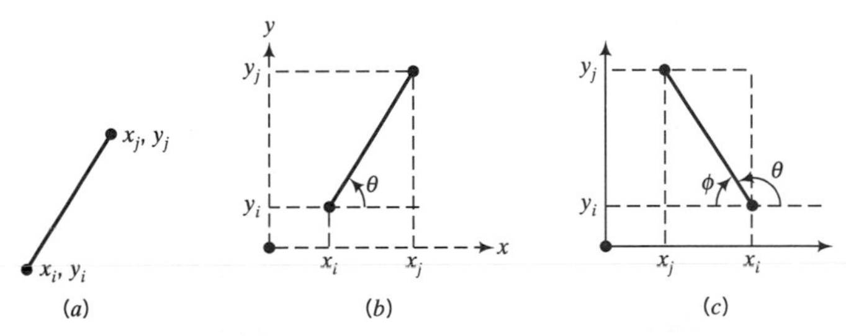

Figure 7.4.a Two markers located in coordinate space. **b** Indication of the ith marker positioned leftward and jth marker positioned rightward. **c** Indication of the ith marker positioned rightward and jth marker positioned leftward.

From which the angle of the segment (θ) may be defined (in radians) as:

$$\theta_i = \tan^{-1}\left[\frac{|y_j - y_i|}{|x_j - x_i|}\right] \tag{2}$$

When the jth marker is leftward and the ith marker is rightward (Figure 7.4.c), the angle of the segment (θ) is greater than $\pi/2$ (up to π). The segment angle (θ) extends into quadrant 2 of a polar coordinate system, so that a segment angle (ϕ) may be defined (in radians) for the acute angle:

$$\phi_i = \tan^{-1}\left[\frac{|y_j - y_i|}{|x_j - x_i|}\right] \tag{3}$$

The segment angle (θ) may then be defined (in radians) as the obtuse angle:

$$\theta_i = \pi - \phi_i \tag{4}$$

7.2 SEGMENT MASS AND MOMENT

a. Center of Mass

A body segment represents a structure with distributed mass. Figures 7.5.a and 7.5.b indicate that for equal segment lengths (δx) there is a variable segment mass

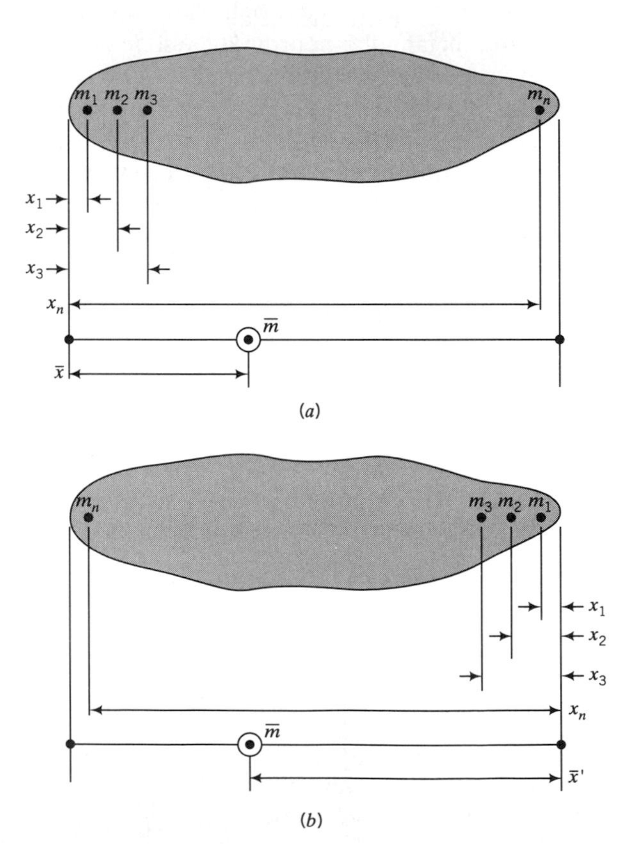

Figure 7.5.a Body segment representing a structure with distributed mass. **b** Variable segment mass for a given segment in Figure 7.5.a.

associated with any given segment length. When the segment is of uniform density, the mass of any given section (m_i) is a function of the volume of that section:

$$m_i = dV_i \tag{5}$$

For a system consisting of n sections, the mass of the entire segment ($\overline{m}$) is given by:

$$\overline{m} = d\sum_{i=1}^{n} V_i \tag{6}$$

The location of the center of mass of a body segment is shown in Figure 7.5.a. Note that the center of mass is located at a specific distance ($\overline{x}$) from the left edge of the segment. Furthermore, each section of the segment exerts its own moment of force proportional to its individual mass times its individual moment arm. In Figure 7.5.a, the center of mass is located at a distance ($\overline{x}$) so that it will produce the same net gravitational moment of force about the left edge as did the original distributed mass. This relationship (Figure 7.5.a) may be expressed as:

$$\overline{m}\,\overline{x} = \sum_{i=1}^{n} m_i x_i \tag{7}$$

Rearranging to solve for the location of the center of mass ($\overline{x}$):

$$\overline{x} = \frac{1}{\overline{m}}\sum_{i=1}^{n} m_i x_i \tag{8}$$

In fact, the center of mass is located at a characteristic point along the segment such that it must create the same net gravitational moment of force about any point along the segment as would occur with the original distributed mass. Therefore, it is also useful to consider the center of mass located at a distance ($\overline{x'}$) from the right edge of the segment (as shown in Figure 7.5.b). The individual section masses (now proceeding rightward to leftward) would then be multiplied by their individual moment arms (now with respect to the right edge of the segment) so that the center of mass would now be located at a distance ($\overline{x'}$) from the right edge of the segment. Mathematically, this would be expressed as:

$$\overline{m}\,\overline{x'} = \sum_{j=1}^{n} m_j x_j \tag{9}$$

$$\overline{x'} = \frac{1}{\overline{m}}\sum_{j=1}^{n} m_j x_j \tag{10}$$

The sum of $\overline{x}$ [from Equation (8)] and $\overline{x'}$ [from Equation (10)] is:

$$\overline{x} + \overline{x'} = \Delta x \tag{11}$$

Table 7.2 represents the anthropometric data for the seven body segments identified in Figure 7.1. The student's attention is directed to the left half of Table 7.2. $\hat{m}$ is the normalized segment mass defined as the ratio of the segment weight (w) to the total body weight (W). The next two columns refer to the normalized center of mass location along the segment axis. $\overline{x}_p$ is the ratio of the proximal end to center of mass length divided by the total segment length. $\overline{x}_d$ is the distal end to center of mass length divided by the total segment length.

Figure 7.6.a identifies a body segment axis as a straight line that connects the ith marker with the jth marker. At any moment in time, the spatial location of

Table 7.2 Anthropometric Data

Segment	$\hat{M}$	$\overline{x_p}$	$\overline{x_d}$	$\overline{\rho_0}$	$\overline{\rho_p}$	$\overline{\rho_d}$	$\overline{h}$
Hand	0.006	0.505	0.495	0.300	0.590	0.580	0.065[a] 0.110[b]
Forearm	0.016	0.430	0.570	0.305	0.525	0.645	0.145
Upper arm	0.028	0.435	0.565	0.320	0.540	0.645	0.190
Foot	0.014	0.500	0.500	0.475	0.690	0.690	0.040[c] 0.150[d]
Foreleg	0.046	0.435	0.565	0.300	0.530	0.645	0.245
Thigh	0.100	0.435	0.565	0.325	0.540	0.655	0.245
Thoracolumbar (thorax and abdomen)	0.355	0.630	0.370	—	—	—	0.300

[a] Gripping hand length
[b] Extended hand length
[c] Vertical foot length
[d] Horizontal foot length

these two markers is identified by x and y coordinates. Before we identify kinematic parameters, it is first necessary to determine the spatial coordinates $(\overline{x_I}, \overline{y})$ of the segment center of mass located along the segment axis. Consequently, it is necessary to identify one of the markers as the distal end of the segment and the other marker as the proximal end of the segment. The following convention is used with respect to human anatomy.

Consider an individual standing erect with legs straight, feet slightly apart, and his arms raised fully extended and stretching outward from the sides of the body. From the frontal view, it might appear as though the person were about to flap their arms and attempt to fly. The convention is then used so that the center of mass of the body (located at approximately the umbilicus) is proximal and that the head, hands, and feet are distal. Therefore, any one of the seven identified body segments (see Table 7.2) will have a proximal end and a distal end with respect to the body's center of mass in the frontal plane. In the sagittal plane, the proximal end may be superior or inferior to the distal end, and the proximal end may either be rightward or leftward of the distal end. For example, with respect to the forearm task (see Example 7.1), the forearm (EW) segment underwent an end-to-end inversion. For the forearm segment, the E marker is the proximal end and the W marker is the distal end. When the forearm segment was flexed past the neutral point

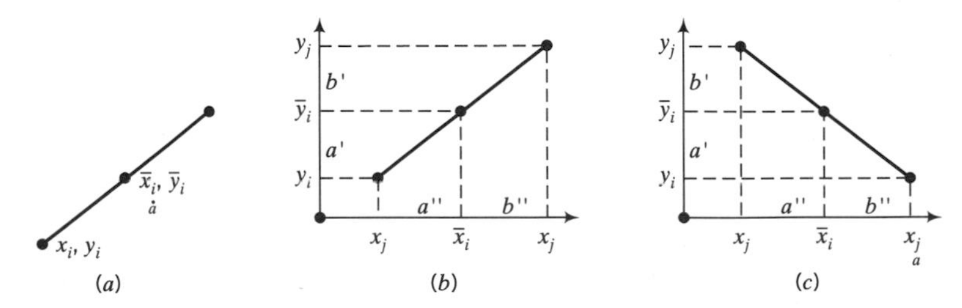

Figure 7.6.a Axial segment distance between the distal-end, proximal-end, and the center of mass. **b** Segment in rectangular coordinate space when the segment angle (θ) is between 0 and 90°. **c** Segment in coordinate space when the segment angle (θ) is between 90° and 180°.

(elbow joint at a 90° angle), the W marker was superior to the E marker as viewed in the sagittal plane. However, when the forearm segment was extended past the neutral point, the W marker was inferior to the E marker as viewed in the sagittal plane. Recall *that the ith marker is always the inferior marker.* In the case of Figure 7.6.a, the *i*th marker has been identified as the distal segment end (D). *The jth marker is always superior*, and in the case of Figure 7.6.a, it has been identified at the proximal end of the segment (P). a is then defined as the axial segment length between the distal end and the center of mass, and b is defined as the axial segment distance between the proximal end and the center of mass location (see Figure 7.6.a).

Figure 7.6.b represents the segment in rectangular coordinate space when the segment angle (θ) is between 0° and 90° (see Section 7.1). *The vertical coordinate* for the center of mass ($\bar{y}_i$) can be obtained with respect to the distal end of the segment or the proximal end of the segment (equations 12, 13, 14). Likewise, *the horizontal coordinate* of the segment center of mass ($\bar{x}_i$) can be obtained with reference to either the proximal or distal segment end; see equations (15), (16), and (17):

$$\bar{y}_i = y_i + a' = y_j - b' \tag{12}$$

$$a' = |y_j - y_j|\bar{x}_d \tag{13}$$

$$b' = |y_j - y_i|\bar{x}_p \tag{14}$$

$$\bar{x}_i = x_i + a'' = x_j - b'' \tag{15}$$

$$a'' = |x_j - x_i|\bar{x}_d \tag{16}$$

$$b'' = |x_j - x_i|\bar{x}_p \tag{17}$$

Figure 7.6.c shows the segment in coordinate space when the segment angle (θ) is between 90° and 180°. Note that the distal end is now rightward and the proximal end is now leftward with respect to the horizontal axis. The vertical coordinate of the center of mass location is still defined by Equation (12). However, the horizontal coordinate of the center of mass location is now defined by Equation (18). Note a', b', a'', and b'' are still defined by equations (13), (14), (16), and (17), respectively.

$$\bar{x}_i = x_i - a'' = x_j + b'' \tag{18}$$

Proceeding from positional data to velocity and acceleration calculations requires more than one moment in time. With respect to kinesiological (cinematographic) data, only the positional data of the various markers can be recorded during the time continuum. Each set of marker positional data is separated in time by a finite time interval, Δt. The calculation of velocity from displacement data then simply requires that we take the finite difference (displacement) data, Δx, and divide by the finite time interval, Δt. Since we require velocity at a particular point in time (at *i*th time), this can be accomplished by subtracting the position at time $i - 1$ from the position at time $i + 1$ and dividing by the corresponding time interval. The vertical velocity for the segment center of mass would then be:

$$v_{y_i} = \frac{\bar{y}_{i+1} - \bar{y}_{i-1}}{2 \cdot \Delta t} \tag{19}$$

The horizontal velocity of the segment center of mass would be:

$$v_{x_i} = \frac{\bar{x}_{i+1} - \bar{x}_{i-1}}{2 \cdot \Delta t} \tag{20}$$

The vertical acceleration of the segment center of mass can also be calculated from three adjacent positional data points (with time interval Δt between any two adjacent sets). Expressing the midpoint method for acceleration at the ith time point:

$$a_{y_i} = \frac{v_{y_i+1/2} - v_{y_i-1/2}}{\Delta t} \tag{21}$$

$$v_{y_i+1/2} = \frac{\overline{y}_{i+1} - \overline{y}_i}{\Delta t} \tag{22}$$

$$v_{y_i-1/2} = \frac{\overline{y}_i - \overline{y}_{i-1}}{\Delta t} \tag{23}$$

With respect to the vertical acceleration of the segment center of mass, substituting equations (22) and (23) into equation (21) and simplifying:

$$a_{y_i} = \frac{\overline{y}_{i+1} - 2\overline{y}_i + \overline{y}_{i-1}}{(\Delta t)^2} \tag{24}$$

Similarly, for the horizontal acceleration of the segment center of mass:

$$a_{x_i} = \frac{\overline{x}_{i+1} - 2\overline{x}_i + \overline{x}_{i-1}}{(\Delta t)^2} \tag{25}$$

Finally, if the angular position (θ) of a segment is also known, the segment angular velocity (ω) and the segment angular acceleration (α) may similarly be determined:

$$\omega_i = \frac{\theta_{i+1} - \theta_{i-1}}{2\,\Delta t} \tag{26}$$

$$\alpha_i = \frac{\theta_{i+1} - 2\theta_i + \theta_{i-1}}{(\Delta t)^2} \tag{27}$$

EXAMPLE 7.2

An industrial firm is developing a wrist brace for vocational bowlers to obtain better wrist support and control when swinging and releasing their bowling ball. A professional bowler is performing this task and his position in time is being filmed in right profile view. Forearm position coordinate data are acquired from an elbow marker (x_E, y_E) and a wrist marker (x_W, y_W), which are 0.323 M apart. During a particular time interval through the task sequence, the following data is acquired:

Time	x_E	y_E	x_W	y_W
0.316	1.877	0.404	1.600	0.239
0.330	1.917	0.408	1.640	0.241
0.344	1.956	0.413	1.682	0.240

Find the nine kinematic parameters for the forearm segment.

SOLUTION 7.2

To find the nine kinematic parameters, proceed systematically.

(i) Orient the segment at the desired time (t = 0.330 s) in coordinate space, indicating marker coordinates and segment angle, θ:

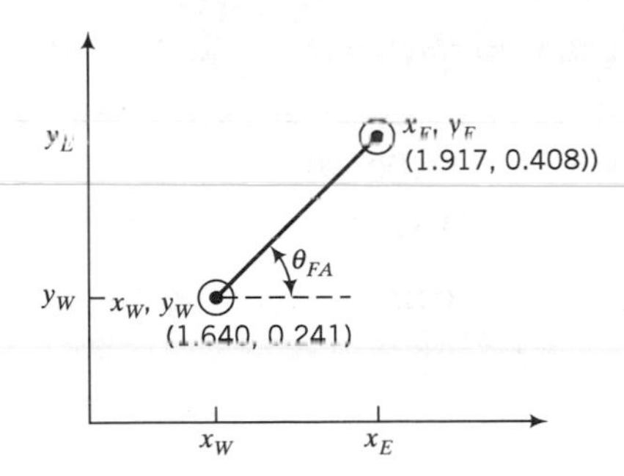

(ii) Identify proximal (P) and distal (D) markers, and indicate center of mass with respect to each end:

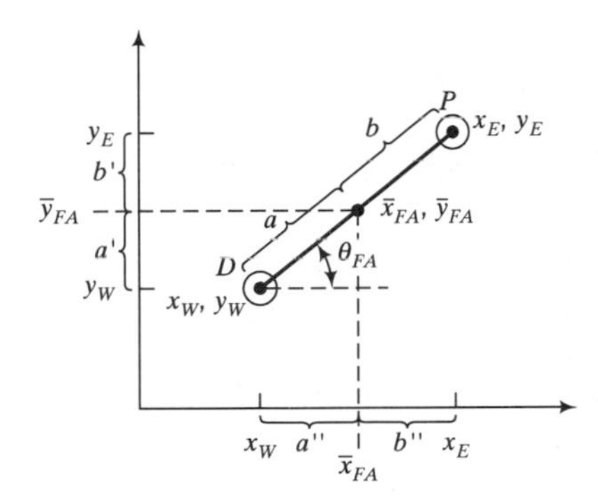

(iii) Solve for θ_i (and also for θ_{i+1} and θ_{i-1}):
For $t = 0.316$ s, from equation (2):

$$\theta_{i-1} = \tan^{-1}\left[\frac{|y_E - y_W|}{|x_E - x_W|}\right]$$

Substituting from the table of data:

$$\theta_{i-1} = \tan^{-1}\left[\frac{.165}{.277}\right] = 0.537 \text{ rad}$$

For $t = 0.330$ s, from equation (2) and the table of data:

$$\theta_i = \tan^{-1}\left[\frac{.167}{.277}\right] = 0.543 \text{ rad}$$

For $t = 0.344$ s, from equation (2) and the table of data:

$$\theta_{i+1} = \tan^{-1}\left[\frac{.173}{.274}\right] = 0.563 \text{ rad}$$

(iv) Solve for $\bar{y}_i$ (and also for $\bar{y}_{i+1}$ and $\bar{y}_{i-1}$):
For $t = 0.316$, from the *distal end* using equations (12) and (13):

$$a' = |y_E - y_W|\bar{x}_d$$
$$\bar{y}_{i-1} = y_W + \alpha'$$

Combining equations (12) and (13), and substituting from Table 7.2 and table of data:

$$\bar{y}_{i-1} = 0.239 + |0.404 - 0.239|(0.570)$$
$$\bar{y}_{i-1} = 0.239 + 0.094 = 0.333 \text{ M}$$

For a check, or as an alternate approach, from proximal end using equations (12) and (14):

$$b' = |y_E - y_W|\bar{x}_P$$
$$\bar{y}_{i-1} = y_E - b'$$

Combining equations (12) and (14), and substituting from Table 7.2 and the table of data:

$$\bar{y}_{i-1} = 0.404 - |0.404 - 0.239|(0.430)$$
$$\bar{y}_{i-1} = 0.4 - 0.071 = 0.333 \text{ M}$$

For $t = 0.330$ s, from the distal end, combining equations (12) and (13) and substituting from Table 7.2 and the table of data:

$$\bar{y}_i = 0.241 + |0.08 - 0.241|(0.570)$$
$$\bar{y}_i = 0.241 + 0.095 = 0.336 \text{ M}$$

For $t = 0.344$ s, from the distal end, combining equations (12) and (13) and substituting from Table 7.2 and the table of data:

$$\bar{y}_{i+1} = 0.240 + |0.413 - 0.240|(0.570)$$
$$\bar{y}_{i+1} = 0.240 + 0.099 = 0.339 \text{ M}$$

(v) Solve for $\bar{x}_i$(and also $\bar{x}_{i+1}$ and $\bar{x}_{i-1}$):

For $t = 0.316$ s, from the distal end and using equations (15) and (16):

$$a'' = |x_E - x_W|\bar{x}_d$$
$$\bar{x}_{i-1} = x_W + a''$$

Combining equations (15) and (16), and substituting data from Table 7.2 and the table of data:

$$\bar{x}_{i-1} = 1.600 + |1.877 - 1.600|(0.570)$$
$$\bar{x}_{i-1} = 1.600 + 0.158 = 1.758 \text{ M}$$

For a check, or as an alternative approach, from the proximal end using equations (15) and (17):

$$b'' = |x_E - x_W|\bar{x}_P$$
$$\bar{x}_{i-1} = x_E - b''$$

Combining equations (15) and (17), and substituting from Table 7.2 and the table of data:

$$\bar{x}_{i-1} = 1.877 - |1.877 - 1.600|(0.430)$$
$$\bar{x}_{i-1} = 1.877 - 0.119 = 1.758 \text{ M}$$

For $t = 0.330$ s, from the distal end, combining equations (15) and (16) and substituting from Table 7.2 and the table of data:

$$\bar{x}_i = 1.640 + |1.917 - 1.640|(0.570)$$
$$\bar{x}_i = 1.640 + 0.158 = 1.798 \text{ M}$$

For $t = 0.344$ s, from the distal end, combining equations (15) and (16) and substituting from Table 7.2 and the table of data:

$$\bar{x}_{i+1} = 1.682 + |1.956 - 1.682|(0.570)$$
$$\bar{x}_{i+1} = 1.682 + 0.156 = 1.838 \text{ M}$$

We now have three of the nine kinematic parameters, that is, the position data:

$$\theta_i = 0.543 \text{ rad}$$
$$\bar{y}_i = 0.336 \text{ M}$$
$$\bar{x}_i = 1.798 \text{ M}$$

In preparation of finding the other six kinematic parameters, prepare a *table of processed data* as follows:

Time	θ (rad)	$\bar{y}$ (M)	$\bar{x}$ (M)
0.316	0.537	0.333	1.758
0.330	0.543	0.336	1.798
0.344	0.563	0.339	1.838
	$\Delta t = 0.14$ s		

(vi) Solve for the rotational kinematic parameters (ω_i, α_i):
From equation (26):

$$\omega_i = \frac{\theta_{0.344} - \theta_{0.316}}{2 \cdot \Delta t}$$

Substituting from the table of processed data:

$$\omega_i = 0.929 \text{ rad/s}$$

From equation (27):

$$\alpha_i = \frac{\theta_{0.344} - 2 \cdot \theta_{0.330} + \theta_{0.316}}{(\Delta t)^2}$$

Substituting from the table of processed data:

$$\alpha_i = \frac{(.564) - (2)(.543) + (.538)}{(.014)^2}$$

$$\alpha_i = \frac{.0160}{.000196} = 81.6 \text{ rad/s}^2$$

(vii) Solve for the vertical translational parameters (v_{yi}, a_{yi}):
From equation (19):

$$v_{yi} = \frac{\bar{y}_{0.344} - \bar{y}_{0.316}}{2 \cdot \Delta t}$$

Substituting from the table of processed data:

$$v_{yi} = 0.214 \text{ m/s}$$

From equation (24):

$$a_{yi} = \frac{\bar{y}_{0.344} - 2 \cdot \bar{y}_{.0330} + \bar{y}_{.0316}}{(\Delta t)^2}$$

Substituting from the table of processed data:

$$a_{yi} = \frac{(.339) - (2)(.336) + (.333)}{(.014)^2}$$

$$a_{yi} = 0$$

(viii) Solve for the horizontal translational parameters (v_{xi}, a_{xi}):
From equation (20):

$$v_{xi} = \frac{\bar{x}_{0.344} - \bar{x}_{0.316}}{2 \cdot \Delta t}$$

Substituting from the table of processed data:

$$v_{xi} = 2.86 \text{ M/s}$$

From equation (25):

$$a_{xi} = \frac{\bar{x}_{0.344} - 2 \cdot \bar{x}_{.0330} + \bar{x}_{.0316}}{(\Delta t^2)}$$

Substituting from the table of processed data:

$$a_{xi} = \frac{(1.838) - (2)(1.798) + (1.758)}{(.014)^2}$$

$$a_{xi} = 0$$

---

b. Mass Moment of Inertia

As we have seen, the location of the center of mass is necessary in order to analyze the translational movement of the body segment through space. In order for the body segment to be in dynamic equilibrium, we need to simultaneously satisfy both the translational and rotational equations:

$$\sum F_y = m \cdot a_y \qquad (28)$$

$$\sum F_x = m \cdot a_x \qquad (29)$$

$$\sum M = I\alpha \qquad (30)$$

Equations (28) and (29) describe the relationship between a linear force (F) and the subsequent linear acceleration (a). For a system in translation, m is the constant of proportionality. For the rotational component of a distributed mass moving in space, equation (30) must be satisfied. m is the moment or torque that creates the resultant angular acceleration (α). I is the constant of proportionality which represents the inertial resistance of the distributed mass to the rotational movements. I is defined as the polar moment of inertia and its value will vary depending upon the specific point about which the axis of rotation is placed. I is a minimum when the axis of rotation is located at the center of mass.

With respect to the distributed mass segment of Figure 7.5.b, recall that the moment of inertia about the right end is:

$$\overline{mx'} = \sum_{j-1}^{n} m_j x_j \tag{9}$$

This equation simply states that for a system with distributed mass, that mass which is located farther from the axis of rotation exerts a proportionately greater moment of inertia than an equivalent amount of mass located nearer the axis of rotation.

Referring to Figure 7.7, the distributed mass of Figure 7.5.a and 7.5.b has now been redrawn to represent a center of mass ($\overline{m}$) located along a segment axis that has a proximal end (P) and a distal end (D). As we have previously noted, x_p is the axial length from the proximal end to the center of mass, and x_d is the axial length from the distal segment end to the center of mass. Let us consider three moments of inertia about this segment, I_o, I_d, and I_p.

I_o is the moment of inertia when the axis of rotation is located at the center of mass. In Figure 7.7, this mass has been subdivided into two equal point masses ($\overline{m}/2$). The location of these two equal point masses is at an axial distance (ρ_0) from the center of mass, so that this axial distance is defined as:

$$\rho_0 = \sqrt{\frac{I_0}{\overline{m}}} \tag{31a}$$

ρ_0 is referred to as the radius of gyration so that when two equal point masses (Figure 7.7) are located the same axial distance (ρ_0), then the same moment of inertia in a plane of rotation about an axis located at the center of mass will be created, as would be the case for the original distributed mass segment.

The moment of inertia at the center of mass (I_0) follows from Equation (31a):

$$I_0 = \overline{m} \cdot \rho_0^2 \tag{31b}$$

The human factors engineer will quickly appreciate that most body segments do not rotate about the center of mass for that segment. Rather, body segments have an axis of rotation about a joint located at one or both ends of the body segment. Therefore, in analyzing the movement of human body segments in space, the moment of inertia can only be taken about a joint axis-of-rotation.

The relationship between the moment of inertia at the center of mass (I_0) and the moment of inertia at *any* axis of rotation along the segment length is expressed by the parallel axis theorem:

$$I = \frac{\overline{m}}{2}(x - \rho_0)^2 + \frac{\overline{m}}{2}(x + \rho_0)^2 \tag{32}$$

where x is the distance between the center of mass and the rotational axis located along the segment line, and $\overline{m}$ is the mass of the segment.

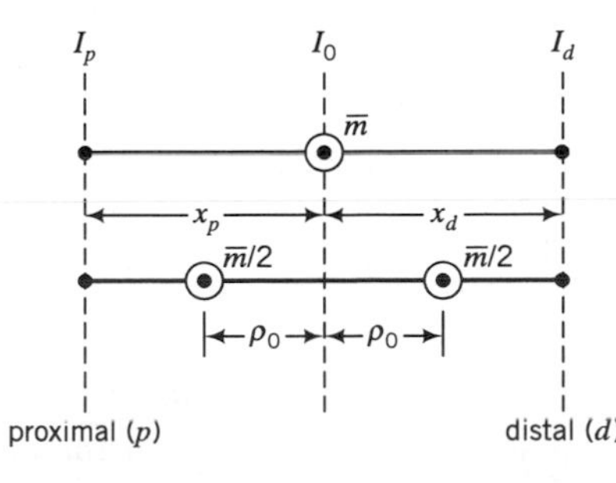

Figure 7.7 Distributed mass of Figure 7.5.a and 7.5.b now represents a center of mass along a segment axis.

Simplifying equation (32) results in:

$$I = \overline{m}\rho_0^2 + \overline{m}x^2 \tag{33}$$

Substituting the equation (31b) into equation (33) results:

$$I = I_0 + \overline{m}x^2 \tag{34}$$

where I_o is the moment of inertia about the center of mass.

Since x can be any axial distance in either direction from the center of mass (providing that new axis of rotation is colinear with the axis for which I_0 was calculated), then we can define the moment of inertia about the distal end of the segment (I_d) by direct substitution into equation (33):

$$I_d = \overline{m}\rho_0^2 + \overline{m}x_d^2 \tag{35a}$$

Likewise, the moment of inertia about the proximal end of the segment (I_p) may be calculated by direct substitution into equation (33):

$$I_p = \overline{m}\rho_0^2 + \overline{m}x_p^2 \tag{36a}$$

Equations (35a) and (36a) can be rewritten as:

$$I_d = \overline{m}(\rho_0^2 + x_d^2) \tag{35b}$$

$$I_p = \overline{m}(\rho_0^2 + x_p^2) \tag{36b}$$

A radius of gyration about the distal segment end (ρ_d) and a radius of gyration about the proximal segment end (ρ_p) can be defined:

$$\rho_d = \sqrt{\rho_0^2 + x_d^2} \tag{37}$$

$$\rho_p = \sqrt{\rho_0^2 + x_p^2} \tag{38}$$

Substituting equation (37) into equation (35b) and also substituting equation (38) into equation (36b) results in more compact mathematical expressions for I_d and I_p:

$$I_d = \overline{m}\rho_d^2 \tag{39}$$

$$I_p = \overline{m}\rho_p^2 \tag{40}$$

Referring once again to Table 7.2, the Table of Anthropometric Data, and the seven body segments that will be analyzed in this chapter, the student's attention is directed to the right half of the table: $\overline{\rho}_0$, $\overline{\rho}_p$, and $\overline{\rho}_d$, represent the *normalized* radii of gyration. When the axis of rotation is located at the center of mass of a body segment, $\overline{\rho}_0$ is the radius of gyration length along the segment axis divided by the total segment axial length. When the axis of rotation is positioned at the proximal end of the body segment, $\overline{\rho}_p$ is the radius of gyration length along the segment axis divided by the total segment axial length. Finally, when the axis of rotation is located at the distal end of the segment, $\overline{\rho}_d$ is the axial length of the radius of gyration divided by the total segment axial length.

EXAMPLE 7.3

For the emergency worker in Example 3.1, calculate the moment of inertia about the shoulder joint (I_s). Treat the upper arm segment and the forearm segment as distributed masses. You may assume that the hand and the "detour sign" are each point masses co-located at 0.70 M from the shoulder joint.

SOLUTION 7.3

Person's Height (H) = 1.75 M
Person's Weight (W) = 700 N
Load Weight (W_L) = 18 N
The system consists of two distributed masses (upper arm and forearm) and two point masses (hand and load being gripped). Begin by drawing a diagram of the system:

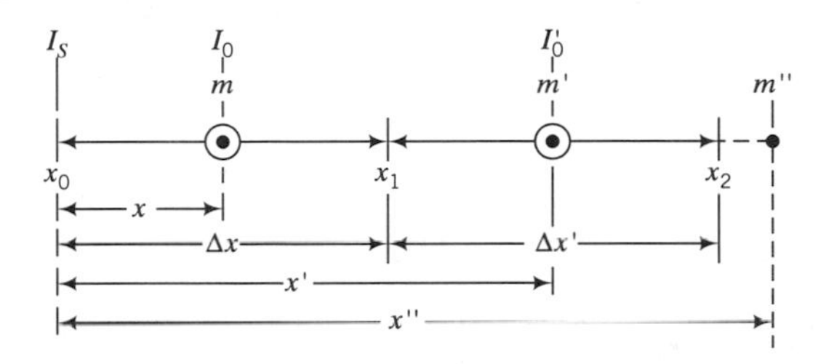

Proceed to calculate (or otherwise identify) all parameters in the diagram above using a systematic method (e.g., proceeding distal to proximal):

(i) Solve for m'', x'':

$$x'' = 0.70 \text{ M (given)}$$

$$m'' = m_L + m_H$$

$$m_L = \frac{18 \text{ N}}{\mathbf{g}} = 1.83 \text{ kg}$$

From Table 7.2:

$$m_H = \left(\frac{700 \text{ N}}{\mathbf{g}}\right)(.006)$$

$$m_H = 0.43 \text{ kg}$$

$$m'' = 1.83 + 0.43 = 2.26 \text{ kg}$$

(ii) Solve for forearm parameters (I'_0, m', and $\Delta x'$):

$$\Delta x' = \bar{h}H$$

Substituting from Table 7.2 and given height:

$$\Delta x' = (.145)(1.75) = 0.254 \text{ M}$$

$$m' = \hat{M}\left(\frac{W}{\mathbf{g}}\right)$$

Substituting from Table 7.2 and given weight:

$$m' = (.016)\left(\frac{700}{9.81}\right) = 1.14 \text{ kg}$$

Finally, solve for I'_0 as follows:

$$\rho'_0 = \bar{\rho}_0 \Delta x'$$

Substituting from Table 7.2:

$$\rho'_0 = (.305)(.254) = .077 \text{ M}$$

$$I'_0 = m' \rho'^2_0$$

Substituting:

$$I'_0 = (1.14)(.077)^2$$

$$I'_0 = .0068 \text{ kg} \cdot \text{M}^2$$

(iii) Solve for the upper arm parameters (I_0, m, and Δx):

$$\Delta x = \bar{h} H$$

Substituting:

$$\Delta x = (.190)(1.75) = 0.333 \text{ M}$$

$$m = \hat{M}\left(\frac{W}{g}\right)$$

Substituting:

$$m = (.028)(71.4) = 2.00 \text{ kg}$$

And solve for I_0 as follows:

$$\rho_0 = \bar{\rho}_0 \Delta x$$

Substituting:

$$\rho_0 = (.320)(.333) = 0.107 \text{ M}$$

$$I_0 = m \rho_0^2$$

Substituting:

$$I_0 = (2.00)(.107)^2$$

$$I_0 = .0029 \text{ kg} \cdot \text{M}^2$$

Solve for the moment of inertia at the shoulder (I_S):
Referring to the system diagram:

$$I_S = I_{UA}(x_0) + I_{FA}(x_0) + I_{M'}(x_0) \tag{i}$$

Using the parallel axis theorem:

$$I_{UA}(x_0) = I_0 + m(x)^2 \tag{ii}$$

$$I_{FA}(x_0) = I'_0 + m'(x')^2 \tag{iii}$$

Treating the hand and sign as point masses:

$$I_{M'}(x_0) = m''(x'')^2 \tag{iv}$$

Substituting equations (ii), (iii), and (iv) into equation (i):

$$I_S = I_0 + I_0' + m(x)^2 + m'(x')^2 + m''(x'')^2$$

Note:

$$x' = \Delta x + \bar{x}_p \Delta x'$$

Substituting from Table 7.2 and the known values:

$$x' = 0.333 + (.430)(.254)$$

$$x' = 0.442 \text{ M}$$

Also:

$$x = \bar{x}_p \Delta x$$

Substituting from Table 7.2 and the known value:

$$x = (.435)(.333) = 0.145 \text{ M}$$

At this point, all parameters in the primary equation are known, so substituting:

$$I_S = .0229 + .0068 + (2.00)(.145)^2 + (1.14)(.442)^2 + (2.26)(0.700)^2$$

$$I_S = .0229 + .0068 + .0421 + .223 + 1.107$$

$$I_S = 1.40 \text{ kg} \cdot \text{M}^2$$

7.3 SEGMENT KINETICS

a. Link-Segment Model

The biomathematical approach for calculating the joint reaction forces (R) and the muscle moments (M) is based upon the link-segment model. At this point, the human factors engineer will have a complete kinematic description (the nine kinematic parameters), the specific anthropometric measurements for the individual or individuals under analysis, and a quantification of the external forces acting upon the body. From these sets of data, the link-segment model will allow the human factors engineer to calculate the joint reaction forces and the muscle moments. This represents the inverse solution (previously referred to in Chapter 3) and is an extremely powerful method for obtaining engineering insight into the net summation of muscle activity which occurs at a joint. Such muscle activity is the final common pathway of the central nervous system drive to the muscles, which represents the final level of human operator control. This information is very useful to the human factors engineer in order to evaluate human-machine system interaction, as well as to evaluate the effects of training, interface modification, or task modification. Obviously, this biomathematical (link-segment) model is necessary since these effects are often obscured in the original kinematic parameters.

The following simplifying approximations are used with respect to the link-segment model:

Approximation 1. Each body segment has a fixed-point mass that is located at the center of mass (which is coincident with the center of gravity in the vertical direction).

Approximation 2. During the course of human body movement, the location of the center of mass along the axial line of each segment remains constant.

Approximation 3. The joints of the body (with respect to upper extremity and lower extremity) are represented as a hinge-type joint.

Approximation 4. During human body movement, the mass moment of inertia (I_0) for a rotational axis located at the center of mass will remain constant along the axis of the body segment.

Approximation 5. The axial segment length between hinge joints (proximal-to-distal end) remains constant during human movement.

Figure 7.8.a depicts the equivalent relationship between the anatomical model and the link-segment model for the upper extremity. The segment masses ($\overline{m}_1$, $\overline{m}_2$, and $\overline{m}_3$) are concentrated at fixed points (Approximation 1). The axial distance from the proximal joint to the center of mass (or from the distal joint to the center of mass) is also constant (Approximation 2). Finally, the lengths of each segment and the moment of inertia (I_{01}, I_{02}, and I_{03}) are constant, (Approximations 4 and 5).

There are five types of forces which act upon the link-segment model. These include the gravitational forces, the ground reaction forces, the external forces, the muscle forces, and the ligament forces. The aforementioned variables represent the external forces acting upon the link-segment model, and are briefly described as follows.

Gravitational Forces. The force of gravity acts vertically downward through the segment's center of mass and is proportional to the mass times the universal gravitational constant.

Ground Reaction Forces. These are external forces which must be quantified and are distributed over an area of the body. Ground reaction forces most commonly act upon the lower extremity, such as a contact area under the foot when walking. In order to represent these forces in the context of a link-segment model, they must interact as a vector quantity (usually measured as a "center of pressure" on a ground-mounted force plate).

External Forces. External forces may act upon any body segment and must also be quantified. With respect to the upper extremity, these forces are also distributed over an area of the body (such as the forces on a hand when using a power tool). As with ground reaction forces, external forces (on either the upper extremity or lower extremity) must be represented as a vector quantity, which may be referred to as the major force vector.

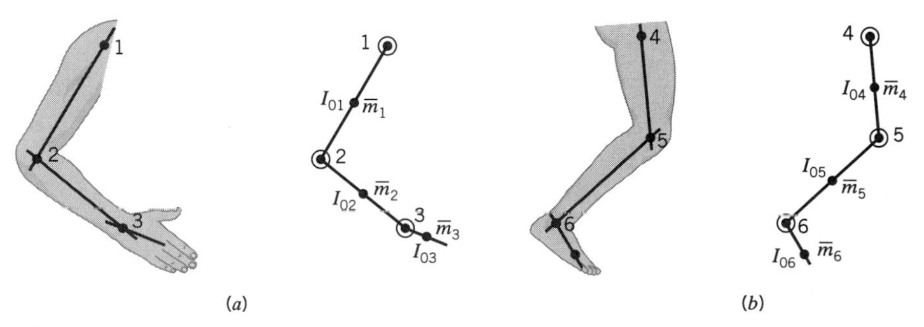

Figure 7.8.a Relationship between the anatomical model and the link-segment model for the upper extremity. **b** Relationship between the anatomical model and the link-segment model for the lower extremity.

Muscle Forces. Muscles insert upon bone (usually near joints) by means of ligamentous insertions. As previously noted (see Chapter 3) muscles insert upon bones as pairs, and at any moment in time one of the muscles may act as the agonist, and the other muscle as the antagonist. The net effect of muscle pair activity at a joint (the difference between the agonist and antagonist activity) can be calculated in terms of a net muscle moment. This is because muscles cross over joints and insert on body segments distal to joints. There is an axial distance between a joint center of rotation and the point of insertion of the muscle along the axis of the distal body segment. When the contracting muscle then exerts a force across the joint, this force is represented as a force vector for that muscle. The moment arm for that muscle is the distance of a line through the joint center of rotation and perpendicular to the muscle line of force. As previously noted, muscles can exert force only by contracting (whether agonist or antagonist) so that when muscle contraction occurs at a given joint, the link-segment model analysis will yield only the net muscle moment (i.e., the difference between the individual agonist muscle moment and the individual antagonist muscle moment).

Ligamentous Forces. When a human body segment approaches the extreme range of motion for its associated joint, passive structures such as ligaments (which are tissues that connect muscle to bone) will add to or subtract from the muscle forces previously described in Item (4). As a general rule, good human factors engineering practice would dictate that ligamentous forces not be appreciable in the performance of standard operational tasks.

The five forces just described constitute all of the external forces acting upon the total body system. As previously noted, these external forces act upon various individual body segments in various ways. However, when proceeding to analyze the link-segment model, the segments themselves are analyzed one at a time. Consequently, reaction forces between these segments must also be calculated. The reaction forces represent internal forces that hold the various segments of the link-segment model together. To accomplish this task, a free-body diagram of each segment must be used. Figure 7.9.a shows the free-body diagram for each segment

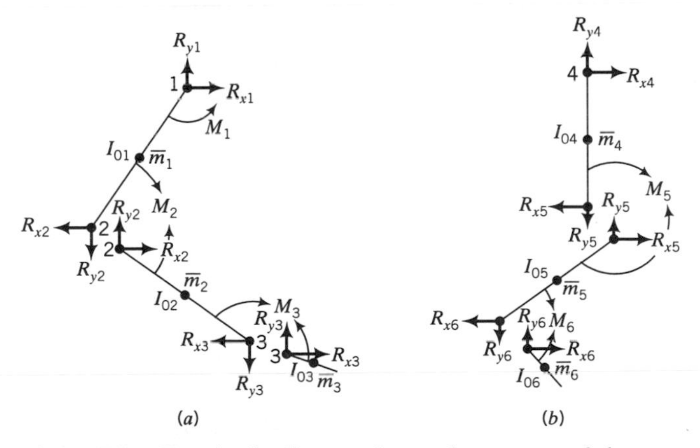

Figure 7.9.a Free-body diagram for each segment of the upper extremity link-segment model. **b** Free-body diagram for each segment of the lower extremity link-segment model.

of the upper extremity link-segment model. Figure 7.9.b shows the free-body diagram for each segment of the lower extremity link-segment model. Conventionally, the segments are disconnected at the joints and the forces that act across the joints are indicated at the distal and proximal ends of each segment.

This analytic approach will permit the human factors engineer to examine each segment and to calculate all unknown joint reaction forces. In order to satisfy Newton's third law, there is an equal and opposite horizontal and vertical force acting at each hinge joint in our link-segment model. Note also that the muscle moments at each joint are of equal magnitude and opposite sign. However, at the distal end of the most distal segment (i.e., the end of the fingers of the hand or the end of the toes of the foot) there is no *muscle* moment for either the upper extremity or the lower extremity. This is because there is no human body joint at these locations and no further distal segment. Recall that muscles can only exert moments by crossing joints and inserting upon a segment distal to the joint. Of course, there may be reaction moments if external forces are applied to the distal end of the most distal segment model.

EXAMPLE 7.4

For the individual performing the task of Example 7.2, position coordinate data is also obtained for the hand segment from the wrist marker (x_W, y_W) and hand marker (x_H, y_H) which are 0.123 M apart. During the same time interval as Example 7.2, the following data is acquired:

Time	x_W	y_W	x_H	y_H
0.316	1.600	0.239	1.573	0.119
0.330	1.640	0.241	1.618	0.120
0.344	1.682	0.240	1.668	0.118

Find the nine kinematic parameters for the hand segment.

SOLUTION 7.4

The general approach to finding the nine kinematic parameters proceeds in the same systematic manner as Solution 7.2 with some useful variations as noted.

(i) Orient the segment:

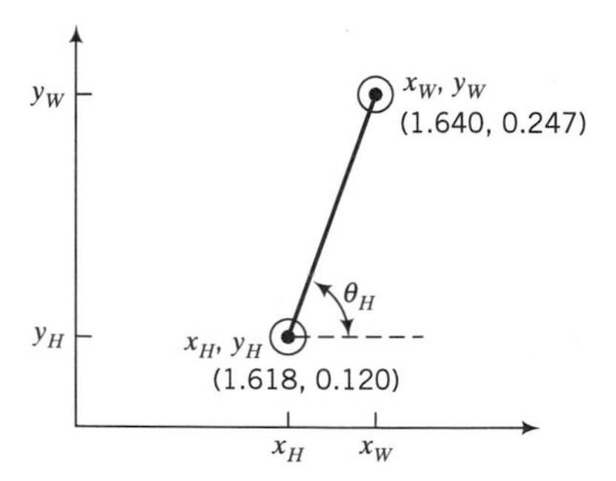

(ii) Identify proximal and distal and indicate center of mass:

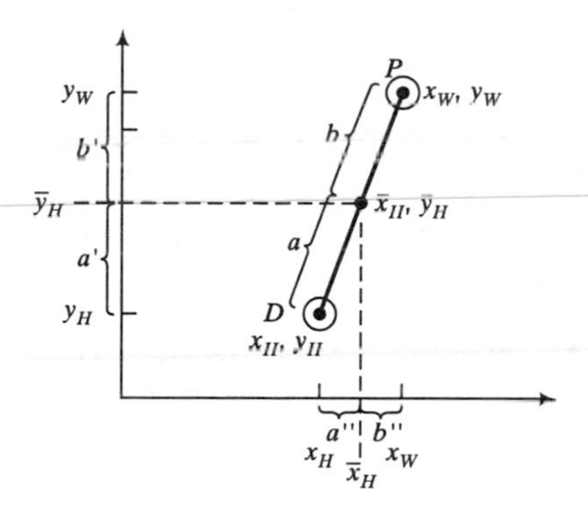

Before proceeding to identify the nine kinematic parameters, prepare a table of formatted data:

Time	y_W	y_H	$\|\Delta y\|$	x_W	x_H	$\|\Delta x\|$
.0316	0.239	0.119	0.120	1.600	1.573	0.027
0.330	0.241	0.120	0.121	1.640	1.618	0.022
0.344	0.240	0.118	0.122	1.682	1.668	0.014

(iii) Solve for θ_i, θ_{i+1}, and θ_{i-1}:

For $t = 0.316$ s, solve equation (2) from the table of formatted data:

$$\theta_{i-1} = \tan^{-1}\left[\left|\frac{\Delta y}{\Delta x}\right|\right] = \tan^{-1}\left[\frac{.120}{.027}\right]$$

$$\theta_{i-1} = 1.349 \text{ rad}$$

For $t = 0.330$ s, solve equation (2) from the table of formatted data:

$$\theta_i = \tan^{-1}\left[\left|\frac{\Delta y}{\Delta x}\right|\right] = \tan^{-1}\left[\frac{.121}{.022}\right]$$

$$\theta_i = 1.391 \text{ rad}$$

For $t = 0.344$ s, solve equation (2) from the table of formatted data:

$$\theta_{i+1} = \tan^{-1}\left[\left|\frac{\Delta y}{\Delta x}\right|\right] = \tan^{-1}\left[\frac{.122}{.014}\right]$$

$$\theta_{i+1} = 1.457 \text{ rad}$$

(iv) Solve for $\bar{y}_i$, $\bar{y}_{i+1}$, and $\bar{y}_{i-1}$:

For $t = .316$ s, from the distal end, combining equations (12) and (13) and substituting from Table 7.2 and the table of formatted data:

$$\bar{y}_{i-1} = y_H + |\Delta y|(\bar{x}_d)$$
$$\bar{y}_{i-1} = 0.119 + (0.120)(0.495)$$
$$\bar{y}_{i-1} = 0.178 \text{ M}$$

For $t = .330$ s, from the distal end, combining equations (12) and (13) and substituting from Table 7.2 and the table of formatted data:

$$\bar{y}_i = y_H + |\Delta y|(\bar{x}_d)$$
$$\bar{y}_i = 0.120 + (0.121)(0.495)$$
$$\bar{y}_i = 0.180 \text{ M}$$

For $t = .344$ s, from the distal end, combining equations (12) and (13) and substituting from Table 7.2 and the table of formatted data:

$$\bar{y}_{i+1} = y_H + |\Delta y|(\bar{x}_d)$$
$$\bar{y}_{i+1} = 0.118 + (0.122)(0.495)$$
$$\bar{y}_{i+1} = 0.178 \text{ M}$$

(v) Solve for $\bar{x}_i$, $\bar{x}_{i+1}$, and $\bar{x}_{i-1}$:

For $t = .316$ s, from the distal end, combining equations (15) and (16), and substituting:

$$\bar{x}_{i-1} = x_H + |\Delta x|(\bar{x}_d)$$
$$\bar{x}_{i-1} = 1.573 + (.027)(0.495)$$
$$\bar{x}_{i-1} = 1.586 \text{ M}$$

For $t = .330$ s, from the distal end, combining equations (15) and (16), and substituting:

$$\bar{x}_i = y_H + |\Delta x|(\bar{x}_d)$$
$$\bar{x}_i = 1.618 + (.022)(0.495)$$
$$\bar{x}_i = 1.629 \text{ M}$$

For $t = .344$ s, from the distal end, combining equations (15) and (16), and substituting:

$$\bar{x}_{i+1} = x_H + |\Delta x|(\bar{x}_d)$$
$$\bar{x}_{i+1} = 1.668 + (.014)(0.495)$$
$$\bar{x}_{i+1} = 1.675 \text{ m}$$

The positional kinematic parameters are as follows:

$$\theta_i = 1.391 \text{ rad}$$
$$\bar{y}_i = 0.180 \text{ M}$$
$$\bar{x}_i = 1.629 \text{ M}$$

At this point, compile a table of processed data:

Time	θ (rad)	$\bar{y}$(M)	$\bar{x}$(M)
0.316	1.349	0.178	1.586
0.330	1.391	0.180	1.629
0.344	1.457	0.178	1.675

$\Delta t = .014$ s

(vi) Solve for ω_i and α_i:

From equations (26) and (27) and substituting from table of processed data:

$$\omega_i = 3.86 \text{ rad/s}$$
$$\alpha_i = 122 \text{ rad/s}^2$$

(vii) Solve for v_{yi} and a_{yi}:

From equations (19) and (24), and substituting from table of processed data:

$$v_{yi} = 0$$
$$a_{yi} = -20.4 \text{ M/s}^2$$

(viii) Solve for v_{xi} and a_{xi}:

From equations (20) and (25), and substituting from table of processed data:

$$v_{xi} = 3.18 \text{ M/s}$$
$$a_{xi} = 15.3 \text{ M/s}^2$$

---

b. Free-Body Diagram of the Link-Segment Model

Newton's second law for three-dimensional (x-, y-, z-axis) coordinate space may be expressed in general form as:

$$\sum F_{x,y,z} = m \cdot a_{x,y,z} \tag{41}$$

$$\sum M_{x-y,y-z,x-z} = I_0\alpha_{x-y,y-z,x-z} \tag{42}$$

Recall from our previous discussion, that we have restricted our system to the sagittal (X-Y) plane, so that this system of six equations may be reduced to three [equations (28), (29), and (30)].

Generally speaking, with respect to the link-segment model, the analysis will proceed from the distal-most segment (usually where the external forces interact with the system) to the more proximal segments (where the internal forces are generated in order to hold the various segments together). The three equations of dynamic equilibrium [equations (28), (29), and (30)] allow the human factors engineer to solve for three unknowns so that the necessary conditions for dynamic equilibrium are satisfied for the system of interest. Figure 7.10.a defines the complete free-body diagram of an individual segment. In general, the human factors engineer will have some information regarding R_{xd} and R_{yd}. These are the reaction forces acting at the distal end of the segment. They may have been determined from a prior analysis of the proximal forces acting upon an adjacent and more distal segment, or they may be the generated reaction forces to external forces acting directly upon the distal end of this particular segment. In addition, the human factors engineer should obtain some quantitative information regarding M_d. This represents the net muscle moment acting at the distal joint of the specific segment. This net muscle moment may have been determined (from a prior analysis) as a reaction moment to the net muscle moment acting at the proximal end of the adjacent and more distal segment. The value of M_d will be zero if there is no distal joint (as would occur at the end of the hand or end of the foot).

The human factors engineer will also obtain experimental data (e.g., kinesiological data) as well as anthropometric data from standard statistical tables (e.g., Tables 7.1 and 7.2). In the usual case, what will remain to be determined is R_{xp}, R_{yp}, and M_p. The first two terms represent the horizontal and vertical reaction forces

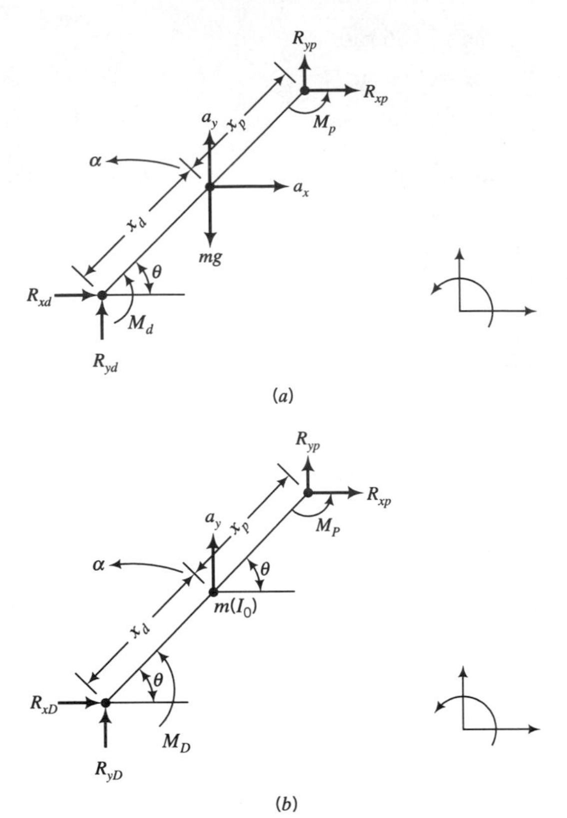

Figure 7.10.a Complete free-body diagram of an individual segment. **b** Representation of the various rotational forces and moments of an individual segment.

(respectively) acting at the proximal joint of the individual segment. The remaining term (M_p) represents the net muscle moment acting upon the individual segment at the proximal joint.

With respect to translational movement in the vertical direction, the following governing equations can be written for Figure 7.10.a:

$$\sum F_y = m \cdot a_y \tag{28}$$

$$R_{yp} + R_{yd} - mg = ma_y \tag{43}$$

With respect to translational movement in the horizontal direction, the following governing equations can be written for Figure 7.10.a:

$$\sum F_x = m \cdot a_x \tag{29}$$

$$R_{xp} + R_{xd} = ma_x \tag{44}$$

With respect to the rotational movement, Figure 7.10.b represents the various rotational forces and moments. If the moments are summed about the center of mass (ΣM_0) of the individual segment, the following governing equations apply:

$$\sum M_0 = I_0\alpha \tag{30}$$

$$M_p + M_d - R_{yd}x_d \cos\theta + R_{xd}x_d \sin\theta + R_{yp}x_p \cos\theta - R_{xp}x_p \sin\theta = I_0\alpha \tag{45}$$

Equations (43), (44), and (45) represent a system of three equations that can now be solved for the three unknowns (R_{yp}, R_{xp}, and M_p) in order to satisfy the condition of dynamic equilibrium of the individual segment.

EXAMPLE 7.5

The individual in Examples 7.2 and 7.4 has a body mass of 85 kg, and grips a bowling ball (mass of ball – 3.41 kg) in the right hand during the swing phase (prior to ground contact).

If the mass of the ball is approximated as a distributed mass (having an ρ_0 equal to .037 m) and the ball's center of mass is co-located at the center of mass of the hand segment:

(a) Find the wrist reaction forces (at the proximal end of the hand segment) and the applied muscle moment at the proximal end of the hand segment.

(b) Find the elbow reaction forces (at the proximal end of the forearm segment) and the applied moment (at the proximal end of the forearm segment).

SOLUTION 7.5(a)

For the hand segment (at the wrist joint), find R_{yp}, R_{xp}, and M_p:
The *translational* FBD for the hand-ball system is:

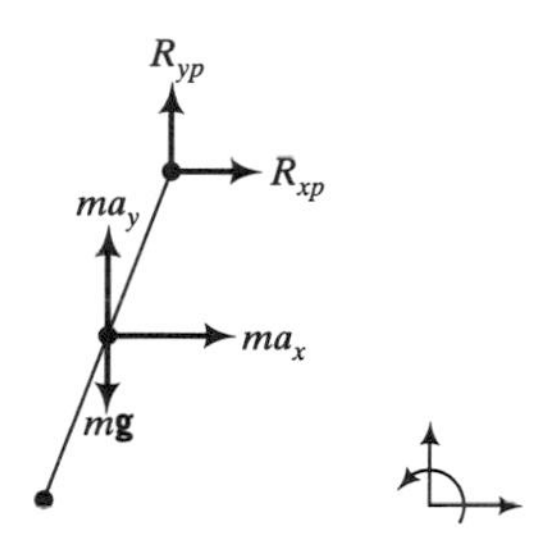

The *rotational* FBD for the hand-ball system is:

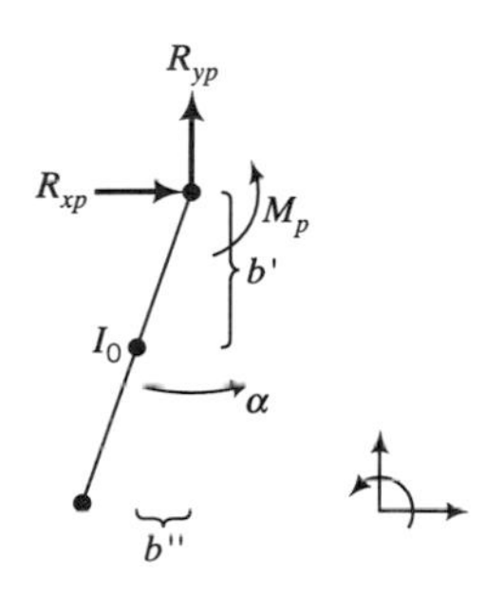

As a starting point, the sign convention is noted and α, R_{yp}, R_{xp}, and g are chosen as positive. M_p is the applied moment around the proximal end of the segment and is chosen positive (which will result in the segment α being positive).

M_d is the reaction moment around the distal end of the segment and is also chosen positive (which also results in the segment α being positive).
Note in the FBD that since the hand-ball system is in the swing phase (prior to ground contact):

$$R_{yd} = R_{xd} = 0$$

Parameters from Example 7.4:

$$a_y = -20.4 \text{ M/s}^2$$
$$a_x = 15.3 \text{ M/s}^2$$
$$\alpha = 122 \text{ rad/sec}^2$$
$$L_H = 0.123 \text{ M (W-H marker distance)}$$

Other parameters are calculated as follows:

(i) Calculation of m:

$$m = m_H + m_B$$

From Table 7.2 and body weight:

$$m_H = (.006)(85) \text{ kg} = 0.51 \text{ kg}$$

Since $m_B = 3.21$ kg, the mass of the hand-ball system is:

$$m = 0.51 + 3.21 = 3.72 \text{ kg}$$

(ii) Calculation of I_0:

First solve for ρ_0 of the hand-ball system:
For the hand segment:

$$(\rho_0)_H = \overline{\rho}_0 L_H$$

Substituting from Example 7.4 and Table 7.2:

$$(\rho_0)_H = (.300)(.123) = .037 \text{ M}$$

Since $(\rho_0)_B = .037$ M, then

$$\rho_0 = (\rho_0)_H = (\rho_0)_B = .037 \text{ M}$$

Then solve for I_0:
Since the hand and ball centers of mass are co-located; for the hand-ball system:

$$I_0 = (I_0)_H + (I_0)_B$$
$$I_0 = m_H \rho_0^2 + m_B \rho_0^2$$

So that:

$$I_0 = m \cdot \rho_0^2$$

Substituting:

$$I_0 = (3.72)(.037)^2 = .0051 \text{ kg} \cdot \text{M}^2$$

(iii) Calculation of b' and b'' (at $t = 0.330$ sec):

$$b' = |\Delta y|(\overline{x}_p)$$
$$b'' = |\Delta x|(\overline{x}_p)$$

Substituting from Example 7.4 and Table 7.2:

$$b' = (.121)(.505) = .061 \text{ M}$$
$$b'' = (.022)(.505) = .011 \text{ M}$$

Now proceed to solve for the hand-ball system in translational dynamic equilibrium. Applying equation (28) to the translational FBD:

$$R_{yp} + m\mathbf{g} = ma_y$$

Substituting and recalling that downwardly directed **g** carries a negative sign:

$$R_{yp} + (3.72)(-9.81) = (3.72)(-20.4)$$

Rearranging and solving:

$$R_{yp} = -39.4 \text{ N (downward)}$$

Applying equation (29) to the FBD:

$$R_{xp} = ma_x$$

Substituting:

$$R_{xp} = (3.72)(15.3)$$

Solving:

$$R_{xp} = 56.9 \text{ N (rightward)}$$

Finally, proceed to solve for the hand-ball system in rotational dynamic equilibrium: Applying equation (30) to the rotational FBD:

$$M_P + (R_{yp})b'' - (R_{xp})b' = I_0(\alpha)$$

Substituting:

$$M_p + (-39.4)(.011) - (56.9)(.061) = (.0051)(122)$$
$$M_p = 0.62 + 0.43 + 3.47 = 4.52 \text{ N} \cdot \text{M}$$

In summary, for the hand segment at the wrist joint:

$$R_{yp} = -39.4 \text{ N (downward)}$$
$$R_{xp} = 56.9 \text{ N (rightward)}$$
$$M_p = 4.52 \text{ N} \cdot \text{M (counterclockwise)}$$

SOLUTION 7.5(b)

For the forearm segment (at the elbow joint), find R_{yp}, R_{xp}, and M_p: The *translational* FBD for the forearm segment is:

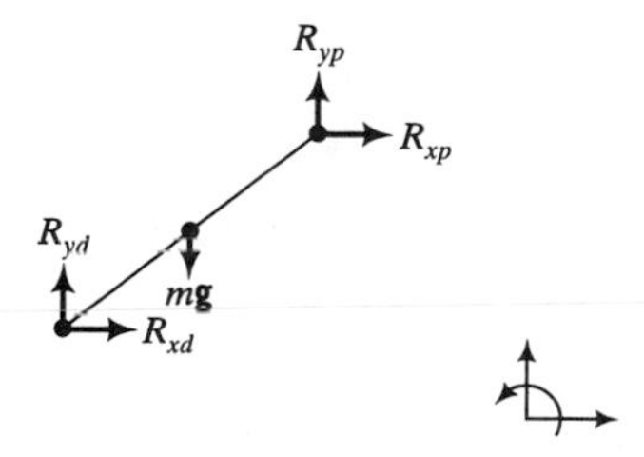

The *rotational* FBD for the forearm segment is:

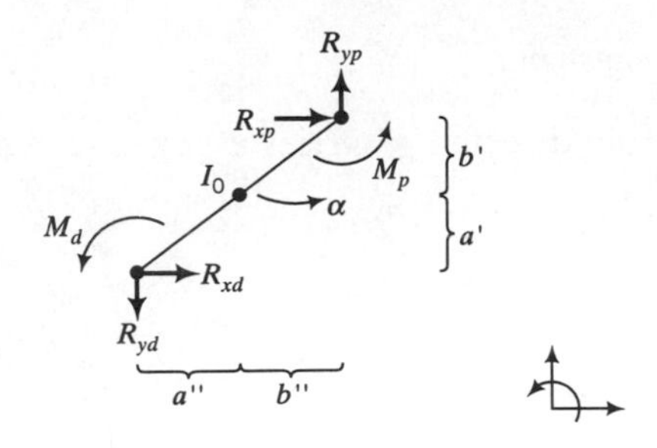

The parameters from Example 7.2:

$$a_y = 0$$
$$a_x = 0$$
$$\alpha = 81.6 \text{ rad/s}^2$$
$$L_{FA} = .323 \text{ M (E-W marker distance)}$$

The parameters calculated in part (a) allow us to define:

$$R_{yd} = -R_{yp} = 39.4 \text{ N}$$
$$R_{xd} = -R_{xp} = -56.9 \text{ N}$$
$$M_d = -M_p = -4.52 \text{ N} \cdot \text{M}$$

Other parameters are calculated as follows:

(i) Calculation of m:
From Table 7.2 and body weight:

$$m = (.016)(85) = 1.36 \text{ kg}$$

(ii) Calculation of I_0:

$$\rho_0 = \bar{\rho}_0 L_{FA}$$

Substituting from Example 7.2 and Table 7.2:

$$\rho_0 = (.305)(.323) = .0985 \text{ M}$$

So that:

$$I_0 = m \cdot \rho_0^2$$

Substituting:

$$I_0 = (1.36)(.0985)^2 = .013 \text{ kg} \cdot \text{M}^2$$

(iii) Determination of a', a'' and b' and b'':

From Example 7.2 (at $t = 0.330$ s):

$$a' = .095 \text{ M}$$
$$a'' = .158 \text{ M}$$

Calculate b' and b'' (at $t = 0.330$ s):

$$b' = |\Delta y|(\bar{x}_p)$$
$$b'' = |\Delta x|(\bar{x}_p)$$

Substituting from Example 7.2 and Table 7.2:

$$b' = (.167)(.430) = .072 \text{ M}$$
$$b'' = (.277)(.430) = .119 \text{ M}$$

Applying equation (28) to the translational FBD:

$$R_{yp} + m(g) + (R_{yd}) = m(a_y)$$

Substituting:

$$R_{yp} + (1.36)(-9.81) + (39.4) = 0$$

Solving for R_{yp}:

$$R_{yp} = 13.3 - 39.4 = -26.1 \text{ N}$$

Applying equation (29) to the translational FBD:

$$R_{xp} + (R_{xd}) = ma_x$$

Substituting:

$$R_{xp} + (-56.9) = 0$$

Solving for R_{xp}:

$$R_{xp} = 56.9 \text{ N}$$

Applying equation (30) to the rotational FBD:

$$M_P + (M_d) - (R_{yd})a'' + (R_{xd})a' + (R_{yp})b'' - (R_{xp})b' = I_0(\alpha)$$

Substituting:

$$M_p + (-4.52) - (39.4)(.158) + (-56.9)(.095) + (-26.1)(.119) - (-56.9)(.072) = (.013)(81.6)$$

Solving for M_p:

$$M_p - 4.52 - 6.23 - 5.41 - 3.11 - 4.10 = 1.06$$
$$M_p = 24.4 \text{ N} \cdot \text{M}$$

In summary, for the forearm segment at the elbow joint:

$$R_{yp} = -26.1 \text{ N (downward)}$$
$$R_{xp} = 56.9 \text{ N (rightward)}$$
$$M_p = 24.4 \text{ N} \cdot \text{M (counterclockwise)}$$

7.4 SEGMENT ENERGETICS

a. Energetic System and Muscle Mechanical Power

Segment energetics refers to the muscle mechanical work which is done by a muscle upon a body segment, and also refers to the muscle mechanical power which is the rate of doing that work. Segment energetics is important because it examines the interaction of the antagonist-agonist pair during human movement. This interaction is accomplished by neural activation of muscle contraction (see Chapter 9). Conse-

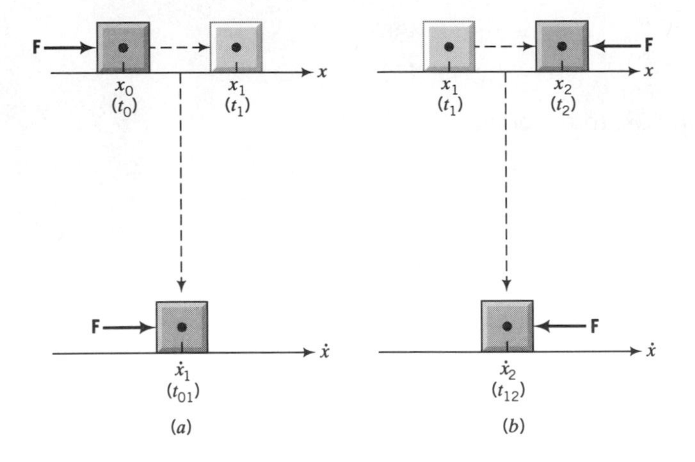

Figure 7.11.a Mechanical work and power having a positive value. **b** Mechanical work and power having a negative value.

quently, patterns of specific muscle contractions reflect the patterns of central nervous system drive to those muscles. This can be very helpful, for example, in designing and evaluating an operational task for the optimal performance, examining a training effect, or evaluating the human subsystem as part of an interaction with a simple technological subsystem (a hand tool).

Recall that mechanical work is a vector quantity and so may have positive or negative values depending upon the direction of the applied forces and the direction of the displacement (Figure 7.11.a). Mechanical power is also a vector quantity, depending upon the direction of the applied force and the displacement rate (Figure 7.11.b). From Newton's second law, it is apparent that positive work represents acceleration work and that negative work represents the deceleration work. Concurrently, positive power is the rate of doing acceleration work, and negative power is the rate of doing deceleration work.

In order to understand segment energetics, we first define the system. Figure 7.12 represents an approximate system of an elbow joint, with somewhat more anatomical detail than was given in Chapter 3. The five elements of the system anatomy are: joint, proximal segment, distal segment, agonist muscle, and antagonist muscle. In Figure 7.12, the center of rotation of the elbow joint has been defined at the intersection of the axial lines through the two adjacent segments. The proximal segment (upper arm) consists of the humerus with the biceps muscle anterior and the triceps muscle posterior (as seen in right profile). The distal segment (forearm) consists of the radius into which the biceps muscle inserts and the ulna into which the triceps muscle inserts. The two muscles represent an agonist-antagonist pair.

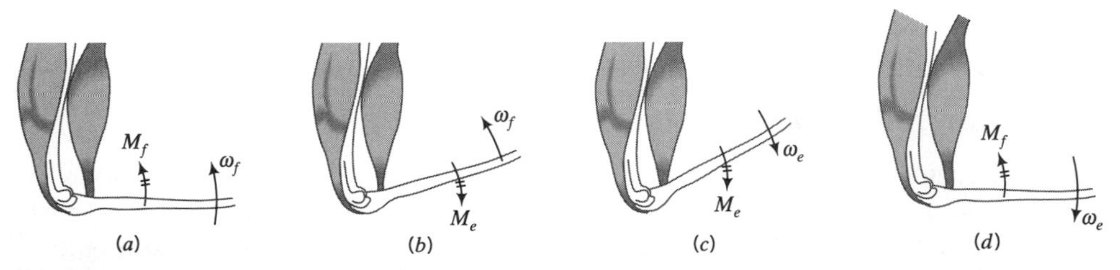

Figure 7.12 Interaction of agonist-antagonist pair during simple flexion-extension of forearm.

The elbow joint itself is located at the distal end of the proximal segment (upper arm) and the proximal end of the distal segment (forearm). Note that the two muscles (by means of their tendons) physically cross over the elbow joint and insert upon the proximal end of the distal segment (forearm). As each muscle contracts it exerts a line of force through the central axis of the muscle. A line drawn through the center of rotation and orthogonal to the muscle line of force represents the moment arm for that muscle about the center of rotation of the segment. The net muscle moment about the joint at any point in time is the sum of the individual agonist muscle moment and the individual antagonist muscle moment. This *net* muscle moment is exerted at the proximal end of the distal segment and is therefore equivalent to the M_p in the free-body diagram presented in Section 7.3.

Continuing our reference to Figure 7.12 the agonist-antagonist muscle moment results in rotation of the distal segment about the joint center. At any point in time (i), the distal segment will then move at some resultant angular velocity ($\boldsymbol{\omega}$). Muscle mechanical power ($\mathbf{P_m}$) is defined as the product of the net muscle moment and the resultant angular velocity:

$$\mathbf{P}_{mi} = M_{pi}\boldsymbol{\omega}_i \tag{46}$$

Note that for the anatomical system of interest, the ith segment is that body segment distal to the joint. As previously noted, the applied moment is the M_p (at the proximal end of the free-body diagram of the segment distal to the joint). Recall that an equal but opposite reaction moment (M_d) occurs at the distal end of the segment proximal to the joint. Angular velocity ($\boldsymbol{\omega}$) is, by definition, uniform over the entire distal segment.

Let us now consider the interaction of an agonist-antagonist muscle pair. All muscles generate force by internal contraction (the shortening of internal contractile elements within the muscle itself). However, the external muscle length may remain constant, shorten, or even lengthen depending upon the nature of the external load upon the muscle. The agonist muscle is considered the "prime mover" so when it generates internal contractile force the *external* length of the muscle shortens and the segment is rotationally displaced in the direction of that shortening. The antagonist muscle opposes the action of the "prime mover" muscle. Consequently, the antagonist muscle generates *internal* contractile force, while the *external* muscle length is increasing with rotational displacement of the segment.

Figure 7.12 examines the agonist-antagonist interaction during a simple flexion-extension movement, which is the *first* half of the "forearm task" described earlier (Figure 7.3). In Figure 7.12.a the forearm (EW) segment is initially parallel to the x-axis of the sagittal plane. The biceps muscle is the agonist exerting a net muscle moment counterclockwise (a flexion moment, M_f). This results in rotational displacement of the forearm (EW) segment in a counterclockwise direction (a flexion angular velocity, $\boldsymbol{\omega}_f$). At this point in time, the forearm segment is accelerating.

As upward flexion of the forearm continues, the net muscle moment at the joint reverses and there is a clockwise extension moment (M_e) as per Figure 7.12.b. This means that the triceps muscle (the antagonist) is exerting an individual muscle moment greater than the individual muscle moment of the biceps. However, the biceps muscle continues to be the agonist since the direction of forearm segment displacement is counterclockwise (at an angular velocity, $\boldsymbol{\omega}_f$). Essentially, the forearm segment continues to move in the direction of biceps muscle shortening. At this point in time, the forearm segment is decelerating and will subsequently come to a complete stop.

After stopping, the forearm segment then undergoes downward extension (Figure 7.12.c). The triceps muscle is now the agonist since it meets the definition of a prime mover. The net muscle moment about the joint center of rotation continues to be clockwise (M_e). However, the forearm segment is now also experiencing rotational displacement in a clockwise direction (at an angular velocity, $\boldsymbol{\omega}_f$). The triceps muscle is the agonist because it is now shortening in the same direction as forearm segment displacement. At this point in time, the forearm segment is accelerating.

Finally, the forearm segment begins to approach its initial, starting position as downward extension continues (Figure 7.12.d). The net muscle moment is now reversed as the biceps muscle exerts an individual muscle moment greater than the individual muscle moment of the triceps. There is now a flexion moment (M_f) that is counterclockwise about the joint center of rotation. However, the forearm segment continues to move in a clockwise direction at an extension angular velocity (ω_e). The biceps remains the antagonist muscle since it is lengthening in a clockwise direction (the direction of rotational displacement). At this point in time, the forearm segment is decelerating and will soon come to a stop.

Figure 7.13 depicts the variation in muscle power during a simple extension-flexion movement, which occurs during the *second* half of the "forearm task" described earlier (Figure 7.3). This figure demonstrates the time-varying nature of the net muscle moment and the segment angular velocity during a task in which the arm is first extended downward and then flexed upward. During the time interval (t_0 through t_1) the net muscle extension moment (M_e) is of the same sign as the segment extension angular velocity ($\boldsymbol{\omega}_e$). This means that the muscle power ($\mathbf{P_m}$) is positive and so describes an acceleration work rate. During the next time interval (t_1 through t_2) the net muscle flexion moment (M_f) is of opposite sign to the segment

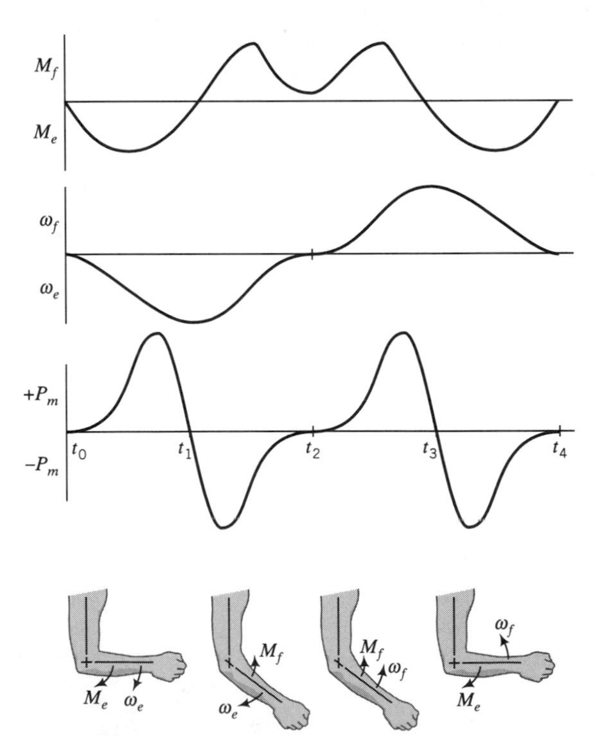

Figure 7.13 Variation in muscle power during simple extension-flexion of the forearm.

angular extension velocity (ω_e). This means that the muscle power ($\mathbf{P_m}$) is negative and so describes a deceleration work rate.

During the third time interval (t_2 through t_3) the net muscle flexion moment (M_f) is of the same sign as the segment angular flexion velocity ($\boldsymbol{\omega}_f$). This would then indicate that the muscle mechanical power ($\mathbf{P_m}$) is positive and so reflects an acceleration work rate. Finally, the last time interval (t_3 through t_4) is characterized by a net muscle extension moment (M_e) which is of opposite sign from the segment angular flexion velocity ($\boldsymbol{\omega}_f$). This would indicate that during the final time interval, the muscle mechanical power ($\mathbf{P_m}$) is negative and so reflects a deceleration work rate.

Taking Figure 7.12 and Figure 7.13 collectively, it is apparent that muscle mechanical power is a mathematical expression of the interaction of an agonist-antagonist muscle pair during a body segment rotational movement about its proximal joint. This agonist-antagonist interaction is a reflection of the pattern of the nervous system final common pathway drive to those muscles that results in the generation of contractile force. Consequently, it is important for the human factors engineer to analyze human movement with respect to joint moments and limb segment angular velocities in order to have an indirect understanding of the nervous system strategies. These strategies represent the time-varying central nervous system drive and the time-varying peripheral nervous system drive that occur during the interaction of the human subsystem with the technological subsystem.

b. Muscle Mechanical Work and Fractional Energetics

Muscle mechanical work may be calculated from the time integral of the muscle power curve (see Figure 7.14). The net work done by an agonist-antagonist muscle pair upon a body segment must be specified during a particular period of time. For example, the muscle mechanical work ($\mathbf{W}_m$), which is done during the time period, t_0 through t_1 (Figure 7.13 and Figure 7.14) is:

$$\mathbf{W}_m = \int_{t_0}^{t_1} \mathbf{P}_{mi} \cdot dt \tag{47}$$

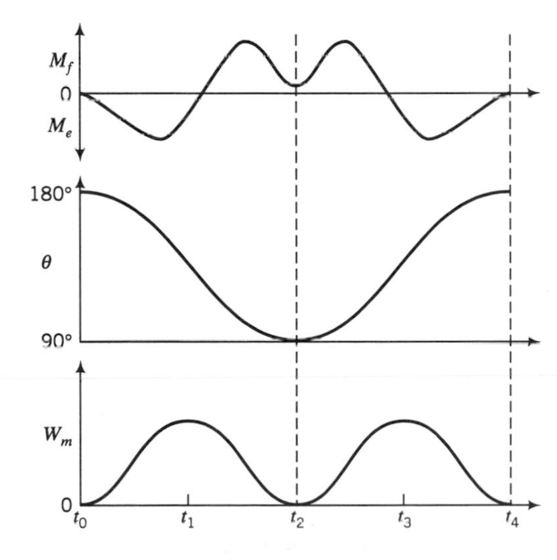

Figure 7.14 Variation in muscle work during simple extension-flexion of the forearm.

Substituting Equation (46) into Equation (47):

$$\mathbf{W}_m = \int_{t_0}^{t_1} \mathbf{M}_{pi}\omega_i \cdot dt \tag{48}$$

Since $\boldsymbol{\omega}_i = d\theta_i/dt$, then substituting into Equation (48) and simplifying:

$$\mathbf{W}_m = \int_{\theta_0(t_0)}^{\theta_1(t_1)} M_{pi} d\theta_i \tag{49}$$

$\mathbf{W}_m$ represents the mechanical work that is performed by an agonist-antagonist muscle pair upon the distal body segment into which both muscles insert. The agonist muscle does work upon the body segment by accelerating it, while the antagonist muscle also does work upon the body segment by decelerating it.

When calculating the muscle mechanical work ($\mathbf{W}_m$), use adjacent zero crossing times (Figure 7.13) when $\mathbf{P}_m$ is reversing its sign to define the upper and lower time limits for the integration. This will allow the human factors engineer to quantify the total positive muscle work (monotonically increasing functions of Figure 7.14 over certain time intervals) separately from a total negative muscle work (monotonically decreasing functions of Figure 7.14 over certain time intervals).

Fractional energetics refers to the fractional distribution of power (F_P) and also to the fractional distribution of work (F_W). Fractional power separates the total power (the sum of the externally generated power and the internally generated power) into an internal power fraction and external power fraction as follows:

$$F_{Pm} = \frac{\mathbf{P}_m}{\mathbf{P}_e + \mathbf{P}_m} \tag{50}$$

$$F_{Pe} = \frac{\mathbf{P}_e}{\mathbf{P}_e + \mathbf{P}_m} \tag{51}$$

Likewise, fractional work separates the total work (the sum of externally generated work and internally generated work) into the fraction due to internal work and the fraction due to external work, as follows:

$$F_{Wm} = \frac{\mathbf{W}_m}{\mathbf{W}_e + \mathbf{W}_m} \tag{52}$$

$$F_{We} = \frac{\mathbf{W}_e}{\mathbf{W}_e + \mathbf{W}_m} \tag{53}$$

Fractional energetics is very useful in the evaluation of a human operator when performing physical work. Fractional energetics provides important parameters as part of the quantitative assessment of physical workload (see Chapter 8). This can be briefly illustrated as follows.

Referring to equations (50)–(53), the sum of fractional power must equal unity, and the sum of fractional work must equal unity:

$$F_{Pm} + F_{Pe} = 1 \tag{54}$$

$$F_{Wm} + F_{We} = 1 \tag{55}$$

Consider these two extreme cases. In the first case, the human operator is performing purely internal work and generating purely internal power (which occurs when moving a body segment in free motion, such as the swing phase of the leg while walking). In this case no external work is being performed, nor is external power being generated so that equations (54) and (55) reduce to:

$$F_{Pm} + \cancelto{0}{F_{Pe}} = 1 \tag{56}$$

$$F_{Wm} + \cancelto{0}{F_{We}} = 1 \tag{57}$$

In the second case, only external work is being performed, and external power generated (such as a person sitting passively in a sled as it proceeds to slide down an incline). In this case the human operator is entirely passive, and not performing any internal work, nor generating any internal power, and equations (54) and (55) reduce to:

$$\cancelto{0}{F_{Pm}} + F_{Pe} = 1 \tag{58}$$

$$\cancelto{0}{F_{Wm}} + F_{We} = 1 \tag{59}$$

The majority of human operator-physical workload situations, however, will involve a human-work interaction that is somewhere between these two extremes. Note that when the internally performed work and internally generated power exactly equal the externally performed work and generated power, then equations (54) and (55) reduce to:

$$F_{Pm} = F_{Pe} = 0.5 \tag{60}$$

$$F_{Wm} = F_{We} = 0.5 \tag{61}$$

In practical application, equations (56)–(61) are never exactly satisfied. Consequently, fractional energetics operationally fall into two broad categories. The first category is when the fractional external power is greater than the fractional internal power and the fractional external work is greater than the fractional internal work. In this case, the human factors engineer should consider three possible subcategories. One, there is an additional external energy source that exceeds the internal energy provided by the human operator. Two, there is an additional internal energy source of the human operator that has not been accounted for in the analysis. Third, there is a combination of an additional external energy source and an additional internal energy source of the human operator.

The second category is when the fractional internal power is greater than the fractional external power and the fractional internal work is greater than the fractional external work. If this is the case, the human factors engineer should first be satisfied that all sources of internal power and work have been accounted for, and that all sources of external power and work have been identified. If this is so, the engineer may then quantitatively examine the energetic efficiency of the human operator and physical workload relationship.

EXAMPLE 7.6

For the system in Example 7.5:

a. Find the muscle mechanical power and the muscle mechanical work for the hand segment.
b. Find the muscle mechanical power and the muscle mechanical work for the forearm segment.

SOLUTION 7.6(a)

From the calculations performed in Examples 7.2, 7.4, and 7.5, construct the following table of calculated parameters:

Segment	$M_p(\theta_2)$	$\omega(\theta_2)$	θ_1	θ_2	θ_3
Hand	4.52	3.86	1.349	1.391	1.457
Forearm	24.4	0.929	0.537	0.543	0.563

For the hand segment:
The muscle mechanical power is obtained by substituting into equation (46):

$$P_m = (4.52)(3.86) = 17.4\text{ W}$$

The muscle mechanical work is calculated over the interval $\theta_1(t = 0.316\text{ s})$ to $\theta_3(t = 0.344\text{ s})$ by approximating that the proximal moment at $\theta_2(M_p)$ is constant over this interval. Consequently, the solution of equation (49) as:

$$\mathbf{W}_m = M_p(\theta_3 - \theta_1) \tag{50}$$

Substituting from the table of calculated parameters:

$$\mathbf{W} = (4.52)(.108) = 0.488\text{ J}$$

SOLUTION 7.6(b)

For the forearm segment:
The muscle mechanical power is obtained by substituting into equation (46):

$$\mathbf{P}_m = (24.4)(0.929) = 22.7\text{ W}$$

The muscle mechanical work performed on the forearm segment is calculated using the same approximations as for the hand segment.
Substituting from the table of calculated parameters into equation (50):

$$\mathbf{W}_m = (24.4)(.026) = 0.634\text{ J}$$

The general principles that have been presented in this chapter can now be applied to each of the seven specific body segments identified in Table 7.2. A three-link, four-segment model of the upper extremity is presented in Table 7.3. Note that the proximal segment of the shoulder joint is orthogonal to the sagittal plane of reference (i.e., the desired line). This proximal segment would be fully visualized in the frontal view. In the first two columns of Table 7.3, the three joints are identified (as hinge-type links) with respect to the distal body segment and the proximal body segment which the joints connect. When the body segment is distal, the movement sign (middle column of Table 7.3) refers to the agonist muscle movement at the proximal end of that segment distal to the joint. The movement sign is defined by the angular change ($\Delta\theta$) of the axis of that segment distal to the joint. A muscle pair acting across each joint is then identified in the fourth column of Table 7.3. Each of the muscles are specifically identified with their *agonist* roles. Each of the muscles (identified in column four) will produce a movement sign (of column three) when functioning as the agonist. Finally, the movement direction (for the distal segment) is specified in the last column, and will occur when the

Table 7.3 Link-Segment Model of the Upper Extremity

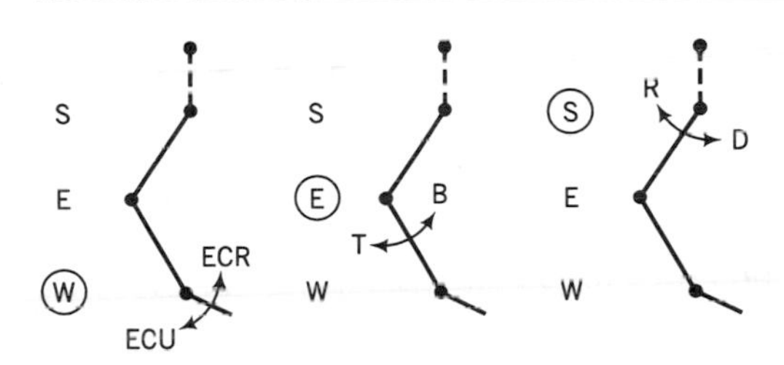

Joint	Body Segment (position)	Movement Sign ($\Delta\theta$)	Agonist Muscle	Movement Direction
Wrist (W)	Hand (distal)	+	Extensor carpi radialis (ECR)	Radial extension
		−	Extensor carpi ulnaris (ECU)	Ulnar extension
	Forearm (proximal)	N.A.	N.A.	N.A.
Elbow (E)	Forearm	+	Biceps (B)	Flexion
		−	Triceps (T)	Extension
	Upper arm (proximal)	N.A.	N.A.	N.A.
Shoulder (S)	Upper arm (distal)	+	Deltoid (D)	Flexion
		−	Rhomboids (R)	Extension
	Shoulder (proximal)	N.A.	N.A.	N.A.

muscle (as indicated in column four) functions as the agonist. It is understood that for any specific movement indicated in Table 7.3, the muscle (of the muscle pair) that is not the agonist will function as the antagonist. It should also be noted that agonist movement sign, muscle identification and movement direction are not applicable (N.A.) with respect to the *proximal* segment of any specific joint. Therefore, the last three columns of Table 7.3 are so indicated.

A one-link, two-segment model for the trunk of the human body is given in Table 7.4. Column headings have been defined and described as previously for Table 7.3. Note, however, that this model of the human trunk is inverted (as

Table 7.4 Link-Segment Model of the Human Trunk

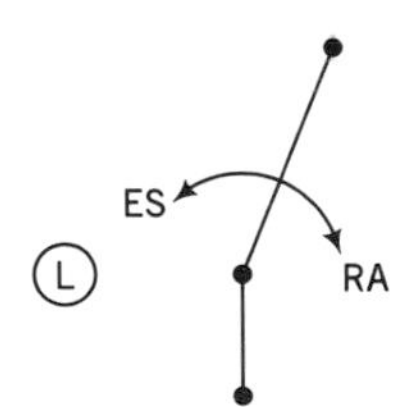

Joint	Body Segment (position)	Movement Sign ($\Delta\theta$)	Agonist Muscle	Movement Direction
Lumbo-sacral (L)	Thoracolumbar spine (distal)	+	Erector spinae (ES)	Extension
		−	Rectus abdominis (RA)	Flexion
	Pelvis (proximal)	N.A.	N.A.	N.A.

Table 7.5 Link-Segment Model of the Lower Extremity

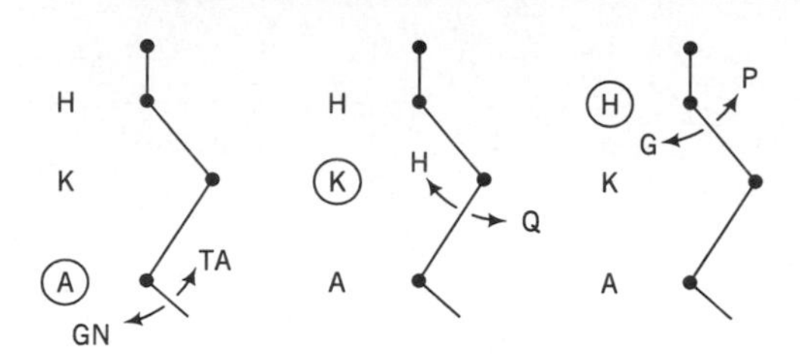

Joint	Body Segment (position)	Movement Sign ($\Delta\theta$)	Agonist Muscle	Movement Direction
Ankle Ⓐ	Foot (distal)	+	Tibialis anterior (TA)	Dorsiflexion
		−	Gastrocnemius (GN)	Plantarflexion
	Foreleg (proximal)	N.A.	N.A.	N.A.
Knee Ⓚ	Foreleg (distal)	+	Quadriceps (Q)	Extension
		−	Hamstring (H)	Flexion
	Thigh (proximal)	N.A.	N.A.	N.A.
Hip Ⓗ	Thigh (distal)	+	Psoas (P)	Flexion
		−	Gluteus (G)	Extension
	Pelvis (proximal)	N.A.	N.A.	N.A.

compared to the model of the lower extremity). Also with respect to the lumbosacral joint, the proximal segment is the pelvis and the distal segment is the thoracolumbar spine.

A three-link, four-segment model of the lower extremity is presented in Table 7.5. Interpretation of the five column headings of Table 7.5 is the same as that already provided in the description of Table 7.3.

This chapter now concludes with the following example.

EXAMPLE 7.7

You are the human factors engineer who performed the data analysis in Examples 7.2, 7.4, 7.5, and 7.6. You have now been assigned to develop the preliminary engineering requirements for the bowler's wrist brace as part of a Stage I systems engineering effort (you may wish to review Section 2.1 at this point).

(a) Interpret the kinematic data of the forearm and hand segment from Examples 7.2 and 7.4.

(b) Interpret the joint reaction forces and net muscle moment data at the wrist and elbow from Example 7.5.

(c) Interpret the muscle mechanical power and work performed on the hand and forearm segments from Example 7.6.

SOLUTION 7.7(a)

Isolated segments of kinematic data should always be viewed in the larger context of a kinematic sequence of events. Interpretation of isolated kinematic data segments should only proceed after the entire temporal sequence of events have been defined, and the HFE has determined the specific segments of the event that deserves focused analysis for the purpose of characterizing the essential requirements of the human-technological system.

In the case of a set of design requirements for a bowler's wrist brace, the HFE would use a four-step process. The sequence of steps would proceed as follows.

First, the engineer would view the entire videotape of our bowler in profile (from which the actual data was obtained) in order to determine the complete sequence of the person-bowling ball interaction. In this case, event phases would be identified:

1. Initial stance with ball firmly gripped in both hands, hold forward of the bowler.
2. Release of one hand and backward swing of the ball as the bowler's body starts moving forward.
3. When the ball is posterior of the body, the point of swing reverses.
4. Forward swing of the ball as the bowler's body continues to accelerate forward.
5. Release of the ball and deceleration of the bowler's body to a stop.

Second, summarize the translational and rotational kinematic parameters for the forearm segment center of mass (from Example 7.2):

	$\bar{x}$	$\bar{y}$	θ
Velocity	2.86 M/s	.214 M/s	.929 rad/s
Acceleration	0	0	81.6 rad/s^2

Schematically indicate the directionality of the translational and rotational kinematic parameters of the forearm center of mass:

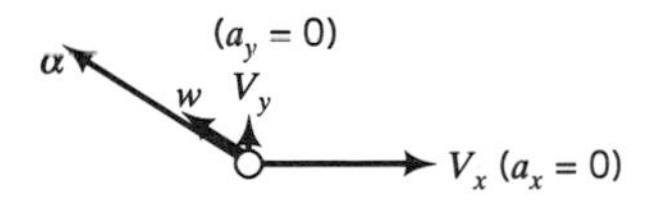

Interpret the movement of the *forearm segment* at this point in time. The center of mass of the forearm is moving at a constant horizontal velocity forward, the vertical velocity is small, and there is a small counterclockwise rotational velocity, which is rapidly accelerating at this point in time. This means that the inclined forearm segment is approaching a more vertical orientation (θ is increasing, as shown below).

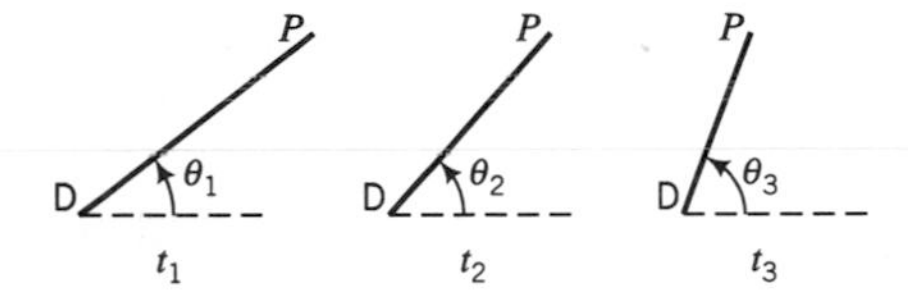

Third, summarize the translational and rotational kinematic parameters for the hand segment center of mass (from Example 7.4):

	$\bar{x}$	$\bar{y}$	θ
Velocity	3.18 M/s	0	3.86 rad/s
Acceleration	15.3 M/s^2	−20.4 M/s^2	122 rad/s^2

Schematically indicate the directionality of those kinematic parameters for the hand center of mass.

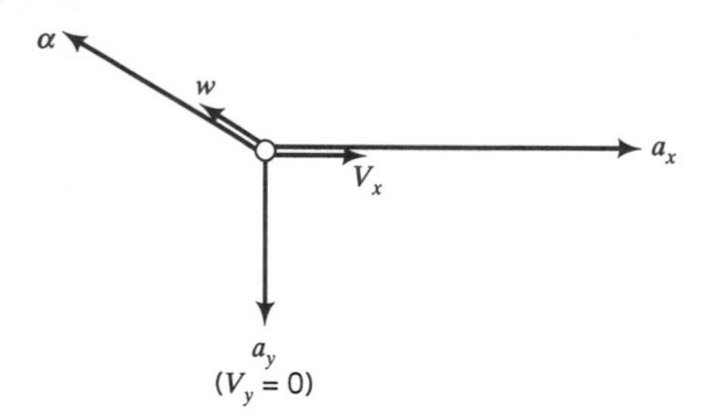

where:

P = proximal end (wrist)
D = distal end (finger-joint)

Interpret the movement of the *hand segment* at this point in time:
The center of mass of the hand is moving forward at an accelerating horizontal velocity, the vertical velocity is zero, and there is a moderate counterclockwise rotational velocity, which is rapidly accelerating at this point in time.

Fourth, interpret the segment-segment interaction with respect to that specific phase of the temporal sequence of events. This is as follows.

With backward swing of the ball [phase (2)], the forearm segment will undergo clockwise rotation (with a decreasing θ) as shown here:

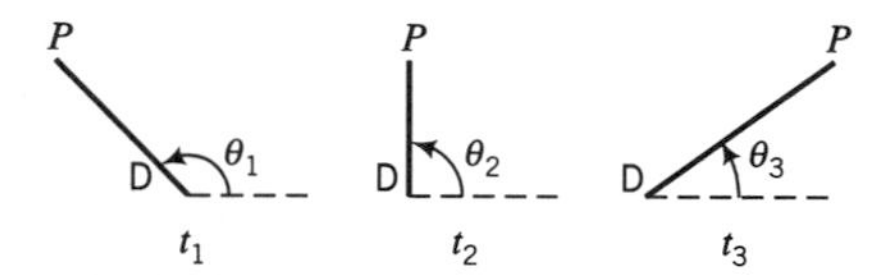

where:

P = proximal end (elbow)
D = distal end (wrist)

Since it has been shown that the forearm segment is approaching a more vertical orientation, with θ increasing, the data segment must be somewhere in Phase 4, which occurs with forward swing of the ball.

The hand segment is also approaching a more vertical orientation (increasing θ) at a positive angular velocity about four times greater than the forearm segment angular velocity. This implies that the hand segment (with respect to the forearm segment) is imparting an additional forward horizontal velocity to the ball (beyond that of the forearm segment velocity).

This is part of the "wrist control" that occurs during a Phase 4 event.

Finally, it is noted that the vertical velocity of both the forearm segment and the hand segment is near or at zero, but the hand segment is now experiencing a moderate negative acceleration. This implies that the data segment is temporally the earlier part of Phase 4, and that the hand (and the forearm also) will subsequently experience an increasingly negative vertical velocity (as the ball, hand and forearm move closer to the ground during the latter part of Phase 4, prior to Phase 5).

SOLUTION 7.7(b)

Interpretation of joint reaction forces and net muscle moment data begins with a proximal-to-distal schematic reconstruction of the force and moment parameters. After this is completed, the interpretation itself, and the specification of design requirements can proceed.

First, perform the segment by segment parameter identification:

i. Since the hand is in the "swing phase," there are no reaction forces at the distal end of the hand segment.

ii. At the proximal end of the hand segment (i.e., at the wrist joint) from Solution 7.5(a):

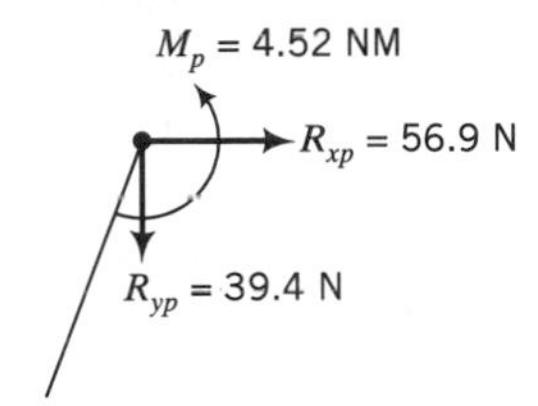

iii. Invoking the Newton's third law at the distal end of the forearm segment (i.e., at the wrist joint):

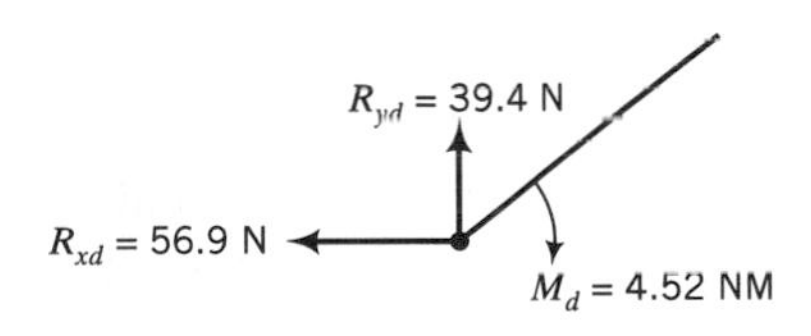

iv. At the proximal end of the forearm segment (i.e., at the elbow joint) from Solution 7.5(b):

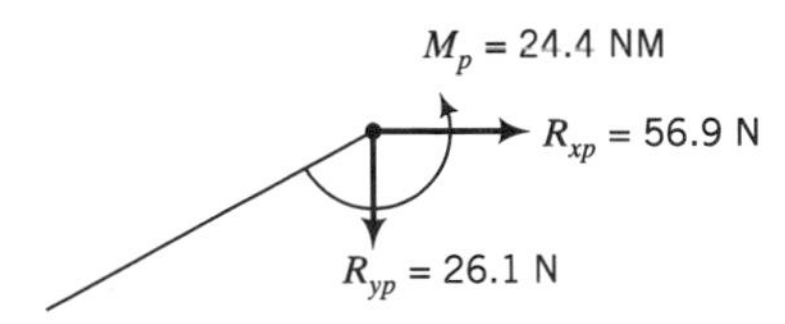

v. At the distal end of the upper arm segment (i.e., at the elbow joint) from Newton's third law:

R_{yd} = 29.1 N

R_{xd} = 56.9 N

M_d = 24.4 NM

Second, schematically diagram the segment-to-segment interaction:

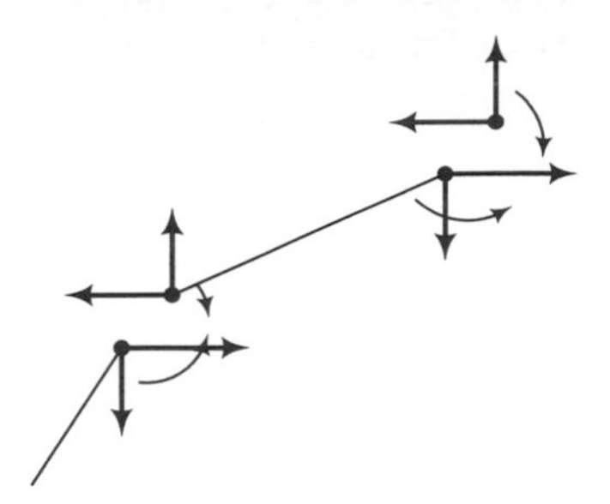

Third, perform the interpretation, which in this case yields the following:

i. The wrist and elbow joints are in tension, as indicated by the direction of the vertical reaction forces at these joints.

This occurs because the joints are in the swing phase (no reaction forces at the distal end of the hand segment) and these joints are loaded by the limb segment weights and the bowling ball weight (all acting vertically downward due to the gravity vector).

Note, however, these tensile forces are not uniform, and that the tensile forces on the wrist (at this point in time) are 150% greater than the tensile forces on the elbow.

A first design requirement, therefore, would need to address the ability of the wrist brace to *attenuate the tensile wrist force with reactive compressive forces.*

ii. The wrist and elbow joints are in shear, as indicated by the direction of the horizontal reaction forces at those joints. Note, in this case that the shear forces are the same at the wrist and elbow joint.

Recall that the wrist shear force is a reaction to the horizontal acceleration of the hand and ball combined mass. This may be a necessary part of ball control.

Therefore, a *second design requirement* for the bowler's wrist brace would specify the capability of the brace to allow a necessary range of wrist shear forces, but provide an upper limit beyond which excessive wrist shear forces would be checked by the upper-limit induced brace reactive shear forces.

iii. The hand is undergoing radial extension at the wrist joint created by the extensor carpi radialis muscle (ECR), which crosses that joint (see Table 7.3).

The forearm is undergoing flexion at the elbow joint induced by the biceps muscle, which crosses the elbow joint (see Table 7.3).

Note that the powerful biceps muscle creates a forearm flexion moment more than five times greater than the hand radial extension moment (caused by the distinctly weaker ECR muscle).

At this point in time, the ECR muscle is generating a radial extension moment (about the wrist) that is resulting in an extremely high rotational acceleration of the hand-ball system. This also may be a necessary part of ball control.

Therefore, a *third design requirement* for a bowler's wrist brace would specify a function of the brace to store elastic energy (during ulnar extension of the hand by the ECU muscles, as per Table 7.3) that could be released during radial extension of the hand in order to assist the vocational bowler in obtaining the extremely high rotational hand acceleration characteristic of a professional bowler.

SOLUTION 7.7(c)

The muscle mechanical power ($\mathbf{P}_m$) and work ($\mathbf{W}_m$) were calculated in Example 7.6, and are summarized below:

Segment	$\mathbf{P}_m$(W)	$\mathbf{W}_m$(J)
Hand	17.4	.488
Forearm	22.7	.634

In order to interpret these values, it is necessary to calculate the external power and work. This is the physical power and work that can be externally quantified (see Chapter 8, which discusses physical workload).

External power ($\mathbf{P}_e$) is the translational power represented by the force-velocity product of the moving bowling ball:

$$\mathbf{P}_e = F_B \cdot V_B \tag{i}$$

External work ($\mathbf{W}_e$) is the translational work defined as the force-displacement product of the moving bowling ball:

$$\mathbf{W}_e = F_B \cdot D_B \tag{ii}$$

Since we are dealing with translational external power and work, there is a horizontal external power (P_{eH}), vertical external power (P_{eV}), horizontal external work (W_{eH}), and vertical external work (W_{eV}) defined as:

$$P_{eH} = m_B \cdot a_x \cdot V_x \tag{iii}$$

$$P_{eV} = m_B \cdot a_y \cdot V_y \tag{iv}$$

$$W_{eH} = m_B \cdot a_x \cdot \Delta x \tag{v}$$

$$W_{eV} = m_B \cdot a_y \cdot \Delta y \tag{vi}$$

We now proceed to solve these four equations.

In Example 7.5, the mass of the ball (m_B) is given as 3.41 kg, and the ball's center of mass is specified as being co-located at the center of mass of the hand segment. Consequently, acceleration, velocity, and displacement data for the bowling ball can be equated to the acceleration, velocity, and displacement of the hand segment.

The following values were summarized in Solution 7.7(a) above:

$$a_x = 15.3 \text{ M/s}^2$$

$$a_y = -20.4 \text{ M/s}^2$$

$$V_x = 3.18 \text{ M/s}$$

$$V_y = 0$$

From the table of processed data given in Solution 7.4:

$$\Delta x = \bar{x}_{0.344} - \bar{x}_{0.316}$$
$$= 1.675 - 1.586 = .089 \text{ M}$$
$$\Delta y = \bar{y}_{0.344} - \bar{y}_{0.316}$$
$$= .178 - .178 = 0 \text{ M}$$

Substituting the above values into equations 7.7(c)(iii)–7.7(c)(vi):

$$P_{eH} = (3.41)(15.3)(3.18) = 166 \text{ W}$$
$$P_{eV} = (3.41)(-20.4)(0) = 0 \text{ W}$$
$$W_{eH} = (3.41)(15.3)(.089) = 4.64 \text{ W}$$
$$W_{eV} = (3.41)(-20.4)(0) = 0 \text{ W}$$

Since the vertical component of external power and work is zero, we need only consider the horizontal component.
Finally, we calculate the fraction of total power and work for the hand segment and the ball itself:
Substituting into equations (50) to (53) results in:

$$F_{Pm} = \frac{17.4}{166 + 17.4} = .095$$

$$F_{Pe} = \frac{166}{166 + 17.4} = .905$$

$$F_{Wm} = \frac{.48}{4.64 + .488} = .095$$

$$F_{We} = \frac{4.64}{4.64 + .488} = 0.905$$

At this point, it is apparent that less than one-tenth of the total power and work is accounted for by the muscle mechanical power and work of the hand segment. Therefore, we then identify additional internal power and work of forearm segment, and recalculate the fraction of total power and work for the hand segment *and* forearm segment (identified as the parenthetical term) and the ball itself:

$$F_{Pm} = \frac{(17.4 + 22.7)}{166 + (17.4 + 22.7)} = .195$$

$$F_{Pe} = \frac{166}{166 + (17.4 + 22.7)} = .805$$

$$F_{Wm} = \frac{(.488 + .634)}{4.64 + (.488 + .634)} = .195$$

$$F_{We} = \frac{4.64}{4.64 + (.488 + .634)} = .805$$

At this point, it is now apparent that about two-tenths of the total power and work is accounted for by the combined hand and forearm segment muscle mechanical power and work. Therefore, the HFE should now systematically consider all the other sources of internal muscle mechanical power and work.

It will then become immediately apparent that both lower extremities of the bowler are actively accelerating the entire body during this phase (4) event [see Solution 7.7(a)]. This will contribute to the overall acceleration of the ball (in addition to the contribution of the right upper extremity). However, further analysis of these other body segments is not really necessary, since the results are not likely to contribute to the design requirements of the bowler's wrist brace.

Finally, since only two tenths of the total power and work has been quantified from internal muscle mechanical power and work, the HFE should also systematically evaluate all other sources of external energy that may not have been accounted for. Reexamination of the videotape will then indicate that Phases 2, 3, and 4 of Solution 7.7(a) represent an approximately pendular motion of the bowling ball. At Phase 3 with swing reversal, there is minimal kinetic energy (the bowler is still moving forward) and maximal potential energy of gravity (vertical height of the ball above the floor). During Phase 4, some of this potential energy will be exchanged for kinetic energy by the pendular lowering of the ball closer to the floor. This would occur even in the presence of a completely passive right upper extremity, that is, analogous to the arm of a swinging pendulum.

As a result of this analysis, another design requirement for the bowler's wrist brace could result from a study of the data segment at Phase 3, the point of swing reversal.

The fourth design requirement would specify that the brace is to maximize ulnar extension at end-posterior swing. This would serve to maximize stored elastic energy (per the third design requirement) and to add a final amount of vertical lift to the ball (in order to maximize the vertical potential energy gradient).

FURTHER INFORMATION

D.B. Chaffin, G.B.J. Andersson, and B.J. Martin: *Occupational Biomechanics*. 3rd Edition. John Wiley. New York. 1999.

Eastman Kodak Company: *Ergonomic Design for People at Work: Volume 1*. Van Nostrand Reinhold. New York. 1983.

C.A. Phillips (ed.): *Effective Upper and Lower Extremity Prostheses*. (AUTOMEDICA 11:1-3). Gordon and Breach. New York. 1989.

G. Salvendy (ed.): *Handbook of Human Factors and Ergonomics*. 2nd Edition. John Wiley. New York. 1997.

D.A. Winter: *Biomechanics and Motor Control of Human Movement*. 2nd Edition. John Wiley. New York. 1990.

PROBLEMS

The following problems represent the application of ergonomic biodynamics as the human operator performs a task. The person is viewed in right profile (side view) and position markers are placed over:

1. The right upper extremity:
   a. The right shoulder (S)
   b. The right elbow (E)
   c. The right wrist (W)
   d. The right hand (H)
2. The right lower extremity:
   a. The right hip (h)
   b. The right knee (K)

c. The right ankle (A)
d. The right foot (F)

Continuous recording of the human operator's movement is obtained and a set of position-time data points recorded:

x_S, x_E, x_W, x_H = horizontal position of the shoulder, elbow, wrist, and hand markers respectively.
y_S, y_E, y_W, y_H = vertical position of the shoulder, elbow, wrist, and hand markers respectively.
x_h, x_K, x_A, x_F = horizontal position of the hip, knee, ankle, and foot markers respectively.
y_h, y_K, y_A, y_F = vertical position of the hip, knee, ankle, and foot markers respectively.

As a human factors engineer, you have decided to analyze a specific moment in time.

i = moment in time at which the ergonomic biodynamic analysis is desired.
$i - 1$ = one time increment (Δt) prior to t.
$i + 1$ = one time increment (Δt) subsequent to t.

You observe that all events are during a repetitive task performance and are cyclical. Each cycle has a Phase A and phase B.

Recall that any external Phase A forces (F_A) or any external Phase B forces (F_B) are in addition to the force of gravity on the body segment.

7.1. A carpenter holds a hand saw in his right hand with his arm bent backward at the start of Phase A when the saw blades engage a piece of lumber (supported between two saw horses). Phase A occurs when the carpenter pushes the hand saw across the lumber (forward stroke) as per Diagram 7.1.A. During Phase B, the carpenter pulls the hand saw back across the lumber (back stroke) returning to the original starting position (see Diagram 7.1.B).

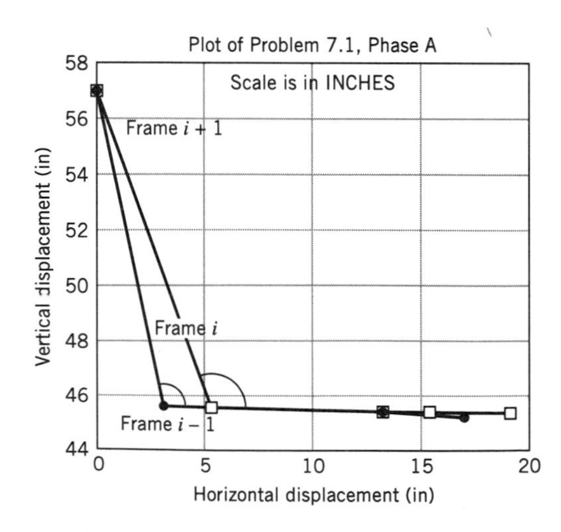

Diagram 7.1A

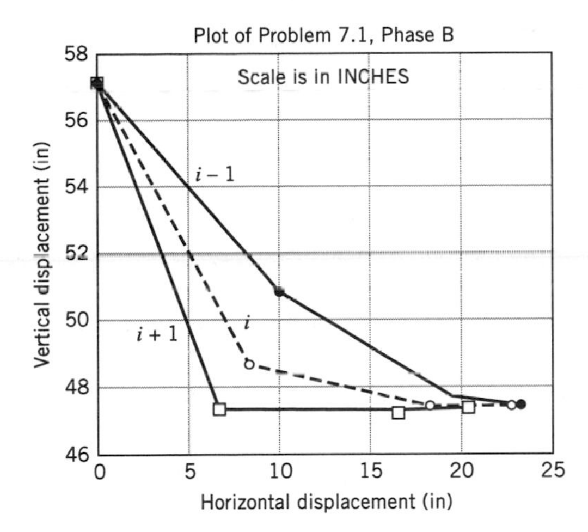

Diagram 7.1B

The person has the following anthropometric dimensions:

1. Linear dimensions:
   a. Vertical (body) height = N/A
   b. Elbow-wrist (E-W) segment length = 0.259 M
   c. Wrist-hand (W-H) segment length = 0.095 M
2. Mass dimensions:
   a. Total body mass (m_B) = 80 kg
   b. Forearm (E-W) segment mass = $(.016)(m_B)$ = 1.28 kg
   c. Hand (W-H) segment mass = $(.006)(m_B)$ = 0.48 kg

The following data is acquired for Phase A of the task (in meters) during a specific moment in time:

Time	x_E	y_E	x_W	y_W	x_H	y_H
$i-1$	.0254	1.150	0.284	1.150	0.380	1.150
i	0.076	1.160	0.335	1.154	0.430	1.151
$i+1$	0.132	1.180	0.389	1.162	0.484	1.154

where $\Delta t = 0.120$ s.

During Phase A, the external force on the hand segment (F_A) is directed leftward, at an angle of 0° to the horizontal axis, is equal to 50 N, and acts at the hand marker (H) location.

a. Calculate the nine kinematic parameters ($\bar{x}, \bar{y}, v_x, v_y, a_x, a_y, \theta, \boldsymbol{\omega}, \boldsymbol{\alpha}$).
   i. For the hand (W-H) segment.
   ii. For the forearm (E-W) segment.

b. Calculate the reaction forces and the applied muscle moment at the proximal end (R_{yp}, R_{xp}, M_p):
   i. Of the hand (W-H) segment.
   ii. Of the forearm (E-W) segment.

c. Calculate the muscle mechanical power and work ($\mathbf{P}_m, \mathbf{W}_m$):

i. For the hand (W-H) segment.
ii. For the forearm (E-W) segment.

d. Interpret the results.

7.2. Repeat Problem 7.1.a for Phase B.

The following data is acquired for Phase B of the task (in meters) during a specific moment in time:

Time	x_E	y_E	x_W	y_W	x_H	y_H
$i-1$	0.254	1.290	0.500	1.210	0.595	1.210
i	0.211	1.240	0.467	1.210	0.575	1.200
$i+1$	0.170	1.200	0.427	1.200	0.522	1.200

where $\Delta t = 0.120$ s.

During Phase B, the external force on the hand segment (F_B) is directed rightward, at an angle of 0° to the horizontal axis, is equal to 50 N, and acts at the hand marker (H) location.

a. Calculate the nine kinematic parameters ($\bar{x}, \bar{y}, v_x, v_y, a_x, a_y, \theta, \boldsymbol{\omega}, \boldsymbol{\alpha}$).

i. For the hand (W-H) segment.
ii. For the forearm (E-W) segment.

b. Interpret the results.

7.3. A luggage handler is initially standing erect, upper arms to the worker's side, forearm and hand extended outward (elbows bent at 90°), as shown in Diagram 7.2. In Phase B, the worker bends her arms down and out and grasps two suitcases on a conveyor belt (while leaning forward only). In Phase A, the worker lifts the suitcases, placing them on a shelf in front of the person so that her arms return to their original starting point at the beginning of Phase B.

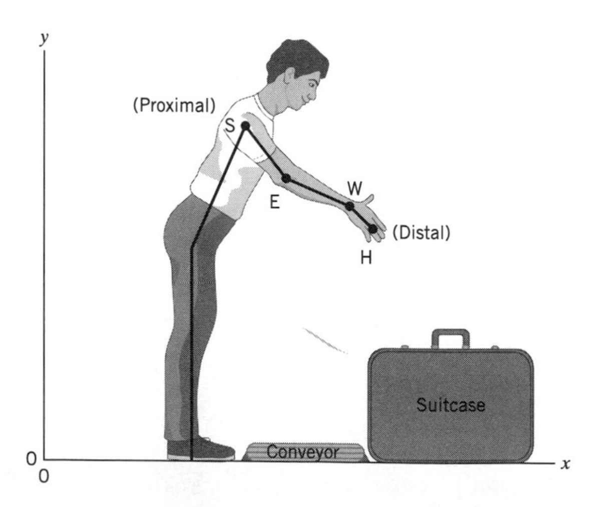

Diagram 7.2

The person has the following anthropometric dimensions:

1. Linear dimensions:

   a. Vertical (body) height = N/A
   b. Elbow-wrist (E-W) segment length = 0.25 M
   c. Wrist-hand (W-H) segment length = 0.09 M

2. Mass dimensions:

   a. Total body mass (m_B) = 65.32 kg
   b. Forearm (E-W) segment mass = (.016)(m_B) = 1.05 kg
   c. Hand (W-H) segment mass = (.006)(m_B) = 0.39 kg
   d. Suitcase (m_{SC}) = 2.27 kg

The following data is acquired for Phase B of the task (in meters) during a specific moment in time:

Time	x_E	y_E	x_W	y_W	x_H	y_H
$i-1$	0.340	1.020	0.532	0.899	0.590	0.870
i	0.341	1.010	0.536	0.885	0.600	0.850
$i+1$	0.358	0.997	0.552	0.879	0.610	0.840

where $\Delta t = 0.043$ s.

During Phase B, there is no external force (except gravity) acting on the hand segment.

a. Calculate the nine kinematic parameters ($\bar{x}, \bar{y}, v_x, v_y, a_x, a_y, \theta, \boldsymbol{\omega}, \boldsymbol{\alpha}$).

   i. For the hand (W-H) segment.
   ii. For the forearm (E-W) segment.

b. Calculate the reaction forces and the applied muscle moment at the proximal end (R_{yp}, R_{xp}, M_p):

   i. Of the hand (W-H) segment.
   ii. Of the forearm (E-W) segment.

c. Calculate the muscle mechanical power and work ($\mathbf{P}_m$, $\mathbf{W}_m$):

   i. For the hand (W-H) segment.
   ii. For the forearm (E-W) segment.

d. Interpret the results.

7.4. Repeat Problem 7.3.a for Phase A.

The following data is acquired for Phase A of the task (in meters) during a specific moment in time:

Time	x_E	y_E	x_W	y_W	x_H	y_H
$i-1$	0.506	0.710	0.666	0.550	0.734	0.495
I	0.510	0.720	0.666	0.565	0.720	0.505
$i+1$	0.498	0.730	0.640	0.585	0.705	0.525

where $\Delta t = 0.043$ s.

During Phase A, the external force (F_A) on the hand segment is the mass of the suitcase (m_{SC} = 2.27 kg) times the gravitational constant (**g** = 9.81 M/s²). This force is directed vertically downward (at an angle of 0° to the vertical axis) and acts at the hand marker (H) location.

a. Calculate the nine kinematic parameters ($\bar{x}, \bar{y}, v_x, v_y, a_x, a_y, \theta, \boldsymbol{\omega}, \boldsymbol{\alpha}$).

   i. For the hand (W-H) segment.

   ii. For the forearm (E-W) segment.

b. Interpret the results.

7.5. A ceramic worker is seated at a "potter rotating table" with his upper leg extended, the knee bent so that the heel of the foot contacts the ground and the ball of the foot contacts the treadle pedal. Phase A *and* Phase B continuously alternate. Phase A (see Diagram 7.3A) begins when the ball of the foot is at its highest point off the ground and pressure is continuously extended downward against the treadle pedal. Phase B (see Diagram 7.3B) begins when the ball of the foot is at its lowest point off the ground and the ball of the foot passively rides the treadle pedal back up to its highest position off the ground.

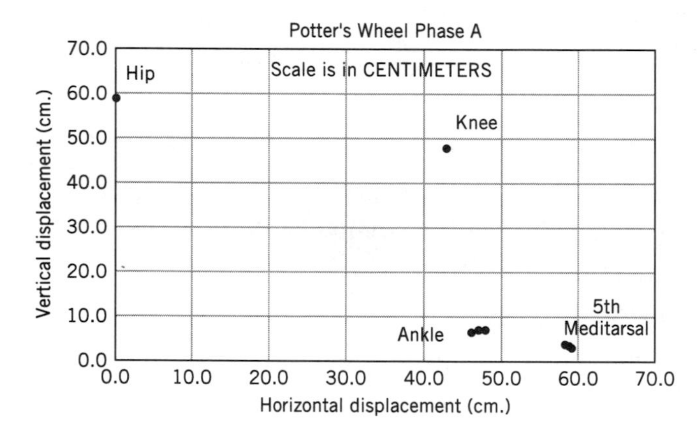

Diagram 7.3A

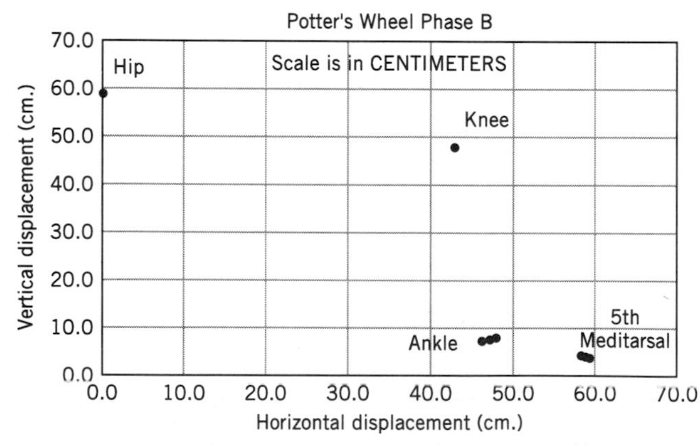

Diagram 7.3B

The person has the following anthropometric dimensions:

1. Linear dimensions:

   a. Vertical (body) height = N/A

b. Knee-ankle (K-A) segment length = 0.425 M
c. Ankle-foot (A-F) segment length = 0.122 M

2. Mass dimensions:

a. Total body mass (m_B) = 65.32 kg
b. Lower leg (K-A) segment mass = (.046)(m_B) = 2.714 kg
c. Foot (A-F) segment mass = (.014)(m_B) = 0.83 kg

The following data is acquired for Phase A of the task (in meters) during a specific moment in time:

Time	x_K	y_K	x_A	y_A	x_F	y_F
$i-1$	0.430	0.485	0.465	0.071	0.585	0.045
i	0.430	0.485	0.472	0.074	0.590	0.042
$i+1$	0.430	0.485	0.480	0.075	0.594	0.040

where $\Delta t = 0.30$ s.

During Phase A, the external force (F_A) is 1.8 N applied by the treadle pedal. The force is directed vertically upward (at an angle of 0° to the vertical axis) and acts at the foot (F) marker location.

a. Calculate the nine kinematic parameters ($\bar{x}, \bar{y}, v_x, v_y, a_x, a_y, \theta, \boldsymbol{\omega}, \boldsymbol{\alpha}$).

i. For the lower leg (K-A) segment.
ii. For the foot (A-F) segment.

b. Calculate the reaction forces and the applied muscle moment at the proximal end (R_{yp}, R_{xp}, M_p):

i. Of the lower leg (K-A) segment.
ii. Of the foot (A-F) segment.

c. Calculate the muscle mechanical power and work ($\mathbf{P}_m, \mathbf{W}_m$):

i. For the lower leg (K-A) segment.
ii. For the foot (A-F) segment.

d. Interpret the results.

7.6. Repeat Problem 7.5 for Phase B.

The following data is acquired for Phase B of the task (in meters) during a specific moment in time:

Time	x_K	y_K	x_A	y_A	x_F	y_F
$i-1$	0.430	0.485	0.480	0.075	0.594	0.040
i	0.430	0.485	0.472	0.072	0.590	0.043
$i+1$	0.430	0.485	0.465	0.071	0.585	0.045

where $\Delta t = 0.30$ s.

During Phase B, the external force (F_B) is 1.8 N applied by the treadle pedal. The force is directed vertically upward (at an angle of 0° to the vertical axis) and acts at the foot (F) marker location.

a. Calculate the nine kinematic parameters ($\bar{x}, \bar{y}, v_x, v_y, a_x, a_y, \theta, \boldsymbol{\omega}, \boldsymbol{\alpha}$).

   i. For the lower leg (K-A) segment.
   ii. For the foot (A-F) segment.

b. Calculate the reaction forces and the applied muscle moment at the proximal end (R_{yp}, R_{xp}, M_p):

   i. Of the lower leg (K-A) segment.
   ii. Of the foot (A-F) segment.

c. Calculate the muscle mechanical power and work ($\mathbf{P}_m, \mathbf{W}_m$):

   i. For the lower leg (K-A) segment.
   ii. For the foot (A-F) segment.

d. Interpret the results.

7.7. A painter stands upright facing a vertical wall. He begins Phase A (see Diagram 7.4A) with arm at side, forearm bent at an angle of 90°, and hand holding a paint roller against the wall. The painter then rolls paint vertically, pushing the roller up the wall (during Phase A). The painter starts Phase B (see Diagram 7.4.B) at the end of this upward roll by then rolling paint (vertically pulling the roller) back down the wall and so returning to the original starting position.

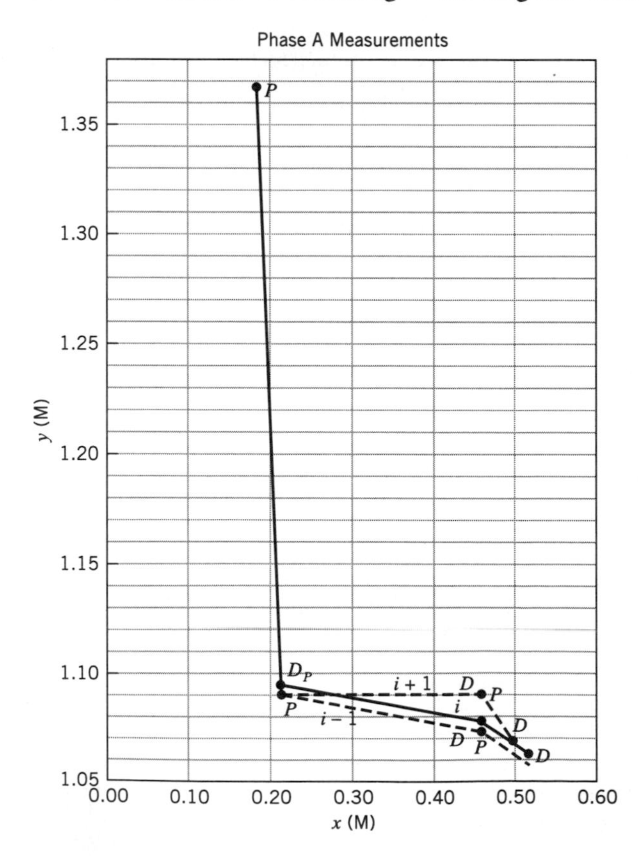

Diagram 7.4A

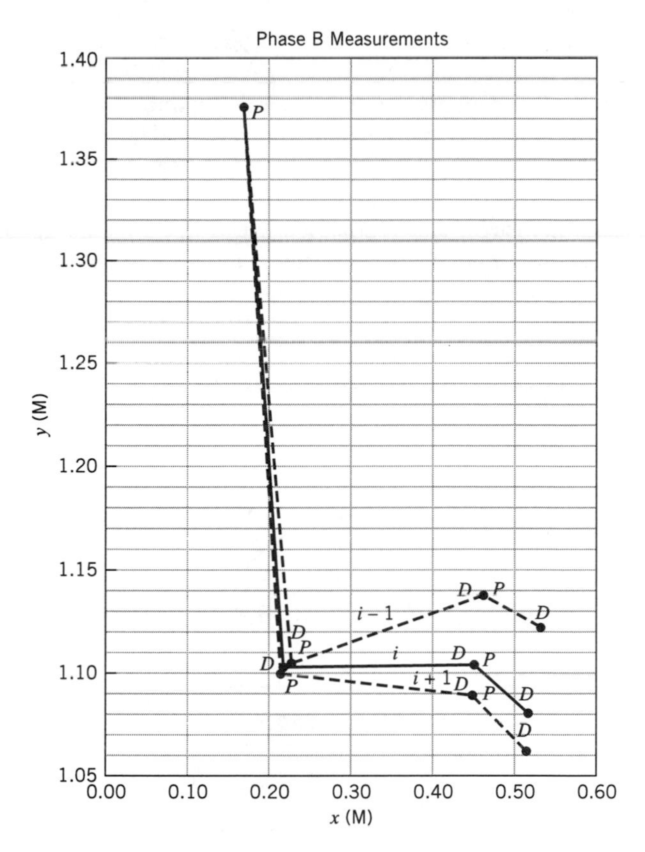

Diagram 7.4B

The person has the following anthropometric dimensions:

1. Linear dimensions:

   a. Vertical (body) height = N/A
   b. Elbow-wrist (E-W) segment length = 0.243 M
   c. Wrist-hand (W-H) segment length = 0.061 M

2. Mass dimensions:

   a. Total body mass (m_B) = 81.818 kg
   b. Forearm (E-W) segment mass = $(.016)(m_B)$ = 1.309 kg
   c. Hand (W-H) segment mass = $(.006)(m_B)$ = 0.491 kg

The following data is acquired for Phase A of the task (in meters) during a specific moment in time:

Time	x_E	y_E	x_W	y_W	x_H	y_H
$i-1$	0.213	1.092	0.455	1.074	0.515	1.058
i	0.213	1.095	0.456	1.079	0.515	1.063
$i+1$	0.213	1.092	0.457	1.090	0.498	1.068

where $\Delta t = 0.067$ s.

During Phase A, there are *two* orthogonal external forces on the hand segment. F_{Ay} is the weight of the roller (4.448 N). The roller is assumed to be a point mass (m_R) located at the center of mass of the hand segment. Because of the gravitational vector (**g**), F_{Ay} acts vertically downward (at an angle of 0 degrees to the vertical axis).

F_{Ax} is the reaction force of the wall against the roller (2.224 N). F_{Ax} acts as a pure horizontal reaction force, directed leftward, and at an angle of 0° to the horizontal axis. F_{Ax} is assumed to act at the center of mass of the hand segment.

a. Calculate the nine kinematic parameters ($\bar{x}, \bar{y}, v_x, v_y, a_x, a_y, \theta, \boldsymbol{\omega}, \boldsymbol{\alpha}$).

   i. For the hand (W-H) segment.
   ii. For the forearm (E-W) segment.

b. Interpret the results.

7.8. Repeat Problem 7.7 for Phase B.

The following data is acquired for Phase B of the task (in meters) during a specific moment in time:

Time	x_E	y_E	x_W	y_W	x_H	y_H
$i-1$	0.225	1.106	0.463	1.139	0.537	1.123
i	0.217	1.105	0.451	1.106	0.517	1.082
$i+1$	0.214	1.101	0.450	1.090	0.515	1.063

where $\Delta t = 0.067$ s.

During Phase B, there are *two* orthogonal external forces on the hand segment. F_{Ay} is the weight of the roller (4.448 N). The roller is assumed to be a point mass (m_R) located at the center of mass of the hand segment. Because of the gravitational vector (**g**), F_{Ay} acts vertically downward (at an angle of 0° to the vertical axis).

F_{Ax} is the reaction force of the wall against the roller (2.224 N). F_{Ax} acts as a pure horizontal reaction force, directed leftward, and at an angle of 0° to the horizontal axis. F_{Ax} is assumed to act at the center of mass of the hand segment.

a. Calculate the nine kinematic parameters ($\bar{x}, \bar{y}, v_x, v_y, a_x, a_y, \theta, \boldsymbol{\omega}, \boldsymbol{\alpha}$).

   i. For the hand (W-H) segment.
   ii. For the forearm (E-W) segment.

b. Interpret the results.

7.9. An individual is seated upon and pedaling a manual bicycle (see Diagram 7.5). For the right leg Phase A begins when the right foot is at the highest point (12 o'clock position) and force is exerted on the pedal continuously (as it rotates clockwise) until the lowest point (6 o'clock position). Phase B begins as the foot passively "coasts" clockwise from 6 o'clock back to 12 o'clock.

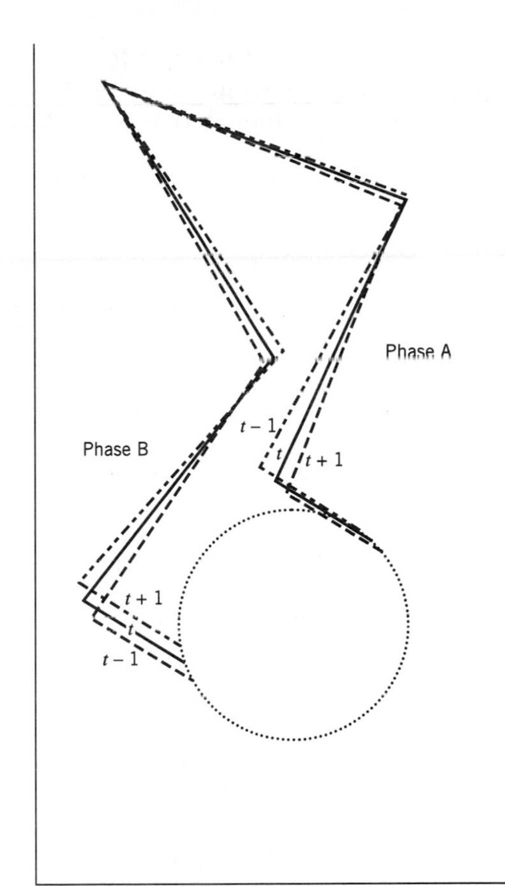

Diagram 7.5

The person has the following anthropometric dimensions:

1. Linear dimensions:

   a. Vertical (body) height = N/A
   b. Knee-ankle (K-A) segment length = 0.405 M
   c. Ankle-foot (A-F) segment length = 0.154 M

2. Mass dimensions:

   a. Total body mass (m_B) = 70.5 kg
   b. Lower leg (K-A) segment mass = (.046)(m_B) = 3.243 kg
   c. Hand (W-H) segment mass = (.014)(m_B) = 0.9871 kg

The following data is acquired for Phase A of the task (in meters) during a specific moment in time:

Time	x_K	y_K	x_A	y_A	x_F	y_F
$i-1$	0.4921	0.8862	0.2966	0.5440	0.4266	0.4610
i	0.4903	0.8916	0.3135	0.5274	0.4435	0.4444
$i+1$	0.4864	0.8961	0.3304	0.5108	0.4604	0.4278

where $\Delta t = 0.025$ s.

During Phase A, the external force (F_A) is the reaction force of the bicycle pedal against the foot (30.0 N). F_A is directed leftward and upward (at a 45° angle to the horizontal axis) and is assumed to act at the foot marker (F) location.

a. Calculate the nine kinematic parameters ($\bar{x}, \bar{y}, v_x, v_y, a_x, a_y, \theta, \boldsymbol{\omega}, \boldsymbol{\alpha}$).

   i. For the lower leg (K-A) segment.
   ii. For the foot (A-F) segment.

b. Calculate the reaction forces and the applied muscle moment at the proximal end (R_{yp}, R_{xp}, M_p):

   i. Of the lower leg (K-A) segment.
   ii. Of the foot (A-F) segment.

c. Calculate the muscle mechanical power and work ($\mathbf{P}_m$, $\mathbf{W}_m$):

   i. For the lower leg (K-A) segment.
   ii. For the foot (A-F) segment.

d. Interpret the results.

7.10. Repeat Problem 7.9 for Phase B.

The following data is acquired for Phase B of the task (in meters) during a specific moment in time:

Time	x_K	y_K	x_A	y_A	x_F	y_F
$i-1$	0.3047	0.6949	0.0743	0.3503	0.2043	0.2673
i	0.3160	0.6872	0.0650	0.3722	0.1950	0.2892
$i+1$	0.3284	0.6807	0.0592	0.3953	0.1892	0.3123

where $\Delta t = 0.025$ s.

During Phase B, the external force (F_A) is the reaction force of the bicycle pedal against the foot (30.0 N). F_A is directed leftward and upward (at a 27° angle to the horizontal axis) and is assumed to act at the foot marker (F) location.

a. Calculate the nine kinematic parameters ($\bar{x}, \bar{y}, v_x, v_y, a_x, a_y, \theta, \boldsymbol{\omega}, \boldsymbol{\alpha}$).

   i. For the lower leg (K-A) segment.
   ii. For the foot (A-F) segment.

b. Calculate the reaction forces and the applied muscle moment at the proximal end (R_{yp}, R_{xp}, M_p):

   i. Of the lower leg (K-A) segment.
   ii. Of the foot (A-F) segment.

c. Calculate the muscle mechanical power and work ($\mathbf{P}_m$, $\mathbf{W}_m$):

   i. For the lower leg (K-A) segment.
   ii. For the foot (A-F) segment.

d. Interpret the results.

7.11. A chemical worker stands erect in front of a table (see Diagram 7.6), arm at side and forearm extended (elbow angle of 90°) and grasps a cylindrical gas canister (initially sitting on the table) in the right hand (removing it from a box of such canisters). Phase A begins and the worker lifts the canister and places it on an overhead conveyor belt (above but in front of the worker). Phase B then begins when the worker's now empty hand retraces the original motion and returns to the box of canisters (and the original starting position).

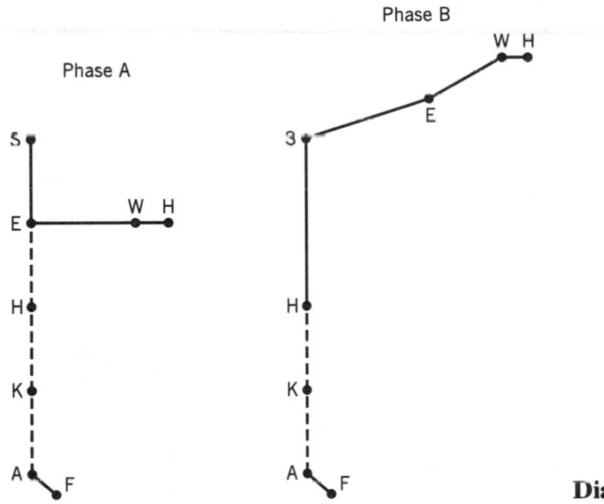

Diagram 7.6

The person has the following anthropometric dimensions:

1. Linear dimensions:

   a. Vertical (body) height = N/A
   b. Elbow-wrist (E-W) segment length = 0.245 M
   c. Wrist-hand (W-H) segment length = 0.07 M

2. Mass dimensions:

   a. Total body mass (m_B) = 85 kg
   b. Forearm (E-W) segment mass = (.016)(m_B) = 1.36 kg
   c. Hand (W-H) segment mass = (.006)(m_B) = 0.51 kg

The following data is acquired for Phase A of the task (in meters) during a specific moment in time:

Time	x_E	y_E	x_W	y_W	x_H	y_H
$i - 1$	1.015	1.100	1.260	1.100	1.335	1.100
i	1.015	1.100	1.259	1.104	1.334	1.105
$i + 1$	1.015	1.100	1.256	1.112	1.331	1.113

where $\Delta t = 0.016$ s.

During Phase A, the external force (F_A) is the weight of the gas canister (W_A). F_A is assumed to be a point mass ($m_{gc} = 0.50$ kg) located at the center of mass of the hand segment. Because of the gravitational vector (**g**), F_A acts vertically downward (at an angle of 0° to the vertical axis).

a. Calculate the nine kinematic parameters ($\bar{x}$, $\bar{y}$, v_x, v_y, a_x, a_y, θ, $\boldsymbol{\omega}$, $\boldsymbol{\alpha}$).

   i. For the hand (W-H) segment.
   ii. For the forearm (E-W) segment.

b. Calculate the reaction forces and the applied muscle moment at the proximal end (R_{yp}, R_{xp}, M_p):

   i. Of the hand (W-H) segment.
   ii. Of the forearm (E-W) segment.

c. Calculate the muscle mechanical power and work ($\mathbf{P}_m$, $\mathbf{W}_m$):

   i. For the hand (W-H) segment.
   ii. For the forearm (E-W) segment.

d. Interpret the results.

7.12. Repeat Problem 7.11 for Phase B.

The following data is acquired for Phase B of the task (in meters) during a specific moment in time:

Time	x_E	y_E	x_W	y_W	x_H	y_H
$i-1$	1.305	1.660	1.430	1.875	1.485	1.925
i	1.304	1.656	1.429	1.871	1.485	1.921
$i+1$	1.302	1.645	1.426	1.866	1.482	1.916

where $\Delta t = 0.016$ s.

During Phase B, there is no external force (except gravity) acting on the hand segment.

a. Calculate the nine kinematic parameters ($\bar{x}$, $\bar{y}$, v_x, v_y, a_x, a_y, θ, $\boldsymbol{\omega}$, $\boldsymbol{\alpha}$).

   i. For the hand (W-H) segment.
   ii. For the forearm (E-W) segment.

b. Calculate the reaction forces and the applied muscle moment at the proximal end (R_{yp}, R_{xp}, M_p):

   i. Of the hand (W-H) segment.
   ii. Of the forearm (E-W) segment.

c. Calculate the muscle mechanical power and work ($\mathbf{P}_m$, $\mathbf{W}_m$):

   i. For the hand (W-H) segment.
   ii. For the forearm (E-W) segment.

d. Interpret the results.

CHAPTER 8

Quantitative Workload Analysis

Work and workload are not interchangeable terms. Work and workload are interrelated with some similarities and some differences which we will now consider. It is useful to view work as the external work performed upon some object or the work that some external object performs upon the human body. In this sense, the physical definition of work applies [e.g., Chapter 4, equations (45), (46), (52), and (53)]. However, when external work interacts with the human body, the individual can experience different amounts of workload for the same amount of external work performed. A simple example is that of a person proceeding to climb up a set of stairs (perhaps from the first floor to the tenth floor at a rate of one floor per minute). Upon arriving at the tenth floor, the individual will have experienced a certain physical workload in response to the performance of a quantity of external work. In this case, the external work would be equivalent to the body weight times the vertical displacement of the body from the first floor to the tenth floor. When the situation is reversed, and the same individual then proceeds downward (from the tenth floor to the first floor) at a rate of one floor per minute, that individual will experience a certain workload upon their arrival on the first floor. The external work in this case will be calculated in a similar manner (as when proceeding up the stairs) so that the magnitude of the external work will be the same whether going up or down the stairs. Using upward positive sign convention, however, ascending the stairs can be defined as positive external work, and descending the stairs can be defined as negative external work. Obviously, while this indicates a difference in the direction of the external work, it does not represent a difference in the magnitude of the external work (see Figure 7.11).

For anyone who has ever performed an equivalent task, they will report that they exerted more effort going up the stairs then they did going down the stairs. Subjectively, the individual will feel that they actually performed more work going up the stairs, then they did going down the stairs. What they are describing is the interaction of the human body with the external work, and so the individual is describing their workload (which can be different) in response to the same magnitude of external work. Obviously, the person's physical workload is related to the external work performed but not simply proportional to it. Therefore, let us examine workload more specifically.

Workload may be defined as the reaction of the human body when performing external work. When the external work is physical work, the bodily reactions consist of physiological adjustments and adaptations required for the performance of that external physical work. When these physiological adjustments and adaptations can be measured and quantified, the human factors engineer has an analytical basis for the quantitative assessment of human physical workload. The acquisition of some essential physiological data was considered in Chapter 5 with respect to the neuromuscular system and the cardiopulmonary system. The quantification of such physiological data will be considered in the first half of this chapter (Sections 8.1 and 8.2). The application of these quantitative relationships will then be considered in the second half of this chapter (Sections 8.3 and 8.4).

Before proceeding to discuss some factors which may affect workload, two other issues should be briefly considered. First, this chapter will examine physical workload defined as the human reaction to external physical work. This chapter will not address mental workload, which is the human reaction to mental work. The interested student is referred to the relevant literature in this area.

A second issue is that physical workload is not simply internal physical work performed by the body in response to external physical work. The student is already familiar with muscle mechanical work and muscle mechanical power (see Chapter 7). As was shown in Example 7.7, relating muscle mechanical work and power to external physical work and power is quite useful in a human factors engineering analysis. However, this is not the same as directly equating internal (muscle mechanical) work to workload. This is because internal work may be viewed as having various forms. One form is muscle mechanical work and power, and it is defined as the work and power performed upon a body segment. Therefore, with respect to the body segment, muscle mechanical work, and power (whether positive or negative) is external to the segment. Another form is muscle contractile work and power, which is internal to the muscle, and it represents the muscle contractile elements acting as chemo-mechanical energy converters. Another form is cardiac work and power necessary for the circulation of blood. Yet another form is pulmonary work and power necessary for respiration.

Using the definition of physical workload given earlier, let us consider various factors that will affect workload. Physical workload is affected by the rate of performing external work. In the stair-climbing example, it is readily apparent that the human workload would change appreciably if the person ascended or descended one floor level every 5 minutes (requiring 5 times longer duration in order to complete a ten-floor ascent and a ten-floor descent). An alternative situation would be to require that the stair-climbing task be performed with a periodicity (cycle) of once per half-hour (10 min ascending, 10 min descending, and ten minutes rest). Over an 8 hour work day, the frequency of the task might range from 1 cycle per day to 16 cycles per day.

Physical workload is affected by the physical condition of the individual. An individual in good physical condition for the task to be performed (such as a marathon runner) would experience less physical workload (for a task requiring both speed and endurance) than an individual poorly conditioned for this specific task (such as a heavyweight wrestler). Physical workload is dependent upon the training and/or skill level of individual. For example, if the handrails in the stairwell were within a comfortable reaching distance, then a person with certain gymnastic training and skill would be able use upper extremity force and movement coordinated with lower extremity force and movement in order to lower the physical workload.

Environmental variables also affect physical workload. Environmental temperature specifically can affect physical workload. Consider the stair-climbing task when performed on a hot summer day when the building air conditioning unit is not operating. In this case, muscle metabolism, thermoregulatory reflexes, and cardiovascular responses must all adjust to the higher ambient temperatures resulting in an increase of the physical workload.

The analysis of physical workload will be presented as follows. Section 8.1 will consider the quantitative analysis of neuromuscular activity with respect to specific system equations. Section 8.2 will consider the quantitative analysis of cardiopulmonary activity again with respect to specific system equations. Section 8.3 will consider the quantitative analysis of static workload with respect to the physiological reactions to minimal-level, mild-level, moderate-level, and high-level static work. Section 8.4 will consider the quantitative analysis of dynamic workload with respect to the physiological reactions to minimal-intensity, mild-intensity, moderate-intensity, and high-intensity dynamic work.

8.1 QUANTITATIVE ANALYSIS OF NEUROMUSCULAR ACTIVITY

a. System Equations (Part I)

The electromyogram (EMG) is a cornerstone for the quantitative analysis of neuromuscular activity. There is a fundamental relationship between the root-mean-square (rms) amplitude of the EMG (A) and a brief (less than 3 s) force impulse (F_δ) when performing an isometric muscle contraction. A muscle contracts isometrically when it exerts force without changing length. For example, gripping a wrench and continuing to tighten a nut that has already been fully tightened, or depressing an automobile brake pedal and continuing to apply force after the pedal is fully depressed. Mathematically, this relationship is as follows:

$$A = f(F_\delta) \tag{1}$$

This amplitude-force relationship may be either linear or nonlinear depending on the specific muscle. Figure 8.1 shows the EMG amplitude relationship to the

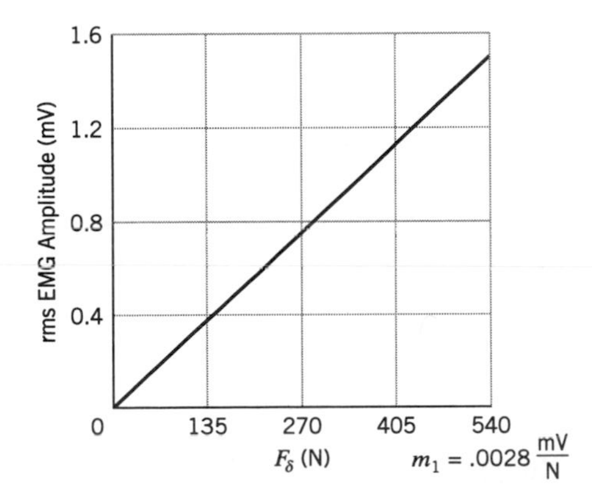

Figure 8.1 EMG amplitude relationship to the muscle impulse force for the hand grip muscles (flexor digitorum).

muscle impulse force for the hand grip muscles (flexor digitorum). The elbow angle is at 90°, and the wrist is in straight alignment with the forearm. This is a linear amplitude-force relationship:

$$A = m_1 \cdot F_\delta \tag{2}$$

where m_1 is a constant slope.

Figure 8.2 demonstrates the EMG amplitude to muscle impulse force relationship for the forearm flexor muscle (biceps). The elbow angle is at 120°, and the force is applied at the distal forearm (wrist). The amplitude-force relationship is nonlinear and characteristic of a power function:

$$A = m_2 \cdot F_\delta^{n_1} \tag{3}$$

where m_2 is an amplitude scale factor, and n_1 is a constant exponent greater than 1.

The A_{mvc} is defined as the maximal EMG amplitude at the maximal voluntary contraction (MVC) force that can be exerted with a brief (less than 3 s) isometric contraction. It is commonly observed that the value of A_{mvc} is not constant (even for the same individual on different days). This is because the rms EMG amplitude is a function of numerous factors including (but not limited to) variations in muscle metabolism, surface electrode position placement, the electrode-skin interface resistance, muscle temperature, and prior muscle fatigue. For any one individual on different days, the A_{mvc} may undergo a variation over a threefold range. Furthermore, the isometric muscle force for brief (less than 3 s) maximal voluntary contraction (F_{mvc}) is quite variable between subjects of different muscle bulk and of different ages. Consequently, it is useful to normalize equations that relate the EMG amplitude to the isometric muscle force impulse. Figure 8.3 is a restatement of Figure 8.1, in which the ordinate has been normalized by A_{mvc} and the abscissa has been normalized by F_{mvc}. Figure 8.3 now represents the fractional amplitude of the EMG during 3-s isometric (impulse) contractions of the same hand grip muscles but now at fractional forces of the F_{mvc} between .125 and 1.00. The mathematical relationship is:

$$\frac{A}{A_{mvc}} = \frac{F_\delta}{F_{mvc}} \tag{4}$$

Note that normalization results in m_1 of equation (2) equal to unity.

Plantar flexion of the foot at the ankle joint results from contraction of the calf

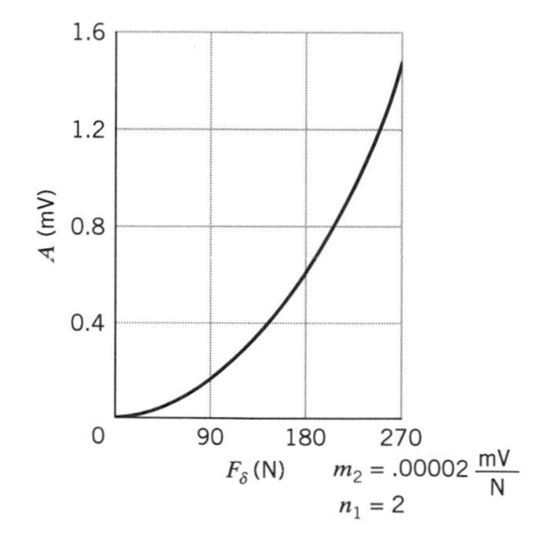

Figure 8.2 EMG amplitude to muscle impulse force relationship for the forearm flexor muscle (biceps).

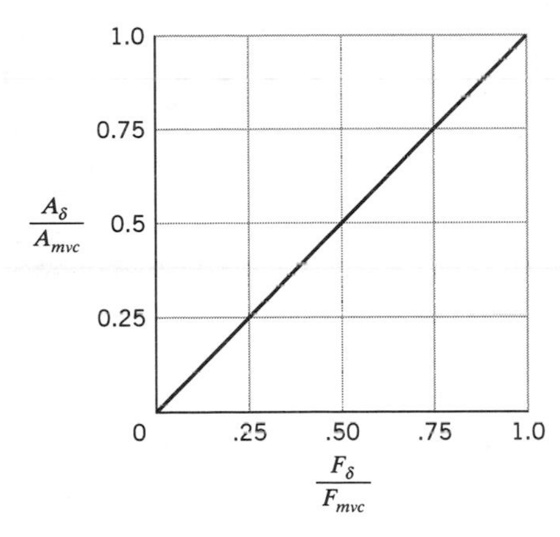

Figure 8.3 Restatement of Figure 8.1, with the ordinate normalized by A_{mvc} and the abscissa normalized by F_{mvc}.

(gastrocnemius) muscle. For this muscle, there is a linear relationship between the amplitude of the surface EMG with brief isometric muscle force impulses as shown in Figure 8.1. In this case the calf muscle contracted isometrically at angle of 90°, and the applied force was at the ball of the foot (F_{mvc} = 600 N). Consequently, Figure 8.3 and equation (4) can equally be applied to the ankle flexor muscle.

With respect to nonlinear amplitude-force relationships, Figure 8.4 indicates the normalized amplitude-force relationship for the forearm flexor muscle (biceps). The mathematical expression of Figure 8.4 is:

$$\frac{A}{A_{mvc}} = \left(\frac{F_\delta}{F_{mvc}}\right)^{n_2} \tag{5}$$

As compared to equation (3), note that the ordinate scale factor (m_2) is now equal to unity. The exponent (n_2) will still have a value greater than 1.

The EMG for all of the above amplitude-force relationships was recorded from electrodes placed on the surface of the skin overlying the respective muscles. Subsequent EMG signal conditioning as well as mathematical processing was performed using the bioelectronic system shown in Figure 5.11. Therefore, the convention used will be that of $A \equiv A_{rms}$ [per Figure 5.11 and equation (51) of Chapter 5]. Furthermore, it is understood that center frequency (f_c) as discussed shortly is also processed per Figure 5.11 and defined by equation (53) of Chapter 5.

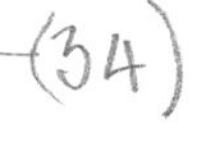

(36)

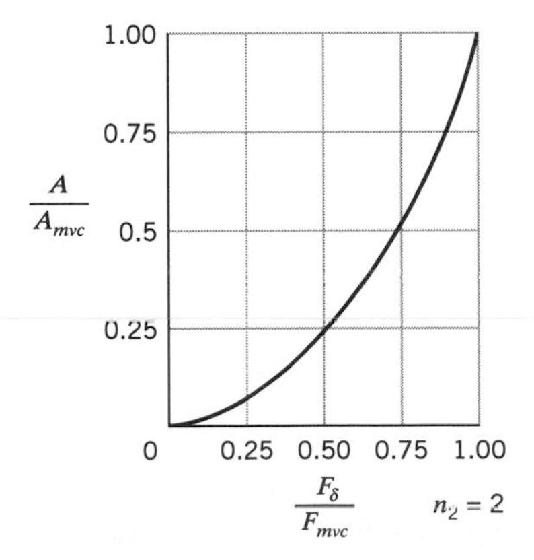

Figure 8.4 Normalized amplitude-force relationship for the forearm flexor muscle.

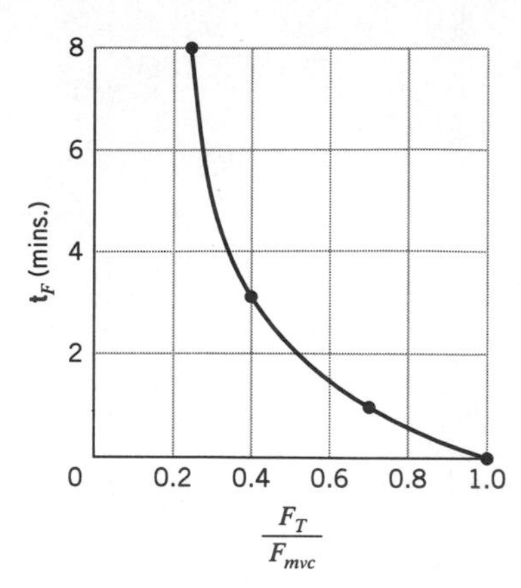

Figure 8.5 Strength-endurance curve applying to all muscles undergoing sustained isometric contractions.

Having considered brief isometric force impulse (F_δ), we proceed to the case of a sustained isometric muscle contraction at a specific target force (F_T). An example of a target force might be the exertion of 200 N of force by the hand grip muscles, which is then sustained for a certain period of time by continuous isometric contraction of the hand grip muscles. Recall that in Figure 8.1, the maximal isometric force (F_{mvc}) was 500 N. By definition, a maximal voluntary isometric muscle contraction cannot be sustained for greater than 3 s and is therefore measured as an isometric force impulse. However, isometric muscle forces less than F_{mvc} can be sustained for longer periods of time before fatigue occurs. In the specific example of 200 N, the sustained isometric contraction would be at a fractional isometric force (F_T/F_{mvc}) of 0.4. Referring to Figure 8.5, it is apparent that the F_T can be maintained for approximately 2 min, after which muscle fatigue will result in the progressive lowering of the hand grip isometric force below this fractional force value. Figure 8.5 also indicates that lower fractional isometric forces can be maintained for longer periods of time and that higher fractional forces are maintained for relatively shorter periods of time. The endurance time is the time-to-fatigue, after which the applied isometric fractional force will decrease. It should be noted that Figure 8.5 implies that fractional isometric forces (F_T/F_{mvc}) up to 0.10 through 0.15 are considered nonfatiguing (see Section 8.3).

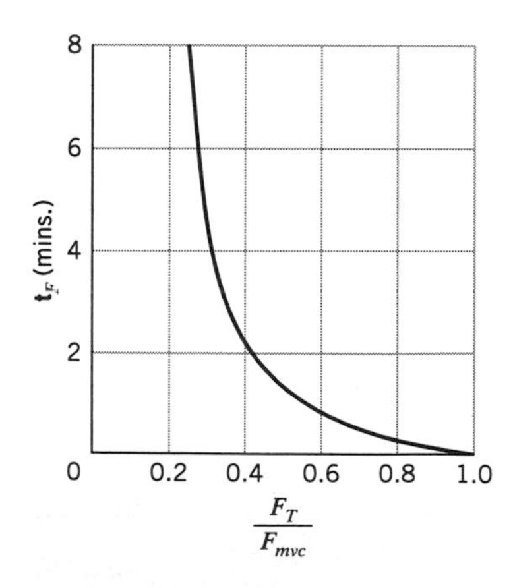

Figure 8.6 Strength-endurance curve for sustained isometric contractions of the biceps muscle.

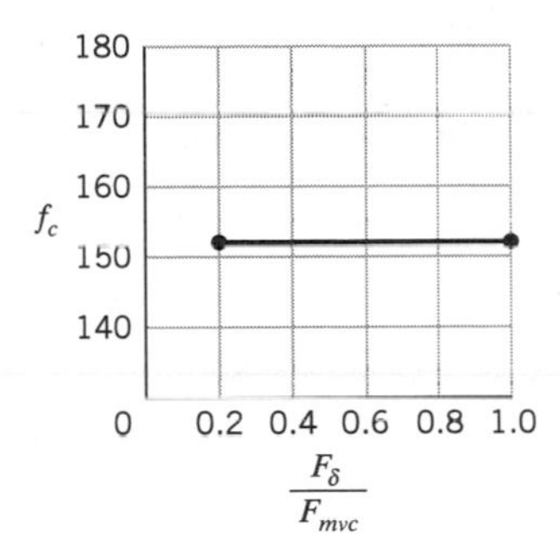

Figure 8.7 Relationship between center frequency (f_c) and fractional isometric force impulses.

The strength-endurance curve of Figure 8.5 applies to all muscles undergoing sustained isometric contractions. It was previously noted that the amplitude-force relationship for the hand grip muscles was linear [equation (2)]. When the amplitude-force relationship is not linear, as was the case for the biceps muscle [see equation (3)], the same strength-endurance relationship is observed. Figure 8.6 indicates the strength-endurance curve for sustained isometric contractions of the biceps muscle. In this case also, fractional forces below 0.10 to 0.15 are considered nonfatiguing.

Before proceeding with further development of the system equations, let us consider the effect of sustained isometric force and muscle fatigue upon the center frequency (f_c). Recall, from Chapter 5, that the Fourier power spectrum is calculated from a fundamental frequency of 4 Hz through approximately 400 to 500 Hz (100 plus harmonics). The center frequency (f_c) was then calculated from that Fourier power spectrum using equations (52) and (53) of Chapter 5. Figure 8.7 indicates the center frequency (in hertz) as a function of fractional isometric force impulses of 0.2–1.0. Note that the center frequency is relatively constant with respect to increasing fractional isometric force impulse. This is in contrast to the response of the rms EMG amplitude (Figure 8.3).

Figure 8.8 indicates the EMG power spectrum recorded for the hand grip muscles at the onset of sustained isometric contraction (solid line) and at the fatigue end point (broken line) when the fractional sustained isometric target force was 0.4. It is apparent that during the fatiguing isometric contraction there is an apparent reduction in the center frequency of the Fourier power spectrum, which can only be qualitatively appreciated in Figure 8.8.

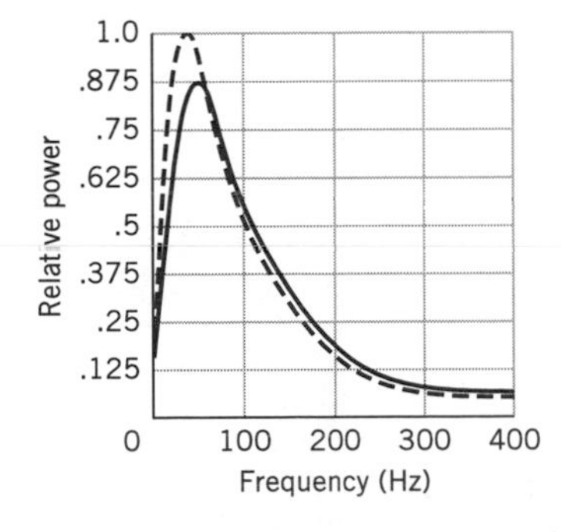

Figure 8.8 EMG power spectrum recorded for the hand grip muscles.

This reduction in the center frequency of the EMG power spectrum with muscle fatigue has been ascribed to the slowing of the conduction velocity of motor unit action potentials on the sarcolema. As previously noted in Chapter 7, the peak amplitude of the MUAP remains constant (with fatigue). The end result will be that when a Fourier analysis is performed on the MUAP of the fatiguing muscle, there will be some reduction of magnitude of the higher frequency harmonics, and some increase in the magnitude of lower frequency harmonics (as compared with the MUAP of nonfatigued muscle).

b. System Equations (Part II)

Let us now consider the quantification of isometric muscle fatigue based upon the frequency and amplitude parameters (f_c and A). When an isometric contraction is sustained to fatigue, there is often a linear rise in the fractional rms EMG amplitude (A/A_{MVC}) of the surface EMG with time (Figure 8.9). This applies to both the hand grip muscles (flexor digitorum) and the foot extensor muscles (gastocnemius). Figure 8.9 indicates sustained isometric contractions of the hand grip muscles at fractional sustained isometric forces of 0.25, 0.40, and 0.70. The abscissa indicates the fractional duration of sustained contraction in which the time at any instant (T_i) is normalized by the specific endurance time (T_F) for the hand grip muscles at a specific fractional isometric target force (see Figure 8.5). The ordinate of Figure 8.10 is the fractional center frequency, which is the center frequency at any moment in time (f_{ci}) normalized by the center frequency obtained for a brief maximal isometric contraction in the unfatigued muscle (f_{cmvc}) immediately prior to the onset of a sustained isometric contraction.

It is immediately apparent that for fractional isometric forces at or below 0.70, there is a linear increase in the fractional rms amplitude of the EMG (Figure 8.9). At the onset of these sustained isometric contractions ($T_i/T_F = 0$) the y-intercept is a fractional rms EMG amplitude near the fractional isometric force that is to be sustained. Although the fractional rms EMG amplitude is substantially higher at the end of each fatiguing contraction, the resultant slope with respect to fractional time is rather linear. These results may be expressed mathematically as:

$$\frac{A_i}{A_{mvc}} = m_5\left(\frac{T_i}{T_F}\right) + \left(\frac{F_T}{F_{mvc}}\right) \qquad (6)$$

Referring to Figure 8.10, the fractional center frequency of the power spectrum of the EMG is initially at unity. However, as the fatiguing sustained isometric contractions are exerted, the fractional center frequency will decrease from 1.0 to approximately 0.75 throughout the duration of the sustained fatiguing isometric

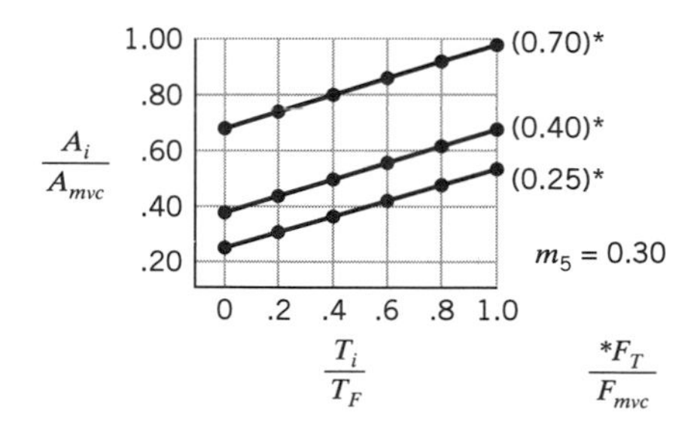

Figure 8.9 Sustained isometric contractions of the hand grip muscles at fractional sustained isometric forces.

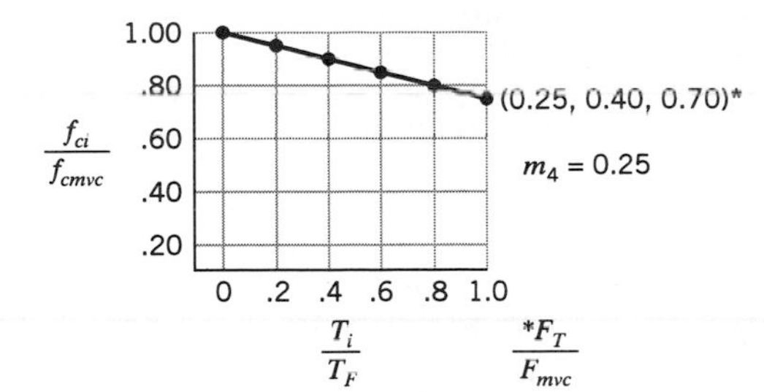

Figure 8.10 Normalized center frequency of the EMG during normalized time to fatigue (at 25%, 40%, and 70% of an MCV).

contractions. Furthermore, this 0.25 fractional decrease in the center frequency is independent of the fractional target isometric force that is exerted and sustained to the fatigue endpoint. Mathematically, this relationship may be expressed as follows:

$$\frac{f_{ci}}{f_{cmvc}} = -m_4\left(\frac{T_i}{T_F}\right) + 1 \tag{7}$$

Operationally, this quantification of neuromuscular activity for static muscle contractions can be restated in a more useful form. It will be seen that the recording and processing of the surface EMG will result in time-varying amplitude and frequency parameters (A and f_c) during the performance of physical work (Sections 8.3 and 8.4). Equations (6) and (7) may then be solved as inverse functions so that the parameter of interest is the time remaining until the onset of muscular fatigue. Consequently, equation (6) may be rearranged to solve for the fractional time to fatigue:

$$\frac{T_i}{T_F} = \frac{1}{m_5}\left(\frac{A_i}{A_{mvc}} - \frac{F_T}{F_{mvc}}\right) \tag{8}$$

Similarly, equation (7) may be rearranged to calculate the fractional time to fatigue:

$$\frac{T_i}{T_F} = \frac{1}{m_4}\left(1 - \frac{f_{ci}}{f_{cmvc}}\right) \tag{9}$$

Equations (6)–(9) are applicable when muscle temperature is 35°C (±3°C). This constraint limits the overall utility of these equations.

In order to practically employ the above system equations for the analysis of static work and static workload, the human factors engineer must consider the effect of both muscle length and muscle temperature when quantifying endurance time and fatigue. The effect of muscle temperature on the F_{mvc} and the effect of muscle length (joint angle) on the F_{mvc} will be considered in Section 8.3. These effects will be described as subsidiary equations in which F_{mvc} will be the dependent variable and either muscle temperature or joint angle will be the independent variable. Such additional mathematical relationships (Section 8.3) are important in quantifying neuromuscular activity with respect to F_{mvc}. However, muscle temperature can dramatically affect both the frequency components of the surface EMG as well as the rms amplitude. Consequently, this section will consider the direct effects of muscle temperature (T) upon the quantification of endurance time (and fatigue) when using f_c and A.

In Chapter 6, the student was introduced to a biothermal model in which core temperature resided in central core tissues, and the skeletal muscle was approximated as a shell tissue (Figure 6.4). It was then noted, that muscle temperature (T_m) could vary widely in response to large fluctuations of surface (skin) temperatures (Example 6.3). This large variation of T_m has significant consequences upon both the rms amplitude and center frequency of the EMG.

There is an effect of muscle temperature on the center frequency of the EMG power spectrum (Figure 8.11.a) and also on the fractional center frequency of the EMG power spectrum (Figure 8.11.b). The fractional rms EMG amplitude is represented in Figure 8.11.c with respect to muscle temperature. The abscissa of Figures 8.11.a, 8.11.b, and 8.11.c is the fractional time during sustained isometric contractions to fatigue at various fractional isometric target forces (0.25, 0.40, and 0.70). The data is for the finger flexors (hand grip muscles) but is reasonably applicable to the foot extensor (gastrocnemius muscle) in which there is a linear relationship between EMG rms amplitude and sustained isometric force. Referring to Figure 8.11.a it is apparent that as muscle temperature changes from 20°C to 40°C, there is a progressive increase in the center frequency (f_c). This is because the conduction velocity of the motor unit action potentials is directly proportional to the temperature of the muscle. An increase in muscle temperature will result in a proportional increase in motor unit action potential conduction velocity. This will result in a reduction in the low frequency components of the MUAP and a general shift to high-frequency components with an increase in muscle temperature. However, during a sustained isometric contraction to the point of fatigue, the reduction of the center frequency of the EMG power spectrum will still be approximately 25% of the initial center frequency for the unfatigued muscle regardless of the

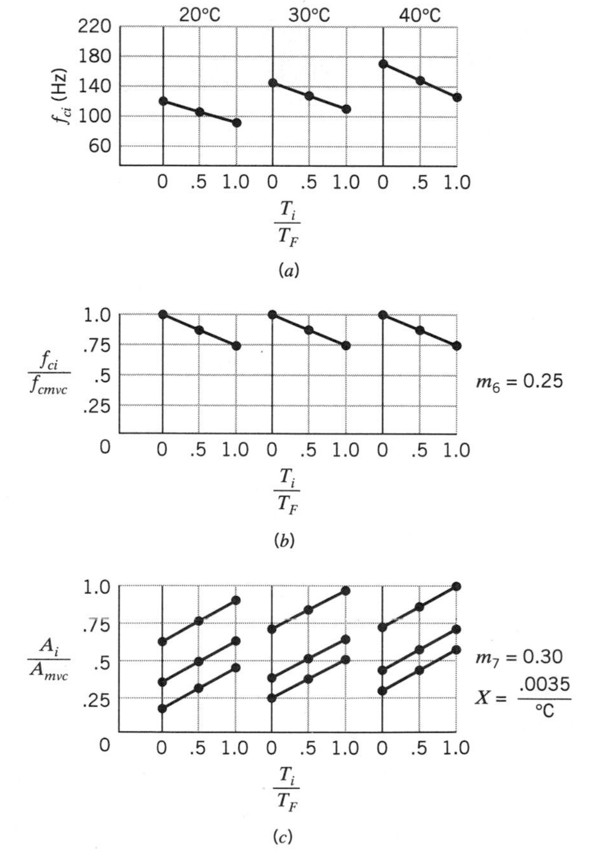

Figure 8.11.a The of muscle temperature on the center frequency of the EMG power spectrum. **b** Muscle temperature effect on the fractional center frequency of the EMG power spectrum. **c** Fractional RMS EMG amplitude.

specific muscle temperature. Consequently, when the fractional center frequency is calculated during a sustained isometric fatiguing contraction (Figure 8.11.b), the result is independent of temperature. Mathematically, the relationships depicted by Figure 8.11.a can be represented as follows:

$$f_{ci} = -[m_{40} + \alpha(T - 40)]\left(\frac{T_i}{T_F}\right) + f_{cmvc40} + \beta(T - 40) \tag{10}$$

where $\alpha = \Delta m/\Delta T$ and $\beta = \Delta f_{cmvc}/\Delta T$.
This mathematical expression is significantly simplified when the fractional center frequency becomes temperature invariant (Figure 8.11.b) as follows:

$$\frac{f_{ci}}{f_{cmvc}} = -m_6\left(\frac{T_i}{T_F}\right) + 1 \tag{11}$$

Finally, Figure 8.11.c considers the fractional rms EMG amplitude and muscle temperature. This information may be summarized mathematically as follows:

$$\frac{A_i}{A_{mvc}} = m_7\left(\frac{T_i}{T_F}\right) + \left(\frac{F_T}{F_{mvc}}\right) + \gamma(T - 40) \tag{12}$$

where $\gamma = (\Delta F_T/F_{mvc})/\Delta T$

Equation (11) may be rearranged to solve for the fractional time to fatigue:

$$\frac{T_i}{T_F} = \frac{1}{m_6}\left(1 - \frac{f_{ci}}{f_{cmvc}}\right) \tag{13}$$

Equation (12) may be rearranged to solve for the fractional time to fatigue:

$$\frac{T_i}{T_F} = \frac{1}{m_7}\left(\frac{A_i}{A_{mvc}} - \frac{F_T}{F_{mvc}} - \gamma(T - 40)\right) \tag{14}$$

The EMG is also quite useful in the quantification of neuromuscular activity during dynamic work. However, it is far more complex to record the electromyogram during dynamic work as compared to static work. Furthermore, the mathematical relationships described earlier are not directly applicable to dynamic work. A significant variable during the performance of dynamic work is that the length of the contracting muscle varies with time. This factor strongly influences the surface EMG in two respects. First, as the muscle contracts the muscle fibers thicken. This results in a change in the motor unit action potential velocity as follows: the MUAP velocity in a stretched muscle (thinner muscle fibers) is significantly higher than for muscle in a contracted state (thicker muscle fibers). Second, during the contraction of muscle, surface EMG electrodes are displaced further away from the central axis of the muscle. This results in a reduction in the amplitude of recorded surface EMG (as compared to the uncontracted muscle).

It has been shown that contracting muscle performs positive muscle work, while lengthening muscle performs negative work (see Chapter 7). One way to evaluate the quantitative importance of the EMG is to examine how well the surface EMG can predict muscle forces during dynamic muscle shortening and lengthening. Both positive and negative work can be performed on an isokinetic muscle-exercising machine (see Chapter 11). An individual can generate a maximum tension while the muscle is lengthening (negative work) or shortening (positive work) at specifically controlled velocities. Under these conditions, the EMG amplitude will remain fairly constant despite the fact that the tension during shortening work is not the same as during lengthening work. However, since the muscle was maximally activated in both cases, it is apparent that the EMG amplitude is useful for quantifying the *state of activation* of the muscle fibers. This is biophysically different from the forces exerted by the muscle upon a body segment.

8.2 QUANTITATIVE ANALYSIS OF CARDIOPULMONARY ACTIVITY

a. System Equations (Part I)

The quantitative evaluation of cardiopulmonary activity requires the acquisition of rate, flow, and pressure information. The appropriate parameters must be defined which satisfy the cardiopulmonary system equations. The appropriate bioelectronic (instrumentation) systems must be identified and utilized to acquire specific measured parameters. The cardiopulmonary system equations can then be applied to obtain additional calculated parameters. This section will address the progressive development of the cardiopulmonary system equations. Let us begin by reviewing some of the essential cardiopulmonary information developed in Chapter 5.

Recall from Section 5.3:

$$f_{HR} = \frac{1}{T} \tag{15}$$

where $T = R\text{-}R$ interval (in seconds), which is obtained from the electrocardiogram (see Figure 5.20). Using a cardiotachometer (see Figure 5.19), the steady-state heart rate (for a 1-min interval) can be obtained. This is related to the period of the cardiac cycle (T) as follows:

$$\bar{f}_{HR} = \frac{60}{T} \tag{16}$$

Note that we use the convention of placing a bar above a parameter to indicate its steady-state value.

From Section 5.3.b, we obtained:

$$\dot{V}_{AIR} = f(t) \tag{17}$$

by using an electropneumogram (see Figure 5.22.a).

Finally, equation (63) of Chapter 5 was derived, and is now restated:

$$\dot{V}_{CO} = \frac{\dot{V}_{O_2}}{C_{AO_2} - C_{VO_2}} \tag{18}$$

Equations (15)–(18) represent some basic rate and flow parameters. However, before proceeding further toward the quantification of cardiopulmonary activity, the human factors engineer will also need pressure information. Such pressure information may be obtained by automatic non-invasive sphygmomanometry. This system may be described as follows.

Most people will have experienced at some point in time an event in which a health care worker (e.g., a doctor or nurse) has taken their blood pressure. This is commonly done with a blood pressure cuff, which is wrapped around the upper arm of the individual. The cuff is then rapidly inflated and subsequently deflated as the health care worker listens with a stethoscope applied below the cuff level. This method of acquiring blood pressure information is referred to as sphygmomanometry. The approach is considered noninvasive since the system is only applied to the surface of the skin. Since a person is physically performing this procedure, it is referred to as *manual* noninvasive sphygmomanometry. Various bioelectronic systems that will perform this procedure repetitively, reliably, and without manual intervention are currently available. These systems are referred to as *automatic* noninvasive sphygmomanometers.

Three major subunits comprise such a system. First, there is an external blood pressure cuff, which has an air bladder sewn within it and a microphone positioned internally between the air bladder and the individual's arm. Second, there is a pneumatic subunit that provides for rapid inflation of air and a slower deflation of air. This inflation/deflation cycle can be repeated at adjustable time intervals. Third, there is an electronic signal detection and processing subunit which translates the vibrations detected by the microphone into a blood pressure output.

The theory of operation is based upon the principle that blood flow through an artery can create a vibration in the elastic arterial wall. These vibrations can then be transmitted through the soft tissues as mechanical pressure waves and be detected by a microphone (which transduces pressure waves into electrical signals). Arterial pressure (P_a) is a time-varying pressure as shown in Figure 5.20. P_a oscillates between a maximum value (referred to as the systolic pressure, P_S) and a minimum value (the diastolic pressure, P_D). The initial rapid inflation of the blood pressure cuff is set to exceed P_S. This rapid inflation phase usually occurs over 1 to 5 s. The subsequent deflation phase occurs at a rate of 2 to 3 mm Hg/s. The initial (rapid) cuff inflation completely occludes the artery below the cuff and the microphone. With subsequent (slow) deflation, cuff pressure then falls to P_S, and some blood flow begins through the compressed artery. This initial flow then sets up vibrations in the elastic arterial wall. Consequently, the first detection of mechanical vibrations with falling cuff pressure is identified by the electronic signal detection and processing unit as the P_S. Further cuff deflation results in an increase in arterial wall vibrations as increased amounts of blood flow through the progressively opening artery. As the cuff pressure decreases to P_D there is a decrease and then cessation of mechanical vibrations in the elastic arterial wall. The electronic signal detection and processing unit proceeds to identify this point of muffling and silence of arterial wall vibrations as P_D. Hemodynamically, blood flow which had been accelerated by the partial narrowing of the artery now becomes uniform and laminar along the course of the artery resulting in a cessation of mechanical vibration in the elastic arterial wall.

Operationally, it is very important to correctly apply the external blood pressure cuff in order to obtain valid and repeatable blood pressure readings. First, the brachial artery should be located by pressing with two fingers about 1 in. above the elbow crease along the inside of the left arm (on a right hand dominant person) or the inside of the right arm (on a left hand dominant person). Proper technique will result in a pulse that can be felt most strongly. Second, the cuff itself should be placed around the arm usually secured with velcro straps, and be sufficiently tight that there is no sliding of the cuff when it is moved up and down or rotated back and forth by a hand applied to the outside of the cuff. Third, the bottom edge of the blood pressure cuff should always be approximated 1 or 2 in. above the elbow and the upper arm remain at the person's side at the same level as the heart. Fourth, the arm muscles must be relaxed, with the palm facing upward, and the elbow should not be hyperextended. Fifth and finally, the person applying the cuff should make certain that there are no kinks in the pressure hose connecting the arm cuff bladder to the pneumatic system subunit of the automatic noninvasive sphygmomanometer.

There are advantages and disadvantages to the use of automatic noninvasive sphygmomanometers, which should be appreciated by the human factors engineer. Some advantages are as follows. First, since they are noninvasive, they are relatively comfortable for the subject, expose the individual to minimal risk, and are usually well accepted by the person. Second, the inflation/deflation cycles can be repeated at regular intervals so that repeated measures of systolic pressure and diastolic pressure can be obtained over time. Third, since the microphone detects pulsatile mechanical vibrations, it can be used to acquire heart rate information (which is a third parameter measurement with some units). There are also some disadvantages. First, since an acoustical microphone is used, all measurements should be acquired in quiet places with low ambient noise levels. High ambient noise will introduce artifact into the microphone signal. Second, the arm from which the measurements are taken must be in a relaxed state. The dominant arm may be active as well as both lower extremities. However, significant muscle contractions or increased tension in the muscles of the arm being measured will effectively uncouple mechanical vibrations in the arterial wall from transmission to the microphone transducer. Third, the arm to which the blood pressure cuff is applied should remain relatively still during the measurement cycle. Otherwise, significant motion artifact can be introduced at the microphone transducer.

It should be noted that these disadvantages apply only during the inflation/deflation pressure cycle during which measurements are taken. Depending upon the type of work and the ambient environment the cycle rate can be adjusted so that there is a sufficiently long inter-cycle interval during which none of the restrictions would apply. As a result of proper operator technique, the use of a good quality system, and the selection of appropriate measurement cycles for the type of work and environment being evaluated, the human factors engineer will obtain important pressure information (P_S and P_D).

The pulse pressure is then defined as follows:

$$\Delta P = P_S - P_D \tag{19}$$

The mean systemic arterial pressure (MSAP) represents the steady state (average) pressure ($\overline{\Delta P}$) during the period of a cardiac cycle (T). It is obvious, that MSAP must be some pressure level above P_D and below P_S (see Figure 5.20). MSAP must be of such a value that the $\overline{\Delta P} \cdot T$ product can be equated to the area under the $P_a = f(t)$ curve:

$$\overline{\Delta P} \cdot T = \int_0^T P_a \cdot dt \tag{20a}$$

Rearranging equation (20a) results in the definition of mean systemic arterial pressure ($\overline{\Delta P}$):

$$\overline{\Delta P} = \frac{1}{T} \int_0^T P_a \cdot dt \tag{20b}$$

Based upon extensive empirical observation, it has been found that the mean systemic arterial pressure may be approximated as:

$$\overline{\Delta P} = P_D + \frac{\Delta P}{3} \tag{21}$$

An alternative expression may be derived by substituting equation (19) into equation (21):

$$\overline{\Delta P} = P_D + \frac{P_S - P_D}{3} \tag{22a}$$

$$\overline{\Delta P} = \frac{2}{3}P_D + \frac{1}{3}P_S \tag{22b}$$

Development of the cardiopulmonary system equations must also consider the definition of $\dot{V}_{O_2}$ and an examination of its relationship to $\dot{V}_{AIR}$. Pulmonary airflow is referred to as *ventilation* (identified as $\dot{V}_{AIR}$). $\dot{V}_{AIR}$ is a cyclical phenomenon composed of two alternating phases.

$\dot{V}_{AIR}$ (inspired) has approximately 20.8% oxygen by volume. Therefore, we may define the fraction of inspired air containing oxygen as:

$$F_{IO_2} = 0.208 \tag{23}$$

$\dot{V}_{AIR}$ (expired) has approximately 15.8% oxygen. Consequently, the fraction of expired air containing oxygen is:

$$F_{EO_2} = 0.158 \tag{24}$$

The difference between F_{IO_2} and F_{EO_2} is the oxygen used by the body. This occurs when oxygen enriched arterial blood (A_{O_2}) flows from the heart and lungs through the human body. Oxygen released to the tissues and venous blood, partially depleted of oxygen ($\dot{V}_{O_2}$), returns to the heart and lungs. This oxygen, which is used by the body, is defined as the oxygen consumption ($\dot{V}_{O_2}$):

$$\dot{V}_{O_2} = (F_{IO_2} - F_{EO_2})\dot{V}_{AIR} \tag{25}$$

The concentration of oxygen in the arterial blood is approximated as:

$$C_{AO_2} = \frac{20 \text{ ml O}_2}{100 \text{ ml blood}} \tag{26}$$

The concentration of oxygen in the venous blood is approximated as:

$$C_{VO_2} = \frac{15 \text{ ml O}_2}{100 \text{ ml blood}} \tag{27}$$

The arterial-venous (A-V) oxygen difference is defined as:

$$\Delta C = C_{AO_2} - C_{VO_2} \tag{28}$$

The cardiac output ($\dot{V}_{CO}$) is obtained by substituting equation (28) into equation (18):

$$\dot{V}_{CO} = \frac{\dot{V}_{O_2}}{\Delta C} \tag{29}$$

$\dot{V}_{O_2}$ and $\dot{V}_{CO}$ are time-varying parameters with the units of milliliters per second.

b. System Equations (Part II)

The performance parameter important to the human factors engineer is the minute oxygen consumption ($\overline{\dot{V}}_{O_2}$). This is the steady-state value of $\dot{V}_{O_2}$ over a 1-min interval, with the units of liters per minute.

At this point, equations (18) and (29) represent a time-varying cardiac output ($\dot{V}_{CO}$). However, these equations are expressions of the Fick equation (derived in Chapter 5) and are valid only for a steady-state cardiac output ($\overline{\dot{V}}_{CO}$). Steady-state cardiac output is calculated (in part) from the steady-state minute oxygen consumption.

In order to define the steady-state oxygen consumption ($\overline{\dot{V}}_{O_2}$), it is necessary to define T_R, the period of a respiratory cycle (in seconds). T_R consists of an inspiratory flow phase (breathing in) and as expiratory flow phase (breathing out). The respiratory frequency f_R is:

$$f_R = \frac{1}{T_R} \tag{30}$$

The steady-state respiratory rate during a 1-min interval:

$$\overline{f}_R = \frac{60}{T} \tag{31}$$

The volume of air inhaled in one respiratory cycle (ΔV_{AIR}) is calculated from the airflow ($\dot{V}_{AIR}$) as:

$$\Delta V_{AIR} = \frac{1}{2}\int_0^{T_R} \dot{V}_{AIR} \cdot dt \tag{32a}$$

In equation (32a) we approximate both the inspiratory flow and the expiratory flow as similar in magnitude and duration so that the volume of air inhaled (ΔV_{AIR}) is one-half the total airflow during the course of a respiratory cycle (T_R).

Since $\dot{V}_{AIR}$ is in milliliters per second, then ΔV_{AIR} of equation (32a) is in milliliters. A steady-state volume of air inhaled per breath ($\overline{\Delta V}_{AIR}$) may be calculated as follows. Ventilation (airflow) is continuously integrated over a period of time (e.g., 15 s), which will include two or more respiratory cycles (of period, T_R). As per equation (32a) the integral is scaled by one-half to account for inspiratory flow only (and exclude expiratory flow) and the value dimensioned in liters. This results in:

$$\overline{\Delta V}_{AIR} = \left[\frac{L}{1000\text{ ml}}\right]\left[\frac{T_R}{30\text{ s}}\right]\int_0^{15\text{ s}} \dot{V}_{AIR} \cdot dt \tag{32b}$$

For steady state, we calculate the minute ventilation ($\overline{\dot{V}}_{AIR}$), which is the volume of air inhaled per one minute interval:

$$\overline{\dot{V}}_{AIR} = \overline{\Delta V}_{AIR} \cdot \overline{f}_R \tag{33}$$

Substituting equation (33) into equation (25) results in the minute oxygen consumption ($\overline{\dot{V}}_{O_2}$), which is the desired steady-state term:

$$\overline{\dot{V}}_{O_2} = (F_{IO_2} - F_{EO_2})\overline{\dot{V}}_{AIR} \tag{34a}$$

Substituting equations (23) and (24) into equation (34a) results in:

$$\overline{\dot{V}}_{O_2} = (.05)\overline{\dot{V}}_{AIR} \tag{34b}$$

Recall from equations (26), (27), and (28) that ΔC is the A-V oxygen difference (in milliliters) per 100 milliliters of blood volume. For steady state, we approximate a $\overline{C}_{AO_2}$ and $\overline{C}_{VO_2}$ so that there is a constant A-V oxygen difference (in liters of oxygen) per liter of blood volume:

$$\overline{\Delta C} = \overline{C}_{AO_2} - \overline{C}_{VO_2} \tag{35}$$

Alternatively $\overline{\Delta C}$ may be calculated directly from ΔC (equation (28)) as follows:

$$\overline{\Delta C} = \left(\frac{\text{L}}{1000\text{ ml}}\right)\left(\frac{1000\text{ ml}}{\text{L}}\right)\Delta C \tag{36}$$

Finally, for the steady-state cardiac output ($\overline{\dot{V}}_{CO}$) in liters per minute, substitute equations (34) and (35) into equation (29):

$$\overline{\dot{V}}_{CO} = \frac{\overline{\dot{V}}_{O_2}}{\overline{\overline{\Delta C}}} \tag{37}$$

A final set of system equations are useful in order to quantify cardiopulmonary activity. These are cardiovascular equations that couple cardiac (heart) function to systemic circulation. The development of these system equations begins with a definition of stroke volume. Referring to Figure 5.20, the amount of blood ejected from the heart into the systemic circulation during a single R-R interval is the difference between the end diastolic volume (V_{ED}) and the end systolic volume (V_{ES}). This is defined as the stroke volume (ΔV_S):

$$\Delta V_S = V_{ED} - V_{ES} \tag{38}$$

ΔV_S is a time-varying parameter that is subject to cycle-to-cycle (beat-to-beat) variations as the heart ejects blood into the systemic circulation.

Recall that the steady-state cardiac output ($\overline{\dot{V}}_{CO}$) has been defined by the Fick principle, which describes an oxygen transport phenomenon (see Chapter 5). This derivation was based upon the approximation that the rate of blood flow through the lungs was equal to the rate of blood flow through the heart. This is a perfectly reasonable approximation because of the series circuit arrangement of the right heart, lungs and left heart (see Figure 5.21). An alternative expression for the steady-state cardiac output, however, is used in the Ohm's law model of the circulation. In this case, $\overline{\dot{V}}_{CO}$ is defined as a hemodynamic phenomenon and represents the blood flow exiting the left heart and entering the systemic circulation. In this case:

$$\overline{\dot{V}}_{CO} = \overline{\Delta V}_S \cdot \overline{f}_{HR} \tag{39a}$$

So that the steady-state stroke volume ($\overline{\Delta V}_S$) is:

$$\overline{\Delta V}_S = \frac{\overline{\dot{V}}_{CO}}{\overline{f}_{HR}} \tag{39b}$$

The Ohm's law model for the systemic circulation states that the hemodynamic blood flow is proportional to the mean arterial pressure ($\overline{\Delta P}$) and inversely proportional to the peripheral resistance (R):

$$\overline{\dot{V}}_{CO} = \frac{\overline{\Delta P}}{R} \tag{40a}$$

Rearranging equation (40a) results in a definition of the peripheral resistance (R) of the systemic circulation:

$$R = \frac{\overline{\Delta P}}{\overline{\dot{V}}_{CO}} \tag{40b}$$

An operationally useful statement of the Ohm's law model is obtained by rearranging equation (40b) as follows:

$$\overline{\Delta P} = R \cdot \overline{\dot{V}}_{CO} \tag{40c}$$

Because the Ohm's law model is based on hemodynamic parameters that couple cardiac (pumping) function with arterial (flow) circulation, we can define cardiovascular energetic equations. Stroke work (W_S) is the pressure-volume work that occurs when the heart must generate a mean systemic arterial pressure ($\overline{\Delta P}$) in order to eject a stroke volume (ΔV_S) per heart beat:

$$W_S = \overline{\Delta P} \cdot \Delta V_S \tag{41}$$

Substituting steady-state stroke volume ($\overline{\Delta V_S}$) from equation (39b) into equation (41), defines the steady-state stroke work:

$$\overline{W_S} = \overline{\Delta P} \cdot \overline{\Delta V_S} \tag{42}$$

The steady-state power output (P_O) of the heart is then defined as:

$$P_O = \overline{\Delta P} \cdot \overline{\dot{V}}_{CO} \tag{43}$$

8.3 QUANTITATIVE ANALYSIS OF STATIC WORKLOAD

a. Physiology of Static Work

The first part of this subsection will consider additional factors related to human strength and endurance when performing static work. This will include development of the remaining system equations [equations (44)–equation (47)] not discussed previously. The second half of this subsection will examine the cardiopulmonary responses to static workload with respect to resting (minimal), mild, moderate, and high intensity static work.

Let us consider some additional factors that will alter the strength of a static muscle contraction. Once such factor is the temperature of the muscle. Recall that most muscles of the body are anatomically located as shell tissues. Unlike the core tissues (in which the temperature is closely regulated to 37°C), the temperature of a shell tissue (such as muscle) may vary over a significantly wider range. The biothermodynamic model (and governing equations) for skeletal muscle as a shell tissue was developed in Chapter 6.

Variations in the muscle temperature alter the strength of that muscle in performing static work. As shown in Figure 8.12, the maximal static strength of muscle (F_{MVC}) remains relatively constant over a temperature range of 30°C to 40°C, as denoted by the F_{MVC} at 35°C ($F_{MVC \cdot 35}$). When muscle temperature is lowered from 30°C to 20°C there is a pronounced fall in the maximal isometric strength of the

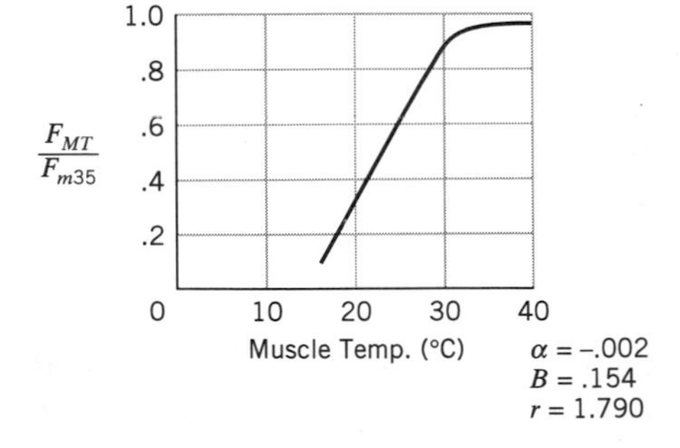

Figure 8.12 Variations in the muscle temperature alter the strength of that muscle in performing static work.

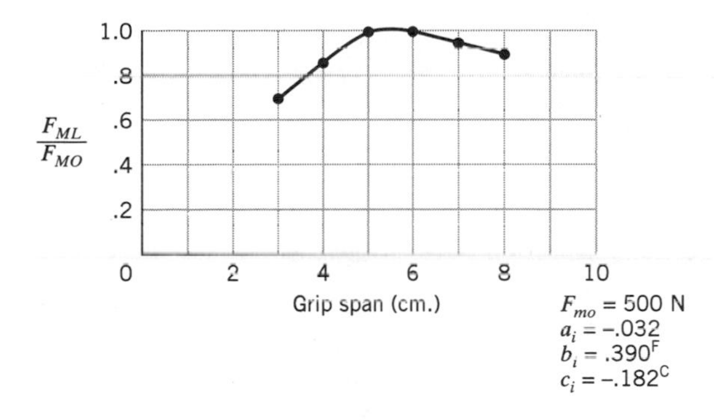

Figure 8.13 Maximum isometric strength at a specific finger flexor length when normalized by $F_{MVC}(l_{F\cdot OPT})$ and the grip spans.

contracting muscle. The relationship between maximal isometric strength at a specific temperature (F_{MVCT}) when normalized by $F_{MVC\cdot 35}$ and the muscle temperature (T) is shown in Figure 8.12 and may be expressed as follows:

$$\frac{F_{MVCT}(30° \leq T \leq 40°)}{F_{MVC\cdot 35}} = 1 \tag{44a}$$

$$\frac{F_{MVCT}(15° \leq T \leq 30°)}{F_{MVC\cdot 35}} = \alpha T^2 + \beta T + \gamma \tag{44b}$$

The fall in the temperature-strength relationship (between 20°C and 30°C) is related to the inhibition of muscle metabolism over this lower temperature range. Muscle metabolism itself is highly dependent upon changes in temperature because (in part) of the temperature-activity relationship of the muscle enzymes.

Another factor that alters the static strength of muscle is its length. This length-tension relationship for skeletal muscle is well established and important in muscle control dynamics (see Chapter 9). This length-tension relationship has direct application to practical static work physiology, since each individual muscle of the human body has one length at which it can develop its maximum isometric strength. This optimal length (l_O) of the muscle is a function of the muscle length itself and the body joint (or joints) over which the muscle crosses. One important example is the hand grip muscles in which the greatest static strength occurs when the finger flexors are at an optimal length ($l_{F\cdot OPT}$), so that there is between 4 and 5 cm of handle separation (the gripping length, l_G) when using a hand-grip dynamometer. Figure 8.13 shows the maximum isometric strength at a specific finger flexor length [$F_{MVC}(l_F)$] when normalized by $F_{MVC}(l_{F\cdot OPT})$ and the grip spans (l_G). The governing equation for this length-force relationship is:

$$\frac{F_{MVC}(l_F)}{F_{MVC}(l_{F\cdot OPT})} = a_1 \cdot l_G^2 + b_1 \cdot l_G + c_1 \tag{45}$$

A second example is shown in Figure 8.14 for the human biceps muscle. This figure represents the maximal isometric strength at a specific biceps length [$F_{MVC}(l_B)$] when normalized by the F_{MVC} at the optimal biceps length ($l_{B\cdot OPT}$) as a function of the elbow joint angle (θ_E) which is the angle between the upper arm and the forearm. This force-angle relationship may be expressed as:

$$\frac{F_{MVC}(l_B)}{F_{MVC}(l_{B\cdot OPT})} = a_2 \cdot \theta_E^2 + b_2 \cdot \theta_E + c_2 \tag{46}$$

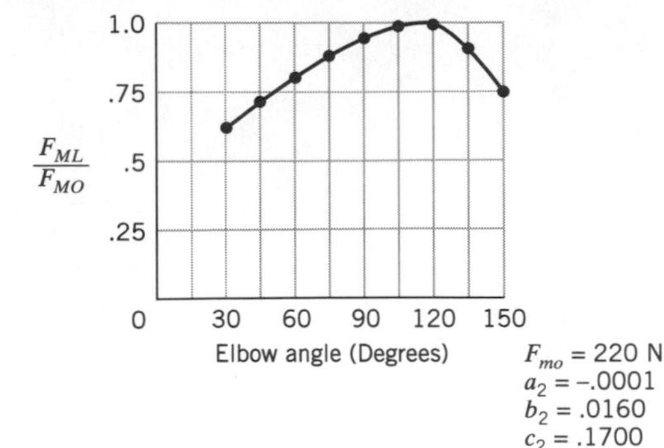

Figure 8.14 Maximum isometric strength at a specific finger flexor length when normalized by $F_{MVC}(l_{F \cdot OPT})$ and the biceps muscle.

A third important example is that of the calf (gastrocnemius) muscle. Figure 8.15 shows the variation in isometric calf muscle strength at a specific calf muscle length $[F_{MVC}(l_C)]$ as a function of ankle angle (θ_A), which is the angle between the foreleg and the foot. This force-angle relationship is:

$$\frac{F_{MVC}(l_c)}{F_{MVC}(l_{C \cdot OPT})} = a_3 \cdot \theta_A^2 + b_3 \cdot \theta_A + c_3 \tag{47}$$

Let us now briefly consider physiological adjustments that affect the force-endurance relationship. As the force generated by an isometrically contracting muscle increases, the intramuscular pressure also increases. Since there is concurrent increased metabolic demand (with the increase in muscle force), eventually the muscle blood flow cannot meet the metabolic requirements. Consequently, the muscle begins to fatigue. The temperature of a muscle performing static work significantly affects the isometric endurance. There is an optimal core muscle temperature above and below which endurance to static work will decrease. This optimal core muscle temperature (for isometric *endurance*) is approximately 28°C.

The cardiopulmonary responses associated with static work reflect the human reaction to isometric muscle contraction. When isometric contractions are sustained at muscle forces that are non-fatiguing (forces that are below 10% to 15% of the muscle's maximum strength), heart rate and blood pressure rise to a certain level and are then maintained throughout the duration of the static work. This represents

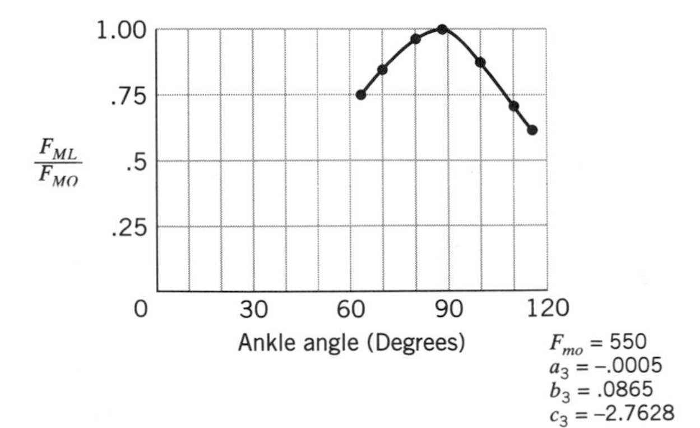

Figure 8.15 Depicts the variation in isometric calf muscle strength at varying ankle angles.

a steady-state response in which the blood pressure will rise 10 to 20 mm Hg, and the heart rate will rise 10 to 20 beats per minute. Static work at 20% or higher of a muscle's maximum strength is fatiguing. Depending upon the type of work performed, the blood pressure may rise as high as 180 to 200 mm Hg. This dramatic rise in blood pressure affects both the systolic pressure and the diastolic pressure during the time course of the fatiguing isometric muscle contractions.

The cardiopulmonary responses to fatiguing static work are not only of greater magnitude but of a different pattern then of those observed during dynamic work. For example, during dynamic work (see Section 8.4), it is common to observe that the systolic pressure will increase, the diastolic pressure will decrease, and the peripheral resistance will decrease during the time course of the fatiguing dynamic work. By contrast, during the course of fatiguing static work, both the systolic pressure and the diastolic pressure will increase and there will be an increase in the peripheral resistance.

Cardiac output increases modestly during fatiguing isometric work. A resting cardiac output of 5 L/min will usually only increase to about 8 L/min over the duration of fatiguing static work.

The muscle mass will influence the magnitude of the blood pressure response to fatiguing static work. When an isometric muscle contraction is sustained to fatigue, the magnitude of the blood pressure response is the same (regardless of the force exerted by the muscle) as long as that *same* muscle is doing the fatiguing static work. However, when large muscle masses are performing fatiguing static work for given periods of time, there is a larger blood pressure response than that found for small muscle masses.

In contrast to the blood pressure response that occurs during fatiguing isometric work, the heart rate response during fatiguing isometric contractions is much more modest. During dynamic work the heart rate may increase to a level above 200 beats/min. However, during static work the heart rate rarely exceeds 120 beats/min.

Oxygen consumption ($\dot{V}_{O_2}$) during fatiguing static work also increases modestly. However, ventilation ($\dot{V}_{AIR}$) can increase markedly. Some individuals will actually experience a distinct hyperventilation. Although heart rate and blood pressure increase continuously throughout the time course of fatiguing static work, there is little change in $\dot{V}_{AIR}$ or $\dot{V}_{O_2}$ until approximately halfway into the duration of fatiguing static work. After that, these parameters increase and may increase markedly.

In summary, the strength and endurance responses associated with resting (minimal), mild, moderate and high intensity static work are presented in Table 8.1. The pulmonary responses associated with resting (minimal), mild, moderate, and high intensity static work are presented in Table 8.2. Cardioenergetic responses are shown in Table 8.3.

Table 8.1 Strength and Endurance Assessment of Static Work

Activity Level	$\frac{F_T}{F_{MVC}}$	$F_{MVC}(T)$ N	$F_{MVC}(l, T)$	$\frac{F_T}{F_{MVC}(l, T)}$	$\frac{A_i}{A_{MVC}}$	$\frac{f_{ci}}{f_{CMVC}}$	$T_F(A)$ min	$T_F(f_c)$ min
Minimal (seated)	0	—	—	—	—	—	—	—
Mild	<.10–.15	480	416	.060	—	—	—	—
Moderate	>.15–.40	405	284	0.352	0.40	0.913	3.0	2.9
High	>.40–.90	450	450	0.500	0.667	0.853	1.35	1.28

Table 8.2 Pulmonary Assessment of Static Work

Activity Level	$\frac{F_T}{F_{MVC}}$	$\overline{f_R}$ cycles/min	$\overline{\Delta V_{AIR}}$ L	$\overline{\dot{V}_{AIR}}$ L/min	$\overline{\dot{V}_{O_2}}$ L/min
Minimal (seated)	0	12	0.42	5.04	0.252
Mild	<.10–.15	12	0.50	6.00	0.300
Moderate	>.15–.40	18	0.39	7.02	0.351
High	>.40–.90	32	0.25	8.00	0.400

Table 8.3 Cardioenergetic Assessment of Static Work

Activity Level	$\frac{F_T}{F_{MVC}}$	$\overline{f}_{HR}$ beats/min	$\frac{P_S}{P_D}$ mm Hg	ΔP mm Hg	$\overline{\Delta P}$ mm Hg	$\overline{\Delta C}$ LO_2/L	$\overline{\dot{V}}_{CO}$ L/min	$\overline{\Delta V_S}$ L	R PRU	$\overline{W_S}$ J	P_0 W
Minimal (seated)	0	70	120/75	45	90	0.05	5.04	0.072	1.07	0.86	1.01
Mild	<.10–.15	82	130/85	45	100	0.05	6.00	0.073	1.00	0.97	1.33
Moderate	>.15–.40	95	181/130	51	147	0.05	7.02	0.074	1.26	1.45	2.29
High	>.40–.90	110	238/181	57	200	0.05	8.00	0.073	1.50	1.95	3.53

b. Analysis of Static Work

EXAMPLE 8.1

Resting (Minimal) Static Work

A receptionist is seated at an information desk, hands folded and answering questions:

(a) Comment upon muscle strength and endurance for this static work:

(b) Regarding pulmonary assessment of this static workload, if the following are measured:

$$T_R = 5\ \text{s}$$
$$\hat{\dot{V}}_{AIR}\ \text{(peak instantaneous pulmonary airflow)} = .264\ \text{L/s}$$

Calculate $\overline{f_R}$, $\overline{\Delta V}_{AIR}$, $\overline{\dot{V}}_{AIR}$ and $\overline{\dot{V}}_{O_2}$.

(c) Regarding cardioenergetic assessment of this static works, if the following are measured:

$$T\text{(R-R interval)} = .857\ \text{s}$$
$$P_S\text{(systolic pressure)} = 120\ \text{mm Hg}$$
$$P_D\text{(diastolic pressure)} = 75\ \text{mm Hg}$$

Then approximate $\overline{\Delta C}$, and calculate f_{HR}, ΔP, $\overline{\Delta P}$, $\overline{\dot{V}}_{CO}$, $\overline{\Delta V}_s$, R, $\overline{W}_s$, and P_O.

SOLUTION 8.1 (a)

Since only minimal postural support is required for any specific muscle group, $F_T/F_{mvc} \leq .1$, and endurance is unlimited.

SOLUTION 8.1(b)

Substituting into (31):

$$\bar{f}_R = \frac{60}{T_R} = \frac{60}{5} = 12\,\frac{\text{cycles}}{\text{min}}$$

Approximate $\dot{V}_{AIR} = f(t)$ as a uniform series of sine waves (inspiration alternating with expiration) with equal positive and negative peak air flow ($\hat{\dot{V}}_{AIR}$) for inspiration and expiration, respectively.

For $\dot{V}_{AIR}$ (inspiration) between $t(0)$ and $t = T_R/2$:

$$\dot{V}_{AIR} = \hat{\dot{V}}_{AIR} \cdot \sin\left(\frac{2\pi}{T_R} \cdot t\right) \tag{i}$$

For ΔV_{AIR} (inspiration), equation (32a) may be written:

$$\Delta V_{AIR} = \int_0^{\frac{T_R}{2}} \dot{V}_{AIR} \cdot dt \tag{ii}$$

Substituting (i) into (ii):

$$\Delta V_{AIR} = \int_0^{\frac{T_R}{2}} \hat{\dot{V}}_{AIR} \cdot \sin\left(\frac{2\pi}{T_R} \cdot t\right) dt \tag{iii}$$

Solving (iii):

$$\Delta V_{AIR} = \hat{\dot{V}}_{AIR}\left[-\frac{T_R}{2\pi} \cdot \cos\left(\frac{2\pi}{T_R} t\right)\right]_0^{\frac{T_R}{2}}$$

$$\Delta V_{AIR} = (\hat{\dot{V}}_{AIR})\left[\frac{T_R}{2\pi}\right][-(-1) - (-(1))]$$

$$\Delta V_{AIR} = \frac{\hat{\dot{V}}_{AIR} \cdot T_R}{\pi} \tag{iv}$$

Because of the approximation of a uniform series of sine waves:

$$\Delta V_{AIR} = \overline{\Delta V}_{AIR} \tag{v}$$

Substituting (v) into (iv):

$$\overline{\Delta V}_{AIR} = \frac{\hat{\dot{V}}_{AIR} \cdot T_R}{\pi} \tag{vi}$$

Substituting the given T_R and given $\hat{\dot{V}}_{AIR}$ into (vi):

$$\overline{\Delta V}_{AIR} = \frac{(.264)(5)}{\pi} = .420\text{ L}$$

Substituting into equation (33):

$$\hat{\dot{V}}_{AIR} = \overline{\Delta V}_{AIR} \cdot \bar{f}_R = (.420)(12) = 5.04\text{ L/min}$$

Substituting into equation (34b):

$$\bar{\dot{V}}_{O_2} = (.05)\bar{\dot{V}}_{AIR} = (.05)(5.04) = .252\text{ L/min} \tag{vii}$$

SOLUTION 8.1(c)

Approximate $\overline{\Delta C}$ as follows: First substituting equations (26) and (27) into (28) for ΔC:

$$\Delta C = 5 \frac{\text{ml } O_2}{100 \text{ ml blood}}$$

Then subsequently, substitute into equation (36) for $\overline{\Delta C}$:

$$\overline{\Delta C} = \left(\frac{\text{L}}{1000 \text{ ml}}\right)\left(\frac{1000 \text{ ml}}{\text{L}}\right)\left(\frac{5 \cdot \text{ml } O_2}{100 \text{ ml blood}}\right)$$

$$\overline{\Delta C} = \left(\frac{\text{L}}{1000 \text{ ml}}\right)\left(\frac{50 \cdot \text{ml } O_2}{\text{L}}\right)$$

$$\overline{\Delta C} = \frac{.05\text{L}_{O_2}}{\text{L}} \tag{i}$$

Substitute into equation (16):

$$\bar{f}_{HR} = \frac{60}{T} = \frac{60}{.857} = 70 \text{ beats/min}$$

Substitute into equation (19):

$$\Delta P = P_S - P_D = 120 - 75 = 45 \text{ mm Hg}$$

Substitute into equation (21):

$$\overline{\Delta P} = P_D + \frac{\Delta P}{3} = 75 + \frac{45}{3} = 90 \text{ mm Hg}$$

Substitute 8.1(b)(vii) and (i) into equation (37):

$$\overline{\dot{V}}_{CO} = \frac{\overline{\dot{V}}_{O_2}}{\overline{\Delta C}} = \frac{.252}{.05} = 5.04 \text{ L/min}$$

Substitute into equation (39b):

$$\overline{\Delta V}_s = \frac{\overline{\dot{V}}_{CO}}{\bar{f}_{HR}} = \frac{5.04}{70} = .072 \text{ L}$$

We calculate R in terms of peripheral resistance units (PRUs):

$$1 \text{ PRU} = \frac{1 \text{ mm Hg}}{1 \text{ ml/s}}$$

A blood pressure of 1 mm Hg results in 1 ml/s of blood flow when the vascular resistance is 1 PRU.

To calculate R, convert $\overline{\dot{V}}_{CO}$ (L/min) to $\overline{\dot{V}}'_{CO}$ (ml/s):

$$\overline{\dot{V}}'_{CO}\left(\frac{\text{ml}}{\text{s}}\right) = \left(\frac{1000 \text{ ml}}{\text{L}}\right)\left(\frac{\text{min}}{60 \text{ s}}\right)\left[\overline{\dot{V}}_{CO}\left(\frac{\text{L}}{\text{min}}\right)\right]$$

$$\overline{\dot{V}}'_{CO} = \left(16.67 \frac{\text{ml} \cdot \text{min}}{\text{L} \cdot \text{s}}\right)\overline{\dot{V}}_{CO} \tag{ii}$$

where $\overline{\dot{V}}'_{CO}$ is in ml/s, and $\overline{\dot{V}}_{CO}$ is in L/min.

Now define R in terms of PRUs:

$$R(\text{PRU}) = \frac{\overline{\Delta P}}{\overline{\dot{V}'}_{CO}} \tag{iii}$$

Substituting into equation (ii):

$$\overline{\dot{V}'}_{CO} = (16.67)(5.04) = 84\,\frac{\text{ml}}{\text{s}}$$

Substituting into equation (iii):

$$R = \frac{\overline{\Delta P}}{\overline{\dot{V}'}_{CO}} = \frac{90}{84} = 1.07\ \text{PRU}$$

For the energetic parameter, stroke work, we use joules of energy:

$$\overline{W}_s = \left(133.3\,\frac{\text{Pa}}{\text{mm Hg(torr)}}\right)\left(.001\,\frac{\text{M}^3}{\text{L}}\right)\overline{\Delta P}\cdot\overline{\Delta V}_s$$

$$\overline{W}_s(J) = (.1333)\overline{\Delta P}\cdot\overline{\Delta V}_s \tag{iv}$$

where $\overline{\Delta P}$ is in mm Hg, and $\overline{\Delta V}_s$ is in liters.
Substituting into equation (iv):

$$\overline{W}_s = (.1333)\overline{\Delta P}\cdot\overline{\Delta V}_s = (.1333)(90)(.072) = .86\ \text{J}$$

For the energetic parameter cardiac output power, P_0, we use units of watts (J/s). So in a manner similar to stroke work:

$$P_0(W) = \left(.1333\,\frac{\text{Pa}\cdot\text{M}^3}{\text{mm Hg}\cdot\text{L}}\right)\left(\frac{\text{min}}{60\ \text{s}}\right)\overline{\Delta P}\cdot\overline{\dot{V}}_{CO}$$

$$P_0(W) = (.00222)\overline{\Delta P}\cdot\overline{\dot{V}}_{CO} \tag{v}$$

where $\overline{\Delta P}$ is in mm Hg, and $\overline{\dot{V}}_{CO}$ is in liters/min.
Substituting into equation (v):

$$P_0 = (.00222)\overline{\Delta P}\cdot\overline{\dot{V}}_{CO} = (.00222)(90)(5.04) = 1.01\ \text{W}$$

EXAMPLE 8.2

Mild Intensity Static Work

A technician is standing at a counter top, continuously gripping a joystick (with a pistol grip) in order to manipulate a remote arm that is handling radioisotopes:

(a) Evaluate the muscle strength and endurance for this static work, if the following are measured.

Pistol grip measurements:

8 cm H × 4 cm L × 2 cm W

T_{AMB} = room temperature (25°C)

F_T (continuous grip force) = 25 N

Also measured (prior to the work session):

F_{MVC} [on a hand-grip dynamometer with a 5.5-cm grip length, at ambient room temperature (25°C)] = 480 N

A_{MVC} = 1.8 mV

f_{CMVC} = 160 Hz (measured at F_{MVC}, above).

(b) Regarding pulmonary assessment for this static work load, if the following are measured:

$$T_R = 5\text{ s}$$
$$\hat{V}_{AIR} = .314\text{ L/s},$$

calculate the customary pulmonary parameters.

(c) Regarding the cardioenergetic assessment of this static work, if the following are measured:

$$T = .732\text{ s}$$
$$P_S = 130\text{ mm Hg}$$
$$P_D = 85\text{ mm Hg}$$

and approximating from equations 8.1(c)(i):

$$\overline{\Delta C} = .05\,\frac{\text{L}_{\text{O}_2}}{\text{L}},$$

then calculate the customary cardioenergetic parameters.

SOLUTION 8.2(a)

Since the technician (and his muscles) are in an environmentally ambient room, we may approximate hand-grip muscle temperature:

$$T_m \approx 35°\text{C}$$

(See Chapter 6 for details.)

The hand-grip dynamometer is set at optimal grip length (between 5 and 6 cm), and the measurements are taken in an environmentally neutral setting, so that:

$$\begin{aligned} F_{MVC}(l_{OPT}, T_{35}) &= 480\text{ N} \\ A_{MVC}(l_{OPT}, T_{35}) &= 1.8\text{ mV} \\ f_{CMVC}(l_{OPT}, T_{35}) &= 160\text{ Hz} \end{aligned} \tag{i}$$

The dimensions of the pistol grip indicates that the joy-stick grip length (l_G):

$$l_G = 4.0\text{ cm} \tag{ii}$$

Recalling equation (44a), and Figure 8.12:

$$\frac{F_{MVCT}(30° \leq T \leq 40°)}{F_{MVC35}} = 1$$

So that we do not need to temperature correct F_{MVC}.
Substituting the indicated coefficients of Figure 8.13 into equation (45):

$$\frac{F_{MVC}(l_F)}{F_{MVC}(l_{F\cdot OPT})} = -.0321 l_G^2 + .3907 l_G - .1829$$

Rearranging and indicating that muscle temperature correction (although not needed) has been accounted for:

$$F_{MVC}(l_F, T_{35}) = F_{MVC}(l_{F\cdot OPT}, T_{35})[-.0321 l_G^2 + .3907 l_G - .1829] \qquad \text{(iii)}$$

Substituting (i) and (ii) into equation (iii):

$$F_{MVC}(l_F, T_{35}) = [480 \text{ N}][.8663] = 415.8 \text{ N}$$

Finally, solve for the sustained fractional static strength:

$$\frac{F_T}{F_{MVC}(l_F, T_{35})} = \frac{25 \text{ N}}{415.8 \text{ N}} = .060$$

Since the fractional static strength is less than 0.10 to 0.15, it is considered nonfatiguing. Therefore, no further analysis (using A_{MVC} or f_{CMVC}) needs to be done.

SOLUTION 8.2(b)

Substituting the given T_R into equation (31):

$$\bar{f}_R = 12 \text{ cycles/min}$$

Making the same approximations for airflow as in Solution 8.1(b)(ii), substitute into equation 8.1(b)(vi):

$$\overline{\Delta V}_{AIR} = \frac{\hat{\dot{V}}_{AIR} \cdot T_R}{\pi}$$

$$\overline{\Delta V}_{AIR} = 0.500 \text{ L}$$

Substituting into equation (33):

$$\bar{\dot{V}}_{AIR} = (.500)(12) = 6.0 \text{ L/min}$$

Substituting into equation (34b):

$$\bar{\dot{V}}_{O_2} = (.05)(6.0) = .300 \text{ L/min}$$

SOLUTION 8.2(c)

From equation (16):

$$\bar{f}_{HR} = 82 \frac{\text{beats}}{\text{min}}$$

From equation (19):

$$\Delta P = 45 \text{ mm Hg}$$

From equation (21):

$$\overline{\Delta P} = 100 \text{ mm Hg}$$

From equation (37):

$$\bar{\dot{V}}_{CO} = 6.0 \frac{\text{L}}{\text{min}}$$

From equation (39b):

$$\overline{\Delta V_s} = .073 \text{ L}$$

From equation 8.1(c)(ii):

$$\overline{\dot{V}'_{CO}} = 100 \frac{\text{ml}}{\text{s}}$$

From equation 8.1(c)(iii):

$$R = 1.00 \text{ PRU}$$

From equation 8.1(c)(iv):

$$\overline{W_s} = 0.97 \text{ J}$$

From equation 8.1(c)(v):

$$P_0 = 1.33 \text{ W}$$

EXAMPLE 8.3

Moderate Intensity Static Work

A student truck driver (when learning in a truck cab simulator) finds it necessary to continually depress a heavy-duty clutch as the student driver manually shifts through a series of heavy gears (with the right hand and arm) from zero truck velocity to normal highway cruising speed. The applied force at the ball of the left foot is 100 N, and the ankle angle is 110°.

It is an early winter morning, and the truck cab simulator heater is broken, so that T_{AMB} = 15°C. A thermistor is affixed to the surface of the left calf muscle, so that the skin temperature (T_S = 20°C) is detected. A biothermodynamic model then allows you to approximate T_m = 25°C [see Figure 6.4 and Example 6.3(c)].

Prior to starting to train in the simulator that day, the student driver was first evaluated at the HFE lab, where the following were recorded:

F_{MVC} (lab) = 500 N (using an ankle-foot dynamometer with an ankle angle of 90°, when the muscle temperature (T_m) was 35°C).

A_{MVC} = 2.0 mV

f_{CMVC} = 150 Hz (measured at the F_{MVC} (lab) above)

After starting up the truck cab simulator, at $t_0 = 0$ the clutch is depressed and surface electrodes continually record the EMG of the left calf muscle during this static work. The following data are then noted:

$$A_i = 0.8 \text{ mV @ } T_i = 1.0 \text{ min}$$
$$f_{ci} = 137 \text{ Hz @ } T_i = 1.0 \text{ min}$$

(a) After 1 min of clutch depression, how much longer can the driver maintain F_T before fatigue sets in and the clutch is released.

(b) Regarding pulmonary assessment for this static workload, if the following are measured:

$$T_R = 3.33 \text{ s}$$
$$\hat{V}_{AIR} = .368 \text{ Liters/s}$$

Calculate the customary pulmonary parameters.

(c) Regarding cardioenergetic assessment of this static work, if the following are measured:

$$T = 632 \text{ s}$$
$$P_S = 181 \text{ mm Hg}$$
$$P_D = 130 \text{ mm Hg}$$

and approximating that $\overline{\Delta C} = .05(L_{O_2}/L)$, then calculate the customary cardioenergetic parameters.

SOLUTION 8.3(a)

Correct F_{MVC} for the muscle temperature (T_m).
Substituting coefficients from Figure 8.12 into equation (44b):

$$\frac{F_{MVCT}}{F_{MVC \cdot 35}} = -.002T^2 + .154T - 1.790$$

Equating (with respect to temperature):

$$F_{MVC \cdot 35} = F_{MVC}(\text{lab}) = 500 \text{ N}$$

Using the approximation of $T_m = 25°\text{C}$ (above), and rearranging for F_{MVCT}:

$$F_{MVCT} = [500][-.002(25)^2 + .154(25) - 1.790]$$
$$F_{MVCT} = [500][.81] = 405 \text{ N}$$

Correct F_{MVC} for the ankle angle, θ_A.
Substituting coefficients from Figure 8.15 into equation (47), and indicating that temperature correction of F_{MVC} has occurred:

$$\frac{F_{MVC}(l_G, T_{25})}{F_{MVC}(l_{GOPT}, T_{25})} = -.0005\,\theta_A^2 + .0865\,\theta_A - 2.7628$$

Equating (with respect to length):

$$F_{MVC}(l_{G \cdot OPT}) = F_{MVCT}(\text{lab}) = 500 \text{ N}$$

So it follows that (with respect to temperature):

$$F_{MVC}(l_{G \cdot OPT}, T_{25}) = F_{MVCT} = 405 \text{ N}$$

Substituting the given ankle angle ($\theta_A = 110°$), and rearranging for $F_{MVC}(l_G, T_{25})$:

$$F_{MVC}(l_G, T_{25}) = [405][-.0005(110)^2 + .0865(110) - 2.7628]$$
$$F_{MVC}(l_G, T_{25}) = [405][.702] = 284 \text{ N}$$

Solve for the sustained fractional static strength, where the applied force at the ball of the foot (100 N) represents the target force (F_T):

$$\frac{F_T}{F_{MVC}(l_G, T_{25})} = \frac{100 \text{ N}}{284 \text{ N}} = .0352 \qquad \text{(i)}$$

Since this fractional static strength is greater than 0.15, it will be fatiguing. Consequently, we must solve for the time remaining to fatigue (ΔT_F). This is done by one of two methods (the rms amplitude method or center frequency method).
Solve for ΔT_F using the rms amplitude method: The following sequential process should be used:
Substitute the coefficients of Figure 8.11.c into equation (14):

$$\frac{T_i}{T_F} = \frac{1}{(.30)}\left(\frac{A_i}{A_{MVC}} - \frac{F_T}{F_{MVC}(l_G, T_{25})} - (.0035)(T - 40)\right) \tag{ii}$$

where F_{MVC}[eq. (14)] $= F_{MVC}(l_G, T_{25})$.
From the given EMG data:

$$\frac{A_i}{A_{MVC}} = \frac{0.8\,\text{mV}}{2.0\,\text{mV}} = 0.4 \tag{iii}$$

Recalling that T in equation (ii) is the muscle temperature (T_m = 25°C), and substituting equations (i) and (iii) into equation (ii):

$$\frac{T_i}{T_F} = (3.333)[(0.40) - (0.352) - (.0035)(-15)]$$

$$\frac{T_i}{T_F} = (3.333)(0.1005) = 0.335$$

Since T_i = 1 min:

$$T_F = \frac{1}{.335} = 3.0\,\text{min}$$

So that the time remaining to fatigue (ΔT_F) is:

$$\Delta T_F = T_F - T_i = 2.0\,\text{min}$$

Solve for ΔT_F using the center frequency method.
The sequential approach is to first solve for T_i/T_F using f_c, and then solve for ΔT_F:
Substitute the coefficient of Figure 8.11.b into equation (13):

$$\frac{T_i}{T_F} = \frac{1}{(.25)}\left(1 - \frac{f_{ci}}{f_{CMVC}}\right)$$

Substitute for the remaining variable using the given EMG data:

$$\frac{T_i}{T_F} = (4.0)\left(1 - \frac{137}{150}\right)$$

$$\frac{T_i}{T_F} = (4.0)(.0867) = .347$$

Since T_i = 1 min:

$$T_F = \frac{1}{.347} = 2.9\,\text{min}$$

So:

$$\Delta T_F = T_F - T_i = 1.9\,\text{min}$$

SOLUTION 8.3(b)

Substituting into equation (31):

$$\bar{f}_R = 18 \frac{\text{cycles}}{\text{min}}$$

As per Solution 8.1(b)(ii), substituting into equation 8.1(b)(vi):

$$\overline{\Delta V}_{AIR} = \frac{\hat{\dot{V}}_{AIR} T_R}{\pi}$$

$$\overline{\Delta V}_{AIR} = .390 \text{ L}$$

$\overline{\dot{V}}_{AIR}$ per equation (33) is:

$$\overline{\dot{V}}_{AIR} = (.390)(18) = 7.02 \text{ L/min}$$

$\overline{\dot{V}}_{O_2}$ per equation (34b) is:

$$\overline{\dot{V}}_{O_2} = (.05)(7.02) = .351 \text{ L/min}$$

SOLUTION 8.3(c)

From equation (16):

$$\bar{f}_{HR} = 95 \frac{\text{beats}}{\text{min}}$$

From equation (19):

$$\Delta P = 51 \text{ mm Hg}$$

From equation (21):

$$\overline{\Delta P} = 147 \text{ mm Hg}$$

From equation (37):

$$\overline{\dot{V}}_{CO} = 7.02 \frac{\text{L}}{\text{min}}$$

From equation (39b):

$$\overline{\Delta V}_s = .074 \text{ L}$$

From equation 8.1(c)(ii):

$$\overline{\dot{V}}'_{CO} = 117 \frac{\text{ml}}{\text{s}}$$

From equation 8.1(c)(iii):

$$R = 1.26 \text{ PRU}$$

From equation 8.1(c)(iv):

$$\overline{W}_s = 1.45 \text{ J}$$

From equation 8.1(c)(v):

$$P_0 = 2.29 \text{ W}$$

EXAMPLE 8.4

High Intensity Static Work

A carpenter/handyman is standing on a ladder (which in turn is leaning against a wall) in an environmentally ambient room. The carpenter leans backwards, extending his left arm straight out and his left hand firmly grips a ladder step (40 cm L × 6 cm W × 2 cm H). The carpenter then hyperextends his neck to look up at the ceiling and proceeds to firmly grip a hammer and tap pre-positioned nails into the ceiling.

The left arm hand-grip muscles maintain a sustained grasp on the ladder step exerting a continuous gripping force (F_T) of 225 N. The right arm hammers for 45-s intervals and then is relaxed at the carpenter's side for 15 s while the right arm cuff pressure cycles to obtain blood pressure readings.

Measured in the HFE laboratory just prior to performing this task:

F_{MVC} (lab) = 450 N (using a hand-grip dynamometer with a 6.0 cm grip length, when the muscle temperature (T_m) was 35°C).

A_{MVC} = 1.2 mV

F_{CMVC} = 170 Hz (measured at F_{MVC} (lab) above).

After initially leaning full back, and starting the first hammer strike at $t_0 = 0$, the left hand exerts its full grip and surface electrodes continuously record the EMG of the left finger flexor muscles during this static work.

The following data are then noted:

$$A_i = 0.8 \text{ mV} @ T_i = 0.75 \text{ min}$$
$$f_{ci} = 145 \text{ Hz} @ T_i = 0.75 \text{ min}$$

(a) After the 45 s of left hand gripping, how much longer can the carpenter maintain F_T before fatigue sets in and he must again lean forward toward the ladder and release or lower the gripping tension?

(b) Regarding pulmonary assessment for this static workload, if the following are measured:

$$T_R = 1.875 \text{ s}$$
$$\hat{V}_{AIR} = .419 \text{ L/s},$$

calculate the customary pulmonary parameters.

(c) Regarding cardioenergetic assessment of this static work, if the following are measured:

$$T = .545 \text{ s}$$
$$P_S = 238 \text{ mm Hg}$$
$$P_D = 181 \text{ mm Hg}$$

and approximating that $\overline{\Delta C} = .05(L_{O_2}/L)$, then calculate the customary cardioenergetic parameters.

SOLUTION 8.4(a)

Correct F_{MVC} for temperature and length effects.

Since the carpenter (and his muscles) are in an environmentally ambient room, we approximate $T_m \approx 35$°C [see Example 6.3(a)].

Since the hand-grip dynamometer is at an optimal length (5.0 to 6.0 cm), and the muscle temperature in the HFE lab is 35°C:

$$F_{MVC}(l_{FOPT}, T_{35}) = 450\text{ N}$$
$$A_{MVC}(l_{OPT}, T_{35}) = 1.2\text{ mV}$$
$$f_{CMVC}(l_{OPT}, T_{35}) = 170\text{ Hz}$$

The dimensions of the ladder step indicate that the left hand will grip a ladder step width of 6.0 cm:

$$l_G = 6.0\text{ cm}$$

Recalling equation (44a) and Figure 8.12, since the $T_m \approx 35°C$, we do not require temperature correction of F_{MVC}.
Recalling Figure 8.13 and equation (45), since:

$$l_G = l_{OPT} = 6.0\text{ cm}$$

then we do not require muscle length correction of F_{MVC}.
Solve for the sustained fractional static strength:

$$\frac{F_T}{F_{MVC}(l_F, T)} = \frac{F_T}{F_{MVC}(l_{FOPT}, T_{35})} = \frac{225}{450} = 0.50 \qquad \text{(i)}$$

Since the fractional static strength is greater than 0.15, it will be fatiguing.
Solve for ΔT_F using the rms amplitude method.
Since $T_m = 35°C$, we may substitute the coefficient of Figure 8.9 into equation (8):

$$\frac{T_i}{T_F} = \frac{1}{(.30)}\left(\frac{A_i}{A_{MVC}} - \frac{F_T}{F_{MVC}}\right) \qquad \text{(ii)}$$

where F_{MVC} [equation (8)] $= F_{MVC}(l_{FOPT}, T_{35})$.
From the given EMG data:

$$\frac{A_i}{A_{MVC}} = \frac{0.8\text{ mV}}{1.2\text{ mV}} = 0.667 \qquad \text{(iii)}$$

Substituting equations (i) and (iii) into equation (ii):

$$\frac{T_i}{T_F} = (3.33)(.167) = 0.556$$

Since $T_i = 0.75$ min:

$$T_F = \frac{0.75}{0.556} = 1.35\text{ min}$$

Remaining time to fatigue (ΔT_F):

$$\Delta T_F = T_F - T_i = 1.35 - 0.75$$
$$\Delta T_F = 0.60\text{ min}$$

Solve for ΔT_F using the center frequency method:
Since $T_m = 35°C$, we may substitute the coefficient of Figure 8.10 into equation (9):

$$\frac{T_i}{T_F} = \frac{1}{(.25)}\left(1 - \frac{f_{ci}}{f_{CMVC}}\right)$$

Substituting the given EMG data:

$$\frac{T_i}{T_F} = (4.0)\left(1 - \frac{145}{170}\right)$$

$$\frac{T_i}{T_F} = (4.)(.147) = .588$$

Since $f_{ci} = 145$ Hz @ $T_i = 0.75$ min

$$T_F = \frac{0.75}{0.588} = 1.28 \text{ min}$$

Remaining time to fatigue (ΔT_F):

$$\Delta T_F = T_F - T_i = 1.28 - .75 = 0.53$$

$$\Delta T_F = 0.53 \text{ min}$$

SOLUTION 8.4(b)

Substituting into equation (31):

$$\bar{f}_R = 32 \frac{\text{cycles}}{\text{min}}$$

As per Solution 8.1(b)(ii), substituting into equation 8.1(b)(vi):

$$\overline{\Delta V}_{AIR} = \frac{\hat{\dot{V}}_{AIR} T_R}{\pi}$$

$$\overline{\Delta V}_{AIR} = .250 \text{ L}$$

$\bar{\dot{V}}_{AIR}$ per equation (33) is:

$$\bar{\dot{V}}_{AIR} = (.250)(32) = 8.0 \text{ L/min}$$

$\bar{\dot{V}}_{O_2}$ per equation (34b) is:

$$\bar{\dot{V}}_{O_2} = (.05)(8.0) = .40 \text{ L/min}$$

SOLUTION 8.4(c)

From equation (16):

$$\bar{f}_{HR} = 110 \frac{\text{beats}}{\text{min}}$$

From equation (19):

$$\Delta P = 57 \text{ mm Hg}$$

From equation (21):

$$\Delta P = 200 \text{ mm Hg}$$

From equation (37):

$$\bar{\dot{V}}_{CO} = 8.0 \frac{\text{L}}{\text{min}}$$

From equation (39b):

$$\overline{\Delta V_s} = .073 \text{ L}$$

From equation 8.1(c)(ii):

$$\overline{V}'_{CO} = 133 \frac{\text{ml}}{\text{s}}$$

From equation 8.1(c)(iii):

$$R = 1.50 \text{ PRU}$$

From equation 8.1(c)(iv):

$$\overline{W}_s = 1.95 \text{ J}$$

From equation 8.1(c)(v):

$$P_0 = 3.55 \text{ W}$$

8.4 QUANTITATIVE ANALYSIS OF DYNAMIC WORKLOAD

a. Physiology of Dynamic Work

This subsection will consider the human reaction to dynamic work. We shall first consider various factors that affect muscular strength and endurance when performing dynamic work. This subsection will then review the cardiopulmonary responses to dynamic work. The responses will be considered with respect to minimal, mild, moderate, and high dynamic work.

The strength of skeletal muscle during dynamic work will be affected by muscle temperature and muscle length. These relationships have already been described in the section on static work.

Strength is also proportional to the skeletal muscle mass performing the dynamic work. More specifically, the strength of a skeletal muscle is directly proportional to its cross-sectional area. When this cross-sectional area is through the middle of the muscle, the individual muscle fibers may be directly counted (under a microscope).

There are two primary types of skeletal muscle fibers. The fast-twitch (type II) muscle fibers are larger diameter fibers that rely on anaerobic metabolism. The slow-twitch (type I) muscle fibers are the smaller diameter fibers that rely on oxidative metabolism. Skeletal muscle strength is primarily determined by the fraction of type II muscle fibers to the total (type I + type II) muscle fiber population.

Skeletal muscle endurance is dependent (in part) upon the actual amount of force being exerted. In this respect there is a strength-endurance relationship similar to that for static work. However, with respect to dynamic work, endurance is also dependent (in part) upon the displacement over which the force is exerted. Skeletal muscle endurance is therefore related to the magnitude of the dynamic work (load times distance) being performed. The endurance of muscle is also dependent upon the rate of repetition with which the work is done. Consequently, endurance of skeletal muscle is dependent upon the muscle power output (work per unit time) required for the performance of dynamic work.

As noted earlier, when seen in cross section, a skeletal muscle will be composed of a mix of type I and type II muscle fibers. In contrast to strength, skeletal muscle endurance is directly proportional to the fraction of slow-twitch (type I) fibers with respect to the total (type I + type II) muscle fiber population. Consequently, physical conditioning programs that emphasize the hypertrophy of type II fibers when the dynamic work primarily emphasizes strength can be established. Other physical conditioning programs can focus upon the hypertrophy of type I fibers when the specific dynamic work requires significant endurance. Finally, balanced physical conditioning programs can be prescribed that result in the hypertrophy of both type I and type II muscle fibers when the dynamic work requires a tradeoff between strength and endurance.

Table 8.4 compares and contrasts various characteristics of type I muscle fibers and type II muscle fibers. This table is presented only to emphasize the functional difference between the two fiber types. For more detailed information, the interested student is referred to the appropriate muscle physiology textbooks.

The cardiopulmonary response to dynamic work includes central (cardiac and pulmonary) reactions and peripheral (systemic vascular) reactions. There are also neural and hormonal responses which make the overall cardiopulmonary adjustments to dynamic work rather complex. In order to provide a systematic (and yet simplified) approach, this section will consider the cardiopulmonary reactions to mild, moderate and heavy dynamic work.

In mild dynamic work, the blood flow to the skin and viscera remains at the original resting levels. However, there is an increase in skeletal muscle blood flow due to increased sympathetic discharge from the midbrain. In general, sympathetic discharge tends to vasoconstrict and reduce blood flow through the peripheral vessels. However, skeletal muscle has a specialized sympathetic vasodilator system which increases blood flow through the exercising muscle. This results in a decrease in vascular peripheral resistance (R). The heart rate is accelerated, resulting in an

Table 8.4 The Two Primary Types of Skeletal Muscle Fibers

Parameter	Type I (Slow)	Type II (Fast)
Color	Red	White
Diameter	Moderate	Large
Latency	Long	Short
Duration	Long	Short
Myosin isoenzyme ATPase activity	Slow	Fast
Ca++ pumping capacity	Moderate	High
Glycolytic capacity	Moderate	High
Oxidative capacity	High	Low
Use	Postural	Movement
Endurance	High	Low
Physio term	Tonic	Phasic
Number of fibers per motor unit	3–6	120–165
Recruitment	Early	Late
Innervation	Small, slowly conducting, steady pulse train	Large, rapid conducting, bursts of activity
100-meter dash (10 s)	15%	85%
2-mile race (60 min)	80%	20%
Distance run	95%	5%

increase in the cardiac output [see equation (39a)]. This increase in cardiac output approximately balances the decrease in peripheral resistance so that the mean systemic arterial pressure remains relatively constant [per equation (40c)].

When an individual performs moderate dynamic work, vasoactive metabolites are produced and accumulate in the exercising muscles. This results in a dramatic dilation of the blood vessels in the exercising skeletal muscle. As a result, the blood flow through the exercising skeletal muscle can increase to amounts from 10 to 20 times that of the resting (nonexercising) state. Venous return to the right side of the heart can also be increased due to the increased contraction of skeletal muscles around the veins returning blood to the right side of the heart. This occurs because of compression of the veins and the subsequent one-way displacement of the blood toward the thorax. The increase in body temperature (which is produced by the dynamic work) will further dilate blood vessels in the skin in order to offload heat (see Section 6.1.a). The net effect of this vasodilation of skeletal muscle and skin blood vessels is to dramatically decrease the systemic peripheral resistance (R). However, there is marked vasoconstriction in the viscera (stomach, intestines, kidneys, etc.) so that blood flow to these organs is reduced. This tends to somewhat offset the marked decrease in systemic peripheral resistance which occurs during moderate dynamic work.

There is a pronounced cardioacceleratory effect as the midbrain continues sympathetic discharge to the heart itself during moderate dynamic work. This pronounced increase in heart rate contributes to a dramatic increase in cardiac output. This effect is partially offset by a decrease in stroke volume due to a shortening of R-R interval (see Figure 5.20). However, there is still a significant increase in cardiac output [see equation (39a)]. The end result is an increase in the mean systemic arterial pressure with moderate dynamic work. This occurs because the increase in cardiac output is proportionally greater than the reduction in peripheral vascular resistance [as per equation (40c)]. This increase in mean systemic arterial pressure is accompanied by an increase (or widening) of the pulse pressure (ΔP). The systolic pressure rises in relation to the increase in cardiac output. However the diastolic pressure will fall, remain the same, or rise only slightly (compared to the resting, non-exercising level) proportional to the decrease in systemic peripheral resistance. The net effect is a widening of the systolic pressure to diastolic pressure differential [see equation (19)].

When a normal individual performs either mild or moderate dynamic work, the increase in cardiac output is caused by the increase in heart rate. It is not caused by an increase in stroke volume, which either remains the same as the resting level, or can actually decrease. An exception to this is in the well-trained individual who normally maintains a slow heart rate, and who will actually increase their stroke volume with moderate dynamic work.

With heavy dynamic work, there is a significant increase in the return of blood to the right heart. This results in an increase in the ventricular end-diastolic volume, which results in an increase in the stroke volume [see equation (38)]. Since there is already a pronounced increase in heart rate, this increase in stroke volume contributes significantly to a dramatic increase in the cardiac output [as per equation (39a)]. This marked increase in cardiac output with heavy dynamic work significantly outweighs the decrease in systemic peripheral resistance, so that there is a further significant rise in the mean systemic arterial pressure [see equation (40c)].

In mild, moderate and heavy dynamic work, blood flow to the heart increases proportionately. This is because cardiac blood flow is under local control, which

Table 8.5 Strength and Endurance Assessment of Dynamic Work

Activity Level	W_e J	P_e W	$\bar{f}_R$ cycles/min	$\overline{\Delta V}_{AIR}$ L	$\bar{\dot{V}}_{AIR}$ L/min	$\bar{\dot{V}}_{O_2}$ L/min	AI
Minimal (standing)	0	0	12	0.5	6.0	0.30	0.136
Mild	25.2	20.2	16	0.65	10.4	0.52	0.248
Moderate	275	110	25	0.80	20.0	1.0	0.417
High	211	205	40	1.00	40.0	2.0	0.800

responds to the increasing metabolic demands of the contracting heart muscle. At all three levels of dynamic work, the blood flow to the brain is constant. This is because cerebral blood flow is also under local control but has much lower metabolic demands.

The pulmonary response to mild, moderate, and heavy dynamic work is progressive and proportional to the level of the dynamic work. For a young healthy adult male, the tidal volume is approximately 500 ml and the respiratory rate is 12 breaths/min at rest. This results in a minute ventilation of approximately 6 L/min [as per equation (33)]. With very heavy dynamic work, the tidal volume can increase up to 1000 ml and the respiratory rate can increase up to 40 breaths/min so that the minute ventilation with very heavy dynamic work can be 40 L/min. For the young, healthy adult male, the minute oxygen consumption will be 300 ml/min at rest [equations (23), (24) and (34)]. The minute oxygen consumption may increase up to approximately 2 L/min for high-intensity exercise.

The maximal steady-state oxygen consumption ($\bar{\dot{V}}_{O_2 \cdot MAX}$) is the highest value that an individual is able to obtain when performing very strenuous work for a period of time. $\bar{\dot{V}}_{O_2 \cdot MAX}$ is referred to as the aerobic capacity and is a physiological measure of the fitness of the individual for prolonged, strenuous dynamic work.

At some point, the individual will reach their upper limit for the performance of dynamic work and $\bar{\dot{V}}_{O_2 \cdot MAX}$ will plateau. Why does this occur? Heart rate and stroke volume will have reached their peak, some dehydration will occur, and the body temperature will have risen. Since there is increased metabolism of the working muscles, the oxygen extraction increases, and lactic acid is formed. However, with high-intensity, prolonged dynamic work, the oxygen extraction by the working skeletal muscles is limited because of a plateauing of the delivery of blood through the peripheral circulation. This occurs as the heart rate reaches its maximum. The build up of lactic acid, feelings of muscle pain, fatigue, and finally exhaustion limit an individual's ability to continue maximal dynamic work.

Table 8.6 Cardiovascular Assessment of Dynamic Work

Activity Level	P_e W	$\bar{f}_{HR}$ beats/min	P_S/P_D mm Hg	ΔP mm Hg	$\overline{\Delta P}$ mm Hg	$\overline{\Delta C}$ LO_2/L	$\bar{\dot{V}}_{CO}$ L/min	$\overline{\Delta V}_S$ L	R PRU
Minimal (standing)	0	75	125/80	45	95	0.05	6.0	0.080	0.95
Mild	20.2	105	125/80	45	95	0.06	8.67	0.083	0.66
Moderate	110	160	151/85	66	107	0.07	14.3	0.089	0.45
High	205	200	180/90	90	120	0.08	25	0.125	0.29

Table 8.7 Cardioenergetic Assessment of Dynamic Work

Activity Level	W_e J	P_e W	$\overline{W}_S$ J	P_0 W	F_{ws}	F_{we}	F_{po}	F_{pe}
Minimal (seated)	0	0	1.01	1.27	1.0	0	1.0	0
Mild	25.2	20.2	1.05	1.83	0.04	0.96	0.08	0.92
Moderate	275	110	1.27	3.40	0.005	0.995	0.030	0.97
High	211	205	2.00	6.67	0.01	0.99	0.03	0.97

The material presented in this section may be summarized as follows. The strength and endurance responses associated with resting (minimal), mild, moderate, and high intensity dynamic work are shown in Table 8.5. The cardiovascular responses for these four levels of dynamic work intensity are shown in Table 8.6. The cardioenergetic responses are shown in Table 8.7.

b. Analysis of Dynamic Work

EXAMPLE 8.5

Resting (Minimal) Dynamic Work

An office clerk is continuously standing at a counter, receiving single-page forms handed from the other side of the counter and placing them into in/out trays at the front of the clerk. In order to evaluate the required muscular strength and aerobic endurance for this dynamic work, the following are measured:

F_L (the force of the external load) = the weight of the single page of paper (W_L) and is negligible compared to the weight of the arm (W_A).

Δx_L (displacement of the external load) = 0.5 M, the distance moved after receiving the form and placing it in a tray.

f_L = the frequency with which the external work ($F_L \cdot \Delta x_L$) is performed (the period of a single work event (T_L) before the initiation of another work event) = 1 work event per minute.

Also measured at the time of this dynamic work:

$$T_R = 5\text{ s}$$
$$\hat{\dot{V}}_{AIR} = 0.314\text{ L/s}$$

Measured at a prior time in the HFE laboratory, during maximal effort treadmill testing:

$$\overline{\dot{V}}_{O_2\max} = 2.2\text{ L/min}$$

(a) Calculate the external work (W_e), the external power (P_e), the customary pulmonary parameters ($\overline{f}_R$, $\overline{\Delta V}_{AIR}$, $\overline{\dot{V}}_{AIR}$, and $\overline{\dot{V}}_{O_2}$) and the aerobic index ($\overline{\dot{V}}_{O_2}/\overline{\dot{V}}_{O_2\max}$).

(b) Regarding cardiovascular assessment of this dynamic work, if the following are measured:

$$T = 0.800 \text{ s}$$
$$P_S = 125 \text{ mm Hg}$$
$$P_D = 80 \text{ mm Hg}$$

For resting dynamic work (minimal oxygen demand), we may also approximate.

$$\overline{\Delta C} = .05 \frac{\text{L}_{\text{O}_2}}{\text{L}} \tag{v}$$

Calculate $(\bar{f}_{HR}$, ΔP, $\overline{\Delta P}$, $\bar{\dot{V}}_{CO}$, $\overline{\Delta V}_s$, and R.

(c) Regarding energetic assessment of this dynamic work, recall that:

$$W_e = 0$$
$$P_e = 0$$

Calculate $\overline{W}_s$, P_0, fractional work and fractional power.

SOLUTION 8.5(a)

It is apparent that the required muscular strength for the performance of the dynamic work will be directly proportional to the external work (W_e) and external power (P_e), which the skeletal muscles must perform.
Hence we calculate W_e and P_e:

$$W_e = F_L \cdot \Delta x_L \tag{i}$$

Substituting into equation (i):

$$W_e = (0)(.5) = 0$$

Proceeding to calculate P_e:

$$P_e = F_L \cdot \Delta x_L \cdot f_L \tag{ii}$$

Or alternatively:

$$P_e = \frac{F_L \cdot \Delta x_L}{T_L} \tag{iii}$$

Substituting into equation (ii):

$$P_e = (0)(.5)(1) = 0$$

The muscular endurance for dynamic work is proportional to the *aerobic index* (AI). The AI is a normalized oxygen demand, which is defined as:

$$\text{AI} = \frac{\bar{\dot{V}}_{O_2}}{\dot{V}_{O_2\max}} \tag{iv}$$

Since $\dot{V}_{O_2\max}$ represents the maximal oxygen consumption that the human operator can physically achieve, the AI will vary between zero and unity. When the AI is a low fraction, endurance for that dynamic task will be prolonged. When the AI is a high fraction, the endurance will be shortened. At an aerobic index of unity, the minute oxygen consumption equals the maximal-effort oxygen consumption, and no further increase in the work performance can be achieved. Furthermore, at an AI of unity, sustained effort will be brief (unless the individual is very well conditioned), and fatigue will be imminent. Therefore, an AI = 1 is a useful endurance end point.

Since the evaluation of muscular endurance requires a determination of aerobic index, we first calculate the customary pulmonary parameters.

Substituting into equation (31):

$$\overline{f}_R = \frac{60}{T_R} = \frac{60}{5} = 12\,\frac{\text{cycles}}{\text{min}}$$

Making similar approximations about airflow, as previously done in Solution 8.1(b)(ii), we substitute into equation 8.1(b)(vi):

$$\overline{\Delta V}_{AIR} = \frac{\hat{\dot{V}}_{AIR} T_R}{\pi} = \frac{(.314)(5)}{\pi} = 0.500\text{ L}$$

Substituting into equation (33):

$$\overline{\dot{V}}_{AIR} = (.500)(12) = 6.0\text{ L/min}$$

Substituting into equation (34b):

$$\overline{\dot{V}}_{O_2} = (.05)(6.00) = 0.300\text{ L/min}$$

Substituting into equation 8.5(a)(iv):

$$\text{AI} = \frac{\overline{\dot{V}}_{O_2}}{\dot{V}_{O_2\max}} = \frac{03}{2.2} = 0.136$$

SOLUTION 8.5(b)

Substitute into equation (16):

$$\overline{f}_{HR} = \frac{60}{T} = \frac{60}{0.80} = 75\,\frac{\text{beats}}{\text{min}}$$

Substitute into equation (19):

$$\Delta P = P_S - P_D = 125 - 80 = 45\text{ mm Hg}$$

Substitute into equation (21):

$$\Delta P = P_D + \frac{\Delta P}{3} = 80 + \frac{45}{3} = 95\text{ mm Hg}$$

Substituting into equation (37):

$$\overline{\dot{V}}_{CO} = \frac{\overline{\dot{V}}_{O_2}}{\overline{\Delta C}} = \frac{0.30}{.05} = 6.0\text{ L/min}$$

Substituting into equation (39b):

$$\overline{\Delta V}_s = \frac{\overline{\dot{V}}_{CO}}{\overline{f}_{HR}} = \frac{6.0}{75} = .080\text{ L}$$

Substituting into equation 8.1(c)(ii):

$$\overline{\dot{V}}'_{CO} = (16.67)\overline{\dot{V}}_{CO} = (16.67)(6.) = 100\,\frac{\text{ml}}{\text{s}}$$

Substituting into equation 8.1(c)(iii):

$$R = \frac{\overline{\Delta P}}{\bar{\dot{V}}'_{CO}} = \frac{95}{100} = 0.95 \text{ PRU}$$

SOLUTION 8.5(c)

Substituting into equation 8.1(c)(iv):

$$\overline{W}_s = (.1333)\overline{\Delta P} \cdot \overline{\Delta V}_S = (.1333)(95)(.080)$$
$$\overline{W}_S = 1.01 \text{ J}$$

Substituting into equation 8.1(c)(v):

$$P_0 = (.00222)\overline{\Delta P} \cdot \bar{\dot{V}}_{CO} = (.00222)(95)(6.0)$$
$$P_0 = 1.27 \text{ W}$$

Define the fractional cardiac stroke work (F_{WS}) as the fraction of total work (external and cardiac) due to stroke work:

$$F_{WS} = \frac{\overline{W}_S}{\overline{W}_S + W_e} \tag{i}$$

Substituting into equation (i):

$$F_{WS} = \frac{1.01}{1.01 + 0} = 1.00$$

Define the fractional external work (F_{WE}) as the fraction of total work (external and cardiac) due to external work:

$$F_{WE} = \frac{W_e}{\overline{W}_S + W_e} \tag{ii}$$

Substituting into equation (ii):

$$F_{WE} = \frac{0}{1.01 + 0} = 0$$

Define the fractional cardiac output power (F_{PO}) as the fraction of total power (external and cardiac) due to cardiac output power:

$$F_{PO} = \frac{P_0}{P_0 + P_e} \tag{iii}$$

Substituting into equation (iii):

$$F_{PO} = \frac{1.27}{1.27 + 0} = 1.00$$

Define the fractional external power (F_{PE}) as the fraction of total power (external and cardiac) due to external power:

$$F_{PE} = \frac{P_e}{P_0 + P_e} \tag{iv}$$

Substituting into equation (iv):

$$F_{PE} = \frac{0}{1.27 + 0} = 0$$

EXAMPLE 8.6

Mild Intensity Dynamic Work

A field engineer is assisting in the evaluation of a manual "crank" generator, which can be used to power a portable communication set. The engineer sits in a chair with the box-like generator at table (chest) height and holds a crank handle (at each side of the box) in each hand. That person then proceeds to apply tangential force (F_L) to rotate each crank handle (offset 180° from each other). The radius of handle rotation is 0.2 M, F_L = 20 N, and the rotation rate is 360° per 1.25 s. In order to evaluate the muscular strength and aerobic endurance required for this dynamic work, the following are measured:

$$T_R = 3.75 \text{ s}$$
$$\hat{\dot{V}}_{AIR} = .545 \text{ L/s}$$

Measured at a prior time:

$$\overline{\dot{V}}_{O_2\max} = 2.1 \text{ L/min}$$

(a) Calculate the external work and power, the customary pulmonary parameters and the aerobic index.

(b) Regarding cardiovascular assessment of this dynamic work, if the following are measured:

$$T = 0.571 \text{ s}$$
$$P_S = 125 \text{ mm Hg}$$
$$P_D = 80 \text{ mm Hg}$$

Approximate $\overline{\Delta C}$ for this mild-intensity dynamic work, and calculate the customary cardiovascular parameters.

(c) Regarding energetic assessment of this dynamic workload, recall that:

$$W_e = 25.2 \text{ J}$$
$$P_e = 20.2 \text{ J}$$

Calculate the customary energetic parameters ($\overline{W}_S$, P_O, F_{WS}, F_{WE}, F_{PO}, and F_{PE}):

SOLUTION 8.6(a)

In this dynamic task, the external work and power are rotational and performed by both upper extremities.
The displacement (Δx_L) during a 360° handle rotation:

$$\Delta x_L = 2\pi r = 1.26 \text{ M}$$

The external work (W_e) is calculated by substituting into equation 8.5(a)(i):

$$W_e = (20)(1.26) = 25.2 \text{ J}$$

External power (P_e) is calculated by substituting into equation 8.5(a)(iii):

$$P_e = \frac{(20)(1.26)}{(1.25)} = 20.2 \text{ W}$$

The customary pulmonary parameters are:
Substituting into equation (31):

$$\overline{f}_R = \frac{60}{3.75} = 16\frac{\text{cycles}}{\text{min}}$$

Substituting into equation 8.1(b)(vi):

$$\overline{\Delta V}_{AIR} = \frac{(0.545)(3.75)}{\pi} = 0.65 \text{ L}$$

Substituting into equation (33):

$$\overline{\dot{V}}_{AIR} = (.65)(16) = 10.4 \text{ L/min}$$

Substituting into equation (34b):

$$\overline{\dot{V}}_{O_2} = (.05)(10.4) = 0.52 \text{ L/min}$$

The aerobic index is calculated by substituting into equation 8.5(a)(iv):

$$\text{AI} = \frac{0.52}{2.10} = 0.248$$

SOLUTION 8.6(b)

For mild dynamic work (mild oxygen demand), we may approximate:

$$\overline{\Delta C} = .06\frac{L_{O_2}}{L}$$

Recall that equation (35) states:

$$\overline{\Delta C} = \overline{C_{AO_0}} - \overline{C_{VO_2}}$$

With dynamic work (in contrast to static work), $\overline{\Delta C}$ progressively widens as the intensity of the work increases. There is increased oxygen demand of the working skeletal muscles which extract progressively more oxygen from the inflowing blood. The arterial blood oxygen content ($\overline{C}_{AO_2}$) remains relatively constant, but the venous oxygen content ($\overline{C}_{VO_2}$) progressively decreases (with increasing muscular dynamic work). Thus, the widening of $\overline{\Delta C}$ as compared to $\overline{\Delta C}$ with resting-level dynamic work.

We can now proceed to calculate the customary cardiovascular parameters:
From equation (16):

$$\overline{f}_{HR} = \frac{60}{0.571} = 105\frac{\text{beats}}{\text{min}}$$

Substitute into equation (19):

$$\Delta P = 125 - 80 = 45 \text{ mm Hg}$$

From equation (21):

$$\overline{\Delta P} = 80 + \frac{45}{3} = 95 \text{ mm Hg}$$

From equation (37):

$$\bar{\dot{V}}_{CO} = \frac{0.52}{.06} = 8.67 \text{ L/min}$$

From equation (39b):

$$\overline{\Delta V_s} = \frac{8.67}{105} = .083 \text{ L}$$

From equation 8.1(c)(ii):

$$\bar{\dot{V}}'_{CO} = (16.67)(8.67) = 145 \frac{\text{ml}}{\text{s}}$$

From equation 8.1(c)(iii):

$$R = \frac{95}{145} = 0.655 \text{ PRU}$$

SOLUTION 8.6(c)

From equation 8.1(c)(iv):

$$\overline{W}_s = (.1333)(95)(.083) = 1.05 \text{ J}$$

From equation 8.1(c)(v):

$$P_0 = (.00222)(95)(8.67) = 1.83 \text{ W}$$

From equation 8.5(c)(i):

$$F_{WS} = \frac{1.05}{1.05 + 25.2} = 0.04$$

From equation 8.5(c)(ii):

$$F_{WE} = \frac{25.2}{1.05 + 25.2} = 0.96$$

From equation 8.5(c)(iii):

$$F_{PO} = \frac{1.83}{1.83 + 20.2} = 0.08$$

From equation 8.5(c)(iv):

$$F_{PE} = \frac{20.2}{1.83 + 20.2} = 0.92$$

EXAMPLE 8.7

Moderate Intensity Dynamic Work

A firm is evaluating the workload for a security guard in a high-rise industrial complex. It is anticipated that part of the guard's routine patrol will involve climbing up and down flights of stairs between clock timer check-in boxes. The clock-in time between adjacent boxes on the routine is part of the workload evaluation.

An experienced security guard (mass = 70 kg) is at the HFE testing laboratory where a two-step stair rise is positioned in front of the standing guard. At $t = 0$, the guard does a double step-up to the first step (rise = 0.2 M) followed by a second double step up to the second step (rise = 0.2 M). The guard then does a double-step backward down to the first step, followed by a double step down to floor level. The elapsed time for the climb phase is 2.5 s and for the descent phase is also 2.5 s. In order to evaluate the required muscular strength, and aerobic endurance for this dynamic work, the following is measured:

$$T_R = 2.4 \text{ s}$$
$$\hat{\dot{V}}_{AIR} = 1.047 \text{ L/s}$$

Measured at a prior time:

$$\overline{\dot{V}}_{O_2\max} = 2.4 \text{ L/min}$$

(a) Calculate the external work and power, the customary pulmonary parameters and the aerobic index.

(b) Regarding cardiovascular assessment of this dynamic work, if the following are measured:

$$T = 0.375 \text{ s}$$
$$P_S = 151 \text{ mm Hg}$$
$$P_D = 85 \text{ mm Hg}$$

Approximate $\overline{\Delta C}$ for moderate (aerobic-type) dynamic work, and calculate the customary *cardiovascular parameters:*

(c) Regarding energetic assessment of this dynamic workload, recall that:

$$W_e = 275 \text{ J}$$
$$P_e = 110 \text{ W}$$

Calculate the customary energetic parameters.

SOLUTION 8.7(a)

In this dynamic task, the external work and power are calculated as follows: Positive work is performed by raising the weight of the body upward a vertical distance of two stair steps:

$$\Delta x_L = (2)(.2M) = .4 \text{ M}$$

$$F_L = W_B = (70 \text{ kg})\left(9.81 \frac{\text{M}}{\text{s}^2}\right) = 687 \text{ N}$$

Substituting into equation 8.5(a)(i):

$$W_e = (687)(.4) = 275 \text{ J}$$

The same magnitude of negative work also occurs when lowering the weight of the body back to ground level.
Positive power is the rate of doing positive work; so substituting into equation 8.5(a)(iii):

$$P_e = \frac{(687)(0.4)}{(2.5)} = 110 \text{ W}$$

The customary pulmonary parameters are:
Substituting into equation (31):

$$\bar{f}_R = \frac{60}{2.4} = 25 \frac{\text{cycles}}{\text{min}}$$

Substituting into equation 8.1(b)(vi):

$$\overline{\Delta V}_{AIR} = \frac{(1.047)(2.4)}{\pi} = 0.80 \text{ L}$$

Substituting into equation (33):

$$\bar{\dot{V}}_{AIR} = (.80)(25) = 20.0 \text{ L/min}$$

Substituting into equation (34b):

$$\bar{\dot{V}}_{O_2} = (.05)(20.0) = 1.0 \text{ L/min}$$

The aerobic index is calculated by substituting into equation 8.5(a)(iv):

$$\text{AI} = \frac{1.0}{2.40} = 0.417$$

SOLUTION 8.7(b)

For moderate dynamic work (moderate oxygen demand), we may approximate:

$$\overline{\Delta C} = .07 \frac{\text{L}_{O_2}}{\text{L}}$$

Recall that $\overline{\Delta C}$ progressively widens (increasing) as the intensity of the dynamic work increases [see Solution 8.6(b)].
The customary cardiovascular parameters are then calculated as follows:
From equation (16):

$$\bar{f}_{HR} = \frac{60}{0.375} = 160 \frac{\text{beats}}{\text{min}}$$

From equation (19):

$$\Delta P = 151 - 85 = 66 \text{ mm Hg}$$

From equation (21):

$$\overline{\Delta P} = 85 + \frac{66}{3} = 107 \text{ mm Hg}$$

From equation (37):

$$\overline{\dot{V}}_{CO} = \frac{1.00}{0.07} = 14.3 \text{ L/min}$$

From equation (39b):

$$\overline{\Delta V_s} = \frac{14.3}{160} = .089 \text{ L}$$

From equation 8.1(c)(ii):

$$\overline{\dot{V}}'_{CO} = (16.67)(14.3) = 238 \frac{\text{ml}}{\text{s}}$$

From equation 8.1(c)(iii):

$$R = \frac{107}{238} = 0.450 \text{ PRU}$$

SOLUTION 8.7(c)

From equation 8.1(c)(iv):

$$\overline{W}_s = (.1333)(107)(.089) = 1.27 \text{ J}$$

From equation 8.1(c)(v):

$$P_0 = (.00222)(107)(14.3) = 3.40 \text{ W}$$

From equation 8.5(c)(i):

$$F_{WS} = \frac{1.27}{1.27 + 275} = 0.005$$

From equation 8.5(c)(ii):

$$F_{WE} = \frac{275}{1.27 + 275} = 0.995$$

From equation 8.5(c)(iii):

$$F_{PO} = \frac{3.4}{3.40 + 110} = 0.03$$

From equation 8.5(c)(iv):

$$F_{PE} = \frac{110}{3.40 + 110} = 0.97$$

EXAMPLE 8.8

High Intensity Dynamic Work

The HFE Laboratory is evaluating United States Forest Rangers as appropriate candidates to join special search-and-rescue teams for rugged and isolated mountainous terrain.

An experimental forest ranger (m_B = 65 kg) is at the HFE laboratory where the ranger is standing on a treadmill inclined at an angle (θ) of 16°. The ranger is wearing a fully equipped backpack (m_P = 10 kg), and at t = 0, the ranger starts running at a treadmill speed (V_T) of 3.5 kph. In order to evaluate the required muscular strength, and aerobic endurance for this dynamic work, the following is measured:

$$T_R = 1.50\text{ s}$$
$$\hat{\dot{V}}_{AIR} = 2.10\text{ L/s}$$

Measured at a prior time:

$$\overline{\dot{V}}_{O_2\max} = 2.5\text{ L/min}$$
$$\Delta x_s\text{ (stride length at 3.5 kph)} = 1.20\text{ M}$$

(a) Calculate the external work and power, the customary pulmonary parameters and the aerobic index.

(b) Regarding cardiovascular assessment of this dynamic work, if the following are measured:

$$T = 0.30\text{ s}$$
$$P_S = 180\text{ mm Hg}$$
$$P_D = 90\text{ mm Hg}$$

Approximate $\overline{\Delta C}$ for heavy (aerobic-type) dynamic work, and calculate the customary cardiovascular parameters.

(c) Regarding energetic assessment of this dynamic workload, recall that:

$$W_e = 211\text{ J}$$
$$P_e = 205\text{ W}$$

Calculate the customary energetic parameters.

SOLUTION 8.8(a)

In this dynamic task, the external work and power are translational and performed by both lower extremities.

Positive work is performed by raising the combined weight of the body and backpack a vertical distance upward (Δx_y) with each stride length (Δx_L). Stride length is defined as the distance (on a treadmill) between that point at which the toes of a foot leave the treadmill [to start a stride] and that point at which the toes of the *same* foot again leave the treadmill (just before the next stride is initiated with that *same* foot).

The stride length (Δx_S) is the hypotenuse of a right triangle since the treadmill walking surface is inclined at an angle, θ = 16°:

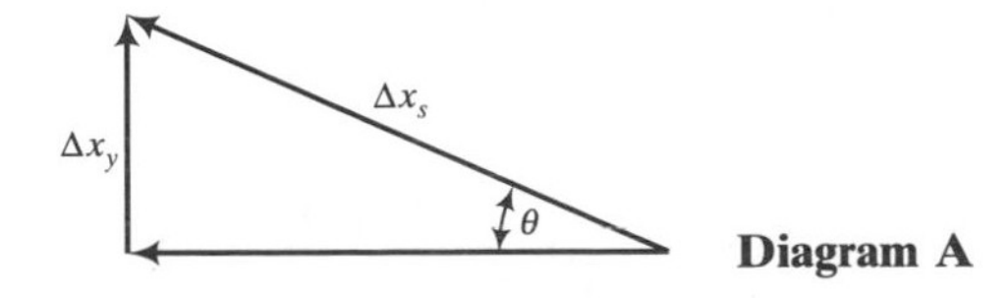

Diagram A

The stride velocity (V_T) is the hypotenuse velocity for the same reasons:

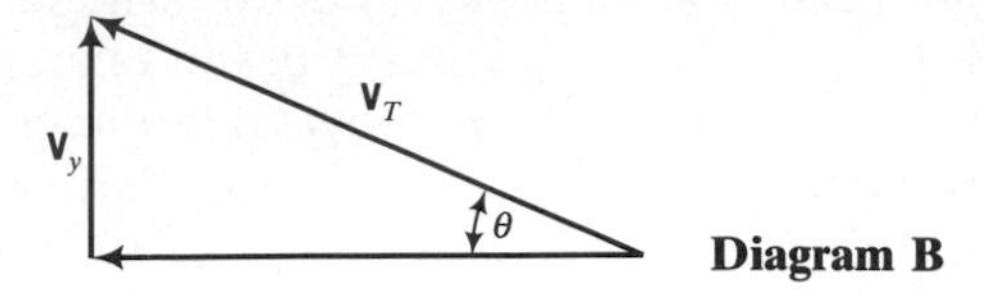

Diagram B

First, calculate the force of the total load:

$$F_L = (m_B + m_P) \cdot g$$
$$F_L = (65 + 10)(9.81) = 736 \text{ N}$$

Next, calculate the vertical distance upward with each stride length:

$$\Delta x_y = \Delta x_S \cdot \tan(16°)$$
$$\Delta x_y = (1.2)(.287) = 0.344 \text{ M} \qquad \text{(i)}$$

Substituting into equation 8.5(a)(i):

$$W_e = (20)(.287) = 211 \text{ J}$$

Positive power is performed by raising the total load (body and backpack) at a constant vertical rate upward (V_y). This is calculated by first considering a stride occurs over a finite time interval, Δt, so that equation (i) can be expressed in terms of velocity:

$$\frac{\Delta x_y}{\Delta t} = \frac{\Delta x_S}{\Delta t} \cdot \tan(16°\text{C}) \qquad \text{(ii)}$$

Since we know the constant velocity of the treadmill (V_T = 3.5 kph), then:

$$\frac{\Delta x_S}{\Delta t} = V_T$$

and

$$\frac{\Delta x_y}{\Delta t} = V_y$$

so that we can rewrite equation (ii) as:

$$V_y = V_T \cdot \tan(16°) \qquad \text{(iii)}$$

Positive external power is then:

$$P_e = F_L \cdot V_y \qquad \text{(iv)}$$

So substituting equation (iii) into equation (iv):

$$P_e = F_L \cdot V_T \cdot \tan(16°) \qquad \text{(v)}$$

Defining V_T:

$$V_T = \left(3.5 \frac{\text{km}}{\text{hr}}\right)\left(\frac{1000 \text{ M}}{\text{km}}\right)\left(\frac{\text{hr}}{3600 \text{ s}}\right)$$
$$V_T = .972 \text{ M/s}$$

And substituting into equation (v):

$$P_e = (736)(.972)(.287) = 205 \text{ W}$$

The customary pulmonary parameters are:
Substituting into equation (31):

$$\overline{f}_R = \frac{60}{1.5} = 40\,\frac{\text{cycles}}{\text{min}}$$

Substituting into equation 8.1(b)(vi):

$$\overline{\Delta V}_{AIR} = \frac{(2.10)(1.50)}{\pi} = 1.00\,\text{L}$$

Substituting into equation (33):

$$\overline{\dot{V}}_{AIR} = (1.00)(40) = 40.0\,\text{L/min}$$

Substituting into equation (34b):

$$\overline{\dot{V}}_{O_2} = (.05)(40.0) = 2.0\,\text{L/min}$$

The aerobic index is calculated by substituting into equation 8.5(a)(iv):

$$\text{AI} = \frac{2.0}{2.50} = 0.80$$

SOLUTION 8.8(b)

For heavy dynamic work (high oxygen demand), we may approximate:

$$\overline{\Delta C} = .08\,\frac{\text{L}_{O_2}}{\text{L}}$$

$\overline{\Delta C}$ continues to progressively increase (widen) as dynamic work intensity increases the oxygen demand [per Solution 8.6(b)]. However, $\overline{\Delta C}$ plateaus at maximal dynamic work effort, so that no further increase will occur beyond this upper limit.
The customary cardiovascular parameters are then calculated as follows:
From equation (16):

$$\overline{f}_{HR} = \frac{60}{0.3} = 200\,\frac{\text{beats}}{\text{min}}$$

From equation (19):

$$\Delta P = 180 - 90 = 90\,\text{mm Hg}$$

From equation (21):

$$\overline{\Delta P} = 90 + \frac{90}{3} = 120\,\text{mm Hg}$$

From equation (37):

$$\overline{\dot{V}}_{CO} = \frac{2.0}{0.08} = 25.0\,\text{L/min}$$

From equation (39b):

$$\overline{\Delta V}_s = \frac{25.0}{200} = .125\,\text{L}$$

From equation 8.1(c)(ii):

$$\overline{\dot{V}}'_{CO} = (16.67)(25.0) = 417\,\frac{\text{ml}}{\text{s}}$$

From equation 8.1(c)(iii):

$$R = \frac{120}{417} = 0.288 \text{ PRU}$$

SOLUTION 8.8(c)

From equation 8.1(c)(iv):

$$\overline{W}_s = (.1333)(120)(.125) = 2.00 \text{ J}$$

From equation 8.1(c)(v):

$$P_0 = (.00222)(120)(25.0) = 6.67 \text{ W}$$

From equation 8.5(c)(i):

$$F_{WS} = \frac{2.00}{2.00 + 2.11} = 0.01$$

From equation 8.5(c)(ii):

$$F_{WE} = \frac{211}{2.00 + 211} = 0.99$$

From equation 8.5(c)(iii):

$$F_{PO} = \frac{6.67}{6.67 + 205} = 0.03$$

From equation 8.5(c)(iv):

$$F_{PE} = \frac{205}{6.67 + 205} = 0.97$$

FURTHER INFORMATION

P. Astrand and K. Rodahl: *Textbook of Work Physiology*. 3rd Edition McGraw-Hill. New York. 1986.

Eastman Kodak Company: *Ergonomic Design for People at Work: Volume 2*. Van Nostrand Reinhold. New York. 1986.

L.A. Geddes and L.E. Baker: *Principles of Applied Biomedical Instrumentation*. 3rd Edition John Wiley. New York. 1989.

A.T. Johnson: *Biomechanics and Exercise Physiology*. John Wiley. New York. 1991.

C.A. Phillips: *Functional Electrical Rehabilitation*. Springer-Verlag. New York. 1991.

PROBLEMS

8.1. a. Write a computational program (e.g., MATLAB) that will solve the system of equations used to analyze *static* workload.

b. Check your computational program by entering the input (given) data of Example 8.2, 8.3, or 8.4. Then compare your computational program result with that specific Example result.

8.2. a. Write a computational program (e.g., MATLAB) that will solve the system of equations used to analyze *dynamic* workload.

b. Check your computational program by entering the input (given) data of Example 8.6, 8.7, or 8.8. Then compare your computational program result with that specific Example result.

8.3. A manufacturer is evaluating a prototype model of a cross-country-ski exercise machine in terms of the human workload that is developed.

A person of body mass (m_b) equal to 85 kg simultaneously slides one ski forward (a stride length of 0.6 M) as they *simultaneously* slide one ski backward (a stride length of 0.6 M). The above occurs over a period of one second and repeats in a continuously cyclical manner. The task is performed on a ski track tilted upward at an angle (θ) of 10° to the horizontal. The coefficient of sliding friction for each ski against the ski track is 0.15.

The HFE laboratory is evaluating this dynamic work. The following are measured:

$$T_R = 1.20 \text{ s}$$
$$\hat{\dot{V}}_{AIR} = 2.60 \text{ L/s}$$

Measured at a prior time:

$$\overline{\dot{V}}_{O_2 max} = 3.0 \text{ L/min}$$

a. Calculate the external work and power, the customary pulmonary parameters, and the aerobic index.

b. Regarding cardiovascular assessment of this dynamic work, if the following are measured:

$$T = 0.28 \text{ s}$$
$$P_S = 180 \text{ mm Hg}$$
$$P_D = 90 \text{ mm Hg}$$

Then by approximating $\overline{\Delta C}$ for heavy (aerobic-type) dynamic work, calculate the customary cardiovascular parameters.

c. Regarding energetic assessment of this dynamic work, calculate the customary energetic parameters.

8.4. A parking attendant is regularly standing up to stamp and hand out tickets to the oncoming cars from one side of the booth. From the other side of the booth for the outgoing traffic, he receives and stamps the ticket plus uses the cash register to determine and store the payments.

The following ergonomic data is recorded:

F_L (the force of the external load) in this case a single ticket = 0 when compared to the arm weight.

Δx_L (displacement of the external load) = 1.2 M

f_L (the period of a single work event before beginning of following event) = 4 min/work event or 0.25 work event/min.

The HFE laboratory is evaluating this dynamic work. The following are measured.

$$T_R = 6 \text{ s}$$
$$\hat{\dot{V}}_{AIR} = 0.295 \text{ L/s}$$

Measured at a prior time:

$$\overline{\dot{V}}_{O_2\max} = 2.05 \text{ L/min}$$

a. Calculate the external work and power, the customary pulmonary parameters, and the aerobic index.

b. Regarding cardiovascular assessment of this dynamic work, if the following are measured:

$$T = 0.83 \text{ s}$$
$$P_S = 120 \text{ mm Hg}$$
$$P_D = 80 \text{ mm Hg}$$

Then by approximating $\overline{\Delta C}$ for minimal (aerobic-type) dynamic work, calculate the customary cardiovascular parameters.

c. Regarding energetic assessment of this dynamic work, calculate the customary energetic parameters.

8.5. A person must assume a continuously crouched position in order to work on the underside of an F16 aircraft (on a runway in the cold early morning hours). This task is being simulated in the HFE laboratory as the person presses the foot down on a force plate. The applied force at the ball of the right foot (F_T) is 100 N, and the ankle angle is 80°. This position is stationary. Since it is a very cold day, the muscle temperature (T_m) is 20°C. Prior to the test, measures were taken on the isometric muscle force for brief maximal voluntary contraction.

These lab measurements are as follows:

$$F_{MVC}\text{ (lab)} = 500 \text{ N (at } T_m = 35°\text{C)}$$
$$A_{MVC} = 2.0 \text{ mV}$$
$$f_{CMVC} = 150 \text{ Hz}$$

At $t = 0$ surface electrodes continually record the EMG of the right calf muscle during this static work. The following data are then noted.

$$A_i = 1.0 \text{ mV @ } T_i = 1.0 \text{ min}$$
$$f_{ci} = 131 \text{ Hz @ } T_i = 1.0 \text{ min}$$

a. After 1 min of this foot press, how much longer can the technician maintain F_T before fatigue sets in?

b. Regarding pulmonary assessment for this static workload, if the following is measured:

$$T_R = 2.0 \text{ s}$$
$$\hat{\dot{V}}_{AIR} = 0.50 \text{ L/s}$$

Approximate $\overline{\Delta C}$ for static-type work, and calculate the customary pulmonary parameters.

c. Regarding cardioenergetic assessment of this static work, if the following are measured:

$T = 0.5$ s
$P_S = 190$ mm Hg
$P_D = 135$ mm Hg

Calculate the customary cardioenergetic parameters.

8.6. The HFE laboratory is evaluating a firefighter while performing a vertical ladder climbing task. A volunteer firefighter ($m_b = 70$ kg) is at the HFE laboratory where the firefighter is climbing a "vertical ladder treadmill," which is at an angle of 90° to the horizontal ground. The ladder step-to-step vertical distance is 0.34 M. The fireman is wearing a fully equipped uniform ($m_p = 11$ kg), and at $t = 0$, the fireman starts climbing the "vertical ladder treadmill" now moving at a speed, $V_T = 0.88$ kph. When the fireman is in dynamic equilibrium with the treadmill speed, his body center of mass is at a constant height above the horizontal ground.

In order to evaluate the human operator workload for this dynamic work, the following are measured:

$T_R = 1.4$ s
$\hat{V}_{AIR} = 2.2$ L/s

Measured at a prior time:

$\overline{\dot{V}}_{O_2\max} = 2.4$ L/min

a. Calculate the external work and power, the customary pulmonary parameters, and the aerobic index.

b. Regarding cardiovascular assessment of this dynamic work, if the following are measured:

$T = 0.26$ s
$P_S = 175$ mm Hg
$P_D = 95$ mm Hg

Approximate $\overline{\Delta C}$ for heavy (aerobic-type) dynamic work, and calculate the customary cardiovascular parameters.

c. Regarding energetic assessment of this dynamic work, calculate the customary energetic parameters.

8.7. An apprentice mechanic continuously grips the handle of a light weight tennis racket on the webbing of which is mounted a circular-shaped angle mirror. It is common that the assistant will grip and hold this device in a prolonged static position to help a master mechanic visualize two different areas of a jet engine at the same time.

a. Evaluate the muscle strength and endurance for this static work, if the following are measured:

Tennis racket grip measurement (cylinder):

Height = 25 cm
Diameter = 5 cm
$T_{AMB} = 25°C$
F_T (continuous grip force) = 50 N

Also measured (at a prior time):

F_{MVC} [on a hand-grip dynamometer with an optimal grip diameter of 5.5 cm (at 25°C ambient temperature)] = 520 N

b. Calculate pulmonary parameters for the following measurements:

$T_R = 4.0$ s
$\hat{\dot{V}}_{AIR} = 0.400$ L/s

Recall that you must approximate $\overline{\Delta C}$ for static-type work.

c. Calculate the customary cardioenergetic parameters for this static work, if the following are measured:

$T = 0.680$ s
$P_S = 150$ mm Hg
$P_D = 90$ mm Hg

8.8. A manufacturer is evaluating its latest version of an exercise rowing machine with respect to the human operator workload when using the rowing machine.

1. Stroke cycle (upper extremity action and lower extremity action occur simultaneously):

Body Region	Action	F	Δx	Δt
Both upper extremities	Pull	90 N	0.8 M	1.5 s
Both lower extremities	Push	90 N	0.8 M	1.5 s

2. Reset cycle (upper extremity action and lower extremity action occur simultaneously):

Body Region	Action	F	Δx	Δt
Both upper extremities	Push	65 N	0.8 M	1.25 s
Both lower extremities	Pull	65 N	0.8 M	1.25 s

a. Calculate the external work and power, the pulmonary parameters, and the aerobic index for the following measurements:

$T_R = 3.5$ s
$\hat{\dot{V}}_{AIR} = .0895$ L/s

Measured at a prior time:

$\hat{\dot{V}}_{O_2 max} = 1.9$ L/s

b. Calculate the cardiovascular parameters for the following measurements:

$T = .420$ s
$P_S = 150$ mm Hg
$P_D = 80$ mm Hg

It will be necessary to approximate $\overline{\Delta C}$ for moderate (aerobic-type) dynamic work.

c. Calculate the energetic parameters.

8.9. A rock climber is being evaluated on an indoor climbing wall for a position on a mountain climbing team. The climber is instructed "free-climb" to a specific spot, hammer in a piton, and attach a safety line. The climber is in good physical shape but is nervous. And as the climber reaches the spot she slips but does not fall, causing a rise in her pulse and blood pressure. The climber jams a piton into a crack, holds onto the rocks with her left hand and starts to hammer in the piton with her right hand. Her left hand grasps the rock face well and exerts a gripping force (F_T) of 125 N. Her right hand hammers for 15 s and stops. Then her right arm is relaxed and her blood pressure is taken.

Lab measurements are as follows:

$$F_{MVC}\,(\text{lab}) = 500\ \text{N}$$
$$A_{MVC} = 1.4\ \text{mV}$$
$$f_{CMVC} = 182\ \text{Hz}$$

While hammering in the piton the left hand exerts its full grip and the EMG of the left finger flexor muscles is recorded during this static work.

$$A_i = 0.9\ \text{mV} \ @\ T_i = 0.25\ \text{min}$$
$$f_{ci} = 149\ \text{Hz} \ @\ T_i = 0.25\ \text{min}$$

a. After the 15 seconds of the left hand gripping, how much longer could the rock climber maintain F_T before fatigue sets in and the climber must release or lower the gripping tension?

b. Regarding pulmonary assessment for this static workload, if the following is measured:

$$T_R = 1.128\ \text{s}$$
$$\hat{V}_{AIR} = 0.512\ \text{L/s}$$

Then approximate $\overline{\Delta C}$ for heavy (dynamic-type) work and calculate the customary pulmonary parameters.

c. Regarding cardioenergetic assessment of this static work, if the following are measured:

$$T_R = 0.327\ \text{s}$$
$$P_S = 190\ \text{mm Hg}$$
$$P_D = 142\ \text{mm Hg}$$

Calculate the customary energetic parameters.

8.10. A person is being evaluated while operating a "paddle-boat," which the manufacturer hopes will replace the commonly used row boat. The person is sitting in the boat and each hand is gripping a handle (connected to a circular wheel in front of the person's chest). As the person rotates the handles, the wheel turns and is so connected to a central shaft that a propeller at the rear of the boat rotates at high speed. This task is currently being performed in an above-ground pool (where the boat is tethered in the center) at the HFE laboratory annex.

The person applies a tangential force to each crank handle (F_T = 22 N) and the radius of rotation (hands-to-wheel center distance) is 0.22 M. Finally, the rotation rate is 360° revolution every 1.22 seconds.

The following are measured:

$$T_R = 3.4 \text{ s}$$
$$\hat{\dot{V}}_{AIR} = 0.592 \text{ L/s}$$

Previously measured:

$$\hat{\dot{V}}_{O_2 max} = 2.7 \text{ L/min}$$

a. Calculate the external work and power, the pulmonary parameters, and the aerobic index.

b. Regarding cardiovascular assessment for this dynamic workload, if the following are measured:

$$T = 0.530 \text{ s}$$
$$P_S = 141 \text{ mm Hg}$$
$$P_D = 87 \text{ mm Hg}$$

Approximate $\overline{\Delta C}$ for moderate (dynamic-type) work and calculate the customary cardiovascular parameters.(dynamic-type) work and calculate the customary pulmonary parameters.

c. Regarding cardioenergetic assessment of this dynamic work, calculate the energetic parameters.

Part III-B

Human Control Engineering

Chapters 9 and 10 that follow represent an area of HFE practice that is identified as Human Control Engineering. The following is a general introduction to this practiced area.

Control engineering (more precisely referred to as control systems engineering) is the engineering practice of the analysis and design of control systems. The student has already been introduced to the definition and concept of a "system" in Section 1.1 of this text. A *control* system is an interrelated arrangement of physical elements that are organized so as to command, direct or regulate either itself or another system.

Control systems engineering is practiced by every engineering discipline and includes industrial and systems engineers, mechanical and manufacturing engineers as well as aeronautical and aerospace engineers. In many of these engineering applications, the term "control system" and "automatic control system" are used interchangeably since the nature and arrangement of the various physical elements are such that the control system behaves as an automaton.

Human control engineering (more precisely referred to as human control systems engineering) is the engineering practice of the analysis and design of human control systems. It focuses upon human capabilities and limitations with respect to the appropriate design of human-technological control systems. Human control engineering is practiced by human factors engineers whether they are industrial, mechanical, aeronautical or systems engineers by discipline. For such engineers, the control system is *not* an automatic control system, but rather a human control system. The nature and arrangement of the various subsystems include a human subsystem in addition to a technological subsystem. The resultant control system is not automatic but rather has human characteristics.

The individual chapters that now follow represent specific areas of human control engineering. Neuromuscular control systems (Chapter 9) represent the *combination* of neuromuscular biophysics *and* computational simulation as applied to the analysis and design of human control systems. Human operator control (Chapter 10) represents the *combination* of information theory *and* control theory as applied to the analysis and design of human control systems.

Neuromuscular Control Systems

Human operator control (both static and dynamic) is of central importance to the human factors engineer because it addresses an individual's ability to perform in a large variety of operational situations. This chapter (Chapter 9) and the companion chapter (Chapter 10) will collectively address human operator control. Chapter 9 will introduce the student to the necessary anatomical, physiological, biophysical, and biomathematical relationships that are the foundations of human neuromuscular performance (whether static or dynamic). This chapter will then specifically focus upon computer simulation of the human neuromuscular control system when performing various tasks that involve: (1) static (isometric) muscle force control; and (2) dynamic muscle velocity control.

The first part of this chapter (Section 9.1) reviews neuromuscular control fundamentals. The material in Section 9.1 is applicable to subsequent development of the neuromuscular control systems of Sections 9.2, 9.3, 9.4, and 10.1.

9.1 NEUROMUSCULAR CONTROL FUNDAMENTALS

Figure 9.1 demonstrates the general schematic flow for a human operator control system. Analysis of Figure 9.1 begins by noting that there are three essential subsystems: the human operator subsystem, the technological subsystem, and the environmental subsystem. The human operator subsystem may be represented by a central nervous processor (CNP) which has bi-directional interaction with the neuromuscular apparatus (NMA). The human operator output is then represented by various types of musculoskeletal movement. The preceding is described more fully in Figure 9.2 The human operator output represents input to the technological subsystem. This occurs by means of a human-system interface (HSI). The technological subsystem itself is represented by an HSI that has bi-directional interaction with the technological plant. The output of the technological subsystem is a direct result of the technological plant operation. Note that the human operator and the resultant operator output do not directly interact with the technological plant itself.

Regarding the human operator subsystem, it is apparent that the operator is driven by input. Figure 9.1 indicates that there are two sources for this human operator input. One source is the technological subsystem output. In this case a loop is closed between the human operator subsystem and the technological subsystem. In those cases in which the human operator subsystem does not receive input from the technological plant output, the human-technological interaction may be considered as open-loop. A second source of human operator input is by means of the environment external to the human operator and the technological system. This external environment represents an environmental subsystem in that it is composed

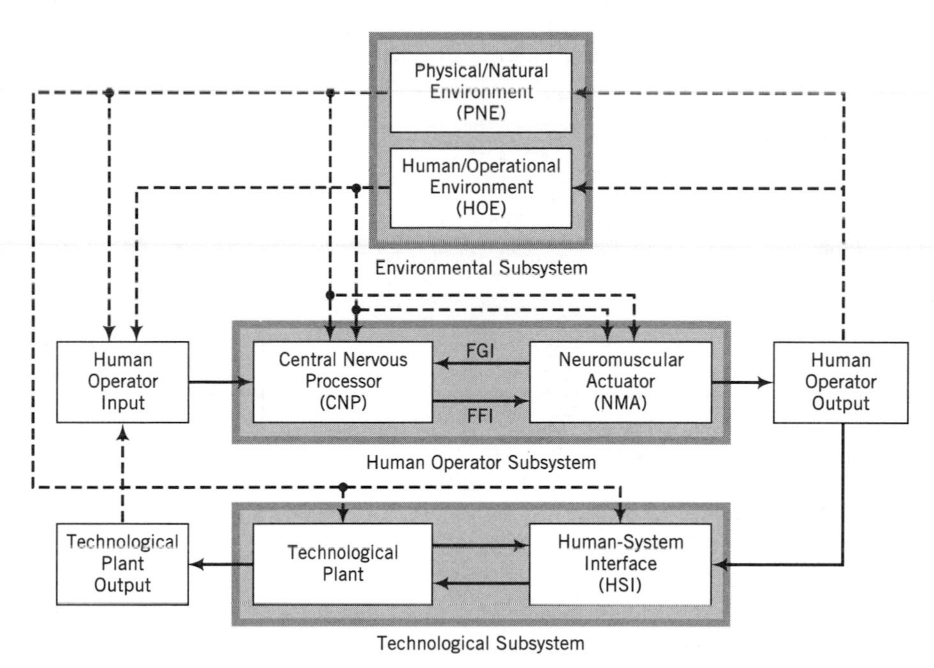

Figure 9.1 General schematic flow for a human operator control system.

of a number of individual elements (see Figure 9.3). The human operator output (which interacts with a technological subsystem) occurs in an external environment and so directly interacts with elements of the environmental subsystem. There are two subsequent outcomes of this interaction. The first outcome is that the human operator is provided with additional inputs. This may represent a form of closed-loop feedback so that human operator control is affected. Alternatively, elements of the environmental subsystem may simply be altered so that the human operator receives inputs that represent alteration of the external environment. In either case, the human operator inputs may directly affect activity of the central nervous processor (CNP), may directly affect the neuromuscular apparatus (NMA), or may simultaneously affect both the CNP and the NMA. This environmental subsystem, human operator input, and CNP/NMA interaction are described more fully with respect to Figure 9.3.

The environmental subsystem interacts with the technological subsystem in two ways. In one case there is indirect interaction because the environmental subsystem and human operator subsystem interaction may alter the human operator output, which in turn represents altered input to the technological system. A second output

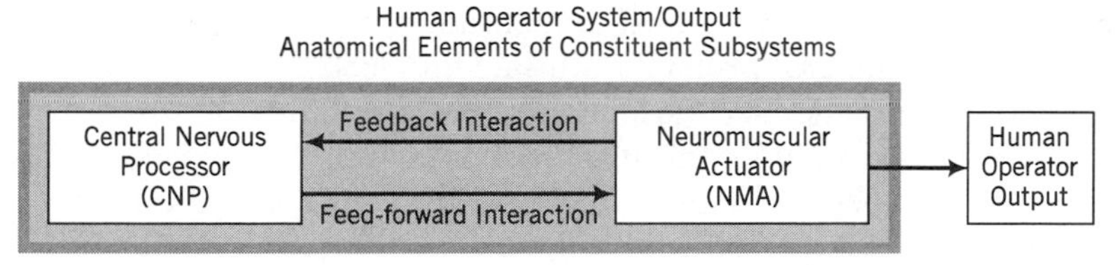

Figure 9.2 Human operator system/output: anatomical elements of constituent subsystems.

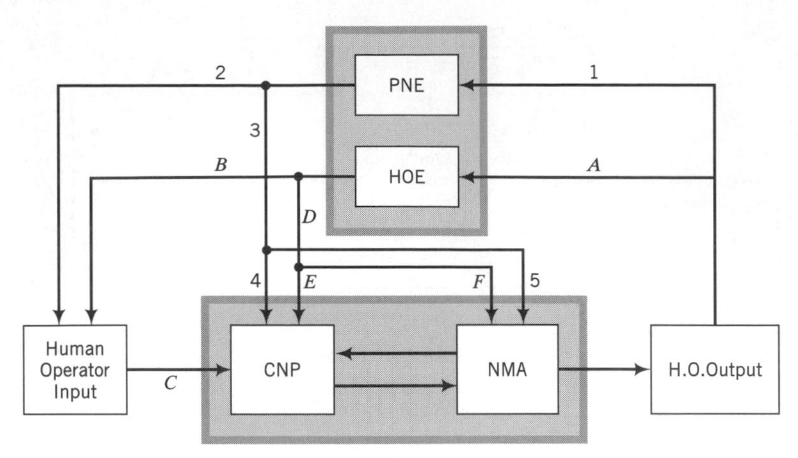

Figure 9.3 External environmental subsystem as related to human operator input and CNP/NMA.

from the environmental subsystem is represented by a direct interaction with the technological subsystem. As a result, the elements of the external environment may directly act upon the human system interface (HSI), or the technological plant, or simultaneously interact with both elements of the technological subsystem. When the human operator output interacts with the environmental subsystem and there is then subsequent interaction of the external environmental with the technological subsystem, another pathway for closed-loop control is developed. The human operator output can also be uncoupled from the environmental subsystem in which case there is no interaction. In this situation the environmental subsystem will interact with both the human operator subsystem and the technological subsystem in an open-loop manner. A central tenet of human factors engineering is that the human operator subsystem, technological subsystem, and environmental subsystem (depicted in Figure 9.1) may be described and modeled mathematically, and that the resultant human operator control model may subsequently be analyzed using the methods of classical control theory.

a. Anatomical System Elements

The human factors engineer will understand that the human operator subsystem represents a set of interacting anatomical elements. The anatomical elements (sensory organs, nerves, muscles, and bones) perform physiological functions which operate according to biophysical laws and may be described with biophysical equations. The resultant engineering control system is basically a mathematical model in a form in which controls system theory may be applied. In this context, the development of the biomathematical model (and the ultimate engineering control system) follows the general principles outlined in Section 2.2. This process begins by defining and understanding those anatomical elements that represent essential features of the human operator control system. Figure 9.2 depicts a subsection of Figure 9.1 that includes the human operator subsystem and the human operator output block. Since we have isolated the human operator for more detailed consideration, Figure 9.2 denotes the human operator as a system itself. It is then apparent that the human operator control system involves four subsystems and a set of output parameters (per Figure 9.2). These subsystems are the central nervous processor, the feed-forward interaction, the feedback interaction, and the neuromuscular actuator. Each of these subsystems consists of between one and four anatomical elements (Table 9.1). Let us now consider this human operator control model at the level of its constituent anatomical elements.

The central nervous processor subsystem may be conceptualized as having four elements (Table 9.1). These elements have an approximately serial arrangement proceeding from the outside to the inside of the body. The cranial sensory element is a sensory array consisting of two subelements. The special sensors are the sensory organs (receptors) of vision, hearing, taste, and smell. The other subelement is the cranial nerves themselves, which conduct the special sensory information to the brain element and brain stem element for subsequent processing. The cranial sensory array accepts human operator data input via these special sensors. This input may be either direct (open-loop) environmental input, feedback interaction (via the environment) from the human operator output, or feedback interaction (via the technological system) from the technological output. Note that with respect to the cranial sensory element, the feedback interaction is external to the human

Table 9.1 The Anatomical System Elements

A. *CNP (Central Nervous Processor) Elements*
   1. Cranial Sensory Array:
      a. Special Sensors
      b. Cranial Nerves
   2. Brain/Midbrain:
      Central (e.g. cortical) nerves
   3. Brainstem/Cerebellum:
      a. Central nerve synapses
      b. Cerebellar nerves
   4. Spinal cord:
      Central (interneuron) nerves [motor efferent]

B. *Feedback Interaction Elements*
   1. Cutaneous Sensory Array:
      a. Specialized Nerve Endings
      b. Peripheral Sensory Nerves
   2. Deep Tissue Sensory Array:
      a. Specialized Muscle, Tendon and Joint Sensors
      b. Peripheral Sensory Nerves
   3. Spinal cord:
      Central (interneuron) nerves [sensory afferent]

C. *Feed-forward Interaction Element*
   1. Central (interneuron) nerve synapse at spinal (motor) nerve root.

D. *NMA (Neuromuscular Actuator) Elements*
   1. Neuromuscular Apparatus
      a. Peripheral motor nerves
      b. Skeletal muscles
   2. Musculoskeletal Apparatus
      a. Tendons and ligaments
      b. Skeletal bones

operator, since this system does not process feedback interaction that is internal to the human operator. Subsequently, the brain/mid-brain element performs a large amount of higher level data processing. The basic structural element is the central (e.g., cortical) nerve. Such central processing includes functions such as cognition, activation, and perception as examples. The brain stem/cerebellum element continues sequential data processing. This element consists of two anatomical subelements: the central nerve synapses and the cerebellar nerves. Recall that a nerve synapsis is an electrochemical junction that allows neuroelectrical activity to propagate between two adjacent nerves. Some example functions of the brain stem/cerebellum element are data prioritization, integration, and neuromuscular coordination. Sequentially, the spinal cord element is the fourth of the central nervous processor subsystems. The essential anatomical elements are the central nerves referred to as interneurons. These are second order neurons that compose the bulk of the spinal cord, and their physiological function is to propagate electrical activity down the spinal cord between the brain stem and the spinal nerve roots. This is referred to as *motor efferent* activity because the direction of data flow is from the central nervous processor to the neuromuscular actuator.

The feedback interaction subsystem allows for the flow of feedback information that is internal to the human operator, as opposed to external. The feedback interaction subsystem consists of three anatomical elements (Table 9.1). A cutaneous sensory element detects and processes the sensations of touch, pressure, pain, vibration, and temperature, which occur at or below the skin surface. This cutaneous sensory array has two essential anatomical subelements. Specialized nerve endings are specific for a particular cutaneous sensation. Peripheral sensory nerves are the second anatomical subelement that electrically conduct these specific sensations (at or below the skin surface) back to the spinal cord itself (at the spinal nerve root). The second anatomical element is the deep tissue sensory array. This deep tissue element detects position, motion, and force at the deeper cutaneous, tendon, and joint structures. The deep tissue sensory array element performs this function by means of two anatomical subelements. Specialized deeper cutaneous, tendon, and joint sensors act as transducers, which encode these physical parameters into electrical signals. The second anatomical subelement is the peripheral sensory nerves themselves, which transmit the resultant electrical activity back to the spinal cord (at the spinal nerve root). The spinal cord element is the third anatomical element that completes the feedback interaction subsystem. The essential anatomical characteristic of the spinal cord element are central nerves, specifically identified as interneurons. In this particular anatomical element, the interneurons are sensory afferent because they conduct electrical activity from the neuromuscular apparatus (at the spinal nerve root) back to the central nervous processor (at the brain stem).

The feed-forward interaction subsystem is probably the simplest subsystem in the human operator system and consists of one anatomical element (Table 9.1). The essential anatomical characteristic is the nerve synapse. In the feed-forward interaction element the central nerves (motor efferent interneurons) synapse at the spinal (motor efferent) nerve roots. This synaptic connection occurs at the cell body of an alpha motor neuron located within the spinal cord but having an axon that exits as a spinal nerve root. All electrical activity impinging upon this cell body (alpha motor neuron) represents upper motor neuron activity which is integrated by the cell body of the alpha motor neuron prior to being propagated outward from the spinal cord along its motor nerve axon. Recall that this axon will eventually innervate and excite a small population of muscle fibers (which has been previously described in Chapter 5 as a motor unit). Because of the ability of the motor cell body to integrate upper motor neuron activity into a final lower motor neuron response, this feed-forward interaction element has sometimes been referred to as the *final common motor pathway.*

The neuromuscular actuator is the fourth subsystem that constitutes the human operator system. The neuromuscular actuator subsystem may be approximated by means of two anatomical elements (Table 9.1). The neuromuscular apparatus element is the fundamental actuator and its operation is described using the biophysical equations of Section 9.1.b. This neuromuscular apparatus element is composed of two anatomical subelements organized in a roughly serial arrangement. The peripheral motor nerves are one of these anatomical subelements. They originally exit at the spinal nerve roots and after a variable amount of branching and interdigitating, ultimately terminate in the muscle tissue itself. The second anatomical subelement is the skeletal muscles themselves. It should be noted that human operator control refers to voluntary motor control, which is accomplished by the skeletal muscles. Other muscles of the body (such as cardiac muscle and smooth muscle) also are subelements of their respective neuromuscular actuators but are under autonomic (involuntary) motor control. Consequently, the human operator cannot voluntarily control these other muscle tissues.

The musculoskeletal apparatus element is responsible for mechanical transduction of the skeletal muscle force, velocity, and length changes. By means of a system of mechanical levers and hinges, these physical parameters may experience gain or attenuation with respect to the external human operator output. There are two anatomical subelements which accomplish this. The tendons and ligaments represent soft tissue subelements. The tendons attach the muscle directly to skeletal bones. The ligaments extend across joints and connect adjacent bones. This is primarily what holds our skeletal bones together. The second anatomical subelement are the skeletal bones themselves, which represent a hard tissue subelement. As a result, our bodies are able to withstand considerable external force acting upon us, and also to generate considerable applied force (acting upon our environment) because the skeletal bones provide the rigid structural framework analogous to a load carrying beam (see Chapter 3).

The human operator output (per Figure 9.2) is a musculoskeletal phenomenon that may be characterized with respect to the force, velocity, and displacement of the output. Depending upon the specific combination of these three physical variables, a particular musculoskeletal output may be characterized with respect to the particular type of contraction that the actuating skeletal muscle performs. Table 9.2 summarizes these relationships. An isometric control system will be simulated in Section 9.2. An isokinetic (isovelocity) control system will be simulated in Section 10.1.a, and an isotonic control system will be simulated in Section 10.1.b. Control system complexity increases as one progresses from top to bottom of Table 9.2. An auxotonic control system (Virtual Medicine) and a minuthotonic control system (Telerobotics) will also be simulated in Section 10.2.b for the purpose of a general introduction to neuromuscular control systems.

The human operator performs in an environment and this environmental interaction is important for two reasons. First, the environment can directly affect the human operator and more specifically the human operator control subsystems. This can occur independently of the human operator output. Second, the environment is an important subsystem with respect to human operator control feedback. Since human beings are predominantly visual creatures and rely mostly upon visual feed-

Table 9.2 Types of Skeletal Muscle Contraction

Muscle Contraction Type	Physical Parameter		
	Force	Velocity	Displacement
Isometric	C/V	0	[C]
Isokinetic	C/V	[C]	V
Isotonic	[C]	V	V
Auxotonic	[I]	V	V
Minuthotonic	[D]	V	V
Oscillotonic	[V]	V	V

0 = Zero
C = Constant
V = Variable
I = Increasing
D = Decreasing

☐ = Characteristic feature of the contraction type

back, the performance of a specific human operator output activity relies upon human operator data input via the special sensory organ of vision. These relationships are indicated in Figure 9.3, which examines a subsection of Figure 9.1. The environmental subsystem is specifically identified and it is seen that the human operator output can directly act upon this subsystem. Furthermore, when the human operator output is translated to human operator input data, the closed-loop feedback path must pass through the environment as represented by an environmental subsystem. Figure 9.3 also indicates two other ways in which the environmental subsystem may interact with the human operator subsystem. The environment may provide human operator input by itself which is processed as input by the human operator control system. Alternatively, the environment may act directly on the human operator to modify the processing and actuating functions of the human operator system.

Let us now turn our attention to the environmental system elements (Table 9.3). It is immediately apparent that the environmental system of Figure 9.3 is composed of two parallel subsystems. The first subsystem is the physical/natural environment (PNE) subsystem. This is the subsystem that one naturally thinks of when describing a human operator interacting with a technological system in a given environment. Physical quantities such as illumination, noise, vibration, and temperature are then specified as elements of the PNE subsystem. A specific human operator and technological system interaction will usually occur in an environment that may be specified with respect to one or a combination of these elements. Note that the environment subsystem of Figure 9.1 is configured as an environmental system in Figure 9.3. The PNE subsystem will interact with both the human operator system and also the technological system. The PNE subsystem is particularly useful in characterizing external feedback of the human operator output with respect to the special sensory organisms of vision and audition. In such cases, illumination levels and noise levels, respectively, will affect the gain of the PNE subsystem element. Specific categories of PNE subsystem interactions with the human operator CNP subsystem are indicated in Table 9.3. Other PNE subsystem elements, such as vibration and temperature, are particularly important since they can exert direct effects upon the human operator CNP subsystem as well as the human operator NMA subsystem. In these cases, the vibration and/or temperature fluxes directly introduce a perturbation function to the human operator control elements. The

Table 9.3 The Environmental Elements and Interactions

I. Environmental System Elements
  A. Physical/Natural Environment (PNE) Subsystem
    1. Illumination
    2. Noise
    3. Odors/Aroma
    4. Vibration
    5. Temperature
    6. Humidity, etc.
  B. Human/Operational Environment (HDE) Subsystem
    1. Task Specifications
    2. Performance Requirements
    3. Self-Expectation
    4. Peer-Expectation
    5. Instruction
    6. Training, etc.

II. Environment-CNP Interactions
    1. Feedback (1-2-C): Special senses (Vision, Audition, Taste, Smell) *reacting* to H.O. Output
    2. Feedback (A-B-C): Operational Performance requirements and/or expectations reacting to H.O. Output
    3. Direct (3-4): CNP elements (e.g. special senses, brain, brainstem) reacting to physical environment only.
    4. Direct (D-E): CNP elements (e.g. special senses, brain, brainstem) reacting to performance requirements and/or expectations, etc. only.

III. Environmental-NMA Interactions
    1. Feedback (1-3-5): NMA elements (e.g. peripheral nerves, skeletal muscle) reacting to H.O. output upon the physical environment.
    2. Feedback (A-D-F): NMA elements reacting to H.O. output upon the performance/operational environment.
    3. Direct (3-5): NMA elements (e.g. peripheral nerves, skeletal muscles) reacting to physical environment *only.*
    4. Direct (D-F): NMA elements reacting to the performance/operational environment *only.*

interested reader is again referred to Sections 1.2 and 1.3 for a simple example of these concepts.

The human/operational environment (HOE) subsystem is the second part of the environmental system and is composed of elements that are uniquely characteristic of the human operator and the task to be performed (human operator output). The HOE subsystem is what makes the human operator control system characteristically "human" and therefore this subsystem does not interact with the technological system (Figure 9.1). The elements of a physical/natural environment are relatively independent of the human operator and are present in varying degrees when any automatic machine is operating with no human interaction at all. In the physical universe, the various elements of the PNE are often treated as independent variables. In contrast, the various elements of the HOE are much more closely associated with the human operator's capabilities and limitations such that these elements may be characterized as human-dependent variables. Such elements as task specification, performance requirements, self-expectation, and prior training do not readily equate to specific physical quantities or mathematical values. However, they are in the simulation and analysis of human operator control. It is critical for the human factors engineer to understand that the human operator is not an automaton and that a human operator control system is not just another automatic control system. Biophysical equations given in this chapter and control system theory (Chapter 10) constitute the human factors engineering approach to an understanding of neuromuscular control systems and human operator control, respectively. However, effectively interfacing the person to the task (or the human operator to the technological system) requires that the HOE subsystem be incorporated (as shown in Figure 9.3).

The human factors engineering student may rightly inquire how such "qualitative" elements of the HOE subsystem may be related (or even modeled) with respect to the quantitative elements in a computer simulation or a control system transfer function. The answer is as follows. With respect to biophysical equations (computer simulation) the HOE subsystem elements will alter the magnitude and/or time course of the biophysical parameters (e.g., maximal force, maximal velocity, or maximal displacement). With respect to control system transfer functions, the HOE subsystem will alter the magnitude of certain gain elements and/or the duration of certain time delays. When this is appreciated, the human factors engineer will have a basis for incorporating the HOE subsystem elements into the general description of the human operator control system (Figure 9.1). An interesting example of this concept is presented in Section 9.2.c in which an isometric control system is modified in order to incorporate physiological fatigue. If an isometric control system were to be simulated with respect to an automatic control machine, then the simulations of Sections 9.2.a and 9.2.b. would suffice since machines do not fatigue. An automatic control machine may eventually break down or malfunction, but it is unlikely that its control output force will progressively "fatigue" with time. However, physiological fatigue is very characteristic of living creatures, and especially human operators. In Section 9.2.c., the student will see how the "fatigue" curve affects the magnitude and time course of a biophysical parameter so that this "human" quality may be incorporated into the computer simulation. For a systematic listing of the general categories by which the HOE subsystem interacts with the human operator CNP subsystem please refer to Table 9.3. Finally, for these specific categories in which the HOE subsystem interacts with the human operator NMA subsystem, also refer to Table 9.3.

The general control system simulations that are presented in Sections 9.2, 9.3, 9.4, and 10.1 follow the characteristic diagram shown in Figure 9.4. There are two external computational paths and two internal computational paths that are used. An external computational path operates on external (input and output) data in order to control an output parameter for a given set of controller/plant characteristics. An internal computational path operates through internal elements in order to control the controller/plant characteristics. The computational paths for the general computer system simulation may be summarized as follows (Figure 9.4). The external feed-forward path (A-E-F) has a reference parameter x_R as input and a computed parameter as output (x_O). The feed-forward pathway includes a controller and a plant. For the controller used in subsequent sections of Chapters 9 and 10, the spinal cord motor (SCM) element of the central nervous processor (CNP) subsystem is used. For the plant, the neuromuscular apparatus (NMA) element of the neuromuscular actuator (NMAc) subsystem is used.

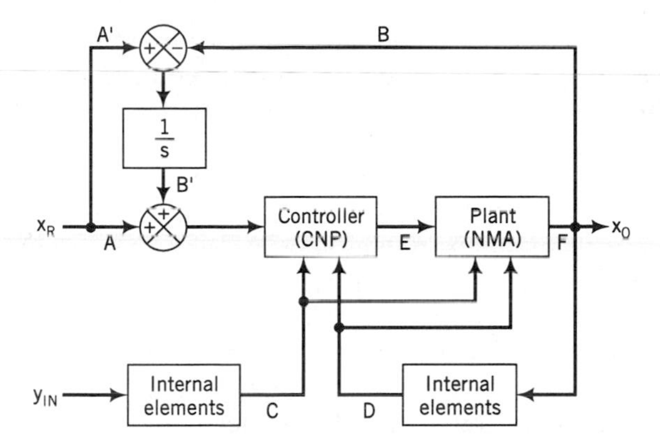

Figure 9.4 Characteristic diagram of the general control system.

The second computational pathway is the external feedback path (B-B'), which is configured as follows. Negative feedback is used so that the output parameter (x_O) is different with the reference parameter (x_R). Integral error (B') is then summated with the reference parameter (x_R) at the feed-forward path. This configuration allows the system to return to the external feed-forward path configuration when the steady-state error is zero. The third computational pathway is the internal feed-forward path (C) of Figure 9.4. Parameter Y_{in} represents input to one or more internal elements, the output of which is used to set one or more controller and/or plant characteristics. The fourth computational pathway is the internal feedback pathway (D) of Figure 9.4. The system output parameter (x_O) provides the input to one or more internal elements, the output of which is used to modify one or more characteristic parameters of the controller element and/or plant element. As indicated in Figure 9.4, these internal elements and their associated feed-forward or feedback pathways are part of the central nervous processor subsystem for the neuromuscular subsystem. It should be noted that either subsystem (CNP or NMA) may have one or more internal feed-forward paths and/or one or more internal feedback paths.

Finally, there are the operational elements of error detection, integral error, reference parameter generation, and reference-error summation (Figure 9.4). For the neuromuscular control systems that are simulated in Sections 9.2, 9.3, 9.4, and 10.1, these operational elements reside within the central nervous processor subsystem. With respect to the human operator control system, these operational elements may be identified with the cranial sensory array element, brain/midbrain element and/or brain stem/cerebellar element.

b. Biophysical System Equations

This section will describe the biophysical relationships that are essential for developing the computer simulation of neuromuscular control systems. Each biophysical relationship is identified by a separate subsection, which is then organized as follows. First, the particular biophysical relationship is identified and described qualitatively. Second, the relationship between the dependent variable and the independent variable is presented graphically, both in direct form and normalized form. Third, the governing biophysical equation which characterizes the specific biophysical relationship is presented. Finally, the operational form of that biophysical equation is developed. The operational equation is the quantitative expression of the specific biophysical relationship that will be used in the subsequent neuromuscular control system simulations. Generally speaking, these operational biophysical equations will be the normalized form of the governing biophysical equation and/or the inverse function of the specific biophysical equation.

Each biophysical relationship represents an operational element within the neuromuscular control system. As will be seen shortly, certain biophysical operational elements are characteristic of the central nervous processor subsystem, while others are characteristic of the neuromuscular actuator subsystem. This chapter and the next will present various neuromuscular control systems as computer simulations. Consequently, these operational elements may be treated as computational elements that represent a fundamental biophysical relationship. As a result, the students can understand how these various elements interact within subsystems and how the subsystems interact with each other in order to achieve control of a specific output parameter. These computer simulations will be executed in MATLAB, and the student should have some familiarity with the development of m-files in that programming environment.

i. The Force-Length Relationship

Historically, one of the first biophysical relationships to be described with respect to skeletal muscle was that the maximum isometric force that could be developed by a muscle (when stimulated) was a function of that muscle's length. This relationship was used for a variety of human skeletal muscles in Chapter 8. The relationship states that every muscle in the human body has an optimal length (L_0) such that when maximally stimulated (100% muscle fiber activation) that muscle will generate its maximal isometric force (P_0^*). Recall that this is an isometric force so that the muscle length does not change during the generation of the force. The relationship further states that at isometric muscle lengths above and below L_0, the maximal isometric force at that length (P_0) will decrease below P_0^*. This is true both for progressively increasing isometric muscle lengths above L_0 and progressively decreasing isometric lengths below L_0 (Figure 9.5.a). It is understood that for each P_0 in Figure 9.5.a the skeletal muscle is maximally stimulated (100% fiber activation) and the muscle length (L) is constant. This relationship can be mathematically described by the force-length equation:

$$P_0 = P_0^* \sin\left[\pi\left(\frac{L}{L_0} - \frac{1}{2}\right)\right] \tag{1}$$

In Chapter 8, the student was introduced to the value of normalizing equation parameters so that when applied to a population of individuals with different physical characteristics, that equation could be more accurately applied. Figure 9.5.b represents the normalized force-length relationship. This may be expressed mathematically by the normalized force-length equation:

$$\hat{P}_0 = \sin\left[\pi\left(\hat{L} - \frac{1}{2}\right)\right] \tag{2}$$

where

$$\hat{P}_0 = \frac{P_0(L)}{P_0^*(L_0)} \tag{3}$$

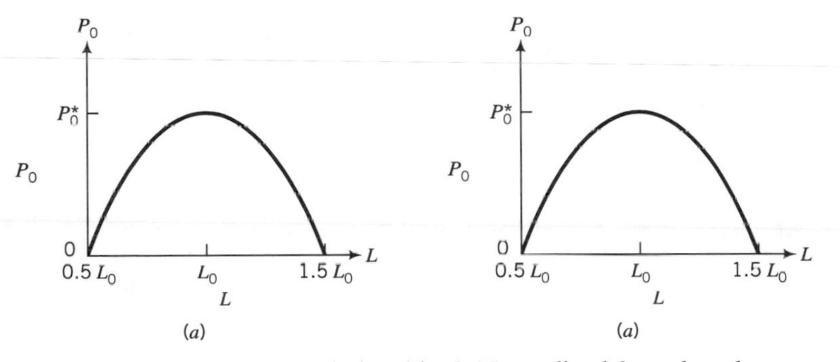

Figure 9.5.a Force-length relationship. **b** Normalized force-length relationship.

and

$$\hat{L} = \frac{L}{L_0} \tag{4}$$

where L is a specific length (e.g., a reference length, an initial length) at which the specific P_0 occurs.

ii. Velocity-Length Relationship

Analogous to the force-length relationship for human skeletal muscle, there is a velocity-length relationship. Although this relationship was not apparent when we addressed quantitative workload assessment (Chapter 8), it is a very important relationship when neuromuscular control is required from the human operator. Together with the force-length relationship, the velocity-length relationship will define the force-velocity-length envelope for a given neuromuscular control system (see Section 9.1.b.iv). The maximal (unloaded) shortening velocity (V_0) is a function of the initial length (L). V_0 is the maximal shortening velocity for a skeletal muscle because there is no external load imposed upon the muscle. Therefore, all of the contractile energy of the muscle is directly translated into shortening velocity. In actual practice, V_0 cannot be experimentally measured but is mathematically extrapolated. This is because real human skeletal muscles must see some small external load in order to be stretched to any specific length (along the L-axis). The actual muscle length (L) at which V_0 occurs is taken as the initial (starting) length for that muscle. Similar to P_0^*, which occurred at a optimal length (L_0), the maxima of the $V_0 - L$ curve also occurs at the optimal length (L_0). The maximal (unloaded) shortening velocity is defined as V_0^*. The biophysical relationship depicted in Figure 9.6.a is expressed quantitatively by the velocity-length equation:

$$V_0 = V_0^* \sin\left[\pi\left(\frac{L}{L_0} - \frac{1}{2}\right)\right] \tag{5}$$

Figure 9.6.b depicts the normalized velocity-length relationship. This relationship may be expressed by the normalized velocity-length equation:

$$\hat{V}_0 = \sin\left[\pi\left(\hat{L} - \frac{1}{2}\right)\right] \tag{6}$$

where

$$\hat{V}_0 = \frac{V_0(L)}{V_0^*(L_0)} \tag{7}$$

and

$$\hat{L} = \frac{L}{L_0} \tag{8}$$

where L is the specific length at which the specific V_0 occurs.

iii. Force-Length Relationship and Velocity-Length Relationship with Time

Equations (1) and (2) for the force-length relationship and also equations (5) and (6) for the velocity-length relationships are independent of time. As such the equations are static and $\hat{L}$ is a specific length that is independent of time. In the vast majority of human operator neuromuscular control systems that are of practical interest to the human factors engineer, the biophysical parameters of skeletal muscle force, velocity, and length are changing with respect to time. The preceding equations can be modified to account for the time variations of these biophysical parameters by defining a time-varying normalized length (L'):

$$L' = \frac{L(t)}{L_0} \tag{9}$$

$$L' = \hat{L}(t) \tag{10}$$

The force-length relationship of Figure 9.5.b may now be expressed as the force-length relationship of Figure 9.7.a. Substituting equation (10) into equation (2), the force-length equation with time variation:

$$\hat{P}_0 = \sin\left[\pi\left(L' - \frac{1}{2}\right)\right] \tag{11}$$

The velocity-length relationship of Figure 9.6.b may now be expressed by the velocity-length relationship of Figure 9.7.b. Substituting equation (10) into equation (6) results in the velocity-length equation with time variation:

$$\hat{V}_0 = \sin\left[\pi\left(L' - \frac{1}{2}\right)\right] \tag{12}$$

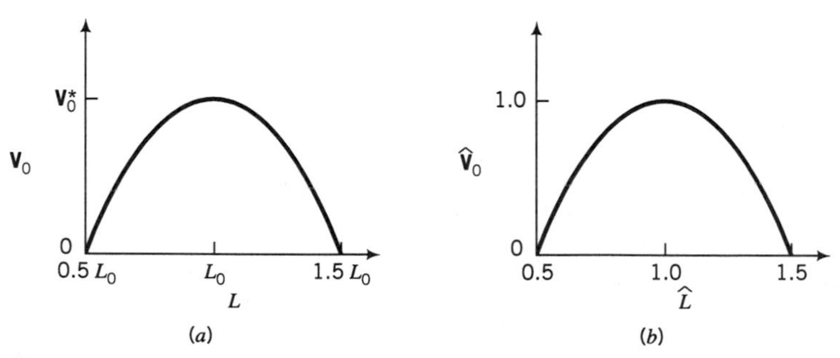

Figure 9.6.a Velocity-length relationship. **b** Normalized velocity-length relationship.

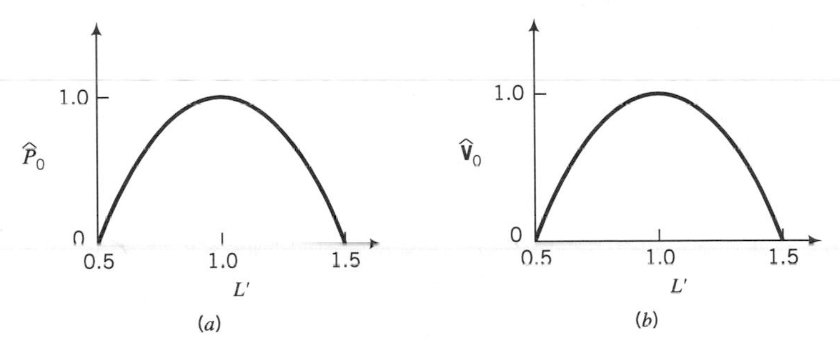

Figure 9.7.a Force-length relationship and velocity-length relationship with time. **b** Normalized force-length relationship and velocity-length relationship with time.

Equations (11) and (12) are valid if we invoke the quasi–steady-state approximation. This approximation requires that as L' varies with time for a reasonably small but discrete time interval, and the mean L' (for that interval) is equivalent to the static $\hat{L}$ when calculating the biophysical variables ($\hat{P}_0$ and $\hat{V}_0$). In other words, the biophysical system is at a steady state, which is equivalent to the static state. Although this is obviously not the exact case in real human operator skeletal muscle (due in part both to viscous effects and frictional effects), it is a useful approximation considering the complexity of the neuromuscular control system.

iv. Force-Velocity-Length Relationship: Part I

The human neuromuscular control system operates within a force-velocity-length domain. In effect, this is an envelope in which the biophysical equations (that interrelate these variables) are valid. This section will describe the coordinate system and two of the envelope boundaries. Referring to Figure 9.8.a, three coordinate axes are identified for the spatial system of reference. The x-axis is skeletal muscle force (P), which is zero at the origin and rightward positive. The y-axis is the velocity of skeletal muscle shortening (V), which is zero at the origin and upward positive. The z-axis represents the length of the skeletal muscle (L), which is at its optimal length (L_0) at the origin.

Using this x-y-z coordinate system, the force-length relationship depicted in Figure 9.5.a would operate in the x-z plane. Translating the origin of Figure 9.5.a ($0.5L_0$) to the origin of Figure 9.8.b (0, L_0) allows us to represent the force-length relationship of Figure 9.5.a on Figure 9.8.b. The velocity-length relationship of Figure 9.6.a is now seen to operate in the y-z plane of the x-y-z coordinate space of Figure 9.8.b. Translating the origin of Figure 9.6.a (in a manner similar to that of Figure 9.5.a) allows us to represent the velocity-length relationship on the y-z plane of Figure 9.8.b.

At this juncture, it will be appreciated that we have now defined an envelope for the neuromuscular control system of a human operator with respect to the force-length plane and the velocity-length plane. Figure 9.8.c represents this same force-velocity-length relationship as a normalized envelope. It is apparent that the force-velocity envelope in the x-y plane is now indicated in Figures 9.8.b and 9.8.c. Consequently, we shall revisit the force-velocity-length relationship in Section 9.1.b.7.

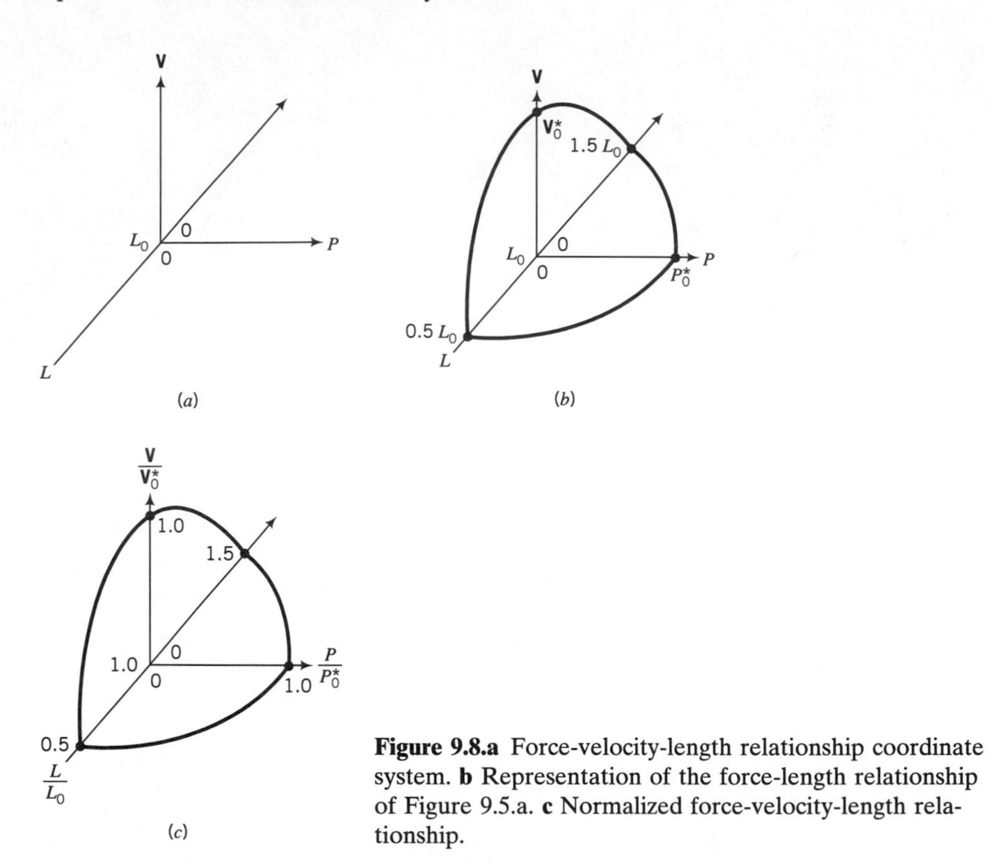

Figure 9.8.a Force-velocity-length relationship coordinate system. **b** Representation of the force-length relationship of Figure 9.5.a. **c** Normalized force-velocity-length relationship.

v. *Force-Velocity Relationship*

The relationship between the external force that a skeletal muscle generates in moving a load and the speed (or velocity) with which that load is moved is recognized as the most fundamental biophysical property of muscle. This relationship is shown in Figure 9.9.a and represents the mechanical output properties of muscle (force and velocity) when skeletal muscle is activated to perform external work. This force-velocity relationship has led many scientists and engineers to view skeletal muscle as a motor. The student will note that when the force in Figure 9.9.a is replaced by the torque that would be generated at a motor shaft and velocity is

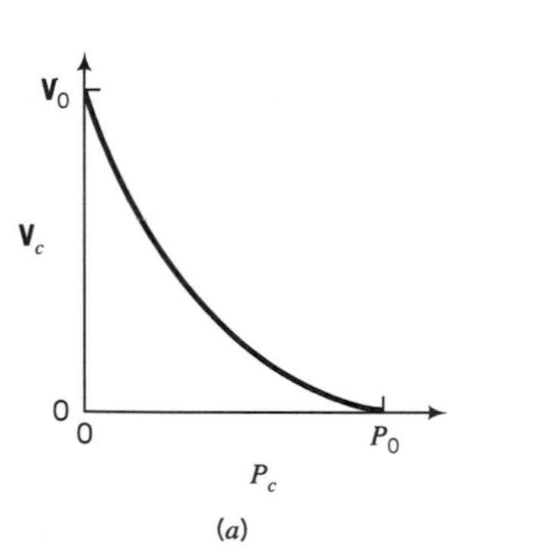

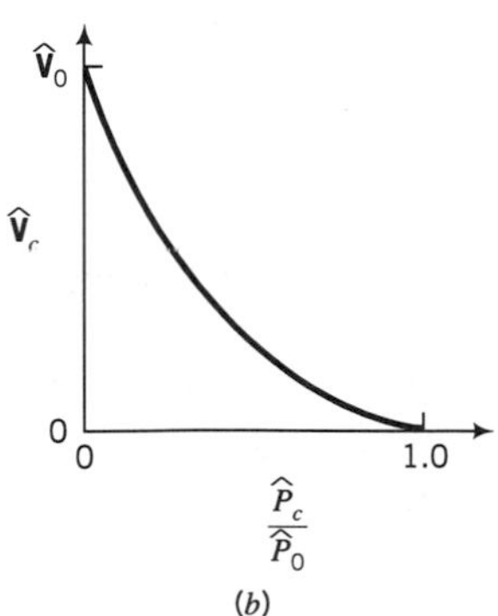

Figure 9.9.a Force-velocity relationship. **b** Normalized force-velocity relationship.

replaced by the angular rotational velocity of the motor shaft, the force-velocity relationship has the same inverse relationship as does shaft rotational velocity versus torque output for a mechanical motor. With this analogy, P_0 would be equivalent to the stall torque output of a motor and V_0 would be equivalent to the zero torque (no load) shaft rotational speed. While this is a useful analogy, there are some important differences that are essential to correctly interpreting the force-velocity relationship of Figure 9.9.a. They will now be considered.

The force-velocity relationship for skeletal muscle, while an inverse relationship, is distinctly nonlinear. This relationship has classically been described as an inverse hyperbolic relationship, and it will be noted that force is treated as the independent variable, while velocity is assigned the dependent variable. Any point along the force-velocity relationship is located at a specific coordinate point (P_c, V_c). P_c represents a constant external load upon the skeletal muscle so that as the skeletal muscle contracts and shortens it moves the external load by generating a constant force. This is known as an isotonic (constant force) type of skeletal muscle contraction (see Table 9.2). V_c is the peak (maximal) external shortening velocity that occurs during the time course of this isotonic muscle contraction. This means that V_c represents a single point in time at which the muscle is maximally activated (100% of the muscle fiber population is being used). This then makes the force-velocity relationship of Figure 9.9.a an apparently static relationship since the time course of events is not indicated. However, since human beings and their skeletal muscles are living systems, it is more realistic to view Figure 9.9.a as an infinite series of steady-state points, each of which occurs over a discrete but finite time interval. This will also allow the human factors engineer to invoke the quasi–steady-state approximation as was done for the force-length relationship and the velocity-length relationship. Finally, this chapter will use the convention that both P and V are subscripted with a "c" when they represent a force coordinate or velocity coordinate respectively along the force-velocity curve (per Figure 9.9.a).

The biophysical relationship expressed in Figure 9.9.a is defined by the force-velocity equation:

$$V_c = V_0\left[1 - \sin\left(\left[\frac{\pi}{2}\right]\left[\frac{P_c}{P_0}\right]\right)\right] \tag{13}$$

Equation (13) may be used as the starting point for developing a normalized force-velocity equation. Begin by normalizing the two velocity terms of equation (13) by V_0^* and normalizing the two force terms of equation (13) by P_0^*:

$$\frac{V_c}{V_0^*} = \frac{V_0}{V_0^*}\left[1 - \sin\left(\left[\frac{\pi}{2}\right]\left[\frac{P_c/P_0^*}{P_0/P_0^*}\right]\right)\right] \tag{14}$$

Define

$$\hat{V}_c = \frac{V_c}{V_0^*} \tag{15}$$

$$\hat{P}_c = \frac{P_c}{P_0^*} \tag{16}$$

Then substituting equations (3), (7), (15), and (16) into equation (14):

$$\hat{V}_c = \hat{V}_0\left[1 - \sin\left(\left[\frac{\pi}{2}\right]\left[\frac{\hat{P}_c}{\hat{P}_0}\right]\right)\right] \tag{17}$$

Equation (17) represents the normalized force-velocity equation and Figure 9.9.b depicts the normalized force-velocity relationship that results.

vi. The Force-Velocity-Length Relationship: Part II

It is apparent in Figure 9.9.a that the velocity intercept is V_0, and the force intercept is P_0. Recall that human skeletal muscle operates within a force-velocity-length relationship and that the envelope in the force-length plane (Figure 9.9.b) is described by equation (1). The envelope in the velocity-length plane of Figure 9.8.b is described by equation (5). Simultaneous solution of these two equations at a common length indicates that for the specific length there is a characteristic P_0 and a characteristic V_0. This is indicated graphically in Figures 9.10.a and 9.10.b. In this example, the common length (L_1) is greater than the optimal length (L_0). P_{01} is then the characteristic P_0 at L_1, and V_{01} is the characteristic V_0 at L_1. This means that if the force-velocity relationship of Figure 9.9.a were to occur at a specific length (e.g., L_1), then its relationship to the force-velocity-length envelope would appear as shown in Figure 9.11.a. If the human operator's skeletal muscle were at its optimal length (L_0), then the force-velocity relationship of Figure 9.9.a would appear in the force-velocity-length relationship as shown in Figure 9.11.b. It is now apparent that we may complete our discussion of the force-velocity-length envelope. The x-y plane (or force-velocity plane) contains the force-velocity relationship and any specific force coordinate and velocity coordinate (P_c and V_c) as described by equation (13).

The preceding example could be repeated using the normalized force-length relationship [equation (2) and Figure 9.5.b], the normalized velocity-length relationship [equation (6) and Figure 9.6.b], and solving for the common length $\hat{L}_1$ as before. Recall that L_1 was greater than the optimal length (L_0) so that the procedure could be repeated a third time and equations (2) and (6) could be solved for a common $\hat{L}_2$ less than the optimal length. If we had selected $\hat{L}_1 = 1.25$ and $\hat{L}_2 = 0.65$, then the resultant normalized force-velocity-length relationship would appear as shown in Figure 9.11.c.

It is now apparent that the human factors engineer can define the surface of a force-velocity-length envelope in which the human neuromuscular control system must operate. This is not surprising! Anthropometric envelopes and performance

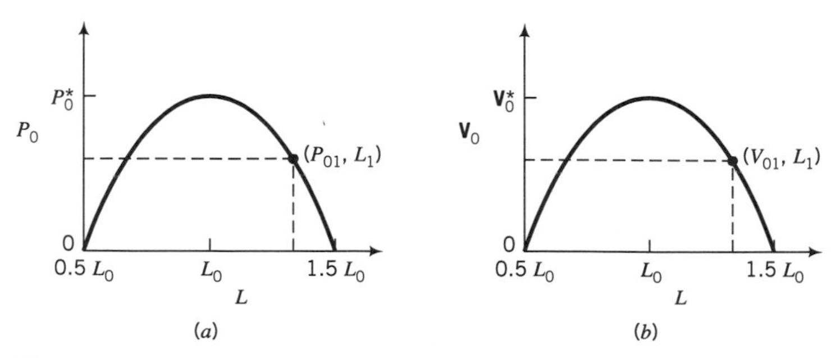

Figure 9.10.a Force-length relationship. **b** Velocity-length relationship.

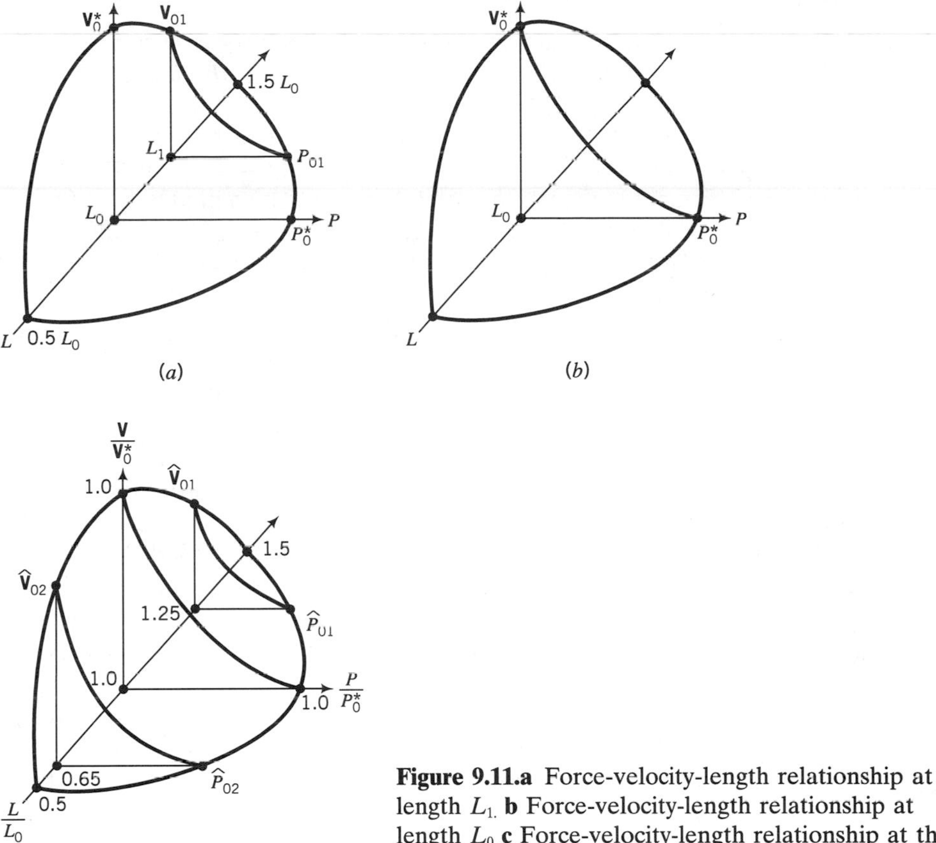

Figure 9.11.a Force-velocity-length relationship at length L_1. **b** Force-velocity-length relationship at length L_0. **c** Force-velocity-length relationship at three different lengths.

envelopes are commonly applied to human operators when performing various tasks. Such anthropometric and performance envelopes are commonly used by the human factors specialist. The human factors engineer understands the biophysical foundations for a particular human operator performance envelope. At this point, the surface of that force-velocity-length envelope has been described with respect to the biophysical equations that govern skeletal muscle. The force-velocity-length relationship may be viewed as a performance envelope within which the human operator will exhibit neuromuscular control.

The surface of the envelope for the force-velocity-length relationship of skeletal muscle has now been defined. Recall that the surface of this envelope represents complete activation of the muscle (100% muscle fiber use). Also recall that skeletal muscle is only one component of the neuromuscular apparatus element. The other component is the peripheral motor nerve. In order to achieve neuromuscular control, the neuromuscular apparatus must operate inside the surface of the envelope. The biophysical equations developed up to this point only characterize the envelope surface. We shall soon see that additional biophysical equations are required that characterize muscle fiber recruitment by means of peripheral nerve stimulation as well as peripheral nerve recruitment by means of central nervous activation. The remainder of this section shall proceed to develop those additional biophysical equations.

vii. Force-Endurance Relationship

A characteristic of human beings who perform continuous sustained muscular work is that they will experience fatigue. This is termed *physiological* fatigue since it represents the inability of the working skeletal muscle to meet the external work demands placed upon the human operator. This is distinct from *mental* fatigue or *psychological* fatigue in which the failure to meet the external workload resides in anatomical elements other than skeletal muscle. The actual cause of muscle fatigue at the biophysical and/or biochemical level is not known. However, it can be operationally quantified as a progressive reduction in the muscle force output over time. This is shown graphically in Figure 9.12.a. For a skeletal muscle at a given length, a characteristic P_0 may be calculated utilizing equation (1). This is represented in Figure 9.12.a by the P_0 at time zero (the y-axis intercept). Recall that this P_0 represents a maximal isometric force (in which one hundred percent of the skeletal muscle fibers are used). Consequently, this P_0 can be maintained for only a brief period (less than 3 s) before it reduces in value. Note that the muscle length is held constant (this is an isometric type contraction) but that now the P_0 decreases as a function of time until it reaches an asymptote of 15% of its original value (Figure 5.12.a). The point of asymptotic approximation is identified T_F, which represents the period of fatigue (the time interval over which fatigue occurs). For all practical purposes, for time beyond T_F there is no further muscle fatigue. Figure 9.12.a is an operational expression of a fatigue curve, and basically is saying that over the course of time the skeletal muscle P_0 is monotonically decreasing.

This fatigue curve allows us to develop the force-endurance relationship for the skeletal muscle of a human operator. Referring to Figure 9.12.b, the y-axis is now simply identified with respect to muscle force which can vary between zero and P_0. For Case 1, the human operator exerts a sustained isometric force of $0.75P_0$

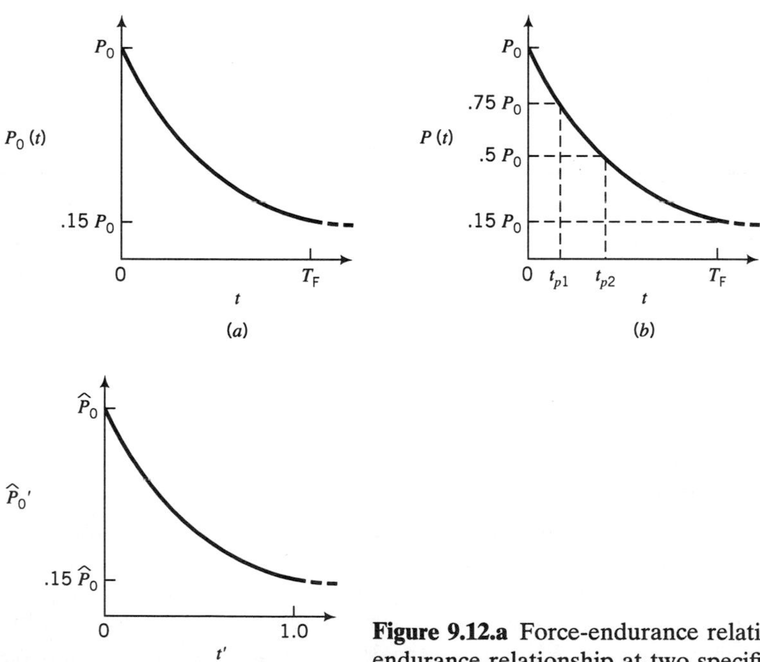

Figure 9.12.a Force-endurance relationship. **b** Force-endurance relationship at two specific endurance times (t_{f1} and t_{f2}). **c** Normalized force-endurance relationship.

beginning at time zero. Referring again to Figure 9.12.b, this isometric force level is maintained until a time (t_{f1}) at which the isometric skeletal force begins to progressively decrease. t_{f1} represents the endurance time at that isometric force level for that human operator. It is as though the human operator desires to maintain a skeletal muscle force of $0.75P_0$ but can do so only for as long as the skeletal muscle's actual P_0 is at or above that value. At that moment in time when the actual skeletal muscle's P_0 falls below $0.75P_0$, the human muscle can no longer maintain the desired isometric force level. For Case 2, the human operator desires to maintain a constant isometric force of $0.5P_0$. This force is generated by the contracting skeletal muscle at time equals zero and continuously maintaining that force until a point in time identified at t_{f2}, which represents the endurance time for that human operator (and the contracting skeletal muscle) when exerting a sustained isometric force of $0.5P_0$. Beyond that point in time, the skeletal muscle's P_0 is continuously decreasing to a value below $0.5P_0$. Figure 9.12.b represents the force-endurance relationship for a human operator. That relationship states that the endurance time for a sustained isometric muscular effort is inversely proportional to the isometric force exerted. The student will note that a quantitative description of that force-endurance relationship requires a mathematical expression for the $P_0(t)$ function of Figure 9.12.a.

The force-endurance equation depicted in Figure 9.12.a is expressed as:

$$P_0(t) = P_0(0)\left[.15 + .85\left(1 - \sin\left[\left[\frac{\pi}{2}\right]\left[\frac{t}{T_F}\right]\right]\right)\right] \tag{18}$$

A normalized force-endurance equation may also be obtained. The mathematical development begins by first normalizing equation (18) with respect to P_0^*:

$$\frac{P_0(t)}{P_0^*} = \frac{P_0(0)}{P_0^*}\left[.15 + .85\left(1 - \sin\left[\left(\frac{\pi}{2}\right)\left(\frac{t}{T_F}\right)\right]\right)\right] \tag{19}$$

The mathematical development continues as follows.
Define

$$\hat{P}_0' = \frac{P_0(t)}{P_0^*} \tag{20}$$

and

$$t' = \frac{t}{T_F} \tag{21}$$

also

$$\hat{P}_0 = \frac{P_0}{P_0^*} = \frac{P_0(0)}{P_0^*} \tag{22}$$

Substituting equations (20), (21), and (22) into equation (19):

$$\hat{P}_0' = \hat{P}_0\left[.15 + .85\left(1 - \sin\left[\left(\frac{\pi}{2}\right)t'\right]\right)\right] \tag{23}$$

Equation (23) may be represented graphically as shown by Figure 9.12.c.

viii. Force-Recruitment Equation

As stated earlier, in order to understand that the neuromuscular control system of a human operator, the human factors engineer must develop biophysical equations for additional elements other than skeletal muscle. This subsection considers the relevant biophysical equation for the peripheral motor nerve component. Recall that this component in combination with the skeletal muscle component will constitute the neuromuscular apparatus. The peripheral motor nerve electrically activates skeletal muscle by generating and conducting bursts of action potentials along the anatomical course of the peripheral motor nerve. Recall that the cell body of the peripheral motor nerve resides in the spinal cord at the level of the spinal nerve root, and the cell body actually generates the action potentials. The *axon* (a long tubular extension of the cell body) will then conduct these action potentials over significant distances (sometimes in excess of 1 M) before finally reaching its destination at a population of muscle fibers within the skeletal muscle itself. The peripheral motor nerve as an electrical conductor only generates and conducts electrical impulses (see Chapter 5). Skeletal muscle is the actual motor that translates these electrical impulses into mechanical parameters (force and velocity). When this occurs, the muscle is said to be reacting to *peripheral motor nerve stimulation*.

In the development of biophysical equations, it is operationally convenient to define a stimulation parameter ($\bar{\$}_p$), which represents a quantity of electrical charge delivered to the skeletal muscle by the peripheral motor nerve. If the electrical impulses arriving at the skeletal muscle from the peripheral motor nerve are viewed as a train of electrical current pulses (that subsequently travel along the individual muscle fibers), then a stimulation parameter can be mathematically represented as the time integral of a current pulse train. This stimulation parameter (the quantity of electrical charge delivered to the skeletal muscle) may be modified by varying the amplitude (or width) of each current pulse (a phenomenon known as *spatial summation*) or by modifying the interpulse interval between rectangular current pulses (a phenomenon known as *temporal summation*). Since these are two independent mechanisms by which the stimulation parameter may be modified it is convenient to normalize the stimulation parameter by a maximum value ($\bar{\$}_{MAX}$). When the stimulation parameter ($\bar{\$}_p$) is a maximum, its normalized value is unity. The maximal stimulation parameter ($\bar{\$}_{MAX}$) is operationally defined as that value of the stimulation parameter at which the maximal isometric force output (P_0) can be obtained (regardless of the relative amount of "spatial" or a "temporal" summation involved). This, of course, requires that the skeletal muscle be held at a constant length during the development of force output.

A typical isometric muscle contraction, in which the force output (P) is plotted as a function of the normalized stimulation parameter ($\$_p/\$_{MAX}$), is shown in Figure 9.13 (upper half). The stimulation parameter on the abscissa is subscripted with a p to indicate that a force output from the skeletal muscle is generated. The isometric output force from the skeletal muscle can vary between zero and P_0 depending upon the amount of normalized stimulation. The upper half of Figure 9.13 is referred to a *force-recruitment relationship* since the progressively increasing isometric force output (as a result of increasing stimulation) is due to the recruitment of increasing numbers of motor fibers within the skeletal muscle.

The force-recruitment equation may be stated as:

$$P = \frac{P_0}{2}\left[1 - \cos\left[\pi\left(\frac{\$_P}{\$_{MAX}}\right)\right]\right] \tag{24}$$

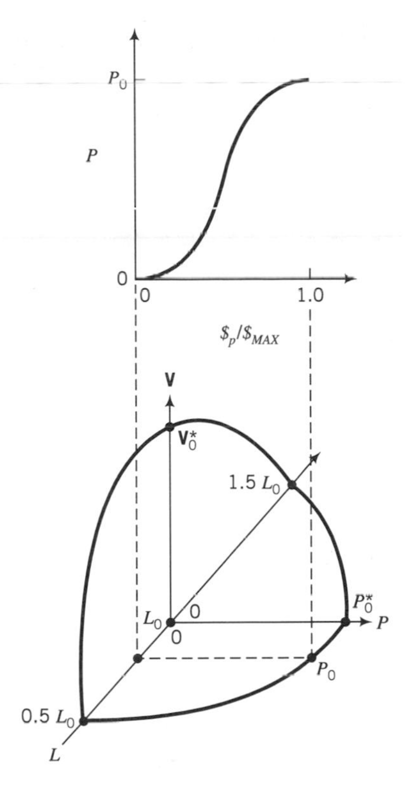

Figure 9.13 Force-recruitment relationship.

ix. Force-Recruitment-Activation Relationship

Recall that the neuromuscular apparatus (as part of the neuromuscular actuator) is basically a plant and that it must interact with elements of the central nervous processor subsystem, which acts as a controller of that plant (see Figure 9.4). At the feed-forward interaction level, this occurs between the interneurons of the spinal cord motor element (central nervous processor subsystem) and the peripheral motor nerve cell body at the spinal nerve root. In effect, the peripheral motor nerve cell body cannot generate electrical impulses (for subsequent conduction along the peripheral nerve axon) unless it is activated by spinal cord motor interneurons. This loss of central nervous processor activation is most dramatically seen in human beings with spinal cord injuries and the resultant skeletal muscle paralysis below the level of the spinal cord injury.

The complex array of nerve fibers impinges upon the cell body of any peripheral motor nerve at the spinal cord root. This complexity is further increased by the fact that synapses are inhibitory, while other of these nerve fiber synapses are excitatory. The situation becomes even more complex when it is realized that some of these interneuron synapses are from interneurons originating at the brain stem level, while other interneuron synapses are from interneurons originating at various spinal nerve root levels. Because of this increasing and layered complexity, the human factors engineer must again resort to defining an operational relationship between peripheral motor nerve stimulation and central nervous processor activation. This may be accomplished by defining a force-activation parameter (A_P). A_p is defined as the ratio of the actual skeletal muscle force output with respect to the maximal skeletal muscle force output (for a constant isometric length). The mathematical development of a force-recruitment-activation equation begins by normalizing equation (24) with respect to P_0:

$$\frac{P}{P_0} = \frac{1}{2} - \frac{1}{2}\cos\left[\pi\left(\frac{\$_P}{\$_{MAX}}\right)\right] \tag{25}$$

The mathematical development then continues as follows.
Define

$$A_P = \frac{P}{P_0} \tag{26}$$

and

$$\bar{\$}_P = \frac{\$_P}{\$_{MAX}} \tag{27}$$

Substituting equations (26) and (27) into equation (25):

$$A_P = \frac{1}{2} - \frac{1}{2}\cos\left(\pi\,\bar{\$}_P\right) \tag{28}$$

Equation (28) may be expressed graphically as shown in Figure 9.14.a.

Finally, the operational feed-forward interaction between central nervous processor activation and neuromuscular apparatus stimulation requires that the relationship between the dependent and independent variable as shown in Figure 9.14.a reverse. Mathematically, this is accomplished by first rearranging equation (28):

$$\cos(\pi\bar{\$}_P) = 1 - 2A_P \tag{29}$$

Solving for $\bar{\$}_P$:

$$\bar{\$}_P = \frac{1}{\pi}\cos^{-1}(1 - 2A_P) \tag{30}$$

Equation (30) is the force recruitment-activation equation which is desired. A graphical representation of equation (30) is shown in Figure 9.14.b.

Referring to the lower half of Figure 9.13, the operational significance of the force-recruitment relationship is apparent. Peripheral motor nerve stimulation of skeletal muscle in an isometric state (constant length) is graphically demonstrated

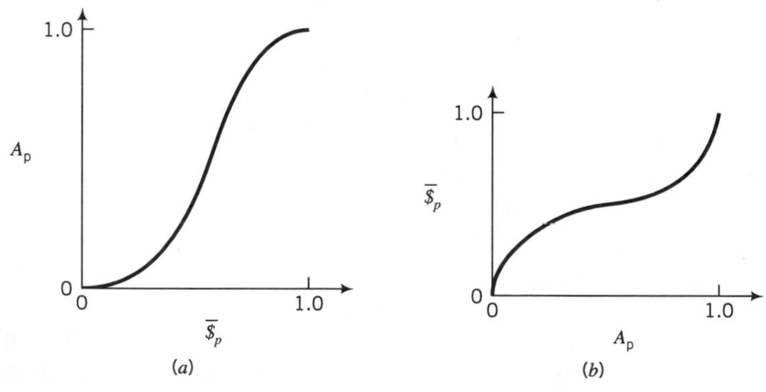

Figure 9.14.a Activation-recruitment relationship. **b** Inverse Activation-recruitment relationship.

in the figure. With respect to the force-velocity-length relationship, specifically in the x-z plane (force-length plane) the force-recruitment relationship modulates the force (P) operating point of human skeletal muscle along the surface of the force-length plane. As Figure 9.13 (lower half) demonstrates, this force modulation may be anywhere from zero to P_0 (at a specific muscle length, L) depending upon the specific level of peripheral motor nerve normalized stimulation (percentage of the skeletal muscle fibers used for force generation).

x. Velocity-Recruitment Relationship

The large majority of human operator neuromuscular control applications of particular interest to the human factors engineer involve dynamic work, rather than static (isometric) work. This requires that the human skeletal muscle develop both force and speed when interacting with external objects. The neuromuscular apparatus, and in particular the peripheral motor nerve interacting with the skeletal muscle, operates in a specific sequence as shown in Figure 9.15. Referring to the upper half of Figure 9.15, the peripheral motor nerve initially generates and conducts electrical impulses that represent a progressively increasing normalized *force* stimulation parameter ($\bar{\$}_p$). The skeletal muscle develops isometric (constant length) force beginning at the origin of Figure 9.15 (upper half) until the necessary force coordi-

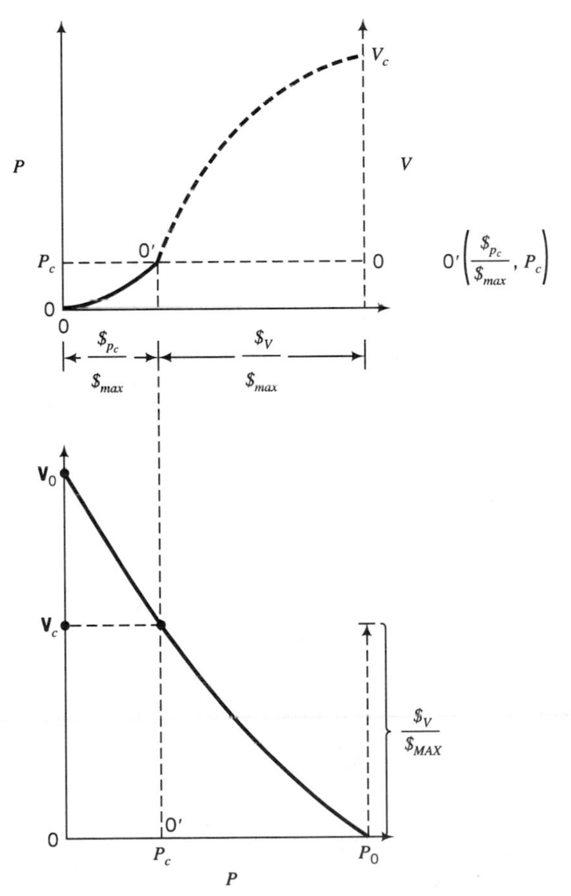

Figure 9.15 Force-recruitment relationship with velocity-recruitment relationship.

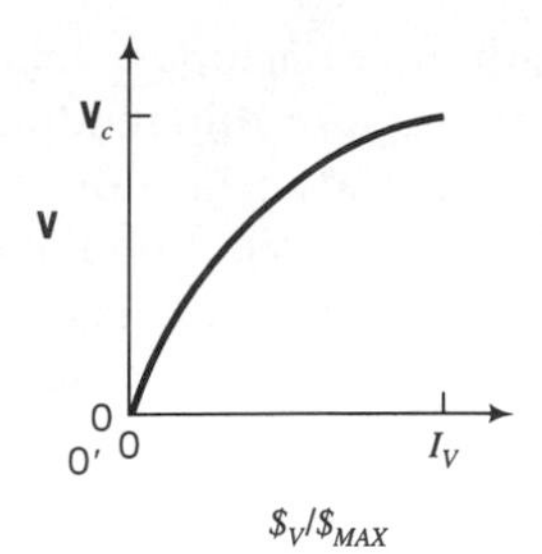

Figure 9.16 Velocity-recruitment relationship.

nate P_c is obtained. At P_c, the applied force of the human operator exceeds the reaction force of the external object and it begins to move. The peripheral motor nerve now continues to generate and conduct electrical impulses, which represent a progressively increasing normalized *velocity* stimulation parameter ($\overline{\$}_v$). For an isotonic (constant force) skeletal muscle contraction, the external speed of muscle shortening *increases* progressively with augmentation in the normalized velocity stimulation parameter, $\overline{\$}_v$ (at a constant external force). When the peripheral motor nerve has been maximally stimulated (100% motor fiber activation of the skeletal muscle) the coordinate velocity (V_c) is obtained. Referring to the lower half of Figure 9.15, this operational sequence is translated to the force-velocity relationship. As the normalized force stimulation parameter progressively increases, the skeletal muscle operating point moves along the x-axis (force axis) from its origin (O) to the coordinate force (P_c) at O'. With subsequent progressive increase in the normalized velocity stimulation parameter, the skeletal muscle operating point moves vertically along the y-axis (velocity axis) up to the coordinate velocity (V_c) on the force-velocity relationship to a point at which the maximal normalized velocity stimulation parameter has obtained. Looking once again at both the upper half and lower half of Figure 9.15, it is apparent that this interaction of the force-recruitment relationship and the velocity-recruitment relationship allows the neuromuscular control system to move the skeletal muscle operating point within the domain of the force-velocity-length relationship.

A graphical presentation of the velocity-relationship is presented in Figure 9.16. Note that the origin (O') of the velocity-recruitment relationship represents a translation of axes with respect to the origin O' of the force-recruitment relationship (upper half of Figure 9.15). The normalized velocity stimulation parameter reaches a maximum at I_V, so that the velocity-recruitment equation may be expressed as follows:

$$V = V_c \cdot \sin\left[\left(\frac{\pi}{2}\right)\left(\frac{\$_V/\$_{MAX}}{I_V}\right)\right] \tag{31}$$

where

$$I_V = 1 - \frac{\$_{Pc}}{\$_{MAX}} \tag{32}$$

Consequently, I_V is a normalized velocity stimulation parameter interval that is less than unity. Since the normalized velocity stimulation parameter may be seen as a differential stimulation parameter (a fraction of stimulation for velocity above that fraction of stimulation for force), we may define a normalized differential stimulation parameter as:

$$\overline{\Delta\$} = \frac{\$_V}{\$_{MAX}} \tag{33}$$

xi. Velocity Recruitment-Activation Relationship

The preceding discussion of the velocity-recruitment relationship has focused upon the neuromuscular apparatus (peripheral motor nerve component interacting with skeletal muscle component) as an element within the neuromuscular actuator subsystem. As we have seen so far, the human operator neuromuscular control system requires that the central nervous processor subsystem have feed-forward interaction with the neuromuscular actuator subsystem. At the lowest level, this involves the spinal cord motor element (interneurons) synapsing at the peripheral motor nerve cell body so that a velocity activation parameter may be operationally defined in a manner analogous to a force activation parameter. It should again be emphasized that the velocity activation parameter (like the force activation parameter) represents the final collective output of the entire central nervous processor subsystem. As a result, higher CNP elements (cranial sensory, brain, and brain stem) ultimately converge upon the final common motor pathway of the neuromuscular control system.

The velocity recruitment-activation equation may be derived mathematically as follows.

Substituting equations (32) and (33) into equation (31) and rearranging:

$$\frac{V}{V_C} = \sin\left[\left(\frac{\pi}{2}\right)\left(\frac{\overline{\Delta\$}}{1 - \dfrac{\$_{Pc}}{\$_{MAX}}}\right)\right] \tag{34}$$

Define

$$A_V = \frac{V}{V_C} \tag{35}$$

Substituting equations (27) and (35) into equation (34) results in:

$$A_V = \sin\left[\left(\frac{\pi}{2}\right)\left(\frac{\overline{\Delta\$}}{1 - \overline{\$}_{Pc}}\right)\right] \tag{36}$$

Equation (36) is one form of the velocity recruitment-activation equation and is represented graphically in Figure 9.17.a. In an actual operational sequence, the

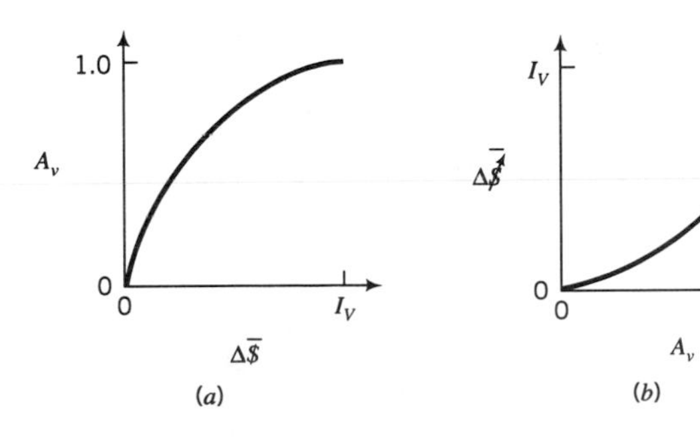

Figure 9.17.a Activation-recruitment relationship. **b** Inverse activation-recruitment relationship.

peripheral motor nerve velocity recruitment is dependent upon spinal cord interneuron activation parameter (A_V) so that the operational form of the velocity recruitment-activation equation is the inverse of equation (36).

Taking the inverse and solving for $\overline{\Delta\$}$:

$$\overline{\Delta\$} = (1 - \overline{\$}_{Pc}) \left[\frac{2}{\pi}\right] \sin^{-1}[A_V] \tag{37}$$

The operational velocity recruitment-activation is illustrated in Figure 9.17.b.

9.2 ISOMETRIC CONTROL SYSTEMS

The systematic consideration of various neuromuscular control systems begins with the force-velocity-length relationship as depicted in Figure 9.18.a. Various configurations of neuromuscular control systems are based upon the relationship of these three parameters. Table 9.4 represents an organizational structure by which specific neuromuscular control systems may be identified. Note that a specific control category is characterized by a unique pattern of the force, velocity, and length parameters. It is important for the human factors engineer to properly assign a specific neuromuscular control system to its characteristic control category. By using such a systematic approach, the engineer will first carefully consider the control task for the human operator and identify the characteristic force, velocity, and length changes during skeletal muscle performance during that task. This is necessary, in part, because of the variation in the complexity, number, and arrangement of the various control system elements with each specific control category. Furthermore, the acquisition of important model parameters will be facilitated by using the appropriate key word variables in database searches of the relative neuromuscular literature.

Referring again to Table 9.4 it is apparent that the characteristic feature of an isometric control system is that the muscle length remains constant regardless of the specific force-time profile. The isometric control category represents the simplest

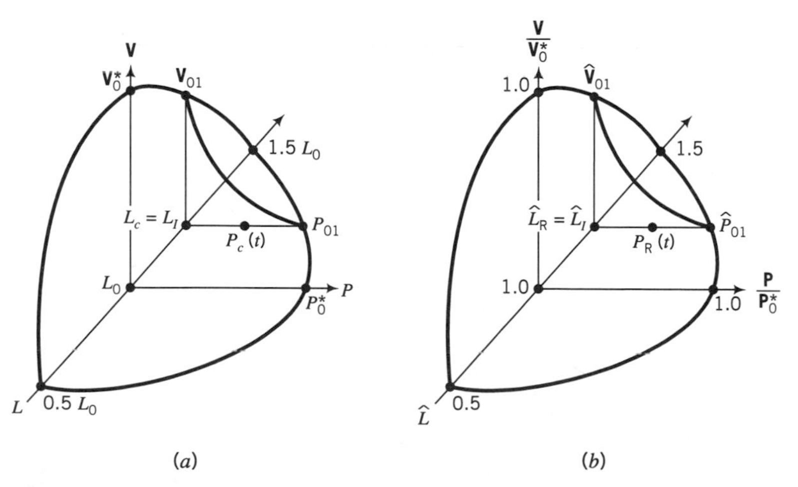

Figure 9.18.a Force-velocity-length relationship for isometric force control. **b** Normalized force-velocity-length relationship for isometric force control.

Table 9.4 Categories of Neuromuscular Control Systems

Control Category	Force	Velocity	Length
I. Isometric[a]		Zero	Constant
A. Isoforce	Constant		
B. Auxoforce	↑		
C. Minuthoforce	↓		
D. Oscilloforce	↑ and ↓		
II. Isotonic[a]	Constant		Varying
A. Isovelocity		Constant	
B. Auxovelocity		↑	
C. Minuthovelocity		↓	
D. Oscillovelocity		↑ and ↓	
III. Auxotonic[b]	Increasing		Varying
A., B., C., D. as per II		as per II	
IV. Minuthotonic[b]	Decreasing		Varying
A., B., C., D. as per II		as per II	
V. Oscillotonic	Increasing and Decreasing		Varying
A., B., C., D. as per II		as per II	

[a] Control Categories covered in Sections 9.2, 9.3, and 9.4.
[b] Control Categories covered in Section 10.1.

form of a neuromuscular control system since, by definition, the velocity of external skeletal muscle shortening is always zero. Note that all four subcategories of an isometric neuromuscular control system will be considered in this chapter. For the remaining neuromuscular control categories, the external skeletal muscle length is varying. It will be apparent from Table 9.4 that the main control categories are identified with respect to the nature of the force parameter. This is because, in classical muscle biophysics, the force-velocity relationship for muscle contraction was so defined that force was identified as the independent variable. Each main category is then subdivided with respect to the specific velocity-time profile that may occur. Recall that velocity refers to the speed of external shortening of the skeletal muscle. Finally, it will be noted that the same four velocity-time profiles are used to identify the subcategories of the other control categories (Table 9.4). This also occurs because of the same historical convention in which velocity is identified as the dependent variable for the force-velocity relationship of skeletal muscle. The specific control categories (other than isometric) are described in Chapter 10 of this text.

a. Constant Force Control System

The first control category to be considered is isometric/constant force. This type of control is required whenever static work is performed and a continuous, non–time-varying force output is desired. The further tightening of an already secured bolt and nut by means of a wrench or the continuous gripping of a pair of pliers as an object is held in place are two common examples of this type of static work effort.

The force-velocity-length relationship for an isometric/constant force control system is depicted in Figure 9.18.a. Note that the coordinate force (P_c) is located at a coordinate length (L_c). The operation of the control system is to maintain P_c as a point on the force-length plane (x-z plane) as a function of time. The normalized force-velocity-length relationship is obtained by normalizing the three respective axes as per equations (3), (4), and (7) as shown in Figure 9.18.b. With respect to Figure 9.18.b, we may then define the following parameters:

$$P_R = \frac{P_c}{P_0^*} \tag{38}$$

$$\hat{L}_R = \frac{L_c}{L_0} \tag{39}$$

$$P' = \frac{P}{P_0^*} \tag{40}$$

The system block diagram for isometric/constant force control is presented in Figure 9.19.a. Each block of this system diagram represents a computational element and the element interconnections represent data flow pathways. Consequently, Figure 9.19.a may be viewed as a flowchart for a computer simulation of the human operator neuromuscular control system. Before proceeding to the identification of specific variables and the element computational equations, let us consider some general characteristics of the system block diagram.

The isometric/constant force control system is divided into two approximate halves by a pair of long vertically dashed lines. The left half of Figure 9.19.a represents the central nervous processor subsystem and the right half represents the neuromuscular actuator subsystem. The two subsystems interact via the feed-forward interface (FFI) and the feedback interface (FBI). Since both FFI and FBI are internal to the human operator (see Figure 9.1) then isometric control (in its simplest form) will depend on internal feedback mechanisms. Isometric control may also depend on external (environmental) feedback when the human operator desires to monitor the result of the isometric force output. One example of this is the human operator interacting with an isometric control stick in order to obtain a displacement output (see Section 9.3.b).

Within the central nervous processor subsystem the controller elements have been identified (G_1 and G_4) as well as the assembly for integral error detection and finally the required force (P_R) and required length (L_R) input parameters. These elements, assembly and parameters collectively define the central nervous processor for the isometric/constant force control system. The neuromuscular actuator subassembly is also defined with respect to its constituent elements (K_P, G_2, and G_{MS}), the force feedback path (via the FBI) and the control parameter output (P). Using the convention of Figure 9.4, the series elements (K_P, G_2, and G_{MS}) constitute the plant that is being controlled. The human factors engineer will note that the elements K_P and G_2 constitute the neuromuscular apparatus and that G_{MS} constitutes the musculoskeletal apparatus that interacts with the external world.

Recall from Chapter 3 that the skeletal system may be simplistically defined as a system of rigid levers acting at fulcrums (which are the joints). Skeletal muscle then inserts, as specific points on these skeletal levers, so that the forces generated at the muscle insertion are subjected to a geometric scale factor with respect to the forces generated at the other end of this skeletal lever (the outside world). For the purposes of understanding neuromuscular control dynamics, it is sufficient to limit our analysis to the output of the neuromuscular apparatus (P'). This simplification will be used for the remainder of this chapter and also for the first section of Chapter 10. In effect, G_{MS} will be approximated as unity but only for the didactic

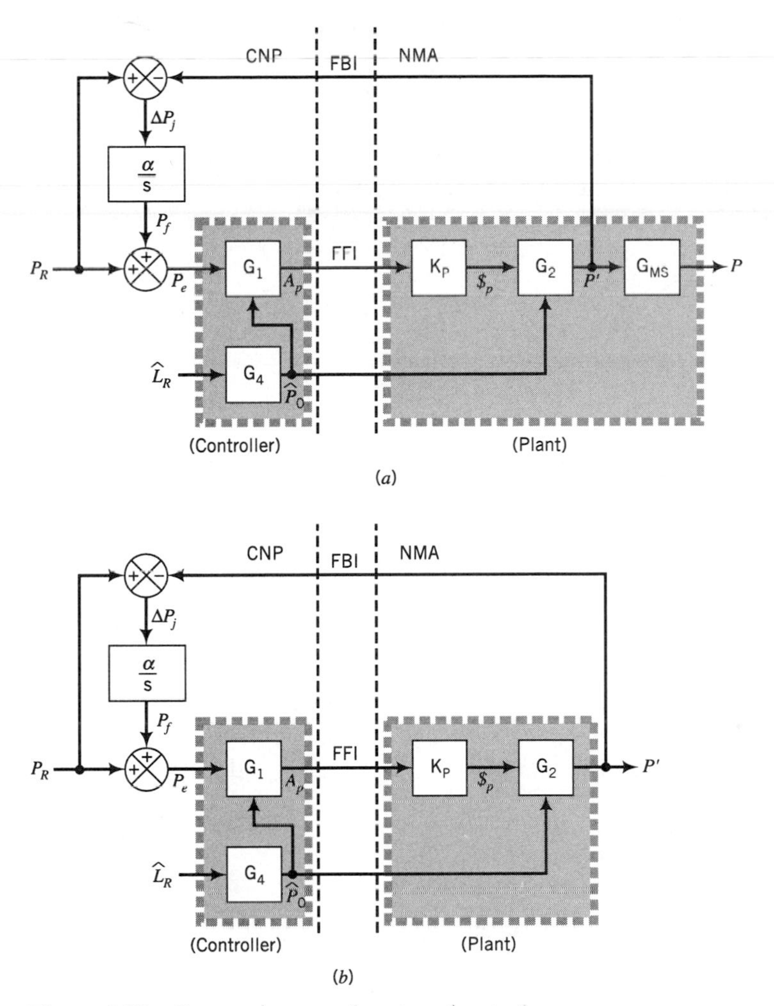

Figure 9.19.a Isometric control system (central nervous processor and neuromuscular actuator). **b** Isometric control system (central nervous processor and neuromuscular apparatus).

purposes of these chapters. Overall system gain is important, however, more important with respect to control system analysis and specifically human operator control. System stability and response time, for example, will certainly be influenced by system gain. Consequently, the effect of the G_{MS} element will be considered in the third section of Chapter 10 where these control theory issues are addressed.

Invoking the approximation that G_{MS} is unity (for didactic purposes only) reduces the system in Figure 9.19.a to that shown in Figure 9.19.b.

For isometric/constant force control, the feedback pathway operates by first calculating differential error as follows:

$$\Delta P_j = P_R - P' \tag{41}$$

The time integral of differential error is then defined as:

$$P_f = \frac{1}{T_f}\sum_{j=1}^{m} \Delta P_j \cdot dt \tag{42}$$

where P_f is the normalized feedback force.

Equation (42) may also be expressed as:

$$P_f = \alpha_f \sum_{j=1}^{m} \Delta P_j \cdot dt \tag{43}$$

where α_f is the force feedback reciprocal time constant:

$$\alpha_f = \frac{1}{T_f}$$

With respect to the feed-forward path of Figure 9.19.b, the normalized error force (P_e) is defined as the output of a summing junction:

$$P_e = P_R + P_f \tag{44}$$

where

P_R = the required normalized force (from the feed-forward pathway);
P_f = the normalized feedback force (from the feedback pathway)

Element G_1 calculates the central nervous processor activation parameter, A_P. This parameter represents the final output of the central nervous processor and is the result of the collective parameters, operations and pathways represented on the left-hand side of Figure 9.19.b. Element G_1 is derived from equation (26) as follows: Normalize equation (26) by P_0^*:

$$A_P = \frac{\dfrac{P}{P_0^*}}{\dfrac{P_0}{P_0^*}} \tag{45}$$

For the input of the element G_1 of Figure 9.19.b:

$$P_e = \frac{P}{P_0^*} \tag{46}$$

Substituting equations (3) and (46) into equation (45):

$$A_P = \frac{P_e}{\hat{P}_0} \tag{47}$$

Element G_1 interacts with element K_P via feed-forward interaction and constitutes the final common motor pathway. Element K_P represents peripheral motor nerve activity, the output of which is the force stimulation parameter, $\$_p$. Recall that the force stimulation parameter represents peripheral motor nerve recruitment of the human's skeletal muscle fibers. Element K_P is defined by equation (30).

Element G_2 represents human skeletal muscle activity. Recall that as more and more skeletal muscle fibers are stimulated, they progressively contribute to the contractile force output of the skeletal muscle. Element G_2 is defined by equation (24) as follows: Normalize equation (24) by P_0^*:

$$\frac{P}{P_0^*} = \left[\left(\frac{1}{2}\right)\left(\frac{P_0}{P_0^*}\right)\right]\left[1 - \cos\left[\pi\left(\frac{\$_P}{\$_{max}}\right)\right]\right] \quad (48)$$

By then substituting equations (3), (27), and (40) into equation (48):

$$P' = \left(\frac{\hat{P}_0}{2}\right)(1 - \cos[\pi\bar{\$}_P]) \quad (49)$$

The computer simulation of the isometric/constant force neuromuscular control system is then completed by substituting $\hat{L}_R$ for $\hat{L}$ of equation (2), so that element G_4 is defined by equation (2).

EXAMPLE 9.1

A laboratory worker (in a standing position) grips (with the finger flexor muscles) a heavy cylinder with one hand. The weight of the cylinder is 0.7 of the P_0^* of the finger flexor muscles, and the coefficient of sliding friction (with a rubber gloved hand) is 1.0. The worker must then lift the cylinder slightly and move it laterally from point A to point B (maintaining the grip) over a 4-s period and then reset the cylinder and release the grip.

Write a computer simulation program for the human operator's neuromuscular control system ($\alpha = 0.25$) when performing this unit step isometric grip function, when:

(a) The cylinder is on a table top so it is gripped at waist height upper arm along the side of the body, elbow bent at 90°, wrist straight. This position allows the length of the finger flexor muscles to be near optimal length so that:

$$L = 1.0L_0$$

and

$$P_0 = 0.7P_0^*$$

Plot the time course of P_R and P' over a 5-s period for this task.

(b) The cylinder is located on an elevated shelf so that the upper arm is raised above shoulder height. The elbow angle is 160° (i.e., forearm extended outward and upward and the wrist cocked downward (ulnar deviation) in order to grip the cylinder. This extends the length of the finger flexor muscles away from optimal length so that:

$$L = 1.35L_0$$

and

$$P_0 = 0.7P_0^*$$

Plot the time course of P_R and P' (over a 5-s period) for this task.

SOLUTION 9.1(a)

An m-file is written in MATLAB for the computer simulation flow chart of Figure 9.19.b. For this program:

$$P_R = 0.7 \text{ (unit step function)}$$
$$\hat{L}_R = 1.0$$
$$\alpha = 0.25$$
$$\Delta t = 5 \text{ s}$$

The program is shown in Figure E9.1.1.

```
% example 9.1(a);
% CA Phillips 15 Sept 1998
clear
% isometric control system
% unit step function
clear
n=51;
r=40;
alpha=0.25;
pi=3.1415926
Psum=0;
Lr=1.0;
Pfeed=0;
Pr=0;
x=1.;
Pohat=sin(pi.*(Lr-0.5));
Ap=Pr./Pohat;
dolhat=(acos(1-2.*Ap))./pi;
Pprim=(Pohat./2.).*(1-cos(pi.*dolhat));
Ppout(1)=Pprim;
for i=2:n
        Pr=.70;
        if i>r
             Pr=0;
        end;
        Prin(i)=Pr;
        x(i)=i;
        Pdelta=Pr-Pfeed;
        Psum=Psum+alpha*Pdelta;
        Pe=Pr + Psum;
        Ap=Pe./Pohat;
        if Ap>1.
               Ap=1.;
        end;
        if Ap<0
               Ap=0;
        end;
        dolhat=(acos(1-2.*Ap))./pi;
        Pprim=(Pohat./2).*(1.-cos(pi.*dolhat));
        Ppout(i)=Pprim;
        Pfeed=Pprim;
end
x=(x-1)/10
```

```
plot (x,Ppout)
hold
plot (x,Prin,'*'
xlabel('time(sec)')
ylabel('Pr  and Pp, *=Pr')
title ('isometric control: unit step')
hold off
```

The plot of the time course of P_R and P' for this task is depicted in Figure E9.1.2.

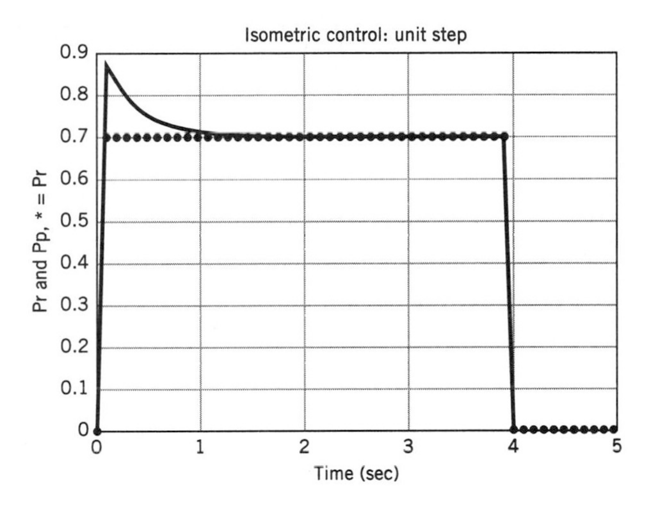

SOLUTION 9.1(b)

An m-file is written in MATLAB for a computer simulation of Figure 9.19.b, with these task parameters:

$$P_R = 0.7 \text{ (unit step function)}$$
$$\hat{L}_R = 1.35$$
$$\alpha = 0.25$$
$$\Delta t = 5 \text{ s}$$

The program is represented in Figure E9.1.3.

```
% example 9.1(b);
% CA Phillips 15 Sept 1998
clear
% isometric control system
% unit step function
clear
```

```
n=51;
r=40;
alpha=0.25;
pi=3.1415926
Psum=0;
Lr=1.35;
Pfeed=0;
Pr=0;
x=1.;
Pohat=sin(pi.*(Lr-0.5));
Ap=Pr./Pohat;
dolhat=(acos(1-2.*Ap))./pi;
Pprim=(Pohat./2.).*(1-cos(pi.*dolhat));
Ppout(1)=Pprim;
for i=2:n
        Pr=.70;
        if i>r
             Pr=0;
        end;
        Prin(i)=Pr;
        x(i)=i;
        Pdelta=Pr-Pfeed;
        Psum=Psum+alpha*Pdelta;
        Pe=Pr + Psum;
        Ap=Pe./Pohat;
        if Ap>1.
               Ap=1.;
        end;
        if Ap<0
               Ap=0;
        end;
        dolhat=(acos(1-2.*Ap))./pi;
        Pprim=(Pohat./2).*(1.-cos(pi.*dolhat));
        if i>r
               Pprim=0;
        end;
        Ppout(i)=Pprim;
        Pfeed=Pprim;
end
x=(x-1)/10
plot (x,Prin,'*')
hold
plot (x,Ppout)
xlabel('time(sec)')
ylabel('Pr  and Pp, *=Pr')
title ('isometric control: unit step')
hold off
```

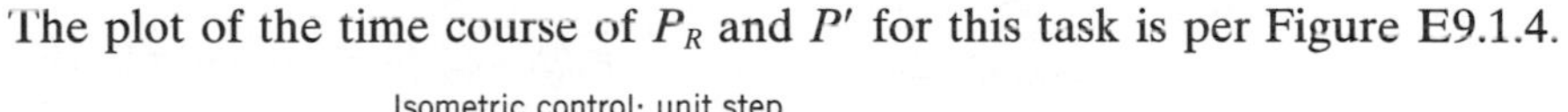

The plot of the time course of P_R and P' for this task is per Figure E9.1.4.

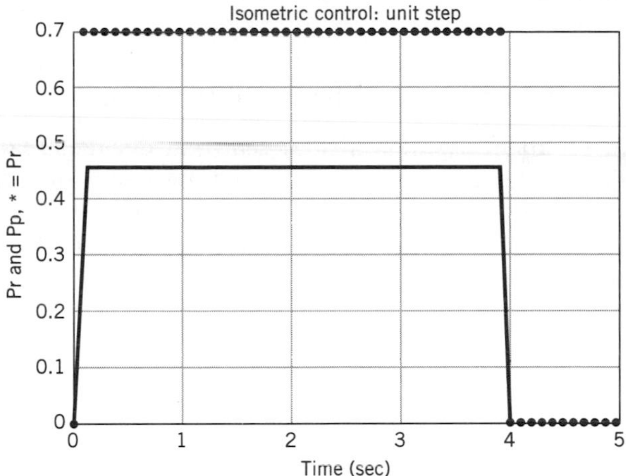

b. Variable Force Control Systems

It is common to think of static work as isometric muscle contractions at a constant force. However, many tasks performed by the human operator, which are of practical interest to the human factors engineer, involve static work in which the isometric muscle contraction generates a time-varying force. One common type of isometric force as a function of time is a progressive, ramp-like increase in isometric force development. Typically, this force-time profile is anticipatory as the human operator prepares to transition from static work to dynamic work (see Example 9.2).

A computer simulation of the isometric/increasing force control system is simply a modification of the isometric/constant force system described in Section 9.2.a. Basically, the required force-time profile, $P_R(t)$, now represents an increasing ramp isometric force function. Recall that this is one of specific input parameters, which is characteristic of the central nervous processor (see Figure 9.19.b). Otherwise, the system elements, control pathways, and other specific parameters remain unchanged.

EXAMPLE 9.2

An elderly person is seated in a chair and prepares to rise from the chair by leaning forward, placing his hands on the arm rests, and initially tensing the thigh muscles at a 0.1 P_0^* force level. Due to age-related muscle weakening, the quadriceps (anterior thigh) muscles must subsequently exert a progressive force increase and must obtain 0.9 of the P_0^* for these muscles in order to proceed from a static (seated) to dynamic (rising) posture.

Write a computer simulation program for this human operator's neuromuscular control system ($\alpha = 0.5$) when performing this ramp isometric leg muscle tensing, when:

(a) The person is seated on a firm and level chair seat, such that his hips' center-of-rotation is horizontally 2 in. above the knee center-of-rotation. The knees are bent near 90°, and the lower legs are perpendicular to the ground. This position allows the length of the quadriceps (anterior thigh) muscles to be near their optimal length; so that:

$$L = 1.0L_0$$

Plot the time course of P_R and P' over a 4-s period for this task.

(b) The person is seated on a soft-cushioned chair seat that declines rearward at a height such that his hips' center-of-rotation is horizontally 2 in. below the knee center-of-rotation. His knees are bent near 90° and his lower legs are perpendicular to the ground. This position results in the length of the quadriceps (anterior thigh muscles) to be away from optimal length so that:

$$L = 1.35L_0$$

Plot the time course of P_R and P' over a 4-s period for this task.

SOLUTION 9.2(a)

An m-file is written in MATLAB for the computer simulation flow chart of Figure 9.19.b. For this program:

$$P_R = \text{ramp function}$$
$$\hat{L}_R = 1.0$$
$$\alpha = 0.5$$
$$\Delta t = 4\text{ s}$$

The program is shown in Figure E9.2.1.

```
% example 9.2(a)
% CA Phillips 15 Sept 1998
% isometric control system
% ramp input function
clear
n=17;
alpha=0.5;
Psum=0;
Lr=1.0;
Pfeed=0;
Pr=.1;
Prin(1)=Pr;
x=1;
PoHat=sin(pi*(Lr-0.5));
Ap=Pr./PoHat;
dolhat=(acos(1-2.*Ap))/pi;
Pprim=(PoHat./2).*(1-cos(pi.*dolhat));
Ppout(1)=Pprim;
for i=2:n
        Pr=Pr+.05;
        Prin(i)=Pr;
        x(i)=i;
        Pdelta=Pr-Pfeed;
      Psum=Psum+alpha*Pdelta;
```

```
        Pc=Psum+Pr;
        Ap=Pe/PoHat;
        if Ap>1.
            Ap=1.;
        end;
        if Ap<0
            Ap=0;
        end;
        dolhat=(acos(1-2*Ap))/pi;
        Pprim=(PoHat/2)*(1-cos(pi*dolhat));
        Ppout(i)=Pprim;
        Pfeed=Pprim;
end
x=(x-1)/4
plot(x,Ppout);
hold;
plot (x,Prin,'*');
xlabel('time(sec)');
ylabel('Pr and Pp, *=Pr');
title ('isometric control: ramp input');
hold off;
```

The plot of the time course for P_R and P' is illustrated in Figure E9.2.2.

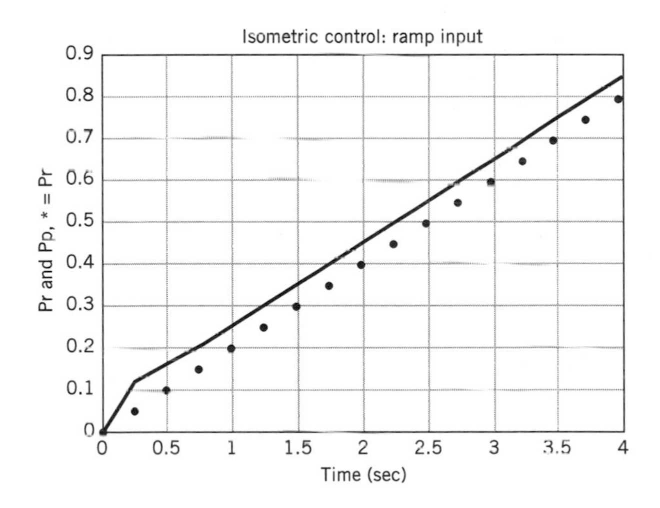

SOLUTION 9.2(b)

An m-file is written for MATLAB for the computer simulation flow chart of Figure 9.19.b. For this program:

$$P_R = \text{ramp function}$$
$$\hat{L}_R = 1.35$$
$$\alpha = 0.5$$
$$\Delta t = 4 \text{ s}$$

The program is shown in Figure E9.2.3.

```
% example 9.2(b)
% CA Phillips 15 Sept 1998
% isometric control system
% ramp input function
clear
n=17;
alpha=0.5;
Psum=0;
Lr=1.25
Pfeed=0;
Pr=.1;
Prin(1)=Pr;
x=1;
PoHat=sin(pi*(Lr-0.5));
Ap=Pr./PoHat;
dolhat=(acos(1-2*Ap))/pi;
Pprim=(PoHat./2).*(1-cos(pi.*dolhat));
Ppout(1)=Pprim;
for i=2:n
        Pr=Pr+.05;
        Prin(i)=Pr;
        x(i)=i;
        Pdelta=Pr-Pfeed;
      Psum=Psum+alpha*Pdelta;
        Pe=Psum+Pr;
        Ap=Pe/PoHat;
      if Ap>1.
            Ap=1.;
      end;
      if Ap<0
            Ap=0;
      end;
        dolhat=(acos(1-2*Ap))/pi;
        Pprim=(PoHat/2)*(1-cos(pi*dolhat));
        Ppout(i)=Pprim;
        Pfeed=Pprim;
end;
x=(x-1)/4;
plot(x,Prin,'*');
hold;
plot(x,Ppout);
xlabel('time(sec)');
ylabel('Pr and Pp, *=Pr');
title ('isometric control: ramp input');
hold off;
```

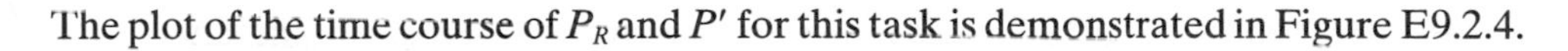

The plot of the time course of P_R and P' for this task is demonstrated in Figure E9.2.4.

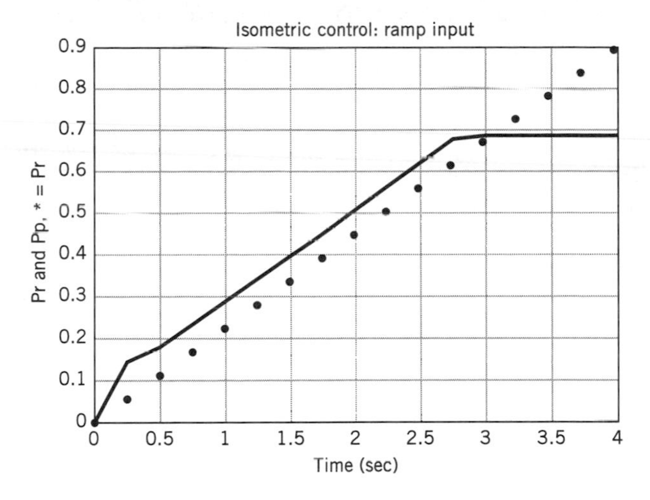

Another important category of isometric control systems with variable force is one in which the isometric force-time profile oscillates between a minimum and maximum value. The isometric force may be an applied force developed by the human operator to perform static work, such as when using a hand tool (see Example 9.3). Alternatively, the isometric force may be a reaction force developed by the human operator in response to an externally applied force. Common examples would be the arm reaction forces when holding a handrail during a bumpy bus ride, holding an arm rest during a turbulent airplane ride, or stabilizing a pneumatic jackhammer while it is operating.

Making the approximation that the isometric time-varying force is periodic and that the force minima and maxima are relatively constant over the time course of the static work will allow the human factors engineer to approximate $P_R(t)$ as a sinusoidal function. It is understood that this represents only a first approximation of the isometric/oscillating force control system, but it is a generally useful approximation for many practical engineering applications of interest to the human factors engineer. As a consequence of this simplifying approximation, the isometric/oscillating force control system becomes a rather straightforward modification of the isometric/isoforce control system. For the isometric/oscillating force control system, all system elements, control pathways and specific parameters other than $P_R(t)$ remain as depicted in Figure 9.19.b. By definition, the isometric skeletal muscle length, $\hat{L}_R$, must remain constant despite significant time-varying fluctuations in the isometric force. For most practical situations, small variations in $\hat{L}_R$ certainly do occur but will have minimal effect over the middle range of $\hat{L}_R$ values (0.75–1.25).

EXAMPLE 9.3

An industrial worker is standing erect and holds a power grinder/sander in a static position so that it rests upon a 2 × 2 in. rectangular bar of metal of 6-ft length. The bar lengths are continuously fed past the worker end-to-end by an automatic conveyor belt. At regular intervals along each bar there is a rough zone that is ground/sanded smooth by the worker increasing the downward force exerted at the top of the grinder/sander. This is by forearm extension at the elbows so that the triceps [upper arm] muscles are utilized. The downward force is exerted in a sinusoidal pattern (of cycle duration, $T = 10$ s) varying between 0.1 of the P_0^* for the triceps muscles and 0.9 of their P_0^*, the higher P_0^* fraction occurring along the rough zone.

Write a computer simulation program for the human operator's neuromuscular control system ($\alpha = 0.2$) when performing this sinusoidal isometric pressing function, when the bar is located on the conveyor belt so that:

(a) The person's upper arms are at the side, the elbow is bent at 90° and the forearms are extended straight out (horizontal to the floor surface). This position allows the length of the triceps (upper arm muscle) to be near its optimal length so that:

$$L = 1.0L_0$$

Plot of the time course of P_R and P' over 1.5 sinusoidal force cycles (15 s).

(b) The person must reach out with the upper arms and lower arms extended outward and downward (at an elbow angle of 150°) in order to reach the top of the metal bar with the sander/grinder. This position results in the length of the triceps muscle to be away from its optimal length so that:

$$L = 0.75L_0$$

Plot the course of P_R and P' over 1.5 sinusoidal force cycles (15 s).

SOLUTION 9.3(a)

An m-file is written in MATLAB for the computer simulation flow chart of Figure 9.19.b. For this program:

$$P_R = \text{sine function}$$
$$L_R = 1.0$$
$$\alpha = 0.2$$
$$\Delta t = 15 \text{ s}$$

The program is presented in Figure E9.3.1.

```
% example 9.3(a)
% CA Phillips 15 Sept 1998
% isometric control system
% sinusoidal input function
clear
n=11;
Psum=0;
Lr=1.0;
alpha=0.2;
Pfeed=0;
Pr=.5;
Prin(1)=Pr;
x=1;
PoHat=sin(pi*(Lr-0.5));
```

```
Ap=Pr./PoHat;
dolhat=(acos(1-2*Ap))/pi;
Pprim=(PoHat./2).*(1-cos(pi.*dolhat));
Ppout(1)=Pprim;
for i=2:16
        Pr=.5+.4*sin(2*pi*i/n);
        Prin(i)=Pr;
        x(i)=i;
        Pdelta=Pr-Pfeed;
        Psum=Psum+Pdelta;
      Pf=alpha*Psum;
        Pe=Pr + Pf;
        Ap=Pe/PoHat;
      if Ap>1.
            Ap=1.;
      end;
      if Ap<0
            Ap=0;
      end;
        dolhat=(acos(1-2*Ap))/pi;
        Pprim=(PoHat/2)*(1-cos(pi*dolhat));
        Ppout(i)=Pprim;
        Pfeed=Pprim;
end
x=x-1;
plot(x,Ppout);
hold;
plot(x,Prin,'*');
xlabel('time(sec)')
ylabel('Pr and Pp, *=Pr');
title ('isometric control: sinusoidal input');
hold off;
```

The plot of the time course for P_R and P' is graphed in Figure E9.3.2.

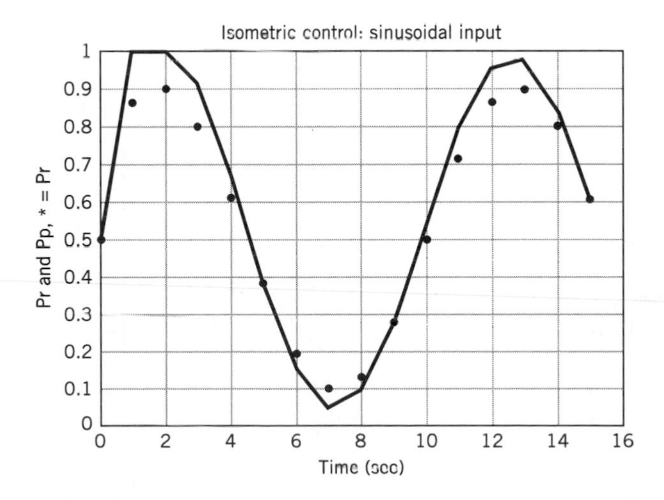

SOLUTION 9.3(b)

An m-file is written for MATLAB for the computer simulation flow chart of Figure 9.19.b. For this program:

$$P_R = \text{sine function}$$
$$L_R = 0.75$$
$$\alpha = 0.2$$
$$\Delta t = 15\ \text{s}$$

The program is per Figure E9.3.3.

```
% example 9.3(b)
% CA Phillips 15 Sept 1998
% isometric control system
% sinusoidal input function
clear
n=11;
Psum=0;
Lr=0.75;
alpha=0.2;
Pfeed=0;
Pr=.5;
Prin(1)=Pr;
x=1;
PoHat=sin(pi*(Lr-0.5));
Ap=Pr./PoHat;
dolhat=(acos(1-2*Ap))/pi;
Pprim=(PoHat./2).*(1-cos(pi.*dolhat));
Ppout(1)=Pprim;
for i=2:16
        Pr=.5+.4*sin(2*pi*i/n);
        Prin(i)=Pr;
        x(i)=i;
        Pdelta=Pr-Pfeed;
        Psum=Psum+Pdelta;
      Pf=alpha*Psum;
        Pe=Pr + Pf;
        Ap=Pe/PoHat;
      if Ap>1.
            Ap=1.;
      end;
      if Ap<0
            Ap=0;
      end;
        dolhat=(acos(1-2*Ap))/pi;
        Pprim=(PoHat/2)*(1-cos(pi*dolhat));
        Ppout(i)=Pprim;
        Pfeed=Pprim;
end;
x=x-1;
plot(x,Prin,'*');
```

```
hold;
plot(x,Ppout);
xlabel('time(sec)');
ylabel('Pr and Pp, *-Pr');
title ('isometric control: sinusoidal input');
hold off;
```

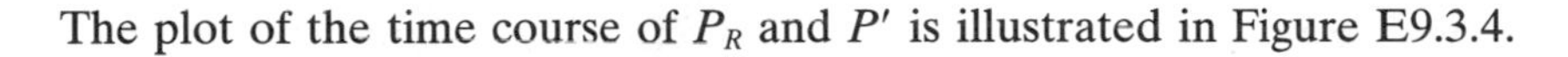
The plot of the time course of P_R and P' is illustrated in Figure E9.3.4.

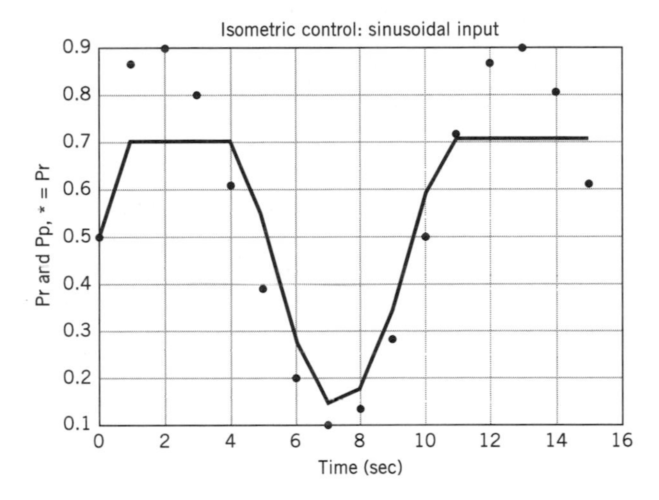

9.3 NEUROMUSCULAR STATIC CONTROL

When a human operator performs static work, force is generated by the operator's musculoskeletal system but there is no external displacement as a result of that generated force. With respect to the skeletal muscle of the human operator, the muscle contraction is referred to as isometric (constant length). Recall that sustained static muscular effort results in a static workload experienced by the human operator. The quantitative analysis of this static workload has been previously presented (see Chapter 8). That analysis was independent of any consideration of the human operator neuromuscular control system that generated the static muscular effort. However, it is equally as important for the human factors engineer to understand the isometric control system when the human operator is performing static work.

a. Isometric Control with Physiological Fatigue

Recall that physiological fatigue is uniquely characteristic of human operator neuromuscular control systems. Indeed, fatigue was one of the characteristics that differentiated human operator control from automatic control systems. Furthermore, recall that an operational definition of physiological fatigue allows the human factors engineer to employ a characteristic biophysical equation [see equation (23)].

When an isometric force is sustained for a prolonged period of time, the human operator experiences skeletal muscle fatigue, which is manifested as a progressive decrease in the skeletal muscle force output. It is important to realize that during the period of sustained isometric force output and during the period of progressive force output declines, the skeletal muscle of the human operator is still performing within the force-velocity-length relationship characterized by Figure 9.11. Physiological fatigue occurs when the isometric force output is sustained for a prolonged period of time. The force output is usually some relatively constant value, and there is no intermittent cycling of force output with some rest period (zero force output). The reason for this is that skeletal muscle recovers from fatigue relatively rapidly with even brief periods of rest (zero force output). A rest interval as short as 10 min will result in a return of skeletal muscle force output from complete fatigue to approximately 70% of the unfatigued force output. A rest interval of 1 hr will allow skeletal muscle to return from a completely fatigued state to over 90% of its original unfatigued force output. Consequently, work-rest cycles are important to the human factors engineer to optimize the human operator endurance time. However, the isometric control system described in this section will consider only the case of a sustained isometric force output by the human operator neuromuscular control system. It is recognized that the design of human-technological systems will never deliberately stress the human operator to the point of complete skeletal muscle fatigue. However, it is instructive to consider an isometric control system with physiological fatigue in order to clearly demonstrate that the human neuromuscular control system is not simply extensions of automatic control theory and the use of a few scaling constants.

For the case of an isometric control system with physiological fatigue, we first modify equation (40):

$$P'_F = \frac{P_F}{P_0^*} \tag{50}$$

where the subscript "F" denotes an isometric neuromuscular control system that is either in a state of fatiguing, or in an unfatigued state that over sustained time has the potential to fatigue.

The system block diagram for an isometric control system with physiological fatigue is a modification of the isometric/constant force control system of Figure 9.19.b. The system block diagram for an isometric control system with physiological fatigue is shown in Figure 9.20. The mathematical development of this system proceeds as follows.

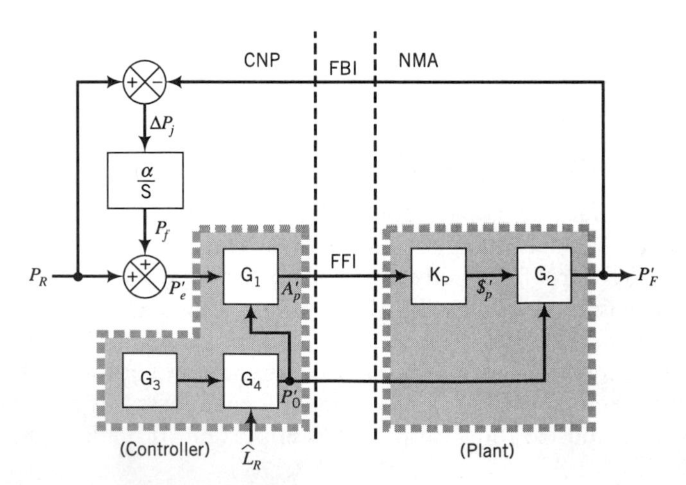

Figure 9.20 Isometric control system with physiological fatigue.

Substituting into equation (41):

$$\Delta P_i = P_R - P'_F \tag{51}$$

Substituting equation (51) into equation (42):

$$P_f = \frac{1}{T_f}\sum_{i-1}^{n} \Delta P_i \cdot dt \tag{52a}$$

which may be written in its alternative form:

$$P_{fi} = \alpha_f \sum_{i=1}^{n} \Delta P_i \cdot dt \tag{52b}$$

where P_f is the normalized feedback force, and α_f is the force feedback reciprocal time constant:

$$\alpha_f = \frac{1}{T_f}$$

Recall that total error (P_e) for the nonfatiguing muscle was expressed as equation (44). Substituting equation (52) for P_f in equation (44) results in the total error (P'_e). For fatiguing muscle exertion:

$$P'_e = P_R + P_f \tag{53}$$

Element G_3 is a clock that generates normalized time (t') as follows:

$$t' = \frac{1}{T_F}\sum_{j=1}^{m} j \cdot \Delta t \tag{54}$$

Since the differential time (Δt) is constant:

$$t' - \frac{\Delta t}{T_F}\sum_{j=1}^{m} j \tag{55}$$

Element G_3 is defined by equation (55).

Element G_4 generates a time varying P_0 that is normalized by P_0^* and progressively decreases as a function of fatigue ($\hat{P}'_0$). Element G_4 is defined by equation (23).

Element G_1 is derived from equation (45) as follows:

$$A'_P = \frac{\dfrac{P}{P_0^*}}{\dfrac{P_0(t)}{P_0^*}} \tag{56}$$

For the input of the element G_1 of Figure 9.20:

$$P'_e = \frac{P}{P_0^*} \tag{57}$$

Substituting equations (20) and (57) into equation (56) results in:

$$A'_P = \frac{P'_e}{\hat{P}'_0} \tag{58}$$

Element G_1 is defined by equation (58).

Element K_P is derived by substituting equation (58) into equation (30):

$$\bar{\$}'_P = \left[\frac{1}{\pi}\right] \cos^{-1}(1 - 2A'_P) \tag{59}$$

where:

$$\bar{\$}'_P = \frac{\$_P}{\$_{\max}} \tag{60}$$

Element K_P is defined by equation (59).

Element G_2 is derived from equation (48) as follows:

$$\frac{P_F}{P_0^*} = \left[\left(\frac{1}{2}\right)\left(\frac{P_0(t)}{P_0^*}\right)\right]\left[1 - \cos\left[\pi\left(\frac{\$_P}{\$_{\max}}\right)\right]\right] \tag{61}$$

Substituting equations (20), (50), and (60) into equation (61):

$$P'_F = \left(\frac{\hat{P}'_0}{2}\right)[1 - \cos[\pi\bar{\$}'_P]] \tag{62}$$

Element G_2 is defined by equation (62).

The computer simulation of the isometric neuromuscular control system with physiological fatigue is completed by setting $\hat{L} = \hat{L}_R$ of equation (2) and element G_4 is defined by equation (23).

EXAMPLE 9.4

A human factors engineer is consulted on the design of a coal car for mining applications. In the anticipated design configuration, the worker rides at the back of the car, facing forward, as the car travels down the mine tracks. The worker pulls on an isometric friction brake as necessary to slow the mining car speed (and also assist in stopping the car).

Three locations for the isometric friction brake handle are being considered by the manufacturer:

(a) Near the middle of the car side in which the operator extends the upper arm outward and downward, the elbow angle at 150°, and the forearm extended outward and downward. The trapezius (posterior shoulder) muscle is extended to near optimal length (L_0) for this muscle.

(*a*) **Diagram A**

(b) About three-quarters back from the front of the car, along the car side, so that the operator holds the upper arm along the side of the body, elbow bent at a 90°

angle, with the forearm extended straight out from the body. The trapezius muscle is now shortened to about 80% of its optimal length (L_0).

(b) **Diagram B**

(c) At the rear of the coal car, to the side of the worker, so that the operator angles the upper arm downward and backward, the elbow angle is 45°, and the forearm is extended forward, but behind the operator. The trapezius muscle is now shortened to about 65% of its optimal length (L_0).

(c) **Diagram C**

Write a computer simulation program for the human operator's neuromuscular control system ($\alpha = 0.5$) when performing a unit step isometric pull (with the posterior shoulder muscle) if the person exerts a P_R of 40% of the trapezius muscle P_0^* and sustains the pull for 15 min, when:

(a) $L_R = 1.0$, Condition a; and plot the time course of P_R and P' for the 15-min pull.

(b) $L_R = 0.8$, Condition b; and plot the time course of P_R and P' for the 15-min pull.

(c) $L_R = 0.65$, Condition c; and plot the time course of P_R and P' for the 15-min pull.

SOLUTION 9.4(a)

An m-file is written in MATLAB for the computer simulation flow chart of Figure 9.20. For this program:

$$P_R = 0.4 \text{ (unit step function)}$$
$$L_R = 1.0$$
$$\alpha = 0.5$$

The program is depicted in Figure E9.4.1.

```
% example 9.4(a)
% CA Phillips 15 Sept 1998
%isometric muscle control
%static (unit step input)
%with physiological fatigue
clear
%timrat is the Delta t used for time iteration
timrat=.02;
n=50;
Psum=0;
Lr=1.0;
Prin=.4;
alpha=0.5;
Pfeed=0;
Pdelta=0;
Pdelarr(1)=Pdelta;
Pr=0;
%G3 time
tprime=timrat;
tpriarr(1)=tprime.*15;
PoHat=sin(pi*(Lr-0.5));
%G4 Po hat prime
Pohapr=PoHat*(.15+.85*(1-sin(pi*tprime/2)));
Pohparr(1)=Pohapr;
Prinarr(1)=Pr;
Ap=Pr/Pohapr;
if AP>1
        Ap=1;
end
if Ap<0
      Ap=0;
end
Aparr(1)=Ap;
dolhat=(acos(1-2*Ap))/pi;
dolharr(1)=dolhat;
Pprim=(Pohapr/2)*(1-cos(pi*dolhat));
Ppout(1)=Pprim;
        for t=2:n
                %G3 time
                tprime=timrat*t;
                tpriarr(t)=tprime.*15;
                %G4 Po hat prime
                Pohapr=PoHat*(.15+.85*(1-sin(pi*tprime/2)));
                Pohparr(t)=Pohapr;
                for i=2:5
                      %i is incremental time during Delta t
                        %Unit Step set to  Prin
                        Pdelta=Prin-Pprim;
                      Psum=Psum + alpha*Pdelta;
                      Pe=Prin + Psum;
                      Ap=Pe/Pohapr;
                        if Ap>1
                                Ap=1;
                        end
```

```
                    if Ap<0
                            Ap-0;
                    end
                        dolhat=(acos(1-2*Ap))/pi;
                        Pprim=(Pohapr/2)*(1-cos(pi*dolhat));
              end
            Prinarr(t)=Prin;
              Pdelarr(t)=Pdelta;
              Aparr(t)=Ap;
              dolharr(t)=dolhat;
              Ppout(t)=Pprim;
      end
plot(tpriarr,Ppout);
hold;
plot(tpriarr,Prinarr,'*');
title('isometric/static control with fatigue');
xlabel('time(min)');
ylabel('Pr and Pp, *=Pr');
hold off;
```

The plot of the time course of P_R and P' is demonstrated in Figure E9.4.2.

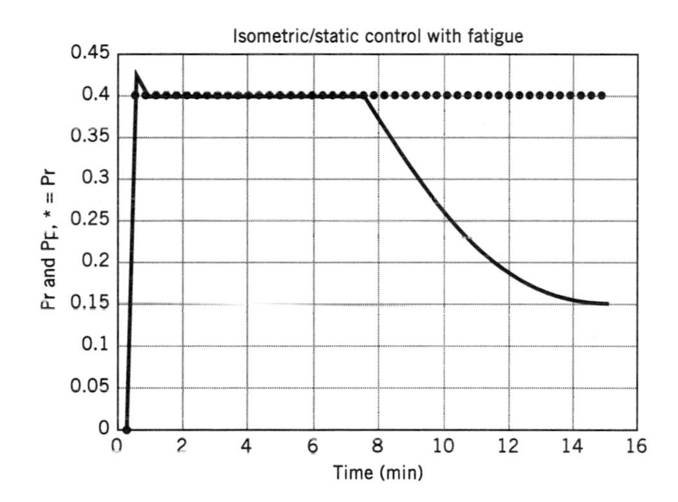

SOLUTION 9.4(b)

An m-file is written for MATLAB for the computer simulation flow chart of Figure 9.20. For this program:

$$P_R = 0.4 \text{ (unit step function)}$$
$$L_R = 0.8$$
$$\alpha = 0.5$$

The program is represented in Figure E9.4.3.

```
% example 9.4(b)
% CA Phillips 15 Sept 1998
%isometric muscle control
%static (unit step input)
%with physiological fatigue
clear
%timrat is the Delta t used for time iteration
timrat=.02;
n=50;
Psum=0;
Lr=.8;
Prin=.4;
alpha=0.5;
Pfeed=0;
Pdelta=0;
Pdelarr(1)=Pdelta;
Pr=0;
%G3 time
tprime=timrat;
tpriarr(1)=tprime*15;
PoHat=sin(pi*(Lr-0.5));
%G4 Po hat prime
Pohapr=PoHat*(.15+.85*(1-sin(pi*tprime/2)));
Pohparr(1)=Pohapr;
Prinarr(1)=Pr;
Ap=Pr/Pohapr;
if AP>1
        Ap=1;
end
if Ap<0
      Ap=0;
end
Aparr(1)=Ap;
dolhat=(acos(1-2*Ap))/pi;
dolharr(1)=dolhat;
Pprim=(Pohapr/2)*(1-cos(pi*dolhat));
Ppout(1)=Pprim;
        for t=2:n
                %G3 time
                tprime=timrat*t;
                tpriarr(t)=tprime*15;
                %G4 Po hat prime
                Pohapr=PoHat*(.15+.85*(1-sin(pi*tprime/2)));
                Pohparr(t)=Pohapr;
                for i=2:5
                      %i is incremental time during Delta t
                        %Unit Step set to  Prin
                        Pdelta=Prin-Pprim;
                      Psum=Psum + alpha*Pdelta;
                      Pe=Prin + Psum;
                      Ap=Pe/Pohapr;
                        if Ap>1
                                Ap=1;
```

```
                end
            if Ap<0
                    Ap=0;
            end
                dolhat=(acos(1-2*Ap))/pi;
                Pprim=(Pohapr/2)*(1-cos(pi*dolhat));
        end
        Prinarr(t)=Prin;
          Pdelarr(t)=Pdelta;
          Aparr(t)=Ap;
          dolharr(t)=dolhat;
          Ppout(t)=Pprim;
    end
plot(tpriarr,Ppout);
hold;
plot(tpriarr,Prinarr,'*');
title('isometric/static control with fatigue');
xlabel('time(min)');
ylabel('Pr and Pp, *=Pr');
hold off;
```

The plot of the time course of P_R and P' is graphed in Figure E9.4.4.

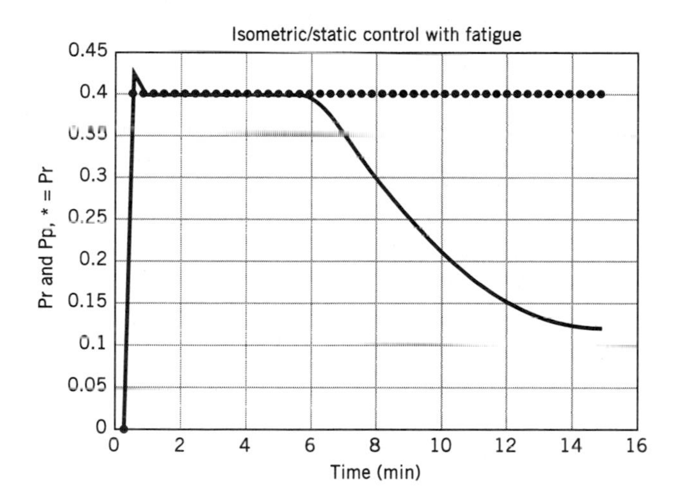

SOLUTION 9.4(c)

An m-file is written in MATLAB for the computer simulation flow chart of Figure 9.20. For this program:

$$P_R = 0.4 \text{ (unit step function)}$$
$$L = 0.65$$
$$\alpha = 0.5$$

The program is printed in Figure E9.4.5.

```
% example 9.4(c)
% CA Phillips 15 Sept 1998
%isometric muscle control
%static (unit step input)
%with physiological fatigue
clear
%timrat is the Delta t used for time iteration
timrat=.02;
n=50;
Psum=0;
Lr=0.65;
Prin=.4;
alpha=0.5;
Pfeed=0;
Pdelta=0;
Pdelarr(1)=Pdelta;
Pr=0;
%G3 time
tprime=timrat;
tpriarr(1)=tprime.*15;
PoHat=sin(pi*(Lr-0.5));
%G4 Po hat prime
Pohapr=PoHat*(.15+.85*(1-sin(pi*tprime/2)));
Pohparr(1)=Pohapr;
Prinarr(1)=Pr;
Ap=Pr/Pohapr;
if AP>1
        Ap=1;
end
if Ap<0
      Ap=0;
end
Aparr(1)=Ap;
dolhat=(acos(1-2*Ap))/pi;
dolharr(1)=dolhat;
Pprim=(Pohapr/2)*(1-cos(pi*dolhat));
Ppout(1)=Pprim;
        for t=2:n
                %G3 time
                tprime=timrat*t;
                tpriarr(t)=tprime.*15;
                %G4 Po hat prime
                Pohapr=PoHat*(.15+.85*(1-sin(pi*tprime/2)));
                Pohparr(t)=Pohapr;
                for i=2:5
                      %i is incremental time during Delta t
                        %Unit Step set to  Prin
                        Pdelta=Prin-Pprim;
                      Psum=Psum + alpha*Pdelta;
                      Pe=Prin + Psum;
                      Ap=Pe/Pohapr;
                        if Ap>1
                                Ap=1;
                        end
```

```
                        if Ap<0
                                Ap=0;
                        end
                          dolhat=(acos(1-2*Ap))/pi;
                          Pprim=(Pohapr/2)*(1-cos(pi*dolhat));
                end
              Prinarr(t)=Prin;
                Pdelarr(t)=Pdelta;
                Aparr(t)=Ap;
                dolharr(t)=dolhat;
                Ppout(t)=Pprim;
        end
plot(tpriarr,Ppout);
hold;
plot(tpriarr,Prinarr,'*');
title('isometric/static control with fatigue');
xlabel('time(min)');
ylabel('Pr and Pp, *=Pr');
hold off;
```

The plot of the time course of P_R and P'_F is presented in Figure E9.4.6.

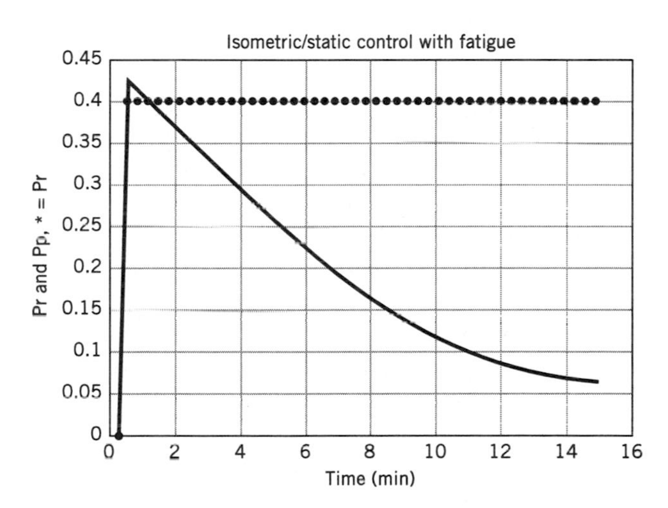

b. Human-Technological Control: Isometric Stick

This subsection will allow us to examine the human operator neuromuscular control system as it interacts with a technological system and the environmental system. Recall from Figure 9.1 that these three systems (human operator, technological, and environmental) interact with each other in various combinatorial configurations depending upon the human operator, the device, the task, and the environment. Figure 9.21 represents the human-technological control system of interest. The human operator system is characterized by the two customary subsystems: a central nervous processor subsystem and a neuromuscular actuator subsystem. Note in Figure 9.21 that the central nervous processor subsystem is represented by all of the

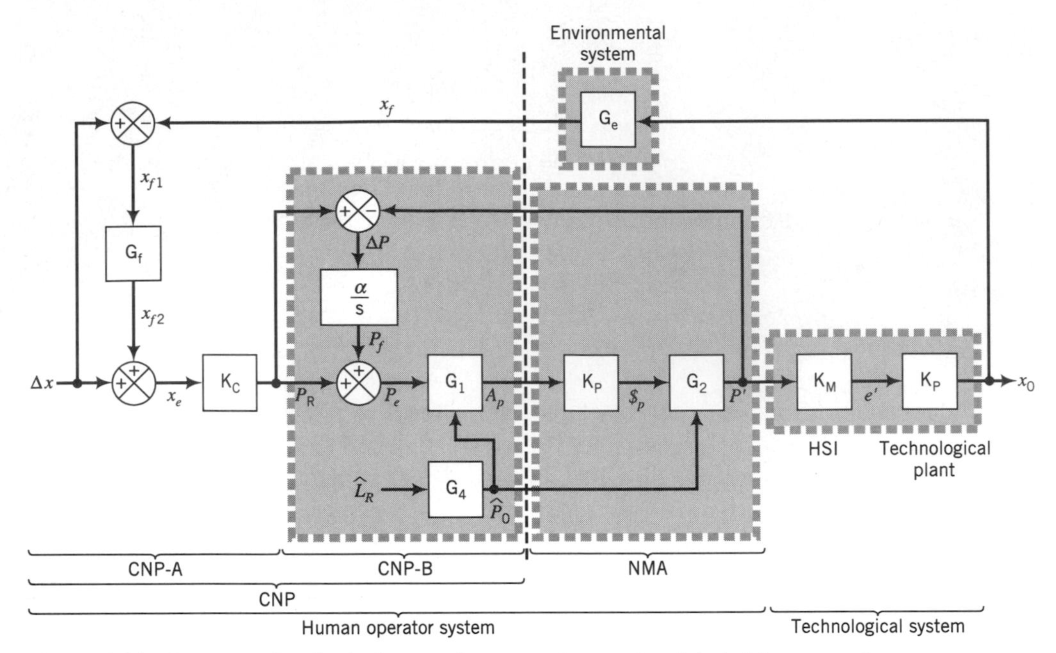

Figure 9.21 Human-technological control system: isometric stick (with zero-order technological plant).

elements, pathways and parameters to the left of the vertical dashed line identified as CNP. The neuromuscular actuator (NMA) is enclosed within the boxed dashed line and consists of the two elements within the neuromuscular apparatus and the feedback pathway internal to the human operator. Recall that implicit in this system block diagram is that the musculoskeletal interface element (G_{MS}) is set to unity and not expressly identified. The human system in Figure 9.21 will operate as an isometric control system.

For human-technological control, it is convenient to divide the central nervous processor subsystem into a CNP-A subsystem and a CNP-B subsystem. The CNP-B subsystem consists of anatomical elements represented by the spinal cord and brain stem. This subsystem has been delineated in Figure 9.21 for two reasons. One, the CNP-B subsystem represents that part of the central nervous processor subsystem that when combined with the CNP-A subsystem accounts for the internal (force) feedback of the human operator. Two, the CNP-B subsystem combined with the NMA subsystem (of Figure 9.22) represents the isometric neuromuscular control system, which was described in Section 9.2. Recall that the isometric neuromuscular control systems of Examples 9.1, 9.2, and 9.3 differed primarily with respect to the definition of the required input isometric force, $P_R(t)$. Otherwise, the governing equations for the CNP-B subsystem and NMA subsystem have previously been developed in Section 9.2. This will make our task of developing the human-technological control system using an isometric stick somewhat more simplified. In effect, what now remains is to develop the appropriate mathematical relationships for the technological system, environmental system, and CNP-A subsystem of Figure 9.21. We shall now proceed to do so, in that respective order.

The technological system is composed of two subsystems: the human-system interface (HSI) and the technological plant. This is based upon the model of the

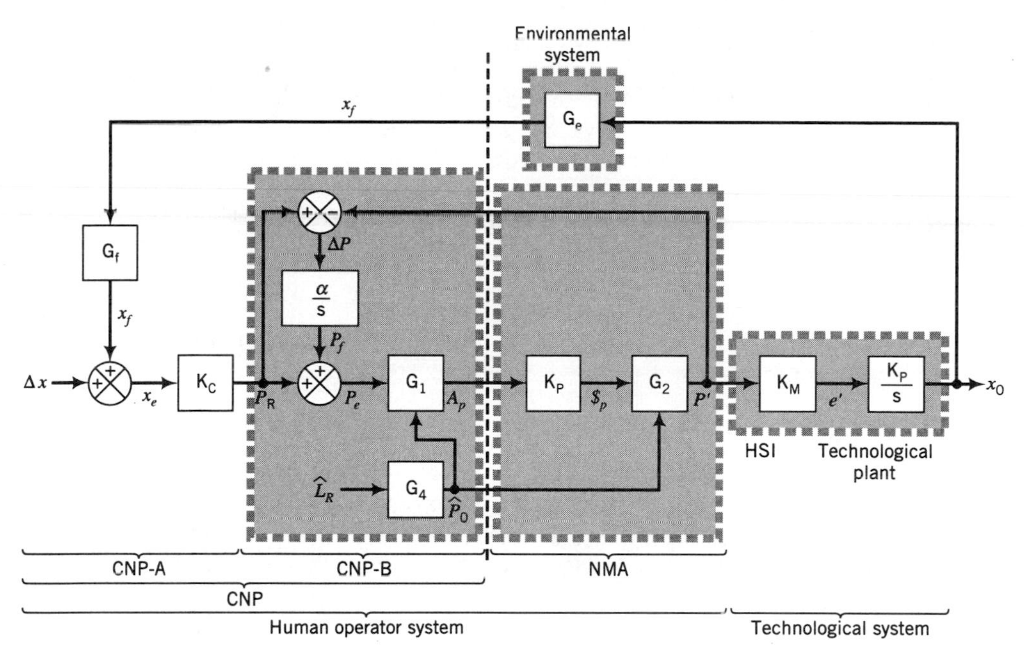

Figure 9.22 Human-technological control system: isometric stick (with first-order technological plant).

technological system presented earlier in this chapter (see Figure 9.1). For the human-technological control system of current interest, the HSI is an isometric control stick. This is a common control device by which the human operator can interact with and exercise control over some specific technological plant. An isometric control stick is basically a transducer, the input to which is human operator isometric force (P'). By definition, this particular control device is characterized as a "stiff stick" because it will only respond to force, and there is no stick displacement regardless of the amount of force applied. The output of the isometric stick will be an electrical voltage that activates the technological plant. Consequently, the isometric stick transducer constant (c) may be defined as follows:

$$c_m = \frac{e}{P} \tag{63}$$

A characteristic of the isometric control stick as a transducer is that there will be some maximal electrical voltage output (e_{max}) when the human operator applies a maximal isometric force, at the optimal muscle length (P_0^*). This maximal transducer output may be expressed as:

$$k_m = \frac{e_{max}}{P_0^*} \tag{64}$$

The isometric stick transducer coefficient (K_M) may then be defined as:

$$K_M = \frac{c_m}{k_m} \tag{65}$$

Substituting equations (63) and (64) into equation (65) yields:

$$K_M = \frac{\left(\dfrac{e}{e_{max}}\right)}{\left(\dfrac{P}{P_0^*}\right)} \tag{66}$$

Define a normalized isometric stick output voltage (e'):

$$e' = \frac{e}{e_{max}} \tag{67}$$

Finally, substituting equations (40) and (67) into equation (66) results in the operational form of K_M:

$$K_M = \frac{e'}{P'} \tag{68}$$

The technological plant may be characterized as a zero-order (Figure 9.21), first-order, or second-order subsystem of the technological system. The order of the technological plant subsystem will significantly influence the human-technological control system performance characteristics. The specific human-technological control system of Figure 9.22 is instructive as it uses a first-order technological plant. Recall that a first-order system generates an output variable that is an integral function of the input variable. Consequently, a first-order technological plant subsystem will introduce a time lag to the human-technological control system. As we shall see in Section 10.3, time lags when combined with time delays will limit the acceptable human-technological control system performance.

The output of the technological plant subsystem is a position (x_0). In computational numerical form, the relationship of output position (x_0) to the input voltage (e') may be expressed as:

$$x_0 = [G_p \cdot dx]\frac{1}{T_P}\sum_{i=1}^{n} e' \cdot dt \tag{69}$$

Define β_p as the reciprocal time constant of integration of the plant:

$$\beta_p = \frac{1}{T_p} \tag{70}$$

Define a technological plant characteristic coefficient (K_P):

$$K_P = G_p \cdot dx \tag{71}$$

Substituting equations (70) and (71) into equation (69):

$$x_0 = K_P\beta_p \sum_{i=1}^{n} e' \cdot dt \tag{72}$$

The environmental system of Figure 9.22 closes the external feedback loop between the technological plant displacement output and the human operator. A human being is a visual creature and a large majority of the sensory information obtained by humans regarding their environment is obtained by means of vision. For the human-technological control system with an isometric stick, we shall represent the external feedback loop of Figure 9.22 as visual feedback. Consequently, the environmental system is represented by a single element (the environmental gain, G_e) and may be expressed as:

$$G_e = \frac{x_f}{x_0} \tag{73}$$

Recall that x_0 is the output displacement of the technological system. Information regarding x_0 is transferred to the human operator via the environment and becomes feedback displacement (x_f) once it enters the human operator system (see Figure 9.22). Under conditions of normal illumination and an unobstructed visual field, one might expect that G_e is approximately unity. Operationally, this approximation is quite useful for computer simulation of the human-technological control system. With decreasing amounts of illumination and/or intermittent obstruction of the visual field, one might expect that G_e becomes less than unity. Consequently, technological system output displacement (x_0) is inadequately transferred to the human operator as feedback displacement (x_f) so that fully compensatory closed loop control does not occur. Finally, the human factors engineer will appreciate that in a completely dark environment or totally obstructed visual field, the human-technological control system reduces to an open-loop system with respect to external feedback through the environment.

The central nervous processor–A (CNP-A) subsystem completes the human-technological control system of Figure 9.22. The anatomical elements of CNP-A subsystem will include the cranial sensory array (visual sense), the cerebrum (decision making), and the brain stem (generalized activation of the neuromuscular control system). CNP-A subsystem data processing begins with the human operator selecting the task to be performed. In the case of current interest, the human operator initially makes a mental decision to effect a change in the technological system output of magnitude Δx. In this case, Δx represents the amount of external displacement to be accomplished and is defined as the distance between the initial technological system output position (prior to any movement) and the final (desired) technological system output position. The human operator then initiates vibrational movement. Central nervous processing continues as the human operator receives input via the cranial sensory array (visual sense) of the actual technological system output position (x_f). The human operator periodically interprets the amount of external displacement that yet remains to be accomplished. This displacement error (x_e) is calculated at the CNP-A level as the difference between the originally desired displacement (Δx) and the actual current displacement position (x_f):

$$x_e = \Delta x - x_f \tag{74}$$

The other characteristic process of the CNP-A subsystem involves the control element, K_c. This element contains the paradigm for the control process. In effect, the human operator has a model, or pattern, at the higher central nervous processor level. This is a form of strategy that the human operator selects in order to accomplish the desired task. At this point, it is appropriate to comment on human strategy before proceeding further.

Strategy is a unique characteristic of living systems when performing tasks. With respect to the human operator, various strategies are often available for completing an assigned task and one or a combination of two or more will ultimately be used in accomplishing the task. Early attempts in the 1940s by control system engineers to model the human operator were based on automatic control theory. For example, in order to model the human operator performing a unit step displacement, a second-order, underdamped, servo-mechanism was developed. Initial enthusiasm was quickly replaced with creeping doubt when it was observed that human tracking response to a unit step did not faithfully follow the automatic control model. Some of the human deviation can be accounted for by neuromuscular fatigue or incorrect perception (interpretation) of sensory input data. A large amount of that deviation, however, is accounted for by the human operator's use of different strategies to acquire a target that undergoes a unit step displacement. Some human operators would generate an initially rapid output response that deliberately overshot the initial amount of unit step displacement and then return to the target acquiring it from above (Figure 9.23.a). Other human operators would execute a series of smaller steps that never overshot the initial displacement but rather acquired the target by coming from below (Figure 9.23.b). Even yet, a third human operator strategy was to use a combination of the first two in which there was an initial rapid (excess displacement) of the original unit step displacement and a subsequent undershooting of the target (during which the human operator would cross over the target path) and then acquisition of the target by a successively smaller undershoot and/or overshoot until target acquisition (Figure 9.23.c).

There are two important lessons to be learned from these early observations. First, the human factors engineer fully understands that the human neuromuscular control system and human operator control, in general, are not simply an extension of automatic control theory. This statement has been made before in this chapter and is once again stated for reinforcement. However, the human factors engineer also realizes that simply because the human operator is a complex system does not mean that human operator control is not capable of control system simulation and analysis. Rather, the human factors engineer understands that by performing a series of simplifying approximations and clearly identifying the system specifications and requirements, a first-order approximation of the human operator control model may be obtained. Since design is an iterative process, the human operator control model can be used with other iterations in the overall human system design process so that additional refinement of the human control system model may be obtained as necessary. In the present case, the human factors engineer must characterize the control element (K_c) with respect to a specific strategy. In essence, the human

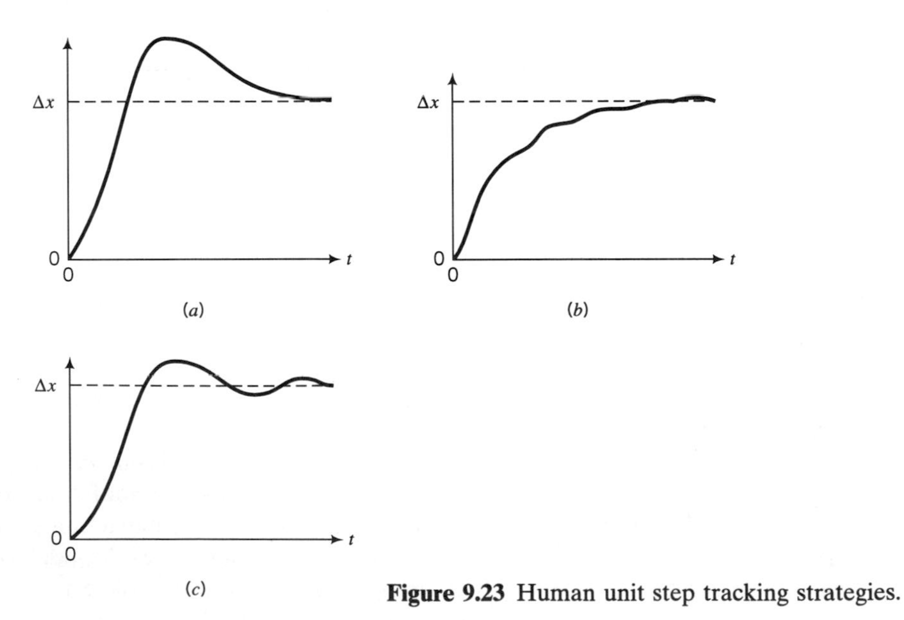

Figure 9.23 Human unit step tracking strategies.

control system model will be valid for that particular strategy. But humans are characteristically inventive and various strategies are possible depending upon the human, the technological system, the environment, and the task. The control element (K_c) may be changed to observe the effects of different control paradigms. Now, on to the task of defining K_c for the current problem of interest.

Begin by treating the control element as a transfer function the output of which is the normalized required force (P_R) that will subsequently be processed by the CNP-B subsystem. The input to the control element will be the displacement error (x_e). This may be expressed as:

$$P_R = K_c \cdot x_e \tag{75}$$

It is then apparent that K_c is the central nervous processor force-displacement coefficient. In effect, it is a proportionality parameter that translates a given amount of displacement into an equivalent amount of force. The paradigm, or model, that immediately comes to mind is that of a mechanical spring. Extending this model, we can define the control element coefficient as:

$$K_c = \frac{\Delta P_A}{\Delta x} \tag{76}$$

Recall that Δx represents the required displacement magnitude (at time equals zero) prior to any human operator movement, and is a constant during the time course of subsequent events [per equation (74)]. ΔP_A has not yet been defined so we proceed to the next step.

Rearranging equation (75) for K_c, and then equating this to equation (76):

$$\frac{\Delta P_A}{\Delta x} = \frac{P_R}{x_e} \tag{77a}$$

Rearranging for ΔP_A:

$$\Delta P_A = \frac{P_R}{\left(\frac{x_e}{\Delta x}\right)} \tag{77b}$$

Define a normalized displacement error (x_e'):

$$x_e' = \frac{x_e}{\Delta x} \tag{78}$$

Substituting equation (78) into equation (77b) and defining P_R' results in definition of the control element parameter (ΔP_A) of the CNP-A subsystem:

$$\Delta P_A = \frac{P_R'}{x_e'} \tag{79}$$

P_R' is the normalized required isometric force (which will vary with time), and obviously the normalized displacement error (x_e') will also vary with time. Equation (79) is a mathematical expression of a specific central nervous processor strategy. Specifically, the ratio of the normalized instantaneous required isometric force (P_R') to the normalized instantaneous displacement error (x_e') should be *constant* (ΔP_A). The human factors engineer understands that other strategies are possible, but that in order to implement a first approximation of the human neuromuscular control system, the engineer must characterize (and mathematically identify) a specific strategy. In other words, certain mathematic models of human operator control are indeed *strategy specific.*

With the generation of P'_R by the CNP-A subsystem data flow then proceeds to the CNP-B subsystem (see Figure 9.22). Note that the CNP-B subsystem is characterized by a normalized required length ($\hat{L}_R$) as an essential subsystem parameter. At this point we have now defined the feed-forward pathways and elements and the external feedback pathway and element that interact around the CNP-A subsystem and the CNP-B subsystem of the isometric neuromuscular control system.

EXAMPLE 9.5

The human-technological control system of Figure 9.22 represents a human operator interacting with a technological device via an isometric control stick. Recall that there is internal force-feedback and external visual feedback from the device to the operator. As a prelude to writing a computer simulation program of this system:

(a) Assemble the set of operational equations necessary to model the block diagram of Figure 9.22.

(b) Given the following:

$$\alpha = 0.1$$
$$\beta_p = 0.1$$
$$K_m = 1.0$$
$$G_e = 1.0$$
$$K_p = 5.0$$

Find K_c.

Recall that K_c is a characteristic parameter of the control element, and the control element is related to the human operator strategy. Indicate the "scaling strategy" used to determine the value of K_c.

SOLUTION 9.5(a)

Referring to Figure 9.22, proceed systematically to follow the computational pathways. Each equation for each computational block is then written with the output variable as the dependent variable.

First, human-feed-forward path (Δx to P_R).

By inspection:

$$x_e = \Delta x - x_f \tag{i}$$

From equation (75):

$$P_R = K_c x_e \tag{ii}$$

Second, human feed-forward path (P_R to P').

Per equation (2):

$$\hat{P}_0 = \sin\left[\pi\left(\hat{L}_R - \frac{1}{2}\right)\right] \tag{iii}$$

By inspection:

$$P_e = P_R + P_f \tag{iv}$$

Per equation (47):

$$A_P = \frac{P_e}{\hat{P}_0} \tag{v}$$

From equation (30):

$$\overline{\$}_P = \left[\frac{1}{\pi}\right] \cos^{-1}(1 - 2A_P) \tag{vi}$$

From equation (49):

$$P' = \left(\frac{\hat{P}_0}{2}\right)(1 - \cos[\pi\overline{\$}_P]) \tag{v}$$

Third, internal feedback path (P' to P_f).
By inspection:

$$\Delta P = P_R - P' \tag{viii}$$

Per equation (43):

$$P_f = \alpha \sum_{j=1}^{m} \Delta P_j \cdot dt \tag{ix}$$

where:

$$\alpha = 0.1 \tag{x}$$

Fourth, technological feed-forward path (P' to x_0).
Rearranging equation (68):

$$e' = K_m P' \tag{xi}$$

Per equation (72):

$$x_0 = K_P \cdot \beta_p \sum_{i=1}^{n} e' \cdot dt \tag{xii}$$

where:

$$\beta_p = 0.1 \tag{xiii}$$

Fifth, external feedback path (x_0 to x_f).
Rearranging equation (73):

$$x_f = G_e x_0 \tag{xiv}$$

Since G_e is given as unity:

$$x_f = x_0 \tag{xv}$$

SOLUTION 9.5(b)

In order to calculate K_c, this control element parameter is defined with respect to a "scaling strategy." This is done as follows.

Recall that the human operator CNP-B and NMA subsystems generate a P' output as a function of a P_R input. Both parameters have been normalized with respect to P_0^* so that the input and output will vary between zero and one.

Now the technological system has an x_0 output in real physical world units of displacement. In effect, the human operator P' output is translated by the technological system to an x_0 output.

External (visual) feedback of x_0 (through the environment) then results in the human CNP-A subsystem sensing x_0 and comparing it with the desired Δx. In order to have a one-to-one proportional comparison, Δx is in the same real-world physical units of displacement as x_0. x_e is the result of that comparison, and so K_c must then "scale" the resultant x_e so that the CNP-A subsystem output (P_R) directly translates to the input P_R of the CNP-B subsystem. The preceding description makes it apparent that the scaling strategy of K_c is to perform the inverse of the scaling already performed by the technological system.

The human factors engineer now has a rationale for defining K_c, and this may be accomplished as follows.
Begin by defining:

$$x_0(t_0) = 0 \tag{i}$$

and

$$x_0(t_f) = \Delta x \tag{ii}$$

For the purposes of scaling, the technological plant may be represented as:

$$x_0(t) = K_p \int e'(t) \tag{iii}$$

At t_f, substituting equation (ii) into equation (iii):

$$\Delta x = K_p \bar{e}' \tag{iv}$$

So that:

$$\bar{e}' = \frac{\Delta x}{K_p} \tag{v}$$

For the HSI element of Figure 9.22:

$$e' = K_m P' \tag{vi}$$

It can be shown that the feed-forward gain (G_{FF}) of the CNP-B subsystem in series with the NMA subsystem (of Figure 9.22) is unity. So that:

$$P' = G_{FF} P_R \tag{vii}$$
$$P' = (1) P_R \tag{viii}$$

So by substituting equation (viii) into equation (vi):

$$e' = K_m P_R \tag{ix}$$

At t_f:

$$\bar{e}' = K_m P_R' \tag{x}$$

So that:

$$P'_R = \frac{\bar{e}'}{K_m} \tag{xi}$$

Substituting equation (v) into equation (xi):

$$P'_R = \frac{\Delta x}{K_m K_p} \tag{xii}$$

With respect to the feed forward pathway of the CNP-A subsystem:

$$P_R = K_c x_e \tag{xiii}$$

At t_0 (when x_0 is zero, as per equation (i)):

$$x_e = \Delta x \tag{xiv}$$

Substituting equation (xiv) into equation (xiii):

$$P'_R = K_c \Delta x \tag{xv}$$

Note that P'_R is now defined as the maximum normalized value of P_R (which occurs at t_0 for the CNP-A subsystem).
Rearranging equation (xv):

$$K_c = \frac{P'_R}{\Delta x} \tag{xvi}$$

Substituting equation (xii) into equation (xvi):

$$K_c = \frac{1}{K_m K_p} \tag{xvii}$$

which is the "scaling strategy" for K_c.
Fourth, substituting given values into equation (xvii):

$$K_c = 0.2$$

Having now derived the governing set of equations and important system parameters for the computer simulation flow chart of Figure 9.22, we may continue with the concluding example for this section.

EXAMPLE 9.6

A forklift operator is seated in the cab and is controlling the forklift position with an isometric control stick. The operator has her upper arm extended somewhat outward, the elbow at an angle of 120°, and the forearm extended straight out (parallel to the earth's surface). The operator grips the control stick handle and pushes (isometrically) straight forward using the pectoralis (anterior shoulder) muscle. By doing so, the forklift operator is able to raise the fork with the isometric control stick. The operator has just engaged a pallet (with a load on it) at ground level using the forks of the lift. The operator now desires to lift the pallet and load a vertical distance, $\Delta x = 2$ M.

(a) Write a computer simulation program of this human-technological control system when performing a unit displacement task using an isometric control stick (as depicted in Figure 9.22).

(b) Plot the time course of:

(i) the displacement (x_0) of the load on the pallet ($x_0 = 0$ at ground level), and the displacement error (x_e) during the performance of this task

(ii) P_R and P' during the performance of this task

SOLUTION 9.6(a)

An m-file is written in MATLAB for the computer simulation flow chart of Figure 9.22. For this program:

$$L_R = 1.0$$
$$\alpha = 0.1$$
$$K_m = 1.0$$
$$\beta_p = 0.1$$
$$K_p = 5.0$$
$$G_e = 1.0$$

The program is per Figure E9.6.1.

```
% example 9.6(a) and 9.6(b)
% CA Phillips 15 Sept 1998
%  Isometric Control System
% w/ first order plant
echo off;
clear;
lr=input('Enter a value for Lr(x)  :');  %lr=1.0 for ex. 9.6
alpha=0.1;
pprime=0;
pf=0;
pi=3.1415926;
Km=1.0;
xprime=0;
eprime=0;
er=0;
Dx=2;
beta=0.1;
Kp=5;
Kc=1/(Km*Kp);  % per equation 9.5(b) (xvii)
Ge=1.0;
for t=1:21;
t2(t)=t-1;
if t==1
     del_x=0;
else
     del_x=Dx;
end;
```

```
xe=del_x-xprime;
xp(t)=xprime;
xerror(t)=xe;
prrec=xe*Kc;
Prprime(t)=prrec;
for x=1:5;
pj=prrec-pprime;
pf=pf+pj;
pp=pf*(alpha);
pe=prrec+pp;
p0hat=sin(pi*(lr-.5));
% G1:
ap=pe/p0hat;

%Kp (for the NMA)
stim=(1/pi)*acos(1-2*ap);
% G2:
pprime=(p0hat/2)*(1-cos(pi*stim));
pprec(t)=pprime;
kmrec(t)=Km*pprime;
% e (Isometric Control Stick):
eprime=Km*pprime;
if eprime>1
    eprime =1;
end
if eprime < 0
   eprime=0;
end
eprec(t)=eprime;
er=er+eprime;
errec(t)=er;
xprime=Kp*(beta)*er;
% note: Kp (for the tech. plant)
end
end
plot(t2,xp);
hold;
plot(t2,xerror,'*');
title('Xprime vs. X error, * = X error');
xlabel('time(sec)');
pause;
hold off;
plot (t2,pprec);
hold;
plot (t2,Prprime, '*');
title('Pprime vs. Pr prime, * = Pr prime');
xlabel('time(sec)');
hold off;
```

SOLUTION 9.6(b)

(i) The plot of the time course of x_0 for Example 9.6(a) is shown in Figure E9.6.2.

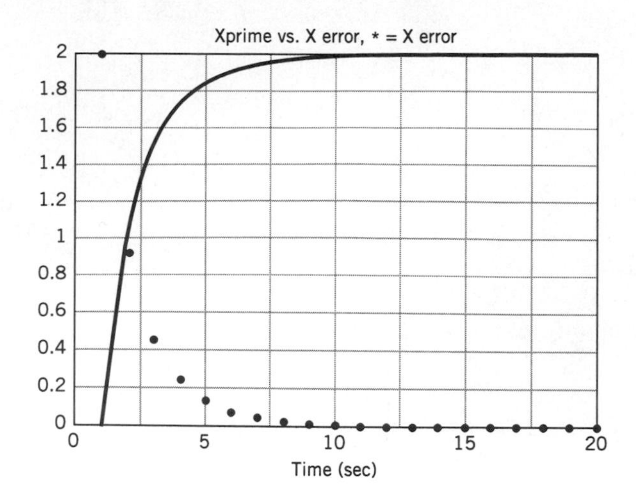

(ii) The plot of the time course of P_R and P' for Example 9.6(a) is illustrated in Figure E9.6.3.

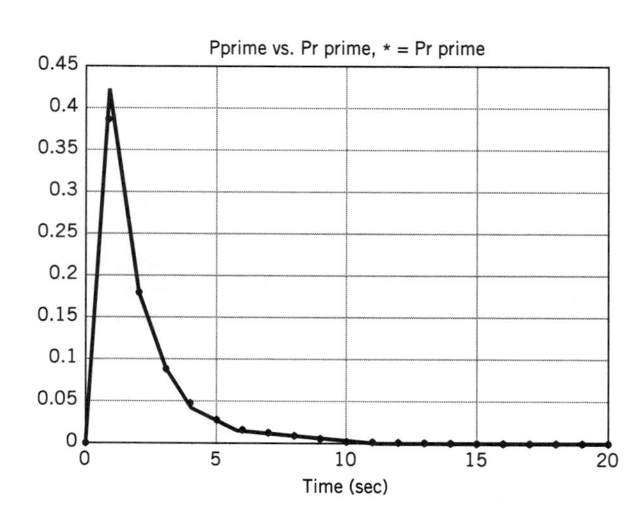

9.4 NEUROMUSCULAR DYNAMIC CONTROL

In order for a human operator to perform dynamic external work, force must be exerted over some specific distance. The resultant force-displacement product represents the dynamic external work performed by the human operator (e.g., see Chapter 8). Recall from Chapter 4 that the applied force may be either constant or variable over the specific displacement distance. The simplest case of external dynamic work is that in which the human operator applies a constant force. If the neuromuscular actuator also generates a constant force (during the displacement), the human's skeletal muscles perform an isotonic contraction. If at the same time, the velocity of skeletal muscle shortening is also constant (during the course of the displacement), the human operator's neuromuscular control system is also performing an isovelocity task. Note that the term *isovelocity* is interchangeably used with the term *isokinetic*. With respect to human factors engineering and to engineering in general, this is not correct since kinetics requires a knowledge of forces that affect movements while velocity, on the other hand, is a kinematic parameter. As a result, isokinetic muscle contractions can occur with any force profile (isotonic, oxotonic, minuthotonic, or oscillotonic) because what is really being said is that the human's muscle is operating in an isovelocity mode. Correct specification of the human operator's neuromuscular control system requires that force, velocity, and length be specified in order to accurately identify the appropriate control category (see Table 9.4). In Section 9.4.a, we shall specifically consider the neuromuscular control system category of isotonic-isovelocity. In Section 9.4.b, we shall consider the human neuromuscular control system interacting with an isotonic control stick.

a. Isovelocity (Isokinetic) Control System

Referring to Figure 9.24.a, we have the characteristic force-velocity-length relationship for a neuromuscular control system performing an isotonic-isovelocity task. Events begin with the human operator's skeletal muscle at an extended length (L_1). This places the operating muscle at a force-velocity relationship defined by V_{01} on the velocity-length curve and by P_{01} on the force-length curve. A target velocity (V_T) is the dependent variable and the skeletal muscle force (P_c) is the independent variable. L_1, V_T, and P_c define the operating point for the human's skeletal muscle, which is located in the three dimensional (force-velocity-length) space. This coordinate point is identified at point A on Figure 9.24.a. As the task proceeds, the human operator's skeletal muscle will generate force by shortening (in the direction of L_0 and L_2). Since the force generated by the muscle is isotonic, P_c will remain constant during this sequence of subsequent length changes. As we shall see shortly, the human operator's neuromuscular control system will then regulate V_T so that this parameter remains relatively constant as the skeletal muscle contracts to progressively shorter lengths. At about midway through this particular operational task, the muscle has shortened to length L_0 so that its operational point B is located at the force-velocity relationship identified by y-intercept V_0^*, and x-intercept P_0^*. At

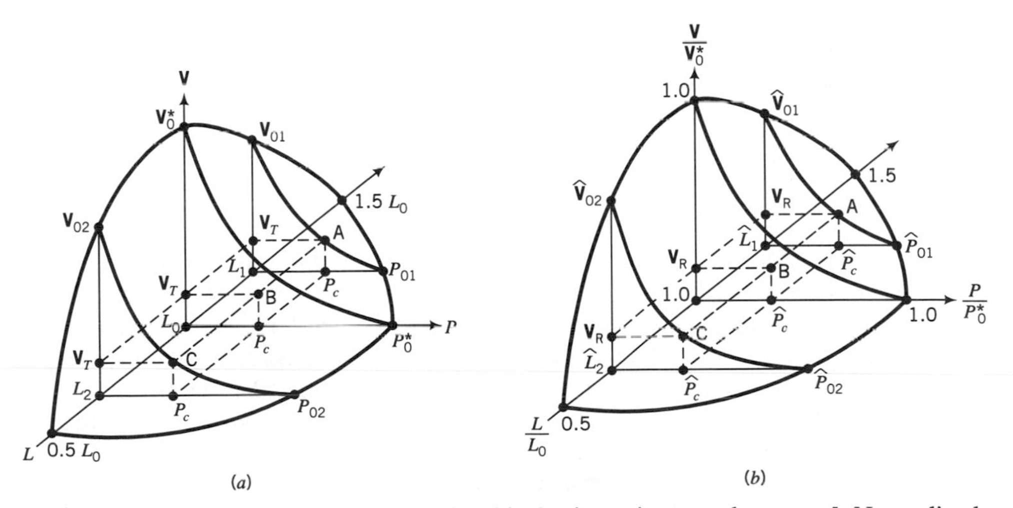

Figure 9.24.a Force-velocity-length relationship for isotonic control system. **b** Normalized force-velocity-length relationship for isotonic control system.

the conclusion of this particular operational task, the human's skeletal muscle has shortened to its minimum length L_2, and its operating point C is then located at a force-velocity relationship identified by V_{02} on the velocity-length relationship and by P_{02} on the force-length relationship. Figure 9.24.a makes it clear that during dynamic external work, the operating point of a human's skeletal muscle moves along a trajectory, the coordinates of which are contained within the three-dimensional force-velocity-length envelope. The basis for the human operator's neuromuscular control system is apparent by examining the specific trajectory of coordinate points A, B, and C. In effect, as the human operator performs a task and skeletal muscle generates force by contracting, it will proceed from a more extended length to a shorter length. During this process, at a biophysical level, the human's skeletal muscle will move through a series of force-velocity templates. The specific operating point on any particular template will then be determined by the amount of central nervous activation and subsequent peripheral nerve stimulation that interacts with the skeletal muscle actuator. Since the human factors engineer already has a system of biophysical equations that describes this process, the engineer now has a basis to develop a computer simulation for the particular control category of interest. We shall now proceed to do this for the isotonic-isovelocity case.

Normalizing the force-velocity-length relationship of Figure 9.24.a by prior equations (3), (4), and (7) results in Figure 9.24.b. With respect to this figure, we may then define a required velocity (V_R):

$$V_R = \frac{V_T}{V_0^*} \tag{80}$$

When the initial length (L_I) is normalized, we define:

$$\hat{L}_1 = \frac{L_I}{L_0} \tag{81}$$

Normalizing the final length (L_F) results in:

$$\hat{L}_2 = \frac{L_F}{L_0} \tag{82}$$

Also with respect to Figure 9.24.b, recall that the normalized coordinate force ($\hat{P}_c$) is defined by equation (16) and that the dependent variable, normalized velocity (V') is defined by:

$$V' = \frac{V}{V_0^*} \tag{83}$$

The computer simulation block diagram for an isotonic-isovelocity neuromuscular control system is shown in Figure 9.25. Despite the increase in the number of computational blocks, the basic pattern for neuromuscular control systems remains the same as described in Figure 9.4. A controller is identified, and in this particular case, it is composed of four computational blocks. A plant consisting of four computational blocks is also identified in this figure. The controller and the plant are arranged sequentially along a feed-forward pathway. Finally, there is a feedback pathway that provides proportional integral control.

Despite the increase in complexity of isovelocity control as compared with isometric control, the basic configuration with respect to anatomical system elements remains unchanged. Comparing Figure 9.19.b (for isometric control) with Figure

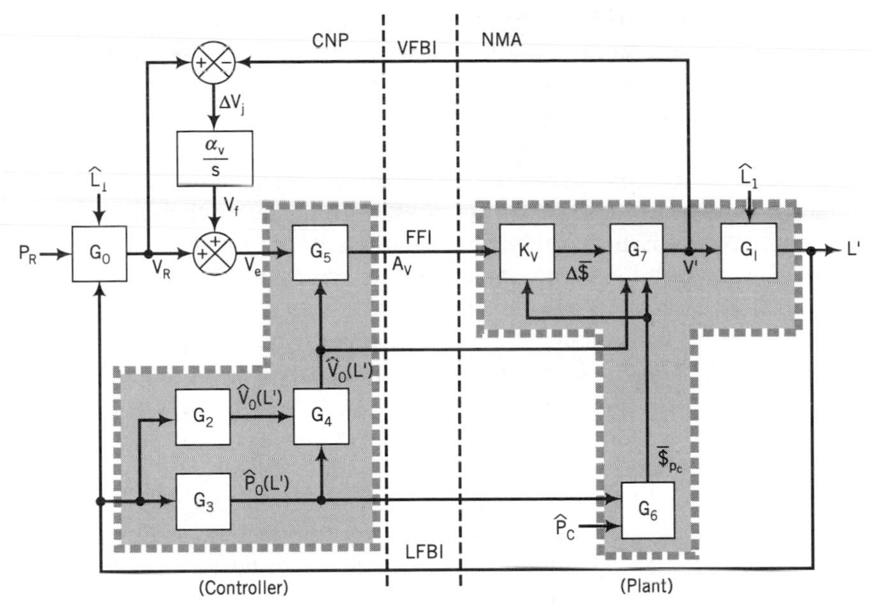

Figure 9.25 Isotonic-isovelocity control system.

9.25 (for isovelocity control) the following characteristics are noted. On the left side of Figure 9.25 is the central nervous processor (CNP) subsystem. The characteristic features of this subsystem are the command parameter V_R, the controller, and the calculation elements of the velocity feedback (VFB) pathway and the length feedback (LFB) pathway. The CNP subsystem interacts with the neuromuscular actuator (NMA) subsystem via the feed-forward interface (FFI). The NMA subsystem is represented on the right-hand side of Figure 9.25. The NMA subsystem is characterized by the isotonic force ($\hat{P}_c$) parameter and the initial length ($\hat{L}_1$) parameter, the plant, the takeoff point for the VFB pathway, and the takeoff point for the LFB pathway. The NMA subsystem interacts with the CNP subsystem via the velocity feedback interface (VFBI) and the length feedback interface (LFBI). Consequently, it is apparent that although the overall complexity of the isotonic-isovelocity neuromuscular control system is significantly increased, the characteristic configuration of the parameters, pathways, and elements is similar to that of the isometric neuromuscular control system (see Figure 9.19.b).

It is readily apparent, however, that there are some configurational differences between the two neuromuscular control systems other than mere complexity. When the neuromuscular control system of the human operator performs dynamic external work, the characteristic output parameter of the human skeletal muscle is muscle length that will change as the skeletal muscle generates force and progressively shortens. In the external world (i.e., outside the human operator) this is represented by an initial position of the human musculoskeletal system and the subsequent time course of displacement is represented by sequentially changing position of that musculoskeletal system. Note, however, that the human operator uses velocity as the required input parameter (V_R) for the neuromuscular control system. Furthermore, the velocity feedback pathway represents the negative feedback pathway by which regulation of the output parameter occurs via integral proportional control. This means that the isotonic (isokinetic) neuromuscular system controls a variable that is one derivative higher than the output variable. This is in marked distinction to the isometric neuromuscular system in which the control parameter (P_R) is the same as the output parameter (P'). With respect to dynamic work, having the control variable one derivative higher than the output variable represents an effective human operator strategy for many tasks of interest to the human factors engineer. A common example is that of operating an automobile. When driving an automobile down a highway the operator may desire to maintain a constant speed. This is accomplished with small variations of the accelerator pedal as the speed of the automobile begins to drop below or rise above the desired speed. When the operator decides to parallel park the vehicle, the car is maneuvered through a sequence of positional changes by means of the operator controlling the relative speed of the vehicle. Other examples of velocity control by the human neuromuscular system include the positional tracking of a flying bird at relatively close range through a pair of binoculars by a bird watcher (see Example 9.7) or the tracking of a shooting star through a narrow field telescope by an astronomer.

The velocity control paradigm depicted in Figure 9.25 is predicated upon an informed decision by the human factors engineer regarding the specific strategy that will be used by the human operator when performing the particular task of interest. It is recognized that alternative strategies may be used by some human operators in which case the paradigm of Figure 9.25 would not be representative. For example, early experiments in the tracking of ramps (which represent a constant amount of positional displacement per unit time) indicated that human operators may use a velocity-matching strategy or alternatively a successive position approximation strategy. In the latter case, a required position (at any point in time) is the input control parameter and the actual operator position is the output parameter. With this particular human strategy, the control parameter is positioned so that a positional neuromuscular control system rather than a velocity neuromuscular control system is used. As we shall see in the next section, when the human operator performs a task involving an isotonic control stick, it is possible to simultaneously use velocity feedback control and position feedback control. Finally, because human operators are human beings and not automatic control devices, it is entirely possible for a human operator to alternate between strategies during the performance of a particular task. As noted previously, this has been observed to occur when the human operator must make an abrupt unit step displacement from a previous position to a new position. The preceding discussion again reminds the human factors engineer to consider the complexity of human neuromuscular control systems so that the engineer will not be overly simplistic with respect to the biophysical modeling of such systems. However, as also noted previously, complexity per se should not deter the human factors engineer from developing a first approximation model of the human operator control system (and strategy!). Since engineering design and development is an iterative process, any first approximation of human operator control will also undergo iteration in parallel with the technological system that is being designed and developed.

A second difference of the isotonic-isovelocity neuromuscular control system when compared to the isometric neuromuscular control system is that the former also involves an internal feedback pathway which is computational rather than regulatory. Referring again to Figure 9.25, the length feedback pathway represents internal feedback of the output variable (L') to the controller. The subsequent result is a sequence of calculations by the controller and by the plant in order to update operational parameter values that characterize the computational elements along the feed-forward pathway. Close inspection of Figure 9.25 as compared to Figure 9.19.b will indicate that it is this internal (computational) feedback pathway and the associated computational elements along that pathway that represent much of the additional complexity associated with an isotonic-isovelocity neuromuscular control system as compared to an isometric neuromuscular control system. The specific details of this length feedback pathway will be considered later in this chapter.

The isotonic-isovelocity neuromuscular control system of Figure 9.25 represents a computer simulation based upon a system block diagram. The following specific discussion will consider in order: (1) the velocity feedback pathway, (2) the feed-forward pathway; and (3) the length feedback pathway. The latter is subdivided into an internal force pathway and an internal velocity pathway.

Referring again to Figure 9.25, the first calculation element in the velocity feedback pathway differences the required velocity (V_R) with the actual output velocity (V') to define a differential velocity (ΔV_j):

$$\Delta V_j = V_R - V_j' \tag{84}$$

This differential velocity represents feedback error, which is subsequently integrated over time according to the following function:

$$V_f = \frac{1}{T_V} \sum_{j=1}^{m} \Delta V_j \cdot dt \tag{85}$$

Operationally, we then define a velocity feedback reciprocal time constant of integration:

$$\alpha_v = \frac{1}{T_V} \tag{86}$$

Substituting equation (86) into equation (85):

$$V_f = \alpha_V \sum_{j=1}^{m} \Delta V_j \cdot dt \tag{87}$$

V_f represents the integral proportional error, which is then summed with the required velocity (V_R) to define the feed-forward error (V_e):

$$V_e = V_R + V_f \tag{88}$$

At this point, data flow is now proceeding along the feed-forward pathway through a sequence of operational elements. This feed-forward pathway (Figure 9.25) then proceeds via G_5 and K_V.

Element G_5 is derived from equation (35) as follows: Normalize equation (35) by V_0^*:

$$A_V = \frac{\dfrac{V}{V_0^*}}{\dfrac{V_c}{V_0^*}} \tag{89}$$

For the G_5 element of Figure 9.25:

$$\frac{V}{V_0} = V_e \tag{90}$$

Substituting equations (15) and (90) into equation (89):

$$A_V = \frac{V_e}{\hat{V}_c} \tag{91}$$

Element K_V is defined by equation (37). Element G_7 is defined from equation (34) as follows:

Substitute equation (27) into equation (34) and rearrange:

$$V = V_c \sin\left[\left(\frac{\pi}{2}\right)\left(\frac{\overline{\Delta\$}}{1 - \overline{\$}_{Pc}}\right)\right] \tag{92}$$

Normalize equation (92) by V_0^*:

$$\frac{V}{V_0^*} = \frac{V_c}{V_0^*} \cdot \sin\left[\left(\frac{\pi}{2}\right)\left(\frac{\overline{\Delta\$}}{1 - \overline{\$}_{Pc}}\right)\right] \tag{93}$$

Substituting equations (15) and (83) into equation (93):

$$V' = \hat{V}_c \sin\left[\left(\frac{\pi}{2}\right)\left(\frac{\overline{\Delta\$}}{1 - \overline{\$}_{Pc}}\right)\right] \tag{94}$$

So far, the feed-forward pathway for this particular neuromuscular control system is similar to the feed-forward pathway for the isometric neuromuscular control system except that P' has been replaced by V'. V' is then translated into L' (the actual output variable, which interacts with the external environment) by computation element G_1. The governing equation for this operational element is derived in the following manner. First, we recognize that the human operator skeletal muscle begins the dynamic external work at a normalized initial starting length ($\hat{L}_1$). Second, during the time course of external dynamic work, the human operator's skeletal muscle will generate force by shortening and the rate of shortening will be reflected by the normalized velocity parameter (V'). The instantaneous normalized muscle length:

$$L' = \hat{L}_1 - \frac{1}{T_L}\sum_{j=1}^{m} V_j' \cdot dt \tag{95}$$

Define a length element reciprocal time constant of integration:

$$\beta_L = \frac{1}{T_L} \tag{96}$$

Substituting equation (96) into equation (95) results in the operational equation for the G_1 element:

$$L' = \hat{L}_1 - \beta_L \sum_{j=1}^{m} V_j' \cdot dt \tag{97}$$

The remainder of the isotonic-isovelocity neuromuscular control system of Figure 9.25 is represented by the length feedback pathway and its constituent elements. Recall that the purpose of this internal feedback pathway is to continuously update characteristic parameters of the feed-forward pathway element as a function of sequential changes in the length of the muscle (L'). The length feedback pathway has two major subdivisions an internal force pathway and an internal velocity pathway. These shall now be considered in that order.

The internal force pathway is a subdivision of the length feedback pathway and is composed of elements G_3 and G_6. Recall that the normalized P_0 ($\hat{P}_0$) varies as a function of muscle length, and that for a particular muscle velocity that length will vary as a function of time (L'). This may be expressed mathematically by restating equations (10) and (11):

$$L' = \hat{L}(t) \tag{10}$$

$$\hat{P}_0(L') = \sin\left[\pi\left(L' - \frac{1}{2}\right)\right] \tag{11}$$

Element G_3 is then defined by equation (11).

Proceeding to element G_6, the isotonic load (P_c) is a characteristic parameter that must be set and then input into the isotonic neuromuscular control system. Setting $P = P_c$ in equation (48):

$$\frac{P_c}{P_0^*} = \left[\left(\frac{1}{2}\right)\left(\frac{P_0}{P_0^*}\right)\right]\left[1 - \cos\left(\pi\left[\frac{\$_{Pc}}{\$_{\max}}\right]\right)\right] \tag{98}$$

Substituting equations (3), (16), and (27) into equation (98):

$$\hat{P}_c = \left[\frac{\hat{P}_0(L')}{2}\right][1 - \cos(\pi\bar{\$}_{Pc})] \tag{99}$$

Rearrange:

$$\cos(\pi \cdot \bar{\$}_{Pc}) = 1 - \frac{2\hat{P}_c}{\hat{P}_0(L')} \tag{100}$$

Solve for the normalized force stimulation parameter ($\bar{\$}_{Pc}$):

$$\bar{\$}_{Pc} = \left(\frac{1}{\pi}\right)\cos^{-1}\left[1 - \frac{2\hat{P}_c}{\hat{P}_0(L')}\right] \tag{101}$$

Element G_6 is defined by equation (101), and used to update elements K_V and G_7 of the feed-forward pathway for the isotonic-isovelocity neuromuscular control system (Figure 9.25). This is necessary because the normalized force stimulation parameter will vary as a function of changing skeletal muscle length.

The other major subdivision of the length feedback pathway is that of the internal velocity pathway. The internal velocity pathway includes element G_2 and element G_4, which are computational elements within the system.

Recall that the normalized $V_0(\hat{V}_0)$ varies as a function of skeletal muscle length. For a specific contractile velocity, the muscle length (in turn) will vary as some function of time (L'). The above is mathematically expressed as equation (12):

$$\hat{V}_0(L') = \sin\left[\pi\left(L' - \frac{1}{2}\right)\right] \tag{12}$$

Element G_2 is defined by equation (12).

Proceeding to element G_4 of the internal velocity pathway, the isotonic coordinate velocity (V_c) is a characteristic parameter that must used to update element G_5 and element G_7 of the feed-forward pathway for this particular isotonic isovelocity neuromuscular control system (Figure 9.25). Element G_4 is defined by equation (17).

EXAMPLE 9.7

A bird watcher is sitting very still and catches sight of a rare bird to the right through the binoculars when the watcher's head is rotated in that direction. Keeping very still otherwise, the bird watcher follows the flight of the bird (by centering it in the binocular viewing field) as the bird flies a circular course (at constant radius from the watcher) from the right side to the left side of the bird watcher. This rare bird flies at a constant speed, and the bird watcher uses the left sternocleidomastoid (SCM) muscle of the neck (the neck "strap" muscle) at an initial length ($L = 1.4L_0$) to shorten and rotate the head leftward to a final length ($L = 0.7L_0$).

In order to accomplish this task, the left sternocleidomatoid muscle shortens at a constant velocity during a seven second interval. The load on this muscle is the weight of the watcher's head (the arms support the weight of the binoculars), which remains constant and represents:

$$P_c = 0.1P_0^* \text{ (for the SCM muscle)}$$

(a) Write a computer simulation program of the human operator neuromuscular control system when performing this isovelocity-isotonic task.

(b) Plot the time course of:

(i) V_R and V' during the performance of this task

(ii) The length of the SCM during the performance of this task

SOLUTION 9.7(a)

An m-file is written in MATLAB for the computer simulation flow chart of Figure 9.25. For this program:

$$\hat{L}_1 = 1.40$$
$$\hat{L}_2 = 0.70$$
$$\hat{P}_c = 0.1$$
$$\Delta t = 7.0$$
$$\alpha = 0.5$$
$$\beta_L = 0.2$$

The program is printed in Figure E9.7.1.

```
% example 9.7(a) and 9.7(b)
% CA Phillips 15 Sept 1998
% Isokinetic Control System
% Unit Step Function

echo off;
clear;
lr=.7;    % requested final length
lhat=1.40;        % these are very sensitive so a pause
pc=.10;     % command is inserted later in case
del_t=7;          % your chosen values don't work
```

```
vprime=0;
lprime=lhat;
vf=0;
pi=3.1415926;
t=0;
alpha=0.5;
sum=0;
beta=0.2;          %suggest beta=1/del_t (approximate)

% G0/calculate
deltalr=lhat-lr;
vinput=deltalr/del_t;

for z=1:11;
t=t+1;
t2(t)=t-1;
% pchat switch
if t==1
   pchat=0;
else
   pchat=pc;
end;
if lprime<=lr
   pchat=0;
end;
pcrec(t)=pchat;
% G0/switch;
if t==1
    vr=0;
else
    vr=vinput;
end;
if lprime<=lr
    vr=0;
end;
vrrec(t)=vr;
for y=1:5;
% G2:
v0hat=sin(pi*(lprime-.5));
v0hatrec(t)=v0hat;
% G3:
p0hat=sin(pi*(lprime-.5));
p0hatrec(t)=p0hat;
% G4:
vchat=v0hat*(1-sin(pi*pchat/(2*p0hat)));
vj=vr-vprime;
delvrec(t)=vj;
vf=vf+vj;
vp=vf*(alpha);
ve=vr+vp;
verec(t)=ve;
% G5:
av=ve/vchat;
if av>1
        av=1;
```

```
end
if av<0
        av=0;
end
avrec(t)=av;
% G6:
stim=(1/pi)*acos(1-(2*pchat/p0hat));
stimrec(t)=stim;
% Kv:
stimbar=(1-stim)*(2/pi)*asin(av);
stimbarrec(t)=stimbar;
% G7:
vprime=vchat*sin((pi/2)*(stimbar/(1-stim)));
velocity(t)=vprime;
% G1:
sum=sum+vprime;
dL=sum*(beta);
lprime=lhat-dL;
if lprime<.5
        pause;  % This stops the program if lprime<.5
                 % i.e. the requested conditions can't be met
end
lprimerec(t)=lprime;
end
end

plot(t2,velocity);
hold;
plot(t2,vrrec,'*');
title('Vprime vs. V requested, * = requested V');
xlabel('time(sec)');
pause;
hold off;
plot(t2,lprimerec);
title('Lprime vs. Time(sec)');
```

SOLUTION 9.7(b)

(i) The plot of the time course of V_R and V' for Example 9.7(a) is as per Figure E9.7.2.

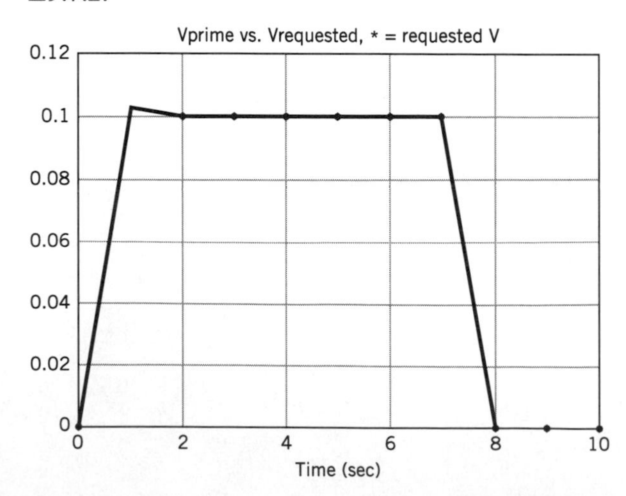

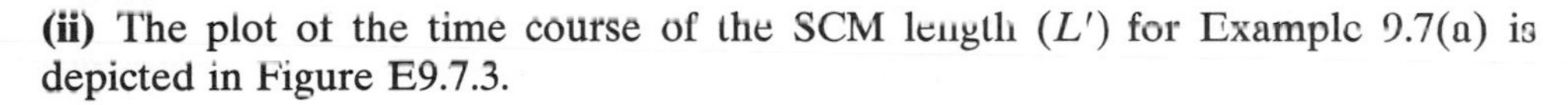

(ii) The plot of the time course of the SCM length (L') for Example 9.7(a) is depicted in Figure E9.7.3.

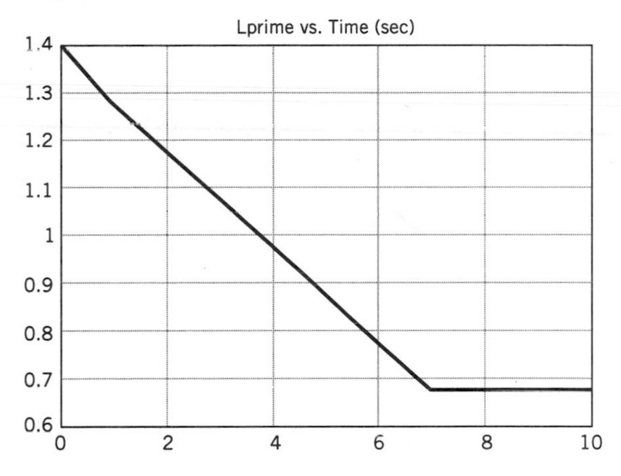

---

b. Human-Technological Control: Isotonic Stick

An isotonic control stick is a displacement-type stick. Movement of the human operator's arm and hand (as the stick is gripped) results in an electrical voltage output proportional to the stick displacement input. Consequently, an isotonic stick is a type of transducer that converts physical displacement into an equivalent electrical voltage. There are two basic types of isotonic control sticks. The first type is a pure-displacement stick in which the resistance of the stick itself to physical displacement is minimal. The term *isotonic* is a misnomer, since such control sticks require almost negligible forces by the human operator in order to displace the stick over its entire control range. The second type of isotonic control stick is a friction-stick. This particular device generates a frictional force upon displacement and the frictional force is constant regardless of the magnitude of the stick displacement or the time rate of that displacement. Consequently, a friction-stick is the semantically correct term for an isotonic control stick. In current design practice, an isotonic control stick of the friction-type may exhibit only a fixed frictional force (not adjustable by the operator) or a variable range of frictional forces (which may be individually set by the human operator prior to control stick use).

Various design parameters are important when using an isotonic control stick to interface a human operator with a technological device. The linearity of the control stick response represents the proportional change of output voltage to a proportional change of input displacement. The stick gain reflects the magnitude of the ratio of stick output voltage to stick input displacement. The zero displacement point is the static position of the control stick at which zero output voltage is obtained. Finally, the stick displacement polarity refers to whether the control stick output voltage is positive or negative over a range of stick displacement. These design parameters are frequently reflected in the derived transfer function for an isotonic control stick element. The following example will illustrate this point.

EXAMPLE 9.8

For the human-technological control system (of Figure 9.26), the human-system interface is an isotonic control stick. The human factors engineer will often find

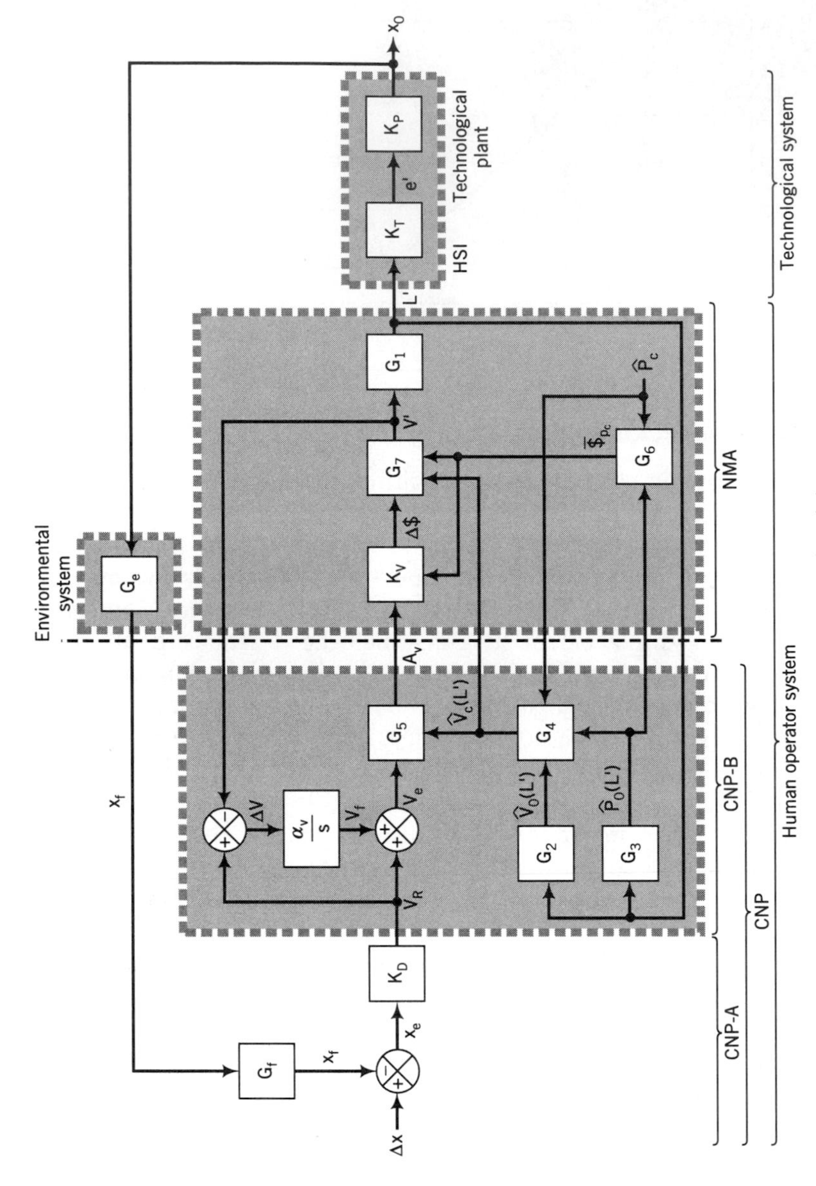

Figure 9.26 Human-technological control system: isotonic stick.

that it is useful to zero the isotonic stick at $\hat{L}_1$ and use the convention that muscle shortening length (L') results in a positive stick output (e'). This requires the engineer to appropriately define the relationship, $e' = f(L')$ for the K_T element of Figure 9.26. Define such a relationship.

SOLUTION 9.8

It is desired to define the relationship:

$$e' = f(L')$$

for the K_T element of Figure 9.26.
This may be accomplished in the following manner.
Begin by graphing the *desired* $e' - L'$ relationship for a linear isotonic displacement control stick:

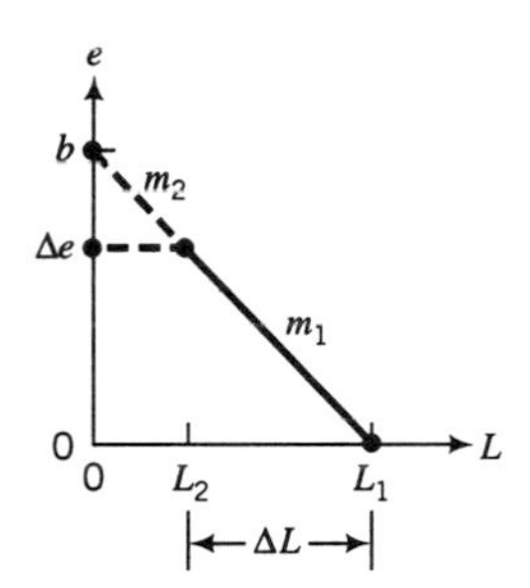

Diagram D

Which is of the form: $e = m \cdot L + b$.
Define m_1: (i)

$$m_1 = \frac{0 - \Delta e}{L_1 - L_2} = -\frac{\Delta e}{\Delta L} \qquad \text{(ii)}$$

Define m_2:

$$m_2 = \frac{\Delta e - b}{L_2 - 0} = \frac{\Delta e - b}{L_2} \qquad \text{(iii)}$$

Since $m_1 = m_2$, we equate equation (ii) and (iii):

$$-\frac{\Delta e}{\Delta L} = \frac{\Delta e - b}{L_2} \qquad \text{(iv)}$$

Solve for b:

$$-\frac{\Delta e \cdot L_2}{\Delta L} = \Delta e - b$$

$$b = \Delta e + \frac{\Delta e L_2}{\Delta L}$$

$$b = \Delta e\left[1 + \frac{L_2}{\Delta L}\right]$$

$$b = \Delta e\left[\frac{\Delta L + L_2}{\Delta L}\right]$$

$$b = \Delta e\left[\frac{L_1 - L_2 + L_3}{\Delta L}\right]$$

$$b = \frac{\Delta e \cdot L_1}{\Delta L} \tag{v}$$

Substitute equation (ii) and (v) into equation (i):

$$e = -\left(\frac{\Delta e}{\Delta L}\right)L + \left[\frac{\Delta e L_1}{\Delta L}\right]$$

$$e = \left[\frac{\Delta e}{\Delta L}\right](L_1 - L) \tag{vi}$$

Rearrange equation (vi):

$$e = [\Delta e]\left[\frac{L_1 - L}{\Delta L}\right] \tag{vii}$$

Define a normalized L ($\overline{L}$):

$$\overline{L} = \frac{L_1 - L}{\Delta L} \tag{viii}$$

So that substituting (viii) into (vii):

$$e = [\Delta e]\overline{L} \tag{ix}$$

Now recall from equation (9):

$$L' = \frac{L}{L_0} \tag{9}$$

Also from equation (81):

$$\hat{L}_1 = \frac{L_1}{L_0} \tag{81}$$

and from equation (82):

$$\hat{L}_2 = \frac{L_F}{L_0} \tag{82}$$

Multiplying the numerator and denominator of equation (viii) by $\frac{1}{L_0}$:

$$\overline{L} = \frac{\frac{L_1}{L_0} - \frac{L}{L_0}}{\frac{L_1}{L_0} - \frac{L_2}{L_0}} \tag{x}$$

Substituting equations (9), (81), and (82) into equation (x):

$$\overline{L} = \frac{\hat{L}_1 - L'}{\hat{L}_1 - \hat{L}_2} \tag{xi}$$

Define a normalized *isotonic* stick voltage output (e') per equation (67):

$$e' = \frac{e}{e_{max}} \tag{67}$$

Define the isotonic stick transfer coefficient (K_T) as:

$$K_T = \frac{e'}{L} \tag{xii}$$

Substituting equation (xi) into equation (xii) and rearranging:

$$e' = K_T \left[\frac{\hat{L}_1 - L'}{\hat{L}_1 - \hat{L}_2} \right] \tag{xiii}$$

Figure 9.26 represents a human operator interacting with a technological device by means of an isotonic control stick. Recall that the human operator represented by the CNP-B subsystem and the NMA subsystem has been characterized in Section 9.4.a. Consequently, Figure 9.26 represents the human operator as an isotonic-isovelocity neuromuscular control system with the following additional interactions: human operator interaction with the technological system; technological system interaction with the human operator CNP-A subsystem (which occurs via the environmental system); and finally, the CNP-A subsystem interaction with the CNP-B subsystem (which collectively define the central nervous processor of the human operator).

It is immediately apparent that this human-technological control system, which uses an isotonic control stick, has distinct similarities to the human-technological control system with an isometric control stick (Figure 9.21) as described in Section 9.3.b. This will become apparent as we define the governing equation for each of the operational elements of Figure 9.26. The description that follows will proceed sequentially from the technological system (which interacts with the human operator system output) to the environmental system (which interconnects the technological system output back to the human operator input) and finally the higher-level human operator central nervous processor (CNP-A) subsystem.

For the case of interest (depicted in Figure 9.26), the technological system is defined by two sequential elements. The human sensory interface (HSI) is an isotonic control stick and represented by element K_T. Recall that the characteristic equation for this operational element is given by equation 9.8(xiii). The next sequential element is the technological plant. It is instructional to represent the plant as a zero-order element. Recall that this was also the case for the human-technological control system utilizing an isometric stick as shown in Figure 9.21. The characteristic coefficient (K_p) of the technological plant (as illustrated in Figure 9.26) may then be defined explicitly by equation (71).

The output of the technological system proceeds to interact with the environmental system by means of a feedback pathway. The human operator (in turn) interacts with the technological system output after such output has passed through the environmental system. The environmental system feedback coefficient (G_e) is defined by equation (73). It should be noted that the prior discussion of the environmental system for human-technological control with an isometric stick (see Section 9.3.b) is equally applicable to the present case of interest, which uses an isotonic control stick.

The human operator task (for the system in Figure 9.26) is to generate a specific output displacement, and so the difference between the starting (initial) position (x_I) and the desired (final) position (x_F) is mentally calculated, and thus Δx is an operational (input) parameter of the human's central nervous processor. Recall that a displacement-type task was modeled for the human operator interacting with an isometric control stick (see Example 9.6). Proceeding along the feed-forward pathway of the higher-level central nervous processor (CNP-A) subsystem a displacement error (x_e) results from the CNP-A subsystem processing of the external (environmental) feedback parameter (x_f). Mathematically, the above operation is defined by equation (74).

The displacement error (x_e) is then sequentially processed by the controller element (K_D). The output of this controller element is the required velocity (V_R) parameter, which represents the interaction between the CNP-A subsystem (higher-level human operator processing) and the CNP-B subsystem of the human operator. The controller element is a very significant computational element within the human-technological control system since it contains the paradigm (a central nervous processor model that (in part) reflects the human strategy used to perform the required task). It is instructive for the human factors engineer to proceed through a systematic analysis of this control element in order to develop an appropriate mathematical equation that most reasonably approximates the human operate paradigm. Such a mathematical formulation might proceed as follows.

The output variable (V_R) is a function of the input variable (x_e):

$$V_R = K_D x_e \tag{102}$$

K_D is the central nervous velocity-displacement coefficient:

$$K_D = \frac{\Delta V_A}{\Delta x} \tag{103}$$

In a manner analogous to equation (79):

$$\Delta V_A = \frac{V_A}{x_e'} \tag{104a}$$

where normalized displacement error (x_e') is defined by equation (78).
Rearranging:

$$V_R = \Delta V_A \cdot x_e' \tag{104b}$$

Substituting equations (78) and (80) into equation (104b):

$$\frac{V_T(L)}{V_0^*(L')} = \Delta V_A \left(\frac{x_e}{\Delta x}\right) \tag{105}$$

Rearrange equation (105) to separate variables:

$$\Delta V_A = \left[\frac{\Delta x}{V_0^*(L_0)}\right]\left[\frac{1}{x_e} \cdot \frac{dL}{dt}\right] \tag{106}$$

Define a dimensionless differential displacement ($d\bar{x}$):

$$d\bar{x} = \frac{dL}{x_e} \tag{107}$$

Define a control element time constant (T_D):

$$T_D = \frac{\Delta x}{V_0^*(L_0)} \tag{108}$$

Substituting equations (107) and (108) into equation (106):

$$\Delta V_A = T_D \cdot \frac{d\bar{x}}{dt} \tag{109}$$

The nature of the control element paradigm is now apparent from equation (109). It was initially apparent from inspection of equation (102) that the central nervous model would involve a differentiator. This occurs because the input variable to the control element is a displacement (x_e) and the output variable (V_R) is a normalized velocity. Equation (109) provides us with additional information regarding the central nervous paradigm. First, the differentiator is operating upon a dimensionless differential displacement which is an incremental change in muscle length (dL) divided by the instantaneous displacement error (x_e). Second, the differentiator has a characteristic time constant (T_D), which represents the magnitude of the required total displacement divided by the theoretical maximal velocity at which that displacement can occur. Recall that the maximal unloaded shortening velocity (V_0^*) occurs at the optimal muscle length (L_0).

EXAMPLE 9.9

The human-technological control system of Figure 9.26 represents a human operator interacting with a technological device via an isotonic control stick. Recall that there is internal velocity feedback and external visual feedback from the device to the operator. Prior to the actual computer simulation program for this system, the human factors engineer is requested to:

(a) Assemble the set of operational equations necessary to model the block diagram of Figure 9.26.

(b) Define the control element coefficient, K_D. Recall that K_D is the characteristic parameter of the control element, and the control element is related to the human operator strategy. Indicate the "scaling strategy" used to determine the value of K_D.

SOLUTION 9.9(a)

Referring to Figure 9.26, proceed systematically to follow the computational pathways. Each equation for each computational block with respect to the output variable.

(i) Feed-forward path (Δx to V_R):

By inspection:

$$x_e = \Delta x - x_f \tag{i}$$

From equation (102):

$$V_R = K_D x_e \tag{ii}$$

(ii) Feed forward path (V_R to V'):

By inspection:

$$V_e = V_R + V_f \tag{iii}$$

Per equation (91):

$$A_v = \frac{V_e}{\hat{V}_c} \tag{iv}$$

From equation (37):

$$\overline{\Delta\$} = (1 - \bar{\$}_{Pc})\left[\frac{2}{\pi}\right]\sin^{-1}(A_v) \tag{v}$$

Per equation (94):

$$V' = \hat{V}_c \sin\left[\left(\frac{\pi}{2}\right)\left(\frac{\overline{\Delta\$}}{1 - \bar{\$}_{Pc}}\right)\right] \tag{vi}$$

From equation (97):

$$L' = \hat{L}_1 - \beta_L \sum_{j=1}^{m} V_j' \cdot dt \tag{vii}$$

where β_L is per equation (96):

$$\beta_L = \frac{1}{T_L} \tag{viii}$$

(iii) Internal velocity feedback pathway (V' to V_f):

By inspection:

$$\Delta V = V_R - V' \tag{ix}$$

From equation (87):

$$V_f = \alpha_v \sum_{j=1}^{m} \Delta V_j \cdot dt \tag{x}$$

where α_v is per equation (86):

$$\alpha_v = \frac{1}{T_v} \tag{xi}$$

(iv) Internal length feedback pathway:

With respect to force-length feedback:
From equation (11):

$$\hat{P}_0(L') = \sin\left[\pi\left(L' - \frac{1}{2}\right)\right] \tag{xii}$$

Per equation (101):

$$\bar{\$}_{Pc} = \left(\frac{1}{\pi}\right)\cos^{-1}\left[1 - \frac{2\hat{P}_c}{\hat{P}_0(L')}\right] \tag{xiii}$$

With respect to velocity-length feedback:
From equation (12):

$$\hat{V}_0(L') = \sin\left[\pi\left(L' - \frac{1}{2}\right)\right] \quad \text{(xiv)}$$

Per equation (17):

$$\hat{V}_c = \hat{V}_0\left[1 - \sin\left(\left[\frac{\pi}{2}\right]\left[\frac{\hat{P}_c}{\hat{P}_0}\right]\right)\right] \quad \text{(xv)}$$

(v) Technological feed-forward path (L' to x_0).

Per equation 9.8(xiii):

$$e' = K_T\left[\frac{\hat{L}_1 - L'}{\hat{L}_1 - \hat{L}_2}\right] \quad \text{(xvi)}$$

By inspection:

$$x_0 = K_p e' \quad \text{(xvii)}$$

where K_p is per equation (71):

$$K_p = G_p \cdot dx \quad \text{(xviii)}$$

(vi) External feedback path (x_0 to x_f):

From equation (73):

$$x_f = G_e x_0 \quad \text{(xix)}$$

SOLUTION 9.9(b)

Define K_D (of Figure 9.26).
For the technological plant:

$$x_0(t) = K_p e'(t) \quad \text{(i)}$$

At t_f (when $x_0(t) = (x)$:

$$\Delta x = K_p \cdot \bar{e}' \quad \text{(ii)}$$

For the HSI element [equation 9.8(xiii)]:

$$e' = K_T\left[\frac{\hat{L}_1 - L'}{\hat{L}_1 - \hat{L}_2}\right] \quad \text{(iii)}$$

At t_f (for the "maximal length of shortening" case, i.e., $L' = \hat{L}_2$):

$$\bar{e}' = K_T \quad \text{(iv)}$$

From equation (102):

$$V_R = K_D \cdot x_e$$

Rearrange:

$$K_D = \frac{V_R}{x_e} \quad \text{(v)}$$

At t_0, $x_e = \Delta x$ (since $x_0 = 0$):

$$K_D = \frac{V_R}{\Delta x} \tag{vi}$$

At t_0 (for the "maximal initial velocity of shortening" case, i.e., $V_R = 1.0$):

$$K_D = \frac{1}{\Delta x} \tag{vii}$$

Substitute equation (ii) into equation (vii):

$$K_D = \frac{1}{K_p \bar{e}'} \tag{viii}$$

Substituting equation (iv) into equation (viii):

$$K_D = \frac{1}{K_p K_T} \tag{ix}$$

This section and this chapter now conclude with the following example.

EXAMPLE 9.10

A crane operator seated in the cab is controlling the crane "neck" (vertical position with an isotonic (displacement) control stick). The upper arm is extended forward from the body, the elbow bent at 130° angle, and the forearm extended straight out (parallel to the earth's surface). This places the trapezius (posterior shoulder) muscle length at $L = 1.40L_0$. The operator has already raised a load in the crane basket (using a pulley system) to a desired height, and now desires to raise the top of the crane neck to a new height which is a vertical distance ($\Delta x = 5$ M) above its previous height.

In order to do so, the operator grips the control stick handle and pulls (isotonically) straight backward using the trapezius (posterior shoulder) muscle. By doing so, the crane operator is able to raise the crane "neck" vertically with the isotonic control stick. The operator pulls backward against a constant isotonic force of the displacement stick (P_c) equal to $.05P_0^*$ of the trapezius muscle.

(a) Write a computer simulation program of this human-technological control system when performing this unit displacement task using an isotonic control stick (as depicted in Figure 9.26).

(b) Plot the time course of:

(i) The force generated ($\hat{P}_c$) during the performance of this task

(ii) The velocity generated (V') and the velocity requested (V_R) during the performance of this task

(iii) The length of the trapezius muscle (L') during the performance of this task

(iv) The vertical displacement (x_0) of the top of the crane neck (where x_0 equals zero prior to the raising of the crane neck) and the displacement error (x_e)

SOLUTION 9.10(a)

An m-file is written in MATLAB for the computer simulation flow chart of Figure 9.26. For this program:

$$\hat{L}_1 = 1.40$$
$$\hat{P}_C = .05$$
$$\Delta x = 5.0$$
$$\alpha = .05$$
$$\beta = 0.1$$
$$K_T = 1.0$$
$$K_p = 10$$
$$G_f = 1.0$$
$$G_e = 1.0$$

The program is represented in Figure E9.10.1.

```
% example 9.10(a) and 9.10(b)
% CA Phillips 15 Sept 1998
%  H.O.-Tech. System Control;
% w/ zero-order plant
%  Unit Step Function: Displacement;
% Isotonic Control Stick;
echo off;
clear;
lhat=1.4;           % these are very sensitive so a pause
pc=.05;             % command is inserted later in case
         % your chosen values don't work;
vprime=0;
xprime=0;
lprime=lhat;
vf=0;
pi=3.1415926;
dx=5;
alpha=0.05;
sum=0;
beta=0.1;
Kt=1.0;
Kp=10;
Kd=1/(Kp*Kt);              %as per eq. 9.9(ix)
Ge=1.0;
deltalr=1.0;               %deltalr=1.5-0.5
for t=1:10;
t2(t)=t-1;
% unit step displacement;
if t==1
   del_x=0;
else
   del_x=dx;
end;
xf=Ge*xprime;
```

```
xe=del_x - xf;
xerror(t)=xe;
if xe<=0
    pchat=0;
else
    pchat=pc;
end;
pcrec(t)=pchat;
vr=Kd*xe;
vrrec(t)=vr;
for y=1:5;
% G2:
v0hat=sin(pi*(lprime-.5));
v0hatrec(t)=v0hat;
% G3:
p0hat=sin(pi*(lprime-.5));
p0hatrec(t)=p0hat;
```

```
% G4:
vchat=v0hat*(1-sin(pi*pchat/(2*p0hat)));
vj=vr-vprime;
delvrec(t)=vj;
vf=vf+vj;
vp=vf*(alpha);
ve=vr+vp;
verec(t)=ve;
% G5:
av=ve/vchat;
if av>1
        av=1;
end
if av<0
        av=0;
end
avrec(t)=av;
% G6:
stim=(1/pi)*acos(1-(2*pchat/p0hat));
stimrec(t)=stim;
% Kv:
stimbar=(1-stim)*(2/pi)*asin(av);
stimbarrec(t)=stimbar;
% G7:
vprime=vchat*sin((pi/2)*(stimbar/(1-stim)));
velocity(t)=vprime;
% G1:
sum=sum+vprime;
dL=sum*(beta);
lprime=lhat-dL;
if lprime<.5
        pause;  % This stops the program if lprime<.5
                 % i.e. the requested conditions can't be met
end
lprimerec(t)=lprime;
```

```
% isotonic control stick;
eprime= Kt*(lhat-lprime)/deltalr;
if eprime>1
    eprime=1;
end;
if eprime<0
    eprime=0;
end;
eprec(t)=eprime;
%Tech. plant (Kp);
xprime=Kp*eprime;
xprec(t)=xprime;
end
end
plot(t2,pcrec);
title('Pchat vs. Time(sec)');
pause;
plot(t2,velocity);
hold;
plot(t2,vrrec,'*');
title('Vprime vs. V requested, * = requested V');
xlabel('Time(sec)');
pause;
hold off;
plot(t2,lprimerec);
title('Lprime vs. Time(sec)');
pause;
plot(t2,xprec);
hold;
plot(t2,xerror,'*');
title('Xprime vs. Xerror, * = Xerror');
xlabel('Time(sec)');
hold off;
```

SOLUTION 9.10(b)

(i) The plot of the time course of $\hat{P}_C$ for Example 9.10(a) is shown in Figure E9.10.2.

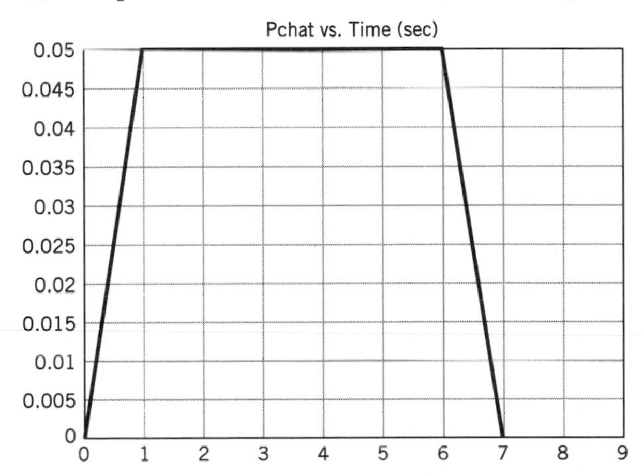

(ii) The plot of the time course of V_R and V' for Example 9.10(a) is graphed in Figure E9.10.3.

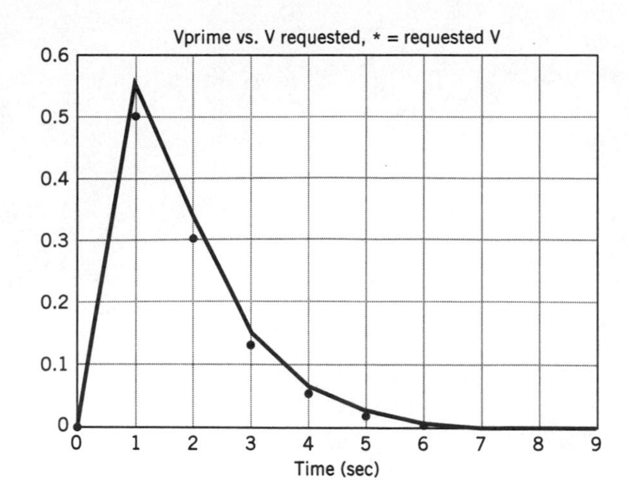

(iii) The plot of the time course of L' for Example 9.10(a) is printed in Figure E9.10.4.

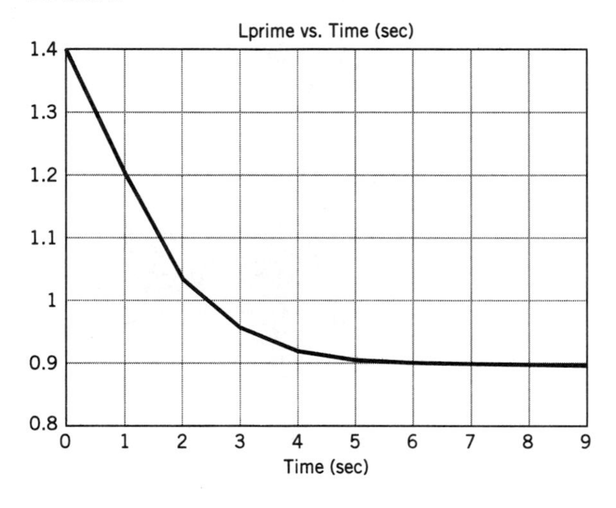

(iv) The plot of the time course of x_0 and x_e for Example 9.10(a) is shown in Figure E9.10.5.

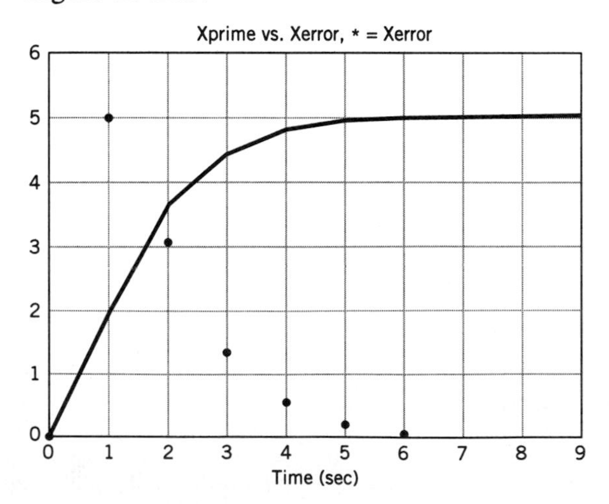

FURTHER INFORMATION

S. Deutsch and A. Deutsch: *Understanding the Nervous System.* IEEE Press. Piscataway, NJ. 1993.

D.M. Etter: *Engineering Problem Solving with Matlab.* 2nd Edition. Prentice Hall. Upper Saddle River, NJ. 1997.

Math Works Inc.: *The Student Edition of MATLAB: User's Guide.* Prentice Hall. Englewood Cliffs, NJ. 1995.

C.A. Phillips: *Functional Electrical Rehabilitation.* Springer-Verlag. New York. 1991.

F.C. Rose, R. Jones and G. Vrbova: *Neuromuscular Stimulation.* Demos. New York. 1989.

PROBLEMS

9.1. P_0^* decreases by 10% per decade of life after age 50. Consider a person (at or below age 50) who is sitting in a chair. The person then contracts the anterior thigh (quadriceps) muscles in order to rise from the chair. The chair has been designed so that the muscle $P_R = 0.5P_0^*$ is necessary for this person to rise. What is the range of $\hat{L}_R$ (minimum and maximum values) for a person to obtain the necessary P_R at:

a. 40 years old
b. 60 years old
c. 80 years old

9.2. (a). For Example 9.1(a), the person now leans over extending the arm *down* to position the cylinder on the rear shelf *below* the counter top. In so performing this task, $\hat{L}_R = 0.70$ and $P_R = 0.70$ for the finger flexor muscles. Plot as a function of time:

(1). P_R and P'
(2). A_P and P'
(3). $\overline{\$}_P$ and P'

(b). For Example 9.1(b), plot P_R and P' as a function of time for:

(1). $P_R = .65$
(2). $P_R = .60$
(3). $P_R = .55$
(4). $P_R = .50$

9.3. Due to associated stiffness of the hip joint from arthritis, the elderly person in Example 9.2(a) is *not* successful in rising from the chair. He then invokes a new chair rising strategy in which he instantaneously activates his anterior thigh muscles to $P_R = 0.90P_0^*$. This strategy works and he starts to rise from the chair. During the 4.5-s time course of rising from seated to erect position, P_R changes as a *decreasing* ramp function. The anterior thigh (quadriceps) muscle still performs an almost *isometric* contraction because the muscle shortening that occurs over the front of the knee joint is offset by proportional muscle lengthening over the front of the hip joint.

Given the above:

Define P_R as a *decreasing* ramp function of the form:

$$P_R(t) = -\left[\frac{P_R(0)}{\Delta t}\right]t + P_R(0)$$

Appropriately modifying the m-file for Example 9.2(a) and plot as a function of time:

a. P_R and P'
b. A_P and P'
c. $\$_P$ and P'

9.4. An experienced assembly line worker is holding a cylindrical tube of caulking glue over an assembly section that moves along a conveyor belt. Initially, he must exert a force with his flexor muscles in his fingers at 0.7 of the P_0^* in order to start the flow of caulking glue, which is to be spread over certain linear distance. As he proceeds through this 4-s task, he has learned that by progressively decreasing the gripping force on the cylindrical tube to 0, a uniform amount of caulking glue material can be distributed over the required linear distance. The cylindrical tube is relatively rigid so that the finger flexor muscles perform an essentially isometric contraction.

Write a computer simulation program for this human operator's neuromuscular control system ($\alpha = 0.25$) when performing this decreasing ramp isometric finger tensing, when:

a. The person is standing and the arm is bent at 120° at the elbow (and the arm-wrist-hand are straight in line) as shown in Diagram 9.1.a. This position allows the finger flexor muscles to be near their optimal length, so that:

$$L = 1.0L_0$$

1. Plot P_R and P' as a function of time.
2. Plot A_P and P' as a function of time.
3. Plot $\$_P$ and P' as a function of time.

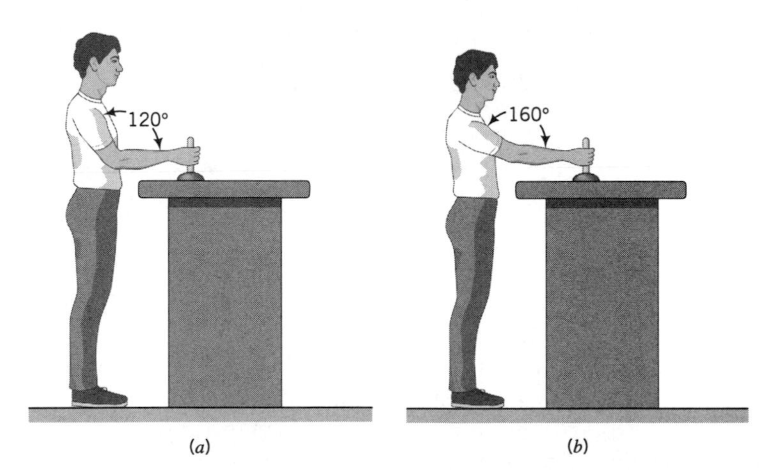

Diagram 9.1

b. The person is standing and the arm is bent at 160° at the elbow in order to reach a second row of assembly sections proceeding along the conveyor belt (see Diagram 9.1.b). This position results in the length of the finger flexor muscles to be away from optimal length so that:

$$L = 1.2L_0$$

1. Plot P_R and P' as a function of time.
2. Plot A_P and P' as a function of time.
3. Plot $\$_P$ and P' as a function of time.

9.5. Regarding Example 9.4, replace the isometric *pull* lever with an isometric *push* lever so that the anterior shoulder (pectoralis) muscles are used.

a. For Example 9.4a [and the diagram of Section 9.4(i)], then:

$$\hat{L}_R = 0.60$$

b. For Example 9.4b [and the diagram of Section 9.4(ii)], then:

$$\hat{L}_R = 0.75$$

c. For Example 9.4c [and the diagram of Section 9.4(iii)], then:

$$\hat{L}_R = 1.0$$

All *other* parameters of Example 9.4 remain the *same*. For part (a), part (b), and part (c) above, separately plot the following as a function of time:

1. P_R vs. P'_F
2. A'_P vs. P'_F
3. $\overline{\$}'_P$ vs. P'_F
4. $\hat{P}'_0$ vs. P'_F

9.6. Modify the m-files that we developed in Solution 9.4.a, Solution 9.4.b, and Solution 9.4.c. in the following manner.

Calculate the endurance time, that is, the time to target failure: $t_F(a)$, $t_F(b)$, and $t_F(c)$. Indicate the t_F for each P_R vs. P'_F plot on the header label of the plot.

9.7. Consider the human technological system of Figure 9.21 (using a zero-order plant). The technological system represented uses an isometric control stick. An example of this system would be an isometric control stick used to operate a stamping press. Because the control stick is isometric, it does not displace when force is applied. Actuation of the device is a function of the force applied to the stick. When the operator places a force on the control stick, the die in a stamp press moves some distance proportional to the force on the stick. When the operator removes the force from the stick, the die returns to its original position.

Given the following:

$$\alpha = 0.1$$
$$K_m = 1.0$$
$$G_e = 1.0$$
$$P_p = 5.0$$
$$\Delta x = 2 \text{ Meters}$$

and defining:

$$G_f = \frac{\beta_f}{s}$$

$$K_c = \frac{1}{K_m K_p}$$

a. Proceed to write a MATLAB m-file for Figure 9.21 of the text.
b. Plot the following as a function of time:

1. x_0 vs. x_e
2. P_R vs. P'

9.8. Regarding Problem 9.7:

a. Modify the block diagram of Figure 9.21 in order to account for physiological fatigue.
b. Modify the m-file developed in Problem 9.7(a) to incorporate physiological fatigue (including fatigue parameters).
c. Plot the following as a function of time:

1. x_0 vs. x_e
2. P_R vs. P'

9.9 For Example 9.7, complete the following for each of the subparts (a)–(g):

i. Identify the specific computational elements of Figure 9.25 for which the two variables have an input-output relationship.
ii. Identify the governing equation for that computational element.
iii. Plot the indicated variables as a function of time.
iv. Explain the result.

The subparts are as follows:

a. V' vs. L'
b. $\Delta\overline{\$}$ vs. V'
c. A_v vs. V'
d. V_R vs. V_e
e. $\hat{P}_0(L')$ vs. L'
f. $\overline{\$}_{Pc}$ vs. $\hat{P}_0$
g. $\overline{\$}_{Pc}$ vs. $\hat{P}_c$

9.10. For Example 9.7, complete the following for each of the subparts (h)–(o):

i. Identify the specific computational elements of Figure 9.25 for which the two variables have an input-output relationship.
ii. Identify the governing equation for that computational element.
iii. Plot the indicated variables as a function of time.
iv. Explain the result.

The subparts are as follows:

h. $\hat{V}_0(L')$ vs. L'
i. $\hat{V}_c(L')$ vs. $\hat{V}_0(L')$
j. $\hat{V}_c(L')$ vs. $\hat{P}_0(L')$
k. $\hat{V}_c(L')$ vs. A_v
l. $\overline{\$}_{Pc}$ vs. $\Delta\overline{\$}$
m. $\hat{V}_c(L')$ vs. V'

n. $\overline{\$}_{Pc}$ vs. V'
o. V_R vs. L'

9.11. An individual is using a motorized wheelchair to help them move from point A to point B, which is a distance of 10.0 M. Operation of this chair is done through an isotonic joystick control (which controls *position,* not velocity). In order for the individual to move the chair forward a specific distance, their wrist flexor muscles must contract to move the joystick backwards (isotonically). Since the user has relatively weak muscle control they like to start the joystick with the wrist flexor muscles at their optimal length ($L_1 = 1.0L_0$) and use very little force ($\hat{P}_c = .025$) to activate this isotonic control stick. Since they would like to move as far as possible with the least amount of force possible, their human sensory interface (HSI) is set at $K_T = 1.5$ and their technological plant (the motorized wheelchair) is set at $K_p = 15$. To further their efforts the feedback gain is set at $G_f = 1.25$. Unfortunately, they are moving over rough terrain, so their environmental gain is only $G_e = 0.80$.

Write the appropriate computer simulation program (based upon Figure 9.26) that represents this problem.

Plot the following as a function of time:

a. V_R vs. V'
b. L'
c. x' vs. x_e

9.12. For Figure 9.22 (when $G_f = 1$), and the associated m-file for Example 9.6.a:

a. Restate the same problem but in which an environmental situation would result in a 15% steady-state overshoot of Δx.
b. Repeat (a) for a 10% steady-state undershoot.
c. Calculate the G_e for (a).
d. Calculate the G_e for (b).
e. Modify the m-file (from Example 9.6.a) with the G_e from (c), keeping all other parameters the same. Then repeat Example 9.6.b.
f. Repeat (e) with the G_e from (d).

9.13. For Figure 9.26 (when $G_f = 1$), and the associated m-file for Example 9.10.a:

a. Restate the same problem but in which an environmental situation would result in a 25% steady-state overshoot of Δx.
b. Repeat (a) for a 20% steady-state undershoot.
c. Calculate the G_e for (a).
d. Calculate the G_e for (b).
e. Modify the m-file (from Example 9.10.a) with the G_e from (c), keeping all other parameters the same. Then repeat Example 9.10.b.
f. Repeat (e) with the G_e from (d).

CHAPTER 10

Human Operator Control

The neuromuscular control system of a human operator is distinctly more complex than the automatic control system of a machine. At this point, the student now has an understanding of human neuromuscular control system fundamentals and both static (isometric) and dynamic (isotonic) control per the preceding Chapter 9. Prior to proceeding with an analysis of human operator control, it is instructive to consider more specifically why the human operator is not simply an automaton. So let us ask the question: What makes human control "human"?

First, the human subsystem will operate within a performance envelope that is significantly more non-linear than most motor actuators. As noted in Section 9.1.b, the neuromuscular actuator of the human operator is defined by a force-velocity-length relationship. Operationally, this translates into a force-velocity envelope within which the human operator skeletal muscle performs in a distinctly nonlinear manner as compared to most motor speed-torque curves. Optimal design of the human-system interface, including detailed specification of design requirements (see Section 10.2.1), requires that the human factors engineer thoroughly understand the operational performance envelope of the neuromuscular actuator.

Second, human operator control involves the element of strategy. Workload may be changed by the use of strategy. For example, a human operator may select a powered hand tool for a particular task as opposed to a manual hand tool. Human operator control may also be significantly affected by strategy. Strategy is a characteristic of the central nervous processor and can significantly alter the central nervous activation of the neuromuscular actuator. At the task-operational level, this would translate into a different central nervous processor strategy for controlling a powered hand tool as compared to the central nervous processor strategy for controlling a manual hand tool. At the element-configuration level, central nervous system processor strategy might alter error detection algorithms. For example, an amateur astronomer may manually track a shooting star in a narrow-field telescope by using a position-updating strategy or a velocity-matching strategy. The use of strategy by a human operator requires the expertise of the human factors engineer because it is this characteristic of strategy that differentiates the human neuromuscular control system from that of an automatic control system (see Sections 9.3.b and 9.4.b). Also recall that fatigue is uniquely characteristic of the human neuromuscular actuator subsystem (Section 9.1.b). Likewise, the characteristic of strategy makes the human central nervous processor subsystem uniquely different (in part) to compensate for the effects of fatigue (see Example 3.1)

Third, human neuromuscular control systems must account for perception as a constituent element in the central nervous processor subsystem. Recall that the cranial sensory array provides data input to the human operator by using the special sensory organs (eyes, ears, and nose) as transducers of electromagnetic, acoustic, and molecular energy (respectively) into electrical impulses conducted along the cranial nerves. Perception is a characteristic element of the central nervous processor subsystem and represents the human interpretation of those sensations. Consequently, perception will affect the human operator's neuromuscular control altering the activation parameter at the final common motor pathway. To the human factors engineer, perception may be viewed as a form of selective filtering of sensory data and also variable gain adjustment to sensory stimuli that characterize the human operator control system. As we shall see in Section 10.1, haptic interface control involves the simulation of specific human perceptions associated with a particular task.

Fourth, recall that the human operator neuromuscular control system interacts with the environmental subsystem in many ways uniquely human. For example, the environmental subsystem provides both direct data input and feedback control to both the central nervous processor subsystem and the neuromuscular actuator subsystem (Table 9.3). Physical variables such as illumination and acoustic noise may dramatically affect human operator neuromuscular control even though the same variables may have a negligible effect on automatic control of a machine. This occurs (in part) because certain environmental variables (such as the level of illumination and the amount of noise) will alter the feedback gain element (G_e) that interacts with the visual sense and the auditory sense. This also occurs (in part) because other environmental variables (such as the uniquely human concepts of self-expectation and peer-expectation) will affect the operational performance of the human control systems. In this regard it is instructive to consider Examples 10.6 through 10.8 (presented in Section 10.3.b), which address a marksmanship training system. As the student works through the human control system analysis for these examples, consider how the various system parameters and coefficients might change for routine target practice as compared to a competitive marksmanship event.

The foregoing statements emphasize that the central nervous processor subsystem is uniquely different from the controller subsystem of an automatic control system. These differences significantly increase the complexity of the simulation and analysis of human operator control. Complexity of the human neuromuscular control system, *per se,* should not deter the human factors engineer from making a reasonable first approximation of that control system.

10.1. VIRTUAL ENVIRONMENTS (VE) AND HAPTIC CONTROL

a. VE and Haptic Control Fundamentals

Virtual reality (VR) can be defined as a synthetic and interactive environment in which a human operator uses appropriate interactive hardware in order to experience both sensory perception and physical activity that simulates a physical environment. This is a rather broad definition of virtual reality, however it does identify the three essential elements of a VR system: (1) a task-oriented human operator; (2) a computer-based information processing system; and (3) appropriate interface hardware (and software) allowing interaction of the human operator with the computer-synthesized environment.

The question is frequently asked: what is the difference between a virtual reality (VR) system and just watching a TV program or going to the movies? The answer is that in a true VR environment, the human being is an active participant, who must interact and change the overall scenario and the imagination of the operator must be affected. In this sense, VR is somewhat analogous to "interactive theater." However, the former is technologically based as opposed to the latter which is socially based. Immersion may be defined as the degree to which the human operator is actively involved with the VR environment. In order to provide a reasonable degree of immersion, a VR system must be real time-oriented and interact with multiple sensory modalities (vision, hearing, touch, etc.). Consequently, an individual would not feel this level of immersion just from watching a TV program or going to a movie. The movie experience however, may be more stimulating (and also immersive), especially were it to engage more of our visual field of view (i.e., cinerama or panorama).

As the human factors engineer considers the design of VR systems one guiding principle should prevail. Whenever there is a decision regarding what effect the engineer wishes to produce, the goal should be to model the phenomenon that is perceived. Specifically, the engineer should model the visual, auditory or tactile signal that seems appropriate for the human operator, and not the actual signal itself. Recall that human stimulation is the physical reaction to an external signal. With respect to a stimulus, the sensory receptor (eye, ear, or skin) acts as a physical transducer. However, human perception is an individual mental interpretation of the stimulus. In other words, perception is a stimulus after it has been processed by the human central nervous processor.

The guiding principle states that the human factors engineer should design *for perception only*. The rationale for this guiding principle is that the large majority of human neuromuscular control system applications involve the human central nervous processor interacting in a coordinated human neuromuscular actuator. On the other hand, a simple reflex action does represent a situation in which an external signal generates a stimulus that does directly interact with the human neuromuscular actuator. However, simple reflex responses are not very useful movements for tasks requiring human operator control.

Teleapplications represent one area for the utilization of VR systems. Teleapplication is defined as a simulated VR environment that is located separately and at a distance from the application location *and* between which there is effective communication. Teleapplications are subdivided into two constituent areas. *Telepresence* refers to a spatially distant VR system in which the virtual environment is a sensory simulation (visual, auditory, olfactory, or tactile). Teleoperation refers to a spatially separated VR system in which the virtual environment is a physical simulation so that manipulation or alteration may occur.

Telerobotics is a specific type of teleapplication in which *both* telepresence and teleoperation are characteristic features. A telerobot is defined as a robot controlled at a distance by a human operator, regardless of the degree of robot autonomy. There are numerous applications for telerobotic devices. These include underwater (inspection, retrieval, etc.), space (assembly, maintenance, etc.) and the earth bound industry (forestry, mining, power line, etc.). Other important applications include process control (nuclear and chemical) as well as the military, medical, and construction industries. Finally, telerobotic systems find application in civil security (fire-fighters, police work, bomb disposal, etc).

Telerobotics differs from VR in several important ways. First, the environments are somewhat unstructured. Second, the human operator must deal with noisy sensory data. Third, energy must be expended by the human operator. This limit of human energy reduces the action of the overall system.

Telerobot systems may be divided into four classes as follows. Class one systems are both power tethered and data tethered. Class two systems are data tethered only. Class three systems are nontethered and telemetry is the communication link. Class four systems are non-tethered and without telemetry (inside deep structures).

Human sensory feedback involves the classical five senses. These are visual, auditory, somatic, olfactory (smell), and gustatory (taste) sensations. Virtual reality systems to date interact with the human operator by means of the visual, auditory and/or somatic sense. The visual sense is used in VR environments most commonly because it is the predominant sensory modality of the human operator. Innovative interactive hardware has been defined by which the human operator may deal with a computer synthesized environment. Such hardware includes visual screens, HUDs (head-up displays), as well as goggles and glasses. The auditory sense has also been routinely used as an interactive sense between the human and the VR environment. Acoustic technology has been defined to allow this interaction and includes microphones, synthesizers, reverberators, and ear phones.

The somatic senses are those nervous system's mechanisms by which sensory information is collected from within the body. These senses are operationally different from the other four special senses (vision, hearing, smell and taste) that collect sensory information from *outside* the body. Somatic senses traditionally have been under-used in VR systems in part because they are diffusely located over the entire body and also because of the large variety of the sensory information collected. This is in counterdistinction to vision and audition, which are anatomically very localized senses and are also of a specific physical phenomenon. However, the somatic senses are particularly important to the human factors engineer for the following reasons. First, appropriate technological hardware currently exists for interaction of the human somatic senses with the VR environment. Such interactive devices include manipulators, data gloves, body suits, thermodes, pins, and pneumatic ladders. Second, the somatic senses are the neurophysiological basis for the design and development of haptic interfaces. Haptic interfaces (discussed later in this chapter) represent a major area for the technological advancement of current VR systems and the human factors engineer is an appropriate individual to address the engineering challenges involved. Consequently, the remainder of this section shall specifically consider haptic interfaces. Visual and auditory interfaces are well treated in various other texts.

As previously stated, an understanding of haptic interfaces begins with the somatic senses, which represent the neurophysiological bases upon which haptic interfaces operate. The somatic senses are traditionally classified into three separate physiological types:

1. The mechanoreceptors, which are stimulated by mechanical displacement of various tissues at the body
2. Thermoreceptors, which are stimulated by heat and cold

3. Nocioreceptors, which represent the human pain sense, and are stimulated by damage to the tissues within the body

The mechanoreceptors provide the operator with the mechanoreceptive somatic sense, which is subclassified into two distinct subcategories. The first category is the tactile sense, which includes touch, pressure and vibration. The second subcategory is the kinesthetic sense, which includes the relative positions and movement rates of different joints of the body.

The neurophysiological basis for the detection and transmission of the mechanoreceptive somatic sense may be separated into four categories: (1) tactile sensations, (2) vibratory sensations, (3) muscle force sensations, and (4) kinesthetic sensations.

Regarding tactile sensations, there are two classes of mechanoreceptors that are involved. The first type (rapidly adapting receptor) reacts with a very brief period of impulse discharge (of only a few hundred milliseconds) and then adapts (ceases impulse discharge). The second type reacts in two phases. Initially there is a burst of activity and then adaptation such that there is a lower but sustained level of impulse discharge. At least six different types of tactile receptors have been identified. These are identified in Table 10.1 with respect to receptor type, tactile stimulus, and perception (interpretation).

Even though the tactile sensations of touch, pressure, and vibration are classified as separate sensations, they are all detected by the same types of tactile sensory receptors (Table 10.1). This results in an interrelationship between the sensations of touch, pressure, and vibration as follows. Touch sensation generally results from stimulation of tactile sensors either in the skin or from those tissues immediately below the skin. Pressure sensation results from mechanical deformation of tactile sensors lying in the deeper tissues below the skin. Vibration sensation is the result of a rapidly repetitive pattern of sensory receptor discharge signal. However, these are still the same receptors used for both touch and pressure (specifically, the rapidly adapting class of receptors).

Muscle force sensation is not the result of special sensory receptors. Certain receptors such as muscle spindles (in skeletal muscle) and golgi tendon apparati (in the tendons) are specialized receptors associated with reflex activity of skeletal muscle. Therefore, the signal output of these specialized receptors acts entirely at the subconscious level. Muscle force sensation is a psychological effect. A common example is that an individual is able to determine how heavy an object is by exerting muscular force upon the object. This type of sensory information is transmitted to the conscious level in two ways. First, there are the usual tactile signals from the deep tissues of the body which are activated (sensory afferent activity). Second, there are signals from the motor cortex of the brain, which indicate the intensity

Table 10.1 Tactile Sensory Receptors

Receptor Type	Tactile Stimulus (Sensation)	Perception (Interpretation)
Pacinian corpuscles	Temporal	Proximity of object
Merkel cells	Spatial	Shape of object
Merkel cells	Amplitude	Force applied to object
Meissner corpuscles	Spatial	Texture
Pacinian corpuscles	Temporal	Hardness
Free nerve endings	Amplitude	Light touch

of motor signals to the skeletal muscles, which are exerting the required force (motor efferent activity).

Kinesthetic sensations represent the third category of the mechanoreceptive somatic sense. Kinesthesia is defined as the conscious awareness of the spatial orientation of the different parts of the body (with respect to one another) and also the rates of movement of those various body parts. Kinesthetic sensations are due to three specialized types of sensory nerve receptors that are located in the joint capsules and ligaments of the body. Recall that a tendon is fibrous tissue that connects a skeletal muscle to a bone. Also recall that a ligament is fibrous tissue that connects one bone to another, essentially holding the bony joint together. The most abundant of these sensory nerve receptors are the Ruffini endings, which are strongly stimulated upon sudden movement of a joint. A second type of specialized sensory nerve ending is the golgi tendon receptors, which are much less abundant but are particularly located in the ligaments about the joint. Third, there are a few Pacinean corpuscles located in the tissues around the joint. These tactile sensors adapt vary rapidly and may therefore assist in signaling the rate of rotation at a joint. It is particularly important to note that the kinesthetic sense relays no information regarding muscle or joint forces. Rather, the kinesthetic sense signals the position, amount of movement (displacement), and rate of movement (velocity) of the limbs at their joints.

The human factors engineer may now ask "how can we couple the human operator to a computer synthesized environment in order to maximize the physical perception of that environment?" The answer is as follows. Consider that the human operator produces output but senses mainly force cues (haptic sense) and position or velocity cues (kinesthetic sense) through some controller that the human factors engineer is going to design. This controller is a human system interface (HSI) since it, in turn, acts with the VR environment in an input-output sense. Recall from Chapter 9 that the human neuromuscular control system, and specifically the neuromuscular actuator react to force (haptic) cues and position/velocity (kinesthetic) cues. We will then define a haptic interface as an HSI controller to be designed specifically with respect to these parameters. Effectively, the design and development of an appropriate haptic interface then provides the maximal VR experience that is desired.

A haptic interface may be operationally defined as a generator of mechanical impedance, that is, a relationship between force and displacement (or velocity). The design and development of effective haptic interfaces requires a knowledge of human force levels when using the fingers of the hand (phalanges) and various grasps (Table 10.2). Appropriate haptic interface design will also require information regarding the human bandwidth for limb motion and sensing (Table 10.3).

Haptic interfaces may be subdivided into three categories. First, ground-based devices are force-reflecting control sticks (joysticks and hand controllers) that are

Table 10.2 Human Force Levels Using Hand Phalanges and Grasps (Newtons)

	Power Grasp	Tip Pinch	Finger Pad	Key Pinch
Male	400	65	91	109
Female	228	45	63	76

Table 10.3 Human Bandwidths for Limb Motion/Sensing

Limb Motion/Sensing	Bandwidth
Unexpected motion	1–2 Hz
Periodic motion	2–5 Hz
Internally generated motion	≤5 Hz
Reflex motion	≤10 Hz
Kinesthetic sensing	20–30 Hz

attached to the floor or other permanent structure. Second, body-based devices are force reflecting exoskeleton devices that are attached to the human operator's forearm or other part of the body. These exoskeleton devices are subclassified into flexible (gloves, suits, etc.) and rigid (linkages, bands, etc.). Third, tactile devices are force reflecting displays that interact at the human skin surface. These tactile displays are further subclassified into three general categories: (1) shape changes (e.g. shape memory actuators, pneumatic actuators, micromechanical actuators, etc.), (2) vibrotactile displays; and (3) electrotactile displays.

Haptic interfaces are designed for a wide variety of VR tasks. It is operationally convenient to divide the tasks in two main categories. Exploratory tasks (exploration) focus upon the extraction of the properties of an object. Exploration represents a sensory dominant task. Manipulatory tasks (manipulation) focus upon a modification of the environment. Manipulation is a motion dominant task. Manipulatory tasks may be further subcategorized based upon the type of task: (1) precision task (e.g., watch repair, threading a needle, etc.), (2) power task (e.g., using a hammer, swinging an axe, etc.), and (3) a hybrid combination of both of these tasks.

The design of a haptic control stick will now be reviewed. The design of a haptic control requires knowledge of the underlying physical principles, an understanding of the various hardware components, and an appreciation of an electromechanical scheme for interfacing these components. The essential element of a haptic control interface is the actuator, and it is this device that determines the specific manner in which force reflection can be obtained.

Understanding of the design of a haptic control stick begins with basic physics. DeLambert's principle was derived from physical mechanics in 1743. Recall from Newton's second law, that a force (F) must be applied to accelerate an object of mass (m) where the mass represents the inertial element. The product of mass times acceleration (ma) is commonly termed the *inertial force.* DeLambert's principle then states that if a force of magnitude ma were applied to the accelerating mass in the direction *opposing* the acceleration, then the system could be analyzed using static equilibrium mechanics. In the case of a haptic control stick, the accelerating force will be that applied by the human operator (F_{HO}) and the opposing force will be that generated by the haptic interface (F_{HI}). This is illustrated at the top of Figure 10.1.

The actuator for this particular haptic control stick is a DC torque motor. Recall that a DC motor is a transducer that converts direct current electrical energy into rotational mechanical energy. The magnetic field B is a steady field and the current I that flows through a conductor is perpendicular to the B field. The force induced on that conductor is proportional to the cross product of the current in the conductor and the strength of the external magnetic field B. This principle is the physical basis

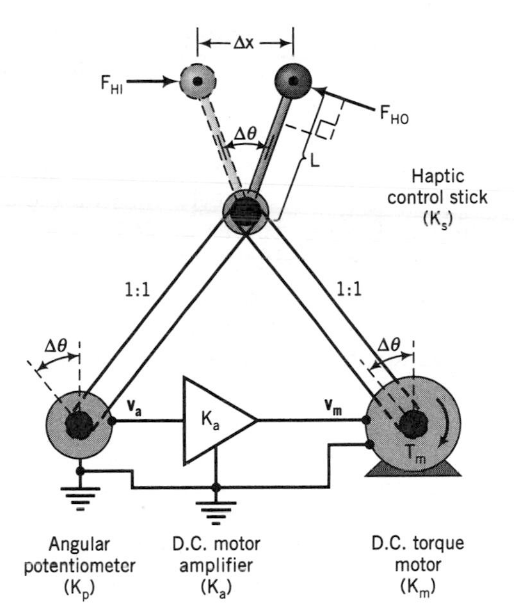

Figure 10.1 Haptic control system for an elastic stick.

for the DC torque motor. The stator is the field circuit on the outside of the motor. The stator remains stationary and is at a constant voltage (E). The rotor itself, which then moves, is the armature circuit that operates in a magnetic field (ϕ). The rotor is connected to the external mechanical load. For a constant ϕ, the torque motor constant (K_m) is proportional to the rotor torque (T_m) output divided by the motor armature current (I_a) input.

The next essential hardware element in a haptic control stick is the sensor (see Figure 10.1). A typical rotational position measurement is accomplished by means of a potentiometer. In the case of an angular potentiometer, a wiper arm is displaced in an angular manner along a circular resistive element. A constant voltage is applied at one point on the resistive element and an adjacent area is connected to ground. The voltage then produced at the wiper arm is the output and is proportional to displacement. In the case of an angular potentiometer, the output voltage produced is proportional to the angular displacement and depends upon a voltage divider effect. When a resistive element reference voltage (V_{ref}) is specified, a potentiometer constant (K_p) may be defined. K_p is the ratio of 2π radians of angular displacement at the wiper arm ($d\theta$) to the reference voltage (V_{ref}) in volts.

The third essential hardware element of a haptic control stick is the DC motor amplifier (as shown in the lower part of Figure 10.1). This DC motor amplifier interconnects the angular potentiometer with the DC torque motor. The input to the DC motor amplifier is the voltage output of the angular potentiometer (V_a). The output of the DC motor amplifier is the motor winding voltage (V_m). The DC motor amplifier serves two purposes. First, it provides the requisite current output so that it acts as an effective driver for the DC torque motor. Second, the DC motor amplifier introduces a motor amplifier constant (K_a). K_a is a dimensionless gain constant (V_m/V_a) that characterizes the particular haptic control stick to be designed. The derivation of K_a in this particular haptic control stick design example will be illustrated later.

The fourth essential feature of a haptic control stick is the control stick itself (as shown in the upper part of Figure 10.1). In the particular design example to follow, it is desired to design an "elastic" control stick. Such a haptic control stick is characterized by a spring constant (K_s). With an elastic control stick, the force that the human operator exerts increases proportionately with the progressive displacement of the stick. Recall that K_s is the constant of proportionality (i.e., linear spring constant) and is the ratio of applied stick force to stick displacement.

Figure 10.1 indicates the interconnection of the control stick with the angular potentiometer and the DC torque motor. The one-to-one (1:1) pulley and cable connections are for illustrative purposes only. Alternatively, the control stick output could be coupled to the sensor and actuator by means of gearing ratios or direct in-line shaft coupling. A 1:1 ratio is used here for mathematical simplicity but could be appropriately designed for either amplification or attenuation of the control stick output.

The operation of the elastic haptic control stick in this particular design example may be described by a series of mathematical equations. These are written with respect to the characteristic system coefficients and represent the first part of the haptic control stick design.

Begin with the human operator exerting a forward force (F_{HO}) on the control stick at a length (L) from the stick center-of-rotation (COR of the stick pulley). In this case, the moment at the stick pulley COR induced by the human operator (M_0) is defined as:

$$M_0 = F_{HO} L \tag{1}$$

Set $x_0 = F(x_0) = 0$. Recall that we are designing an "elastic" haptic control stick, so that at the point when the human operator has displaced the stick forward a certain distance (Δx):

$$K_s = \frac{F_{HO}(\Delta x)}{\Delta x} \tag{2a}$$

and it follows:

$$\Delta x = \frac{F_{HO}}{K_s} \tag{2b}$$

To simplify this design analysis, we shall invoke small angle theory, so that the resultant angular rotation of the stick pulley ($\Delta\theta_s$) is related to stick displacement (Δx) as:

$$\Delta x = L \cdot \Delta\theta_s \tag{3a}$$

Alternatively:

$$\Delta\theta_s = \frac{\Delta x}{L} \tag{3b}$$

Substituting equation (2b) into equation (3b):

$$\Delta\theta_s = \frac{F_{HO}}{K_s \cdot L} \tag{4}$$

Since this particular design is using a 1:1 coupling ratio between the control stick and the angular potentiometer, the resultant angular change at the center of rotation (COR) of the potentiometer shaft ($\Delta\theta_p$) at the potentiometer pulley is:

$$\Delta\theta_p = \Delta\theta_s \tag{5a}$$

The output voltage of the angular potentiometer (V_a) is a function of the potentiometer constant (K_P) which may be derived as follows.

Set

$$\theta_0 = V_a(\theta_0) = 0 \tag{5b}$$

so that at $\Delta\theta_P$:

$$K_p = \frac{\Delta\theta_P}{V_a} \tag{6a}$$

Rearranging:

$$V_a = \frac{\Delta\theta_P}{K_p} \tag{6b}$$

The output voltage of the angular potentiometer is directly connected to the input at the motor amplifier. This amplifier has a characteristic gain constant (K_a), which determines the output voltage from the amplifier (V_m) as follows:

$$K_a = \frac{V_m}{V_a} \tag{7a}$$

Also expressed as:

$$V_m = K_a V_a \tag{7b}$$

Recall that for a constant ϕ, the DC motor constant (K_m) is:

$$K_m(\phi) = \frac{T_m}{I_a} \tag{8a}$$

where:

T_m = rotor torque output (Joules)
I_a = armature current (amperes)

When the armature is not rotating, there is no back EMF and the motor winding current is limited to the applied motor voltage (V_m). For a *shunt-wound* DC motor, this is expressed as:

$$I_a = \frac{V_m}{R_a} \tag{8b}$$

where R_a = resistance of the armature coil winding (ohms).
Substituting equation (8b) into equation (8a) results in an alternative definition of K_m:

$$K_m = \frac{R_a T_m}{V_m} \tag{8c}$$

Rearranging equation (8c) now defines T_m in relation to V_m:

$$T_m = \left(\frac{K_m}{R_a}\right) V_m \tag{8d}$$

T_m is transmitted to the motor rotor pulley, which in turn is transmitted by a 1:1 ratio cable as torque to the pulley of the control stick (T_{CS}). Thus:

$$T_{CS} = T_m \tag{9}$$

T_{CS} is then transmitted as a haptic interface force (F_{HI}), which acts at the end of the control stick of length, L, and acts to displace the control stick rearward (in the opposite direction of F_{HO}):

$$T_{CS} = F_{HI} L \tag{10a}$$

Rearranging:

$$F_{HI} = \frac{T_{CS}}{L} \tag{10b}$$

The next phase of the design process will address two objectives. First, for any given amount of force exerted by the hand of the human operator against the stick (F_{HO}) there should be an equal and opposite amount of force exerted by the haptic interface stick against the hand of the human operator (F_{HI}):

$$F_{HI} = F_{HO} \tag{11}$$

Second, the force experienced by the human operator should be elastic. Specifically, the force at the haptic control stick (whether generated by the human operator or reflected back by the control stick) should linearly increase with increasing amounts of stick displacement:

$$F_{HO} = K_s \, \Delta x \tag{12}$$

These two design objectives may be satisfied by using the above set of characteristic system equations in order to appropriately define the motor amplifier constant (K_a). Mathematically, this requires that K_a be so defined that equations (11) and (12) are simultaneously satisfied. This may be accomplished in the following manner.

Recall that K_a is a dimensionless gain constant defined by equation (7a) and shown schematically in Figure 10.1. In effect, K_a is the overall transfer function for the input-output voltage relationship of the motor amplifier element. The design objective is then to identify a set of subsidiary transfer functions that collectively satisfy both equations (11) and (12). This is indicated using a cascade block diagram in Figure 10.2.

Block 1 is the inverse of equation (6b):

$$\Delta\theta_p = K_p \cdot V_a \tag{6c}$$

Block 2 is the inverse of equation (5a):

$$\Delta\theta_s = \Delta\theta_p \tag{5c}$$

Block 3 is a restatement of equation (3a).

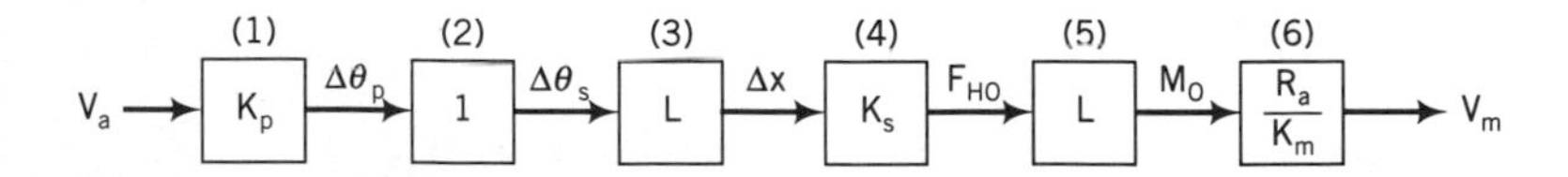

K_s = Spring constant (N/m)
K_p = Potentiometer constant (Radius/Volt)
K_a = Motoramplifier constant (Dimensionless)
K_m = Torquemeter constant (Joule/Amp)

Figure 10.2 Motor amplifier K_a transfer function.

Block 4 is the inverse of equation (2b):

$$F_{HO} = K_s \, \Delta x \tag{2c}$$

At this point, we have satisfied equation (12), which was one of the design objectives. Note that equation (12) is identical to equation (2c).

Block 5 is a restatement of equation (1).

In order to satisfy the other design objective, we begin by restating equation (11) in terms of the moment:

$$F_{HI} L = F_{HO} L \tag{13}$$

Substituting equations (10a) and (1) into equation (13):

$$T_{CS} = M_0 \tag{14}$$

Then substituting equation (9) into equation (14):

$$T_m = M_0 \tag{15}$$

This is simply a restatement of the other design objective. We then analyze Block 6 as follows. Let the voltage output of Block 6 be some as yet unknown voltage (V_0) so that:

$$V_0 = \left(\frac{R_a}{K_m}\right) M_0 \tag{16}$$

Substituting equation (8c) into equation (16), and rearranging:

$$V_0 = \left(\frac{M_0}{T_m}\right) V_m \tag{17}$$

Substituting equation (15) into equation (17) indicates that the other design objective is satisfied when the output of Block 6 is:

$$V_0 = V_m \tag{18}$$

Finally, since we have expressed K_a as a cascade block diagram, we apply block diagram algebra [see Section 10.2.b.i] to Figure 10.2:

$$K_a - \frac{V_m}{V_a} = \frac{K_p \cdot K_s \cdot L^2 \cdot R_a}{K_m} \tag{19}$$

Equation (19) satisfies our remaining design objective.

This review of haptic control interface design is for illustrative purposes only. The haptic control system depicted in Figure 10.1 would be overly simplistic if actually implemented in an operational environment. For example, the design does not account for the time delay associated with the generation of a reflected inertial force. In actual practice, the human operator would displace the control stick in real time and the actual displacement would occur at some velocity. Without accounting for time delay, the generated opposing force would lag behind that of the human operator applied force. Another oversimplification is that damping of the torque motor has not been included in the design analysis. As a consequence, when the human operator abruptly reduces or removes the applied force, the now unopposed haptic force generated by the torque motor would induce an oscillation of the control stick about its zero-force position. However, the above approach is a reasonable first approximation as it provides an interesting example of how a haptic control interface operates and some of the design objectives involved.

b. VE and Haptic Control Applications

One area in which haptic control interfaces may be effectively applied is Virtual Medicine (VM).

EXAMPLE 10.1

A physician enrolled in a general surgery residency program is the human operator in this example in which a VM system being developed for physician training purposes (Figure 10.3).

The physician stands facing a virtual work counter-top, on top of which (and above which) are positioned various surgical instruments which act as interface controls to the VM video display screen at which the physician is looking.

On the video-display screen is a frontal view of a patient's abdomen, and a virtual incision has already been made over the area of the gall bladder. The operative procedure being practiced is the removal of the gall bladder. The skin has already been retracted, and a layer of fascia is now exposed. Fascia is a fibrous tissue that has elastic mechanical properties. Pulling on facia is mechanically equivalent to stretching spring.

A small slit has already been made in the surface of the fascia (with the blade-edge of a scalpel) and the human operator has now inverted the scalpel (so that the blade end is upward). Finally, the operator has now inserted the handle end of the scalpel into the small slit in the fascia (as viewed on the video display unit).

At this point in the virtual machine operation, the surgeon is gripping an actual scalpel handle (with thumb, index, and middle fingers) as if holding an ink pen.

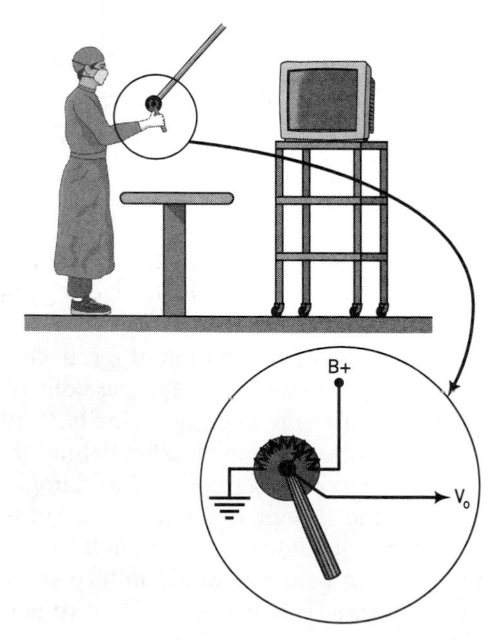

Figure 10.3 Virtual medicine system.

The scalpel handle itself is actually the "control stick" of a human sensory interface (HSI) configured as follows: Extending downward from above the human operator is a microphone boom, but at the end of the boom rod is a control-stick potentiometer. The wiper-arm of this potentiometer is connected to an actual scalpel handle, which serves as the control-stick itself (see Figure 10.3)

The HFE is now evaluating the next stage in the operative procedure known as "blunt dissection." At this stage, the operator will enlarge the slit in the fascia by flexing the hand at the wrist (using the flexor carpi muscles of the forearm) while slightly elevating the forearm itself (using the biceps muscles).

The primary muscle action will be horizontal along the plane of the fascia so that progressive elastic resistance will be encountered as the horizontal displacement of the scalpel handle increases (Δx), and the slit in the fascia is lengthened (see Figure 10.4).

The above procedure will be visualized on the display, and the HFE also will use a haptic interface in which the force will be generated at the scalpel handle that will be proportional to the horizontal lengthening of the opening in the fascia. Consequently, the HFE is now evaluating a haptic interface patterned after an "elastic control stick" for this stage of the operative procedure.

(a) Refer to the block diagram of a human operator-technologcial system interaction in which the HSI is an isotonic control stick (Figure 9.26). Appropriately modify this figure to incorporate a haptic interface, specifically an "elastic" control stick with a spring constant, K_S;

(b) Referring to Figure 10.4, the stiffness coefficient of the fascia (K_F) is obtained from biomaterial mechanical data, and is defined as:

$$K_F = \frac{F_e(\Delta x)}{\Delta x}$$

The HFE may then approximate that the vertical height of the wrist center-of-rotation (c) is a constant vertical distance Δy above the fascial plane (f) and

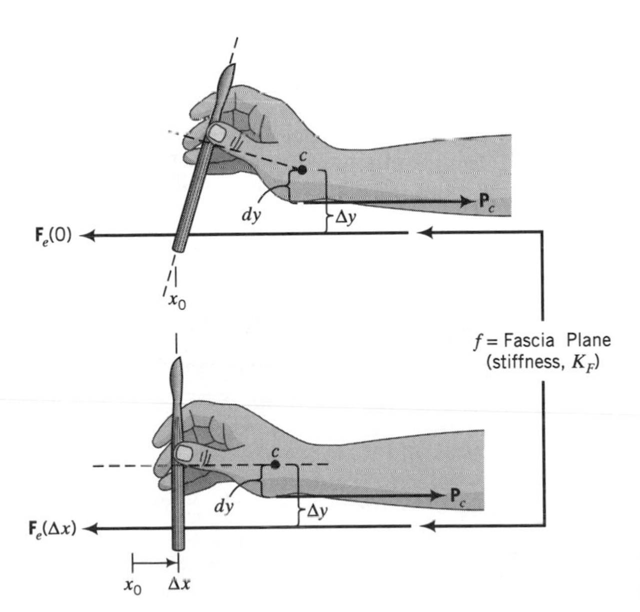

Figure 10.4 Free-body diagram for biomechanical analysis.

also that c is a constant vertical distance (dy) above the line-of-force of the flexor carpi muscles (P_c), as shown in Figure 10.4. Finally, given that:

$$dy = 2.0 \text{ cms}$$
$$\Delta y = 6.0 \text{ cms}$$

(i) Define P_c in relation to F_e

(ii) Define P_c as a function of the fascia stiffness coefficient, K_F.

(c) Referring to the human operator-technological system block diagram developed in part (a) for a haptic interface representing an "elastic" control stick; and given the following values:

$$\Delta x = 5 \text{ cm}$$
$$\hat{P}_C = 0 \text{ (at } t_0 = 0)$$
$$K_D = 0.1$$
$$K_S = 0.5$$

Other parameters as per Example 9.10(b).

(i) Write a computer simulation program for neuromuscular control system when performing a unit displacement task using a haptic (elastic) control stick; and

(ii) Plot the time course of:

(a′) The force generated ($\hat{P}_C$) during the performance of this task

(b′) The velocity generated (V') and the velocity commanded (V_R) during task performance

(c′) The muscle length (L') that occurs during the task

(d′) The displacement (x_0) of the scalpel ($x_0 = 0$ at $t_0 = 0$) and the displacement error (x_e) during the performance of this task.

SOLUTION 10.1(a)

A neuromuscular dynamic control system appropriately modified with a haptic "elastic control stick" is shown in Figure 10.5.

SOLUTION 10.1(b)

(i) Referring to Figure 10.4, define $P_c = f(F_e)$:
From inspection of Figure 10.4, draw the free-body diagram for the upper half and lower half of the figure:

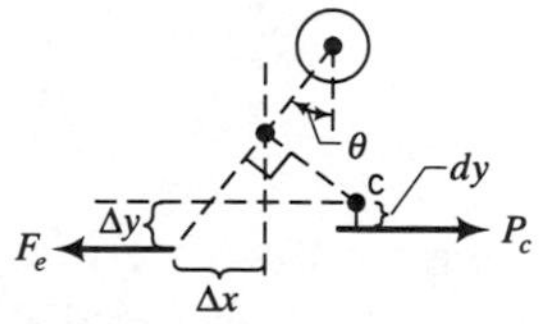

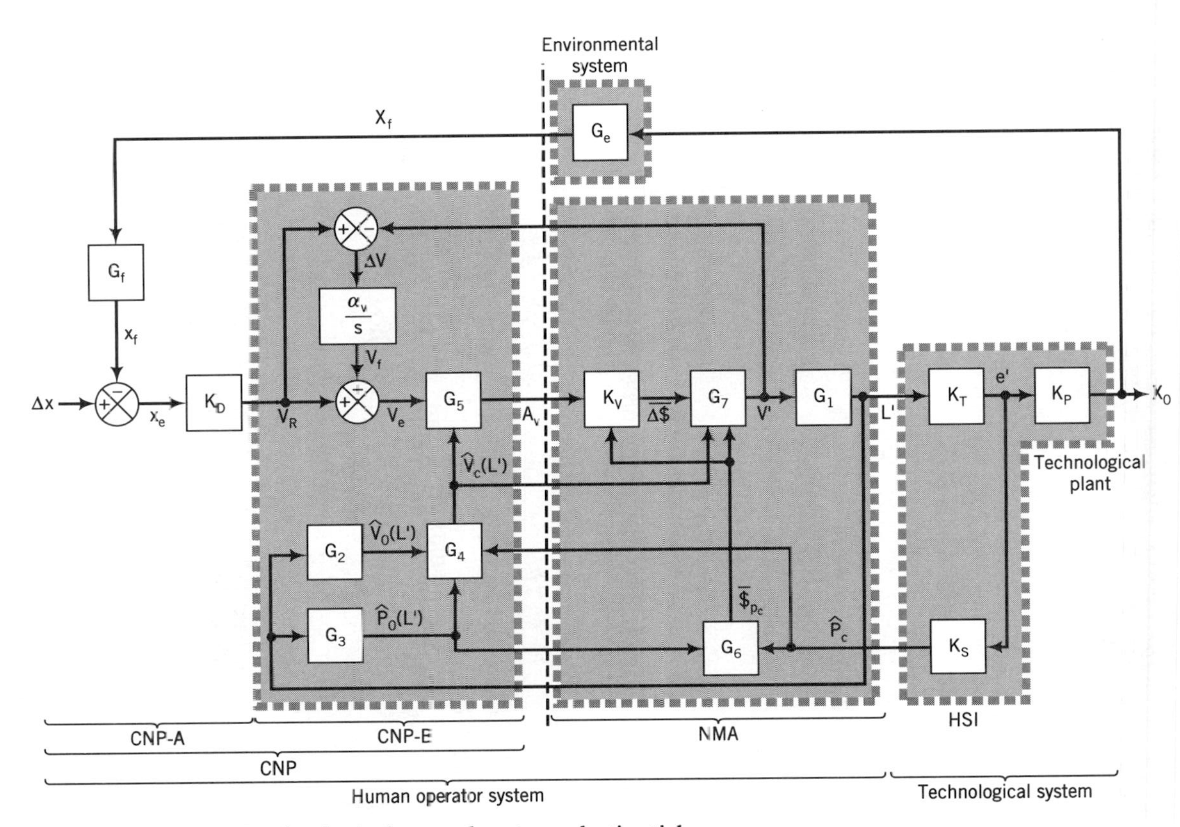

Figure 10.5 Human-technological control system: elastic stick.

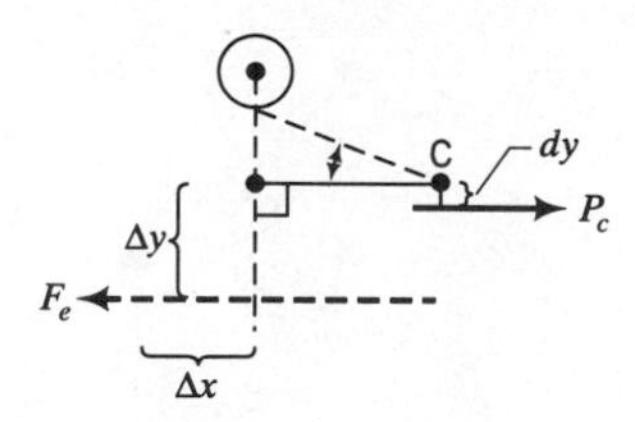

At point C:

$$\sum M_c = 0:$$
$$P_c dy - F_e \Delta y = 0 \tag{i}$$

So that:

$$P_c = \left[\frac{\Delta y}{dy}\right] F_e \tag{ii}$$

Substituting given values:

$$P_c = 3F_e \tag{iii}$$

(ii) Define $P_c = f(K_F)$:
Rearrange the definition of K_F:

$$F_e(\Delta x) = K_F \Delta x \tag{iv}$$

Substitute equation (iv) into equation (ii):

$$P_c = K_F \left[\frac{\Delta x \cdot \Delta y}{dy}\right] \tag{v}$$

SOLUTION 10.1(c)

(i) An m-file is written in MATLAB for the computer simulation flow chart of Figure 10.5. A representative program is shown in Figure E10.1.1.

```
% Example 10.1
% CA Phillips 15 Sept 1998
%  H.O.-Tech. System Control;
%  Unit Step Function: Displacement;
% Elastic Control Stick;
echo off;
clear;
lhat=1.40;
pchat=0;
vprime=0;
xprime=0;
lprime=lhat;
vf=0;
pi=3.1415926;
dx=5;
alpha=0.05;
sum=0;
beta=0.1;
Kd=0.1;
```

```
Kt=1.0;
Kp=10;
deltalr=1.0;              %deltalr=1.5-0.5
Ks=0.5;
Ge=1.0;
for t=1:10;
t2(t)=t-1;
% unit step displacement;
if t==1
     del_x=0;
else
     del_x=dx;
end;
xf=Ge*xprime;
xe=del_x - xf;
xerror(t)=xe;
vr=Kd*xe;
vrrec(t)=vr;
for y=1:5;
% G2:
v0hat=sin(pi*(lprime-.5));
v0hatrec(t)=v0hat;
% G3:
p0hat=sin(pi*(lprime-.5));
p0hatrec(t)=p0hat;
% G4:
vchat=v0hat*(1-sin(pi*pchat/(2*p0hat)));
vj=vr-vprime;
delvrec(t)=vj;
vf=vf+vj;
vp=vf*(alpha);
ve=vr+vp;
verec(t)=ve;
% G5:
av=ve/vchat;
if av>1
        av=1;
end
if av<0
        av=0;
end
avrec(t)=av;
% G6:
stim=(1/pi)*acos(1-(2*pchat/p0hat));
stimrec(t)=stim;
% Kv:
stimbar=(1-stim)*(2/pi)*asin(av);
stimbarrec(t)=stimbar;
% G7:
vprime=vchat*sin((pi/2)*(stimbar/(1-stim)));
velocity(t)=vprime;
% G1:
sum=sum+vprime;
dL=sum*(beta);
lprime=lhat-dL;
```

```
if lprime<.5
        pause;   % This stops the program if lprime<.5
                  % i.e. the requested conditions can't be met
end
lprimerec(t)=lprime;
% elastic  control stick(part 1);
eprime= Kt*(lhat-lprime)/deltalr;
if eprime>1
    eprime=1;
end;
if eprime<0
    eprime=0;
end;
eprec(t)=eprime;
% elastic control stick (part 2);
pchat=Ks*eprime;
pcrec(t)=pchat;
%Tech. plant (Kp);
xprime=Kp*eprime;
xprec(t)=xprime;
end;
end;
plot(t2,pcrec);
title('Pchat vs. Time(sec)');
pause;
plot(t2,velocity);
hold;
plot(t2,vrrec,'*');
title('Vprime vs. V requested, * = requested V');
xlabel('Time(sec)');
pause;
hold off;
plot(t2,lprimerec);
title('Lprime vs. Time(sec)');
pause;
plot(t2,xprec);
hold;
plot(t2,xerror,'*');
title('Xprime vs. Xerror, * = Xerror');
xlabel('Time(sec)');
hold off;
```

(ii) The requested time plots are as follows:

(a′) The generated force ($\hat{P}_C$) is presented in Figure E10.1.2.

(b′) The velocity generated (V') and the velocity requested (V_R) are depicted in Figure E10.1.3.

(c′) The muscle length (L') is shown in Figure E10.1.4.

(d′) The displacement (x_0) of the scalpel and the displacement error (x_e) are illustrated in Figure E10.1.5.

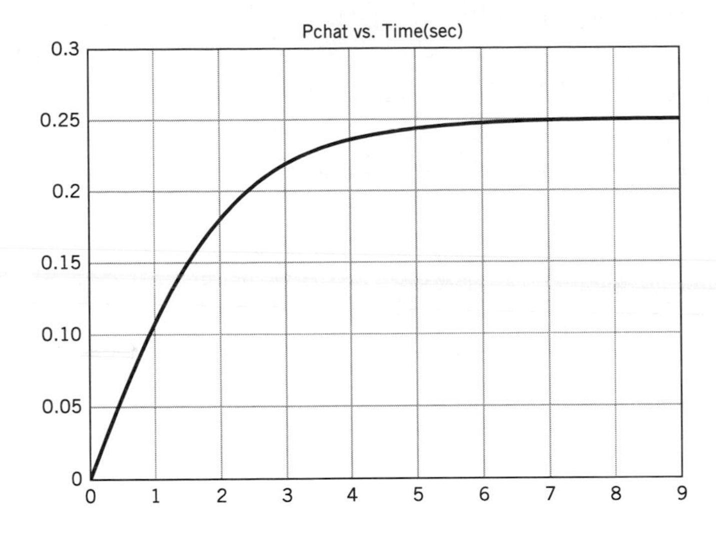

Figure E10.1.2

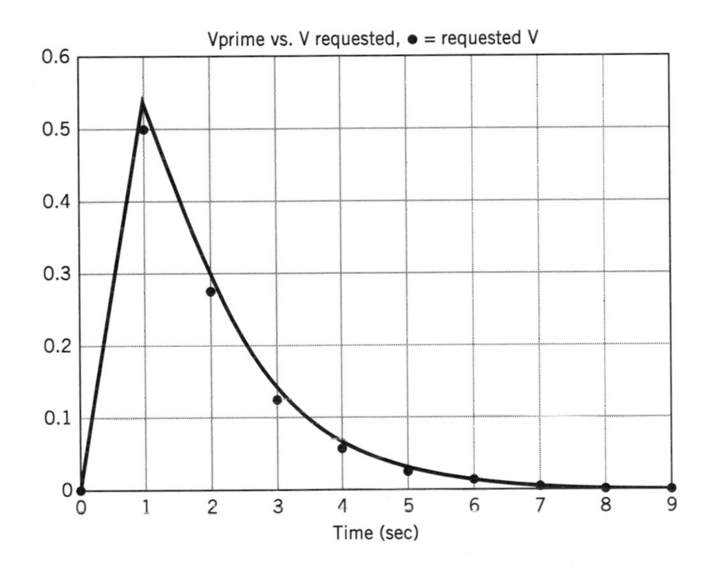

Figure E10.1.3

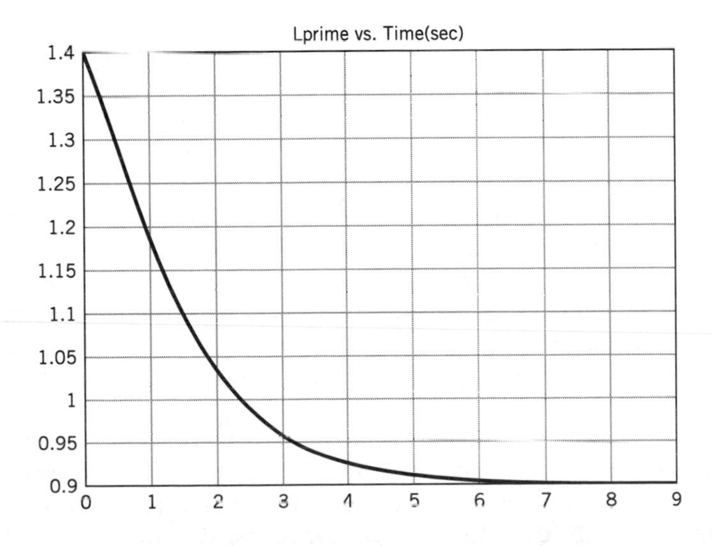

Figure E10.1.4

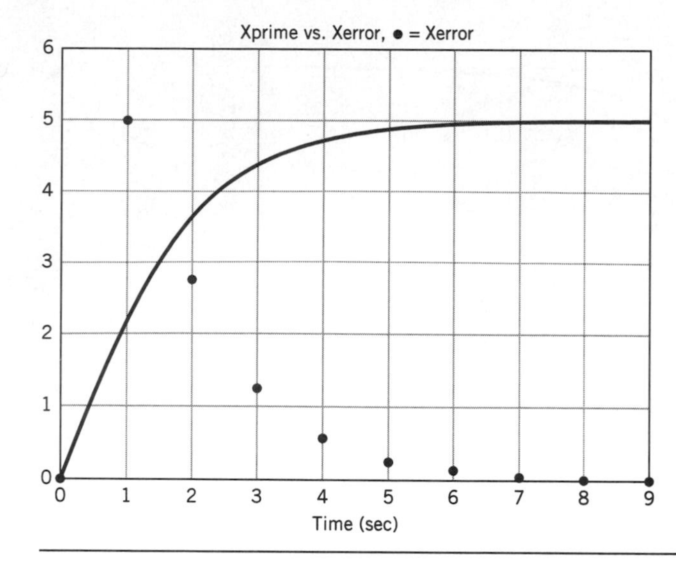

Figure E10.1.5

Telerobotics is another area in which haptic control interfaces are effectively used.

EXAMPLE 10.2

A human operator is sitting at a control console on board a ship at sea. The control console is connected by a tethered cable to an underwater robot in this example of a telerobotic system.

The lower half of the underwater robot rolls on tank treads over the sea bottom and is currently at a salvage site in which material is being removed from the sunken Titanic.

The upper half of the underwater robot has a swiveling head in which the central eye is a video camera lens that transmits a real-time televised image of the robot's field of view back to the video display located at the center of the shipboard control console.

The underwater robot has a torso that swivels independently of the head and has two mechanical arms, one arm each attached to each side of the torso. At this point, both mechanical arms have attached to a locked purser's safe, and already lifted it vertically to the mid-torso level of the underwater robot.

Back at the control console, the robot arm movements are controlled by a panel of thumb and index finger actuated "haptic" switch levers (see Figure 10.6). The flexor pollucus longus (FPL) muscle of the human operator is about to contract, flexing the thumb at the first knuckle joint and depressing the switch lever.

The HFE is now evaluating the next action in the sunken safe removal, in which the underwater robot arms will extend forward, moving the locked safe horizontally away from the torso over a given distance, Δx. This action is necessary in order to place the locked safe on a loading platform (from which it will be hoisted out of the water).

Since water is a viscous medium, a moving object will encounter viscous (rate-dependent) reaction forces. Consequently, the HFE is evaluating a haptic switch lever that will react like a "viscous" control stick. This means that a greater speed of the robot arm movement over the desired horizontal distance (Δx) will result in a greater required force to be applied by the human operator at the controls. These control forces will reflect the reaction forces that the robot arms are encountering in the viscous aqueous environment:

(a) Modify the human operator-technological system block diagram for an isotonic control stick (Figure 9.26) to incorporate a human system interface (HSI) that represents a "viscous" control stick;

(b) Referring to the left side of Figure 10.6, the viscous coefficient (C_v) is used as an amplifier parameter, and defined as:

$$C_v = \frac{F_L(\dot{x})}{\dot{x}}$$

Referring to the right side of Figure 10.6, the HFE may approximate the thumb length ($\overline{AC}$) and also the action of the FPL muscle (P_c) as acting at a distance ($\overline{BC}$) from the thumb joint center-of-rotation (c). Given that θ is the angle-of-inclination (from horizontal) of the thumb length ($\overline{AC}$), and ϕ is the angle of insertion of the *FPL* muscle near the thumb joint (c):

(i) Define P_c in relation to F_L.

(ii) Define P_c as a function of the viscous coefficient, C_v.

(c) Referring to the human operator-technological system developed in part (a) for a haptic interface representing a "viscous" control stick, and given the following:

$$\Delta x = 5 \text{ decimeters } (0.5 \text{ M})$$
$$\hat{P}_C = 0 \text{ (at } t_0 = 0)$$

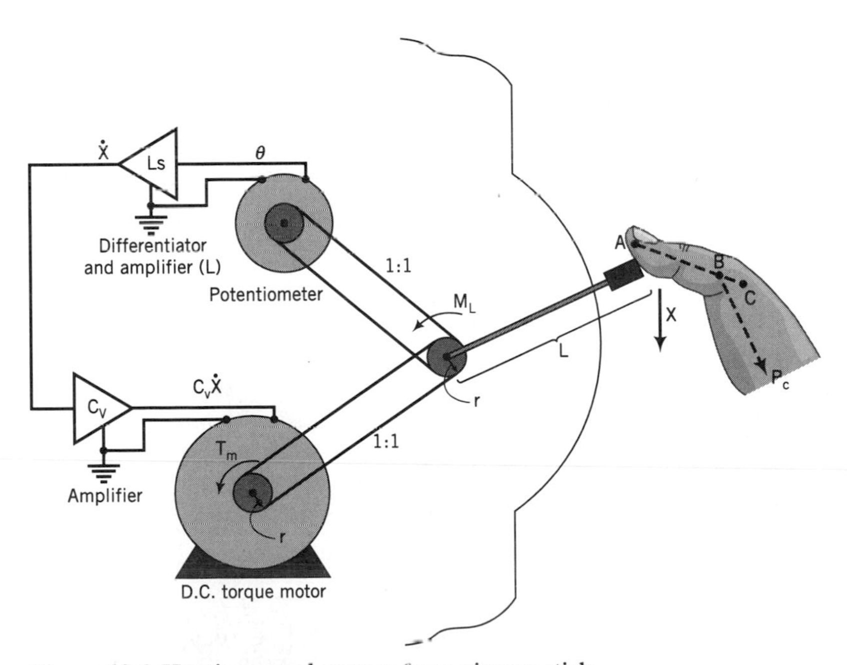

Figure 10.6 Haptic control system for a viscous stick.

$$K_D = 0.1$$
$$C_v = 1.0$$

Other parameters as per Example 9.10(b)

(i) Write a computer simulation program for the neuromuscular control system when performing a unit displacement task using a haptic (viscous) control stick.

(ii) Plot the time course of the following:

(a′) The force generated ($\hat{P}_C$) during the performance of this task

(b′) The velocity generated (V') and the velocity commanded (V_R) during task performance

(c′) The muscle length (L') that occurs during the task

(d′) The displacement (x_0) of the locked safe ($x_0 = 0$ at $t_0 = 0$) and the displacement error (x_e) during the performance of this task.

SOLUTION 10.2(a)

A neuromuscular dynamic control system appropriately modified with a haptic "elastic control stick" is shown in Figure 10.7.

SOLUTION 10.2(b)

(i) Referring to Figure 10.6, define P_c in relation to the lever force, F_L:
From inspection of the right side of Figure 10.6, draw the free-body diagram for the thumb and lever system:

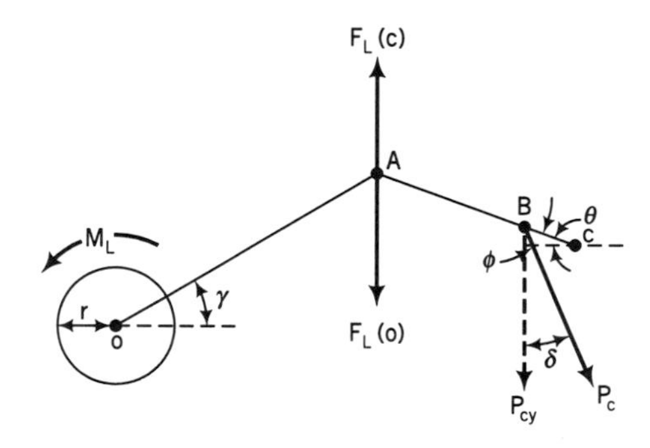

Where: $\overline{OA} = L$
$F_L(c) = F_L$ seen at point c
$F_L(o) = F_L$ seen at point o

Referring to the right half of this diagram:
At point C:

$$\sum M_c = 0: \tag{i}$$
$$P_{cy}\,[\overline{BC}\cos\theta] - P_{cx}[\overline{BC}\sin\theta] - F_L[\overline{AC}\cos\theta] = 0$$

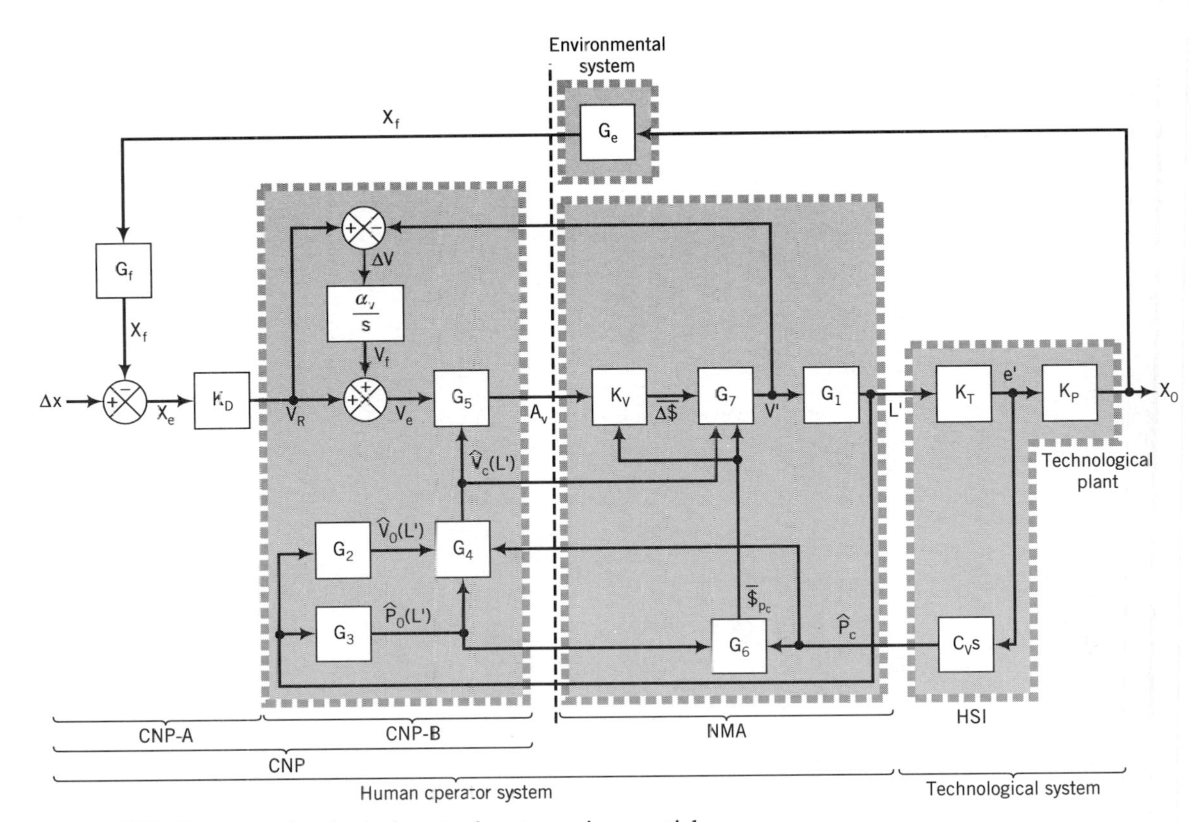

Figure 10.7 Human-technological control system: viscous stick.

Now since:

$$P_{cy} = P_c \cos[\delta] \quad \text{(ii)}$$
$$P_{cx} = P_c \sin[\delta] \quad \text{(iii)}$$

and it can be shown that:

$$\delta = 90 - \theta - \phi \quad \text{(iv)}$$

Then substituting equations (ii)–(iv) into equation (i):

$$P_c[\cos(90 - \theta - \phi)][\overline{BC}\cos\theta] - P_c[\sin(90 - \theta - \phi)][\overline{BC}\sin\theta] - F_L[\overline{AC}\cos\theta] = 0 \quad \text{(v)}$$

Rearrange for $P_c = f(F_L)$:

$$P_c = \frac{F_L[\overline{AC}\cos\theta]}{[\cos(90 - \theta - \phi)][\overline{BC}\cos\theta] - [\sin(90 - \theta - \phi)][\overline{BC}\sin\theta]} \quad \text{(vi)}$$

(ii) Define $P_c = f(C_v)$:
Referring to the left-hand side of the diagram, at o, for $\sum M_o = 0$:

$$M_L - F_L \cdot L \cdot \cos\gamma = 0 \quad \text{(vii)}$$

Rearrange for F_L:

$$F_L = \frac{M_L}{L \cdot \cos\gamma} \quad \text{(viii)}$$

From inspection of the left side of Figure 10.6, it is apparent that:

$$M_L = T_m \quad \text{(ix)}$$

since the radius r (motor pulley) equals r (switch lever pulley).
Restating equation (ix) with respect to the lever switch force:

$$F_L \cdot r = T_m \quad \text{(x)}$$

From the definition of C_v, then the viscous force at the lever (*and* at the motor) is:

$$F_L = C_v\dot{x} \quad \text{(xi)}$$

Substituting equations (ix)–(xi) into equation (viii):

$$F_L = \frac{C_v\dot{x}r}{L \cdot \cos\gamma} \quad \text{(xii)}$$

Substituting equation (xii) into equation (vi) results in the requested function:

$$P_c = \frac{C_v(\dot{x}r\overline{AC}\cos\theta)}{L \cdot \cos\gamma([\cos(90 - \theta - \phi)][\overline{BC}\cos\theta] - [\sin(90 - \theta - \phi)][\overline{BC}\sin\theta])} \quad \text{(xiii)}$$

SOLUTION 10.2(c)

(i) An m-file is written in MATLAB for the computer simulation flow chart of Figure 10.7. This program is indicated by Figure E10.2.1.

```
% Example 10.2
% CA Phillips 15 Sept 1998
%  H.O.-Tech. System Control;
%  Unit Step Function: Displacement;
% Viscous Control Stick;
echo off;
clear;
lhat=1.40;
pchat=0;
vprime=0;
xprime=0;
lprime=lhat;
ep(1)=0;
vf=0;
pi=3.1415926;
dx=5;
alpha=0.05;
sum=0;
beta=0.1;
Kd=0.1;
Kt=1.0;
Kp=10;
deltalr=1.0;            %deltalr=1.5-0.5
Cv=1.0;
Ge=1.0;
for t=1:10;
z=t+1;
t2(t)=t-1;
% unit step displacement;
if t==1
     del_x=0;
else
     del_x=dx;
end;
xf=Ge*xprime;
xe=del_x - xf;
xerror(t)=xe;
vr=Kd*xe;
vrrec(t)=vr;
for y=1:5;
% G2:
v0hat=sin(pi*(lprime-.5));
v0hatrec(t)=v0hat;
% G3:
p0hat=sin(pi*(lprime-.5));
p0hatrec(t)=p0hat;
% G4:
vchat=v0hat*(1-sin(pi*pchat/(2*p0hat)));
vj=vr-vprime;
delvrec(t)=vj;
vf=vf+vj;
vp=vf*(alpha);
ve=vr+vp;
```

```
verec(t)=ve;
% G5:
av=ve/vchat;
if av>1
        av=1;
end
if av<0
        av=0;
end
avrec(t)=av;
% G6:
stim=(1/pi)*acos(1-(2*pchat/p0hat));
stimrec(t)=stim;
% Kv:
stimbar=(1-stim)*(2/pi)*asin(av);
stimbarrec(t)=stimbar;
% G7:
vprime=vchat*sin((pi/2)*(stimbar/(1-stim)));
velocity(t)=vprime;
% G1:
sum=sum+vprime;
dL=sum*(beta);
lprime=lhat-dL;
if lprime<.5
        pause;   % This stops the program if lprime<.5
                  % i.e. the requested conditions can't be met
end
lprimerec(t)=lprime;
% viscous  control stick(part 1);
eprime= Kt*(lhat-lprime)/deltalr;
if eprime>1
    eprime=1;
end;
if eprime<0
    eprime=0;
end;
eprec(t)=eprime;
% viscous control stick (part 2);
ep(z)=eprime;
del_ep=ep(z)-ep(z-1);
pchat=Cv*del_ep;
pcrec(t)=pchat;
%Tech. plant (Kp);
xprime=Kp*eprime;
xprec(t)=xprime;
end;
end
plot(t2,pcrec);
title('Pchat vs. Time(sec)');
pause;
plot(t2,velocity);
hold;
plot(t2,vrrec,'*');
```

```
title('Vprime vs. V requested, * = requested V');
xlabel('Time(sec)');
pause;
hold off;
plot(t2,lprimerec);
title('Lprime vs. Time(sec)');
pause;
plot(t2,xprec);
hold;
plot(t2,xerror,'*');
title('Xprime vs. Xerror, * = Xerror');
xlabel('Time(sec)');
hold off;
```

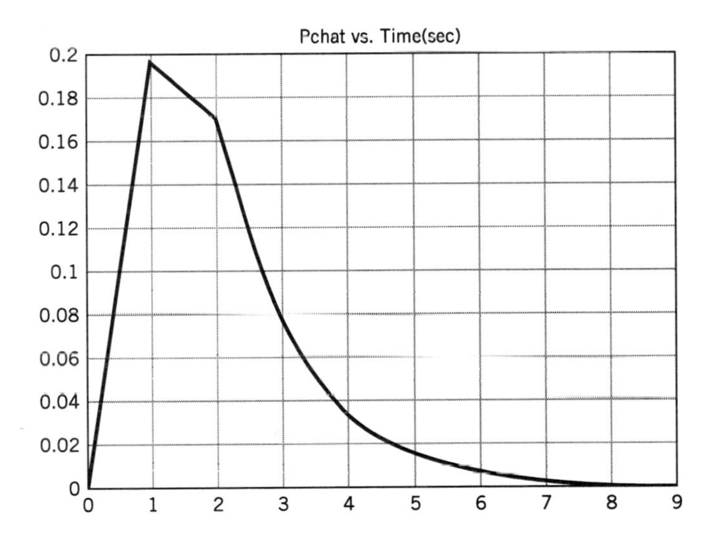

Figure E10.2.2

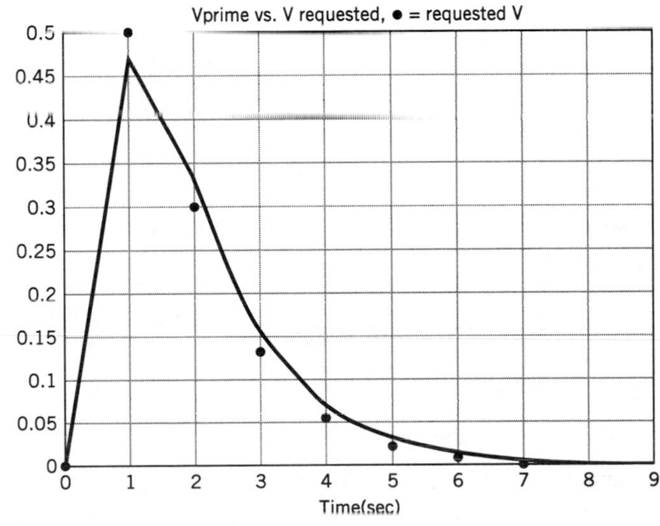

Figure E10.2.3

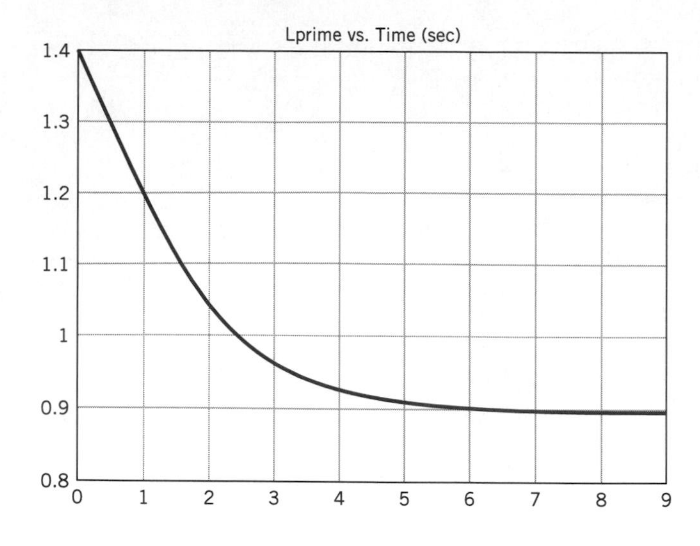

Figure E10.2.4

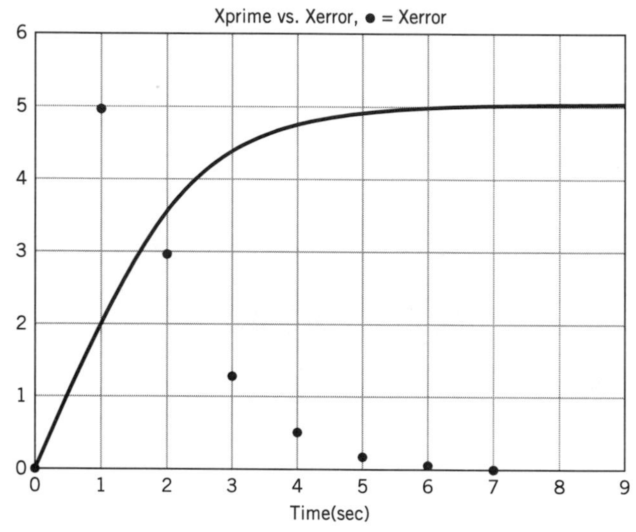

Figure E10.2.5

(ii) The requested time plots are as follows:

(a′) The generated force ($\hat{P}_C$) is presented in Figure E10.2.2.

(b′) The velocity generated (V') and the velocity requested (V_R) are depicted in Figure E10.2.3.

(c′) The muscle length (L') is shown in Figure E10.2.4.

(d′) The displacement (x_0) of the locked safe and the displacement error (x_e) are illustrated in Figure E10.2.5.

10.2 HUMAN OPERATOR INFORMATIC CONTROL

a. Information Theory and Human Action

i. Information Theory and Its Application

Let us begin by discussing the value of information. In the preceding material (Section 10.1), the human factors engineer was able to provide the human operator with additional haptic information (force and position cues) along with the customary visual cues. The basis for doing so was that the additional information was more helpful (in effect, more "immersive") to the human operator than having only the visual cues alone. In order to more rigorously examine this hypothesis, let us consider the concept of information and its application.

Consider the experimental scenario depicted in Figure 10.8. The human operator is seated at a keyboard configured at the top with two semicircles, each of five discrete lights. At the bottom of the keyboard are two sets of pushbuttons (five pushbuttons to a set) such that each pushbutton is located directly under each light. The task to be accomplished is a simple stimulus-response task. The *input stimulus* may be one or more of the lights going on (as depicted in the circles on the panel). The human operator has his ten fingers on each of the ten possible pushbuttons at the lower part of the panel. The *output response* occurs when the human operator presses one or more of the pushbuttons directly below an illuminated light. There are two different methods by which the difficulty of this stimulus-response task may be adjusted. First, the number of possible input lights could be varied (any number between the same single light and all ten of the lights). This would represent a variation in the *spatial* information of the task. Second, the rate at which one or more of the lights is turned on during a given period of time could also be varied. This would represent the *temporal* information of the task. This task may be expressed in terms of information theory.

The amount of information (H) that is present in a set (x) of equally probable alternatives (N) may be expressed as:

$$H(x) = \log_2 N \tag{20}$$

where $H(x)$ is the information content (in bits) of the xth set, and N is the number of equally probable alternatives or choices that the human operator must select.

Consider the simplest task (involving the first set), in which only the same single light will be randomly illuminated; for example, the light underneath the operator's right index finger. In this simple stimulus-response scenario, there is only one choice for the human operator ($N = 1$) so that from equation (20), the information content (H) of this task (first set) is zero. Now consider a slightly more difficult task (involving a second set), in which one of either two lights may be illuminated, always the same pair of lights and each of equal probability; for example, the light below the operator's right index finger and the light below the operator's left index finger. In a task in which only one of the two lights will be illuminated at a time (not both at the same time), then the human operator is faced with one of two alternatives ($N = 2$) and must make one of two choices, so that equation (20) would predict that the information content (H) of this task (second set) is unity (1.0 bit).

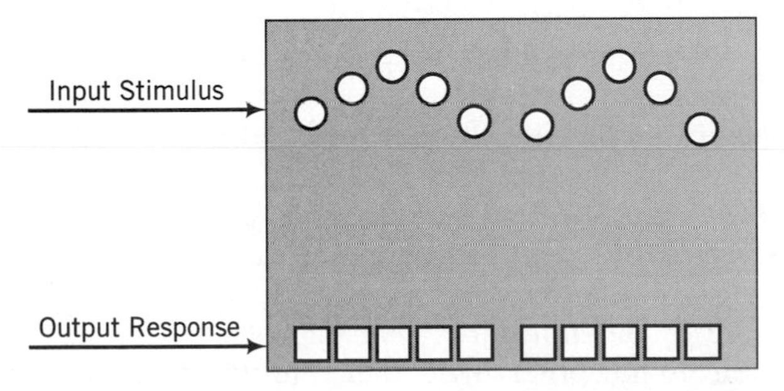

Figure 10.8 Stimulus-response display board.

In other words, one bit of spatial information is defined as an alternative of choice. Obviously, the task difficulty increases with an increase in the information content (H).

The baud rate is the unit of signaling (stimulus) speed. The baud rate may be defined as the ratio of the spatial information (in bits) to the rate of bit occurrence (per second):

$$\text{baud rate} = H/\text{time} \tag{21}$$

where baud rate is in units of bits/second.

Equation (21) states that 1 baud is equivalent to a temporal information rate of one bit (of information) per second. Consequently, the human operator will experience an increase in task difficulty that may also be associated with an increase in the information rate (baud rate).

The relationship between the task difficulty and the information content may be appreciated in this example of an office worker. The office worker is a telephone receptionist seated at a desk and whose job it is to answer telephone calls and forward them to another person in the office complex. This task has both a spatial dimension and a temporal dimension. With respect to spatial information, two telephones on a desk would represent a low bit per stimulus task (1.0 bit per equation (20)). Alternatively, 16 telephones on a desk would represent a high bit per stimulus task (4.0 bits per equation (20)). With respect to the temporal dimension, one call (on two possible phones) per one hundred seconds would represent a low baud rate task (.01 baud per equation (21)). Alternatively, one call every 10 s would represent a high baud rate task (0.1 baud per equation (21)).

The preceding discussion has referred to the information content in which all events are equally probable. Most situations of practical interest to the human factors engineer will involve a number of alternatives in which the probability of occurrence of the various events will differ. A common example would be that of a human operator who is receiving and transcribing international telegraphic code. There are 26 letters in the alphabet so that 26 alternatives of choice (N) represent the set.

However, it is commonly known that in the English language, certain letters of the alphabet appear with much greater frequency than do others. The letter *e* appears most commonly, and the frequency of appearance of the other 25 letters of the alphabet will occur less commonly and with varying probability (e.g., *a* occurs much more frequently in normal English communication than does *q*). As a result, the human factors engineer is very frequently interested in the information content of a set of stimuli, each a specific probability of occurrence. In this situation, the information content (H) is calculated as follows:

$$H(x) = \sum_{i=1}^{N} P_i \left(\log_2 \left[\frac{1}{P_i} \right] \right) \tag{22}$$

where P_i is the probability of the ith alternative in the xth ensemble (set) of N alternatives.

Returning to the example of the light and pushbutton panel (see Figure 10.8), consider the case in which there were two alternatives and each consisted of one light (above the pushbutton of the left index finger and above the pushbutton for the right index finger). If it had transpired that one of the two lights were to illuminate twice as often as the other light, then the information content (H)

for this human operator task (for this set) would be governed by equation (22). Specifically, the information content (H) would be:

$$H(x) = (0.67)(\log_2[1.5]) + (0.33)(\log_2[3])$$
$$H(x) = (0.67)(0.59) + (0.33)(1.59) = 0.92 \text{ bit}$$

This simple example indicates that the information content in a two-choice alternative, each of different probability, is less than the information content of a two-choice alternative, each of equal probability. In effect, when all of N alternatives have equal probability (P_i):

$$NP_i = 1 \tag{23}$$

and $H(x)$ is a maximum. This is shown by first rearranging and then substituting equation (23) into equation (22), and factoring:

$$H(x) = \log_2(N) \sum_{i=1}^{N} P_i \tag{24}$$

so that:

$$H(x) = NP_i \cdot \log_2(N) \tag{25}$$

And by substituting equation (23) into equation (25), it is apparent that equation (22) reduces to equation (1), and thus is the maximum information that can be gleamed from a signal.

In a variety of manual control tasks, the human operator is frequently required to react to a continuously changing input stimulus. In effect, the human operator is responding to a continuously varying forcing function (i.e., command signal). This is a somewhat different situation than described above in which a limited number of discrete system states represent the information content of the task. In the continuous time case, it is typical to describe the baud rate in units of frequency (hertz).

Returning to Figure 10.8, it should be noted that there is an information content associated with the input stimulus (the lights), and another information content associated with the output response (the buttons). The input information rate (bits/s) described by equation (2) is applied at the input stimulus (turning on a light). Recall that task difficulty will increase as a function of the input information rate (the telephone example cited earlier). There is also an output information rate (bits/s) which is determined by applying equation (2) to the *correct* output response of the human operator (pushing the button below the light that has turned on).

In a simple information processing experimental paradigm, the human operator will perform as indicated in Figure 10.9. Note that the input information rate (bits/s) is treated as the independent variable. Task difficulty progressively increases with an increase in the input information rate. The dependent variable is the output information rate (bits/s), which represents the output correct response rate. This relationship is characterized graphically by a continuous curve, which initially has a positive (rising) slope, reaches an inflection point (the channel capacity) and then continues with a negative (decreasing) slope. For the human operator, as long as information arrives on the ascending limb (Figure 10.9), there is effective information processing (output response throughput of the input stimulus). When information arrives too rapidly (the descending portion of Figure 10.9), human operator information processing becomes overloaded and there is inadequate or ineffective

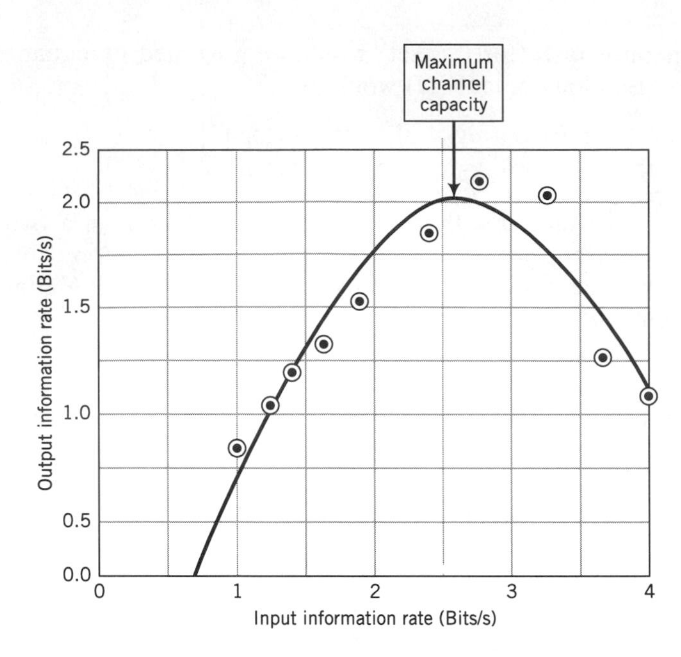

Figure 10.9 Human operator information throughput.

throughput (input stimulus driven output response). The inflection point represents the maximum rate at which the human operator can process information (in an input-output manner) and is termed the "maximum channel capacity."

The human maximum channel capacity may be estimated from Figure 10.9 as approximately 2.3 bits/second. For this particular human operator, the calculated channel capacity is low because of a high level of environmental stress (concurrently large acceleration forces were being experienced by the human operator). Various estimates of the human operator maximum channel capacity have been made based upon stimulus-response experiments (in a nonstressful laboratory environment). It would appear that the human operator can obtain a maximum value of between 6 to 10 bits/s in certain tasks. Other laboratory experiments have involved manual tracking tasks in which the human operator follows a continuously varying demand signal. When the information content is continuously varying (as opposed to discrete), the "maximum channel capacity" may be characterized by another measure termed the *bandwidth.* For a stimulus that is periodic and predictable (e.g., sine wave tracking with hand-arm movement), the human operator bandwidth is between 1 and 2 Hz. For a random and unpredictable signal (e.g., unit step track with hand-arm movement), the human operator bandwidth is below 1 Hz (maximum of 1 Hz). Below and up to these human operator channel maximum capacities (and bandwidths), the human is an adequate and effective information processor (for the particular task and environment specified).

Referring again to Figure 10.9, the maximum channel capacity may be viewed as an "overload" point such that when it is exceeded there is a decrease in information throughput of the human operator. In effect, the human operator is in a condition of "information overload" and the maximum channel capacity (of Figure 10.9) may be considered as the "overload point." There are at least three different ways in which the channel capacity of a human operator may be exceeded. First, the level of difficulty of a control task may be inherently high. As we have seen, the higher task difficulty is associated with a higher input baud rate (Figure 10.9) so that the human operator must process input information at an excessive rate. In this case, human operator performance can only improve by an increase in the channel capacity itself.

A second source of information overload occurs with inexperienced, untrained operators. In a typical operational task, a great deal of the information presented to an operator is either irrelevant or it is redundant. The inexperienced and/or untrained operator is not sufficiently selective, and therefore tends to process more information than is essential for the task performance. Consequently, the human operator experiences information overload and a decline in task performance. In this case, the use of strategy (see Section 9.3) can significantly improve human operator information processing. Recall that strategy is an internal paradigm of the central nervous system processor and creates a model for predictive control of a system. As a result, the skilled (trained) operator may significantly improve task performance by remaining below his maximum channel capacity and not exceeding the overload point.

A third way in which information overload may occur is the situation in which two tasks compete for the attention of the human operator. As a result, there is the simultaneous presentation of two sets of task information, which must be processed by a human operator with a finite channel capacity. Possible human operator reactions may involve overfocusing on one task (with significant degradation of performance of the other task) or dividing attention between the two tasks (and accepting some performance degradation in each of the tasks). Issues with respect to time sharing and attention distribution are the domain of psychological models and outside the scope of this text. For both practical and didactic purposes, this textbook considers only single task performance.

It should be recognized that information theory and its application have some practical limitations. First, most of the equations of information theory are descriptive rather than explanatory so that, per se, they do not provide information regarding the specific mechanics of the information processing scheme. Second, many important psychological mechanisms (such as short-term memory and long-term memory) are not adequately characterized by the equations of information theory. Despite these limitations, the human factors engineer must work with quantitative (analytical) models when addressing the concept of information and human operator informatic control. Once again, as has been a recurring theme in this text, in the initial design and development phase, only a preliminary first-order approximation of the human operator informatic model may be necessary. However, further refinements of the initial model will continue and evolve through the various iterative phases of human system design (see Section 2.1). Although psychological theory may not necessarily be advanced by the human factors engineer, the goal of the working engineer is practical application rather than theoretical development.

ii. Human Response (Action) Time

We act and we react! A characteristic feature of human beings is that they respond to their environment. For the purpose of quantitative analysis, consider the following axiom: "For every response (action time), there is both a reaction (reaction time) and a movement (movement time)." Mathematically, this may be expressed as follows:

$$AT = RT + MT \tag{26}$$

where:

AT = action time (seconds)
RT = reaction time (seconds)
MT = movement time (seconds)

The time dimension of a human operator response during the performance of a task is governed primarily by whether the response is "internally" paced or "externally" paced. In an internally paced response, the human operator sets the pace for the task and so the task is said to be "self-paced." This pace, of course, may be highly variable depending upon the motivation of the human operator, the physical environment, and many other variables. It is these self-paced tasks that occur most commonly in ordinary, everyday task performance, and are of practical interest to the human factors engineer. The "externally" paced task, is one in which the demands of the task set the pace for the human operator response. Playing a video arcade game is an example of an "externally" paced task. These tasks occur most commonly in controlled laboratory experiments that evaluate various aspects of human operator performance but also occur with some regularity in ordinary everyday tasks.

There is a distinct advantage to a self-paced task with respect to human operator informatic control. Specifically, the level of task difficulty (and the associated informational processing rate) may be voluntarily set by the human operator to a level consistent with satisfactory task performance. For example, the human operator may drive his vehicle along a straight course of a four-lane interstate highway at a particular speed. The same human operator would reduce his vehicle speed when driving along a single-lane, winding mountain road. It should be noted that the material presented in this section with regard to human operator response time is based upon externally paced tasks (usually in a controlled experimental laboratory setting).

The response time (AT) may be defined as the minimum time to generate a response output for an input stimulus. Referring to equation (26), the reaction time (RT) equals the response time (AT) when there is zero movement time (MT). Referring to Figure 10.8, consider the case when all ten of the human operator's fingertips are in immediate contact with all ten of the output response pushbuttons. Assuming negligible movement of the pushbutton (when depressed) and also negligible movement of the fingertip, the time interval between a stimulus light illuminating (input) and a responding pushbutton depression (output) would represent the reaction time. This is a common method by which the human operator reaction time is evaluated in the experimental laboratory.

Let us first consider reaction time. *Simple* reaction time refers to the RT when there is only *one* specific stimulus event and the same output response occurs. Since there is no choice between alternatives ($N = 1$), the information content (H) associated with a simple reaction time is zero. There are few ordinary everyday tasks where simple reaction time only is used, because generally few tasks are so simple as to have zero information content. Simple reaction time will range between .15 and .20 s (and .20 s is a generally accepted value). However, there is some variation depending upon a number of factors. The type of stimulus (visual, auditory, or tactile), the operator's state of expectation of the signal, and the individual's age are a few examples.

Most ordinary tasks that are regularly performed by a human operator include some information content. This means that an output response to an event occurs after a period of uncertainty and a choice among alternative actions have been made. As the task difficulty increases, the alternatives of choice (N) increase and so does the information content (H) of the task. Therefore, it is not surprising that the human operator reaction time will also increase with increasing task information content.

Choice reaction time may be mathematically related to the information content of the stimulus event (H_s) by means of the Hick-Hymen law:

$$\text{RT} = a + bH_s \tag{27}$$

where:

a = *simple* reaction time ($H_S = 0$)
b = a proportionality constant analogous to a reciprocal baud rate (s/bit)

$$b = \frac{1}{\beta} \tag{28}$$

where:

β = choice reaction time baud rate (bits per second)

For many operational tasks of practical interest to the human factors engineer, the empirical fitting coefficients a and b may be reasonably approximated, so that an operational form of the Hick-Hymen law is:

$$\text{RT} = 0.20 + (0.15)H_s \tag{29}$$

where RT is the choice reaction time (s).

When the number of alternatives of choice (N) is specified, either equation (20) or equation (22) may be substituted for H_S in equation (27). Recall that equation (20) is selected when the alternatives of choice are equi-probable, and equation (22) is selected when the various alternatives of choice have various probabilities. Consequently, the Hick-Hymen law is very useful for quantifying the human operator reaction time when a complex task (high information content) is performed by the human operator.

Various factors can affect the choice reaction time other than the information content of the task. Such factors include the amount of prior practice by the human operator, use of a warning signal prior to task occurrence, and a compatible stimulus (e.g., the output response push button being located directly under the input stimulus light, as per Figure 10.8).

Let us now consider movement time. A simple form of movement is that of directly moving from point A (the starting point) to point B (the end point). This is referred to as *step tracking*. There are various examples in the real world of step tracking. Examples include the aiming of a camera upon a stationary object, changing lanes while driving on a multilane highway, and reaching over a distance in order to place the finger on a flip switch. In the performance of such tasks, the movement time would be affected by the distance of the movement as well as the accuracy required by the size of the object toward which one is moving. This relationship between the distance moved and the object size represents the difficulty of the movement response. Fitts' law defines an index of movement difficulty (ID) with analogy to information theory:

$$\text{ID (bits)} = \log_2(2A/W) \tag{30}$$

where:

A = distance of movement from start to target center
W = width of the target

Equation (30) has the distance-width ratio (A/W) multiplied by two as an alternative form of:

$$\text{ID (bits)} = \log_2[A/(W/2)] \tag{31}$$

Since A is the distance of movement from the start to the target center, the target can be acquired by overshooting the center, but within the limit of plus $W/2$. However, it is equally possible to acquire the target by undershooting the center, but within the limit of minus $W/2$. In the original experiments that led to Fitts' law, it was demonstrated that for movement between 2 and 16 in. a longer time is required when that movement must end within a small target area compared to when the target area is larger. In these original experiments, the two targets were placed a fixed distance apart and the subject had to tap them alternately and as quickly as possible.

The movement time (MT) may be defined as the time required to physically make a response which begins when the movement is initiated and ends when the target is acquired. The basis for Fitts' law is that the movement time can be predicted from the index of movement difficulty:

$$\text{MT} = a + b\,(\text{ID}) \tag{32}$$

The empirical constants (a and b) in equation (32) depend upon the type and nature of movement involved. Referring to Figure 10.10, the slope of the regression line decreases from arm to finger (as the value of the y-intercept correspondingly increases). In other words, a is a delay constant that depends upon the body member being used. Furthermore, b (s/bit) is a measure of the information handling capacity, analogous to a reciprocal baud rate.

Substituting equation (30) into equation (32) results in the generally used form of Fitts' law:

$$\text{MT} = a + b \cdot \log_2\left[\frac{2A}{W}\right] \tag{33}$$

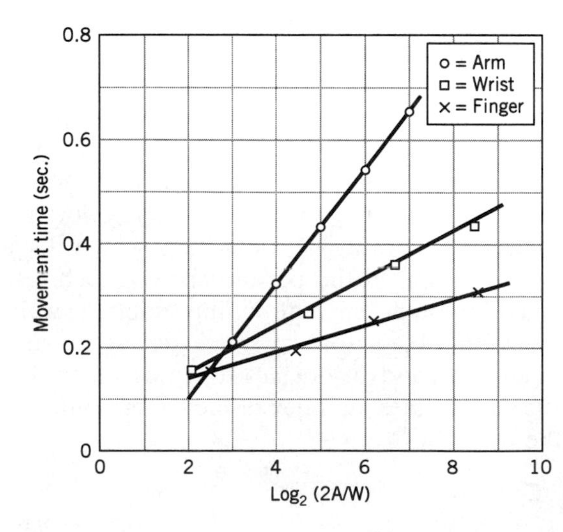

Figure 10.10 Fitt's law for arm, wrist, and finger movement.

There are various types of movements for which Fitts' law does not hold. However, it has been demonstrated that the rule does apply for single prepared movements, provided that they are reasonably large. For these single prepared movements, the times are only about 60% of the times associated with alternating movements.

Fitts' law may be somewhat simplified for practical application (and still retain acceptable accuracy) by making the approximation:

$$a \ll [b \cdot \log_2(2A/W)] \tag{34}$$

Invoking this approximation, equation (33) reduces to:

$$\mathrm{MT} = b \cdot \log_2(2A/W) \tag{35}$$

An operationally useful representation of Fitts' law for the movement time (MT) for the arm, wrist, and finger is depicted in Figure 10.10. The empirical constants a and b depend upon the type of movement the human operator performs. The index of difficulty depends upon the displacement amplitude and target width. Many tasks of practical interest to the human factors engineer involve upper extremity movement by the human operator of either the arm, wrist, or finger. A set of equations may be derived from experimental laboratory data (of Figure 10.10) resulting in three operationally useful forms of Fitts' law:

1. Human operator movement time (MT) for the arm:

$$\mathrm{MT} = .01 + (.105)\log_2\left[\frac{2A}{W}\right] \tag{36}$$

2. Human operator movement time (MT) for the wrist:

$$\mathrm{MT} = .095 + (.045)\log_2\left[\frac{2A}{W}\right] \tag{37}$$

3. Human operator movement time (MT) for the finger:

$$\mathrm{MT} = .105 + (.025)\log_2\left[\frac{2A}{W}\right] \tag{38}$$

Referring to Figure 10.11, a typical human operator output in response to a unit step is presented. Symbols on the amplitude ordinate and time abscissa of Figure 10.11 are identified with respect to the Fitts' law parameters.

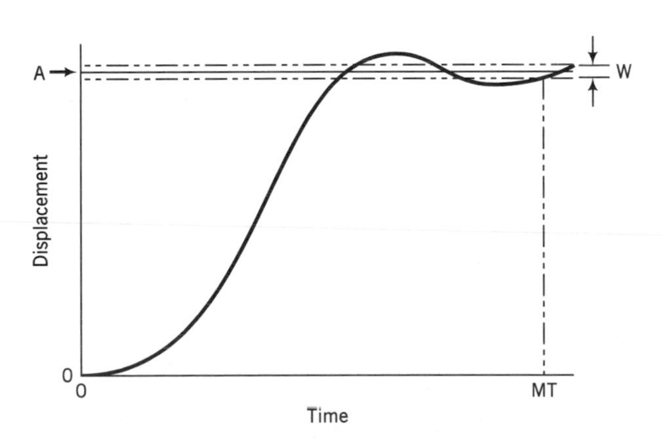

Figure 10.11 Typical human movement response for a unit-step track.

b. Human Operator Informatic Model

i. System Modeling Review

This introductory review will consider three aspects of systems modeling that will be used in the development of the human operator informatic model. These three aspects are: (a) the Laplace transform; (b) block diagram fundamentals; and (c) the transfer functions of systems.

The Laplace Transform. The Laplace transform is an operational technique by which time functions are related to frequency dependent functions of a complex variable. The Laplace transform is a mathematical transformation that is extremely useful for solving linear differential equations with constant coefficients. Such equations characterize linear and time-invariant (LTI) systems.

The Laplace transform is defined as follows. When $f(t)$ is a real function of a real time variable (t) and defined for time greater than zero, then:

$$L|f(t)| = F(s) = \int_{0^+}^{\infty} f(t)e^{-st}\,dt \qquad (39)$$

Equation (39) represents the Laplace transform of $f(t)$. Note that "s" is a complex variable referred to as the *Laplace operator* and "t" is a real variable that represents time.

The purpose of a Laplace transformation is to apply equation (39) to a problem that is stated in the real variable time domain. Application of equation (39) transforms that problem into the complex variable s domain. The solution of the transformed problem may then proceed using simple algebraic techniques. Thus a differential equation is converted into an algebraic equation in s, which is a considerable simplification. However, the solution will be in terms of the Laplace operator (s). Consequently, another mathematical operation is necessary to again transform the problem solution back to the real variable time domain and so obtain the time domain solution. This second transformation is a mathematical operation from the s domain (or frequency domain) into the t domain (or time domain). Such an operation is referred to as the inverse Laplace transform, which may be mathematically defined as follows:

$$L^{-1}|F(s)| = f(t) \qquad (40)$$

The technique of partial fraction expansion provides a rather straightforward approach to simplifying the inverse Laplace transform and may be reviewed by the student as he/she feels necessary. For most control system problems, there is a common subset of Laplace transforms that commonly occur. Consequently, the inverse Laplace transformation may be obtained from tables of transform pairs. A sample table that is applicable to the informatic model developed in this subsection is presented in Table 10.4.

Block Diagram Fundamentals. The Laplace transform allows the engineer to represent the input-output relationships of both linear devices and linear systems by means of transfer functions:

$$\frac{\text{Output}}{\text{Input}} = F(s) \qquad (41)$$

Table 10.4 Table of Laplace Transform Pairs

Number	$F(s)$	$f(t)\ t \geq 0$	Function
1	1	$\delta(t)$	Unit impulse
2	$\frac{1}{s}$	$u(t)$	Unit step
3	$\frac{1}{s^2}$	t	Unit ramp
4	$\frac{1}{s}e^{-bs}$	$u(t-b)$	Unit-step time delay
5	$\frac{1}{s+a}$	e^{-at}	Exponential
6	$\frac{e^{-bs}}{s+a}$	$e^{-a(t-b)}$	Exponential time delay
7	$\frac{1}{s(s+a)}$	$\frac{1}{a}(1-e^{-at})$	—
8	$\frac{e^{-bs}}{s(s+a)}$	$\frac{1}{a}(1-e^{-a(t-b)})$	—

If the output is expressed in terms of the Laplace transform $O(s)$ and the input is expressed in terms of the Laplace transform $I(s)$, then the transfer function $F(s)$ is given by:

$$\frac{O(s)}{I(s)} = F(s) \tag{42}$$

Rearranging equation (42) results in the Laplace transform of the output $O(s)$ in response to an input $I(s)$ as follows:

$$O(s) - F(s)I(s) \tag{43}$$

$F(s)$ represents the transfer function and is characterized by a numerator in the Laplace domain [$N(s)$] and a denominator in the Laplace domain [$D(s)$]:

$$F(s) = \frac{N(s)}{D(s)} \tag{44}$$

One way to express the numerator and denominator of a transfer function is with the factored form of a polynomial expression:

$$F(s) = \frac{(s+z_1)(s+z_2)\cdots(s+z_n)}{(s+p_1)(s+p_2)\cdots(s+p_3)} \tag{45}$$

where z represents the roots of the numerator and are referred to as the *system zeros*. Additionally, the p in equation (45) represents the characteristic roots of the denominator and are referred to as the *system poles*.

The system transfer function represented by equation (45) may be specified to within a constant when all of the system's zeros and poles have been specified. This constant is referred to as the system's gain-factor and is often denoted by either a G or a K.

A block diagram is an abbreviated and graphical representation of a physical system and depicts the functional relationships between the various elements. The utility of a block diagram is that it may be used to characterize and evaluate the

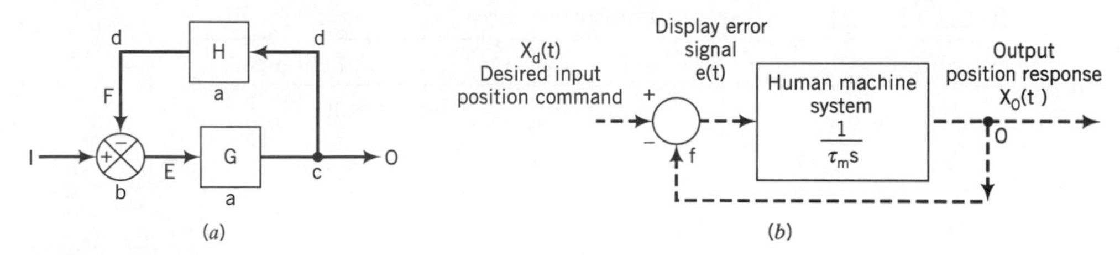

Figure 10.12.a Block diagram elements. **b** Human-machine control system.

contributions of individual elements with respect to the overall system performance. In general, the block itself is characterized by a transfer function of the form shown by equation (42). The block diagram for a specific system consists of two or more blocks and has four characteristic features. These characteristic features are blocks, summing points, takeoff points, and unidirectional arrows. The diagrammatic representation of these features is identified respectively as a, b, c, and d in Figure 10.12.a.

Block diagram algebra refers to a set of derivable transformations expressed as algebraic equations. Block diagram algebra is used in order to simplify the block diagrams of complicated control systems. The algebraic equation permits a block diagram to be transformed into an equivalent diagram that represents a specific mathematical operation. Table 10.5 is a useful table of transformation theorems that are used in the remainder of this section. Note that a specific mathematical operation may involve combining blocks in cascade. Eliminating a feedback loop or removing a block from a feedback loop are two other examples.

Transfer Functions of Systems. As noted above, block diagram algebra (see Table 10.5) is used for the reduction of complicated block diagrams. The final form of the reduced block diagram will result in a characteristic transfer function of that system. Three specific types of transfer functions of systems are of particular interest to the human factors engineer. These are briefly described as follows.

The first type of system transfer function is the canonical form of a feedback control system. Such a system is depicted in Figure 10.12.a. The canonical form consists of two blocks of which one is in the forward signal path (G) and the other is in the feedback signal path (H). There is a single input into a summing node and a single output. It is often useful to reduce this canonical form into a single block with a single input and a single output. Essentially, this process involves a transformation that eliminates the feedback loop and the summing point.

The second type of system transfer function is the unity feedback system. A unity feedback system is defined as that system in which the primary feedback (f) is exactly equal to the controlled output (o). A unity feedback system is depicted in Figure 10.12.b. Also note that setting $H = 1$ in Figure 10.12.a would result in a linear, unity feedback system.

The third type of system transfer function is a system with multiple inputs. In actual engineering practice, it is sometimes necessary to evaluate system performance when two or more inputs are simultaneously applied to different parts of the system. A system transfer function that characterizes the system output with respect to all the various inputs acting together is then desired. As long as the system is linear, such a system transfer function may be derived using the superposition theorem. This procedure is as follows:

Table 10.5 Useful Transformation Theorems

Number	Block Diagram	Equivalent Diagram	Algebraic Equation	Operation
1	I → F_1 → F_2 → O	I → F_1F_2 → O	$O = (F_1F_2)I$	Combining blocks in cascade
2	I → ∓ ⊗ → F_1 → O; F_2 feedback	I → $\frac{F_1F_2}{1 \pm F_1F_2}$ → O	$O = F_1(I \mp F_2O)$	Eliminating a feedback loop
3	Same as 2	I → $\frac{1}{F_2}$ → ∓ ⊗ → F_1F_2 → O	Same as 2	Removing a block from feedback loop
4	I → F → ∓ ⊗ → O; J	I → ∓ ⊗ → F → O; J → $\frac{1}{F}$	$O = FI \mp J$	Moving a summing point ahead of a block
5	I → F → O; I	I → F → O; $\frac{1}{F}$ → I	$O = FI$	Moving a take-off point beyond a block

1. Select one system input $[I_1(s)]$ for evaluation, and set all other system inputs to zero.
2. Transform the resultant block diagram into canonical form by using the appropriate transformations (see Table 10.5).
3. Derive the system output response $[O_1(s)]$ in terms of selected input acting alone, per equation (43).
4. Repeat steps (1), (2), and (3) for each of the remaining system inputs.
5. Add together as an algebraic summation all of the output responses that were derived in (1)–(4):

$$O(s) = F_1(s)I_1(s) + F_2(s)I_2(s) \cdots + F_n(s)I_n(s) \tag{46}$$

This algebraic sum will represent the total system output response when all of the various inputs are acting individually and simultaneously at different parts of the system.

These three specific types of transfer functions of systems are now illustrated with the following example. The interested student is referred to other standard texts if a more detailed review is desired.

EXAMPLE 10.3

Determine the transfer function of each of the following systems (by means of the indicated operations):

a. Reduce the canonical form of a negative feedback system (as shown in Figure 10.12.a) to an equivalent single block by using algebraic equations.
b. Reduce the canonical form of a negative feedback system (as shown in Figure 10.12.a) to an equivalent unity feedback system using Table 10.5.
c. Determine the output $P(s)$ for the following multiple input single output (MISO) system using the five-step method:

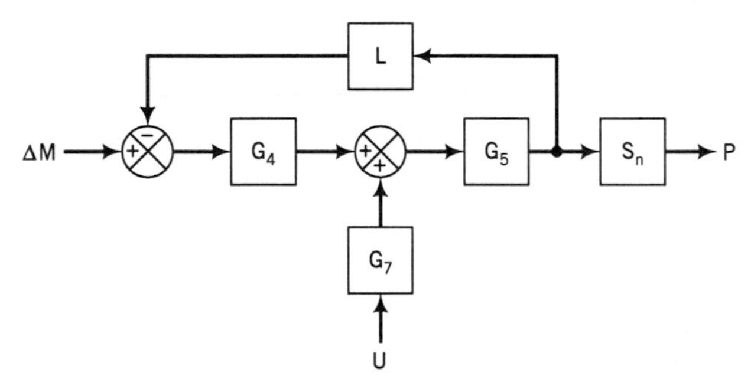

SOLUTION 10.3(a)

The algebraic equations describing the canonical form of the negative feedback system are directly determined by inspection of Figure 10.12.a:

$$O = GE \tag{i}$$

$$F = HO \tag{ii}$$

$$E = I - F \tag{iii}$$

Now proceed to derive the relationship $O = f(I)$:
Substituting equation (iii) into equation (i):

$$O = GI - GF \tag{iv}$$

Substituting equation (ii) into equation (iv):

$$O = GI - GHO \tag{v}$$

Separate variables:

$$O(1 + GH) = GI \tag{vi}$$

And rearranging:

$$O = \left[\frac{G}{1 + GH}\right] I \tag{vii}$$

The requested transfer function is then:

$$\frac{O}{I} = \frac{G}{1 + GH} \tag{viii}$$

SOLUTION 10.3(b)

For *any* feedback system composed of only linear blocks in the feedback signal path, the unity feedback system is obtained by using Transformation 3 of Table 10.5. For Figure 10.12.a, this results in:

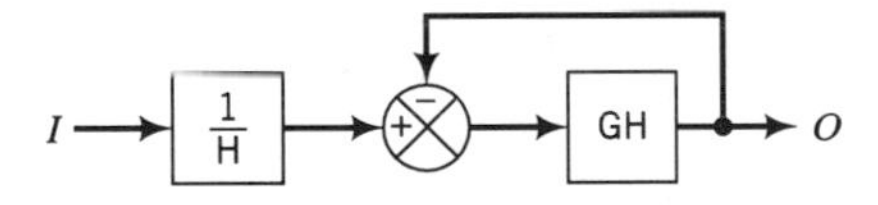

SOLUTION 10.3(c)

Start with the MISO system of Example 10.3(c) and determine the output $P(s)$ as follows:
For $U = 0$, solve for P_M:

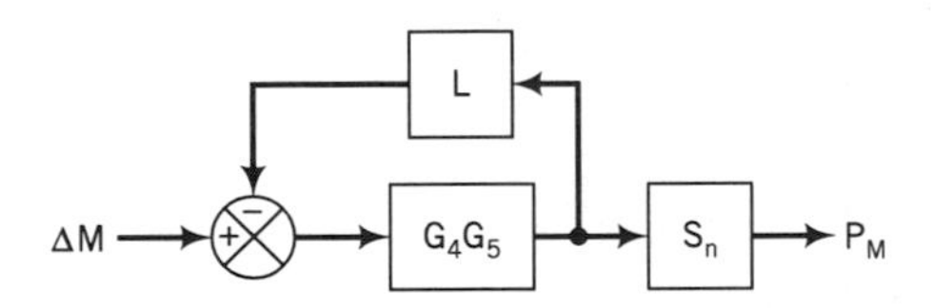

Invoking Transformation 2 of Table 10.5:

ΔM → $\frac{G_4G_5}{1 + G_4G_5L}$ → S_n → P_M

Now solve for P_M by invoking Transformation 1 of Table 10.5:

$$P_M = \left[\frac{G_4 G_5 S_n}{1 + G_4 G_5 L}\right] \Delta M \tag{i}$$

For $\Delta M = 0$, solve for P_u:

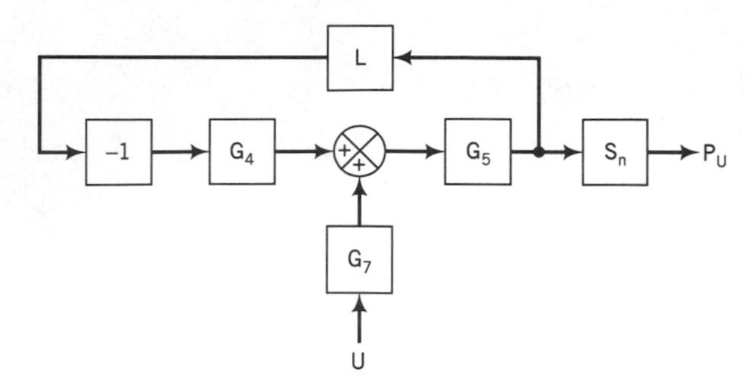

Rearrange as a single input single output (SISO) system:

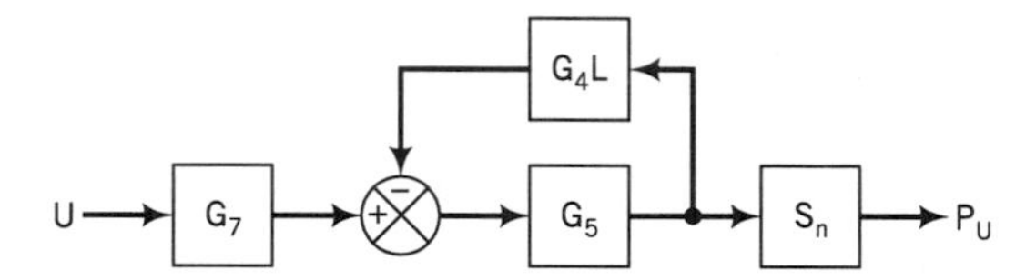

Invoking Transformation 2 of Table 10.5:

Now solve for P_u by invoking Transformation 1 of Table 10.5:

$$P_u = \left[\frac{G_7 G_5 S_n}{1 + G_4 G_5 L}\right] U \tag{ii}$$

Using superposition theorem for a system of linear blocks:

$$P = P_M + P_u \tag{iii}$$

The requested transfer function is then obtained by substituting equations (i) and (ii) into equation (iii):

$$P = \left[\frac{G_4 G_5 S_n}{1 + G_4 G_5 L}\right] \Delta M + \left[\frac{G_7 G_5 S_n}{1 + G_4 G_5 L}\right] U \tag{iv}$$

Finally, equation (iv) may be simplified by factoring:

$$P = \left[\frac{G_5 S_n}{1 + G_4 G_5 L}\right] (G_4 \Delta M + G_7 U) \tag{v}$$

---

ii. Preliminary Model Development

Fitts' law characterizes a specific type of movement in which the human operator attempts to move as rapidly as possible (between points A and point B) *and also* be as accurate as possible (i.e., acquire the target in a minimum amount of time). Such a movement is represented as a single-step tracking task. It is immediately apparent that the speed of movement and the accuracy of movement tend to oppose each other: a movement can be very fast (and lose accuracy) or very accurate (but at a lower speed of movement). Consequently, the movement time predicted by Fitts' law requires a "speed-accuracy" tradeoff. This means that the human operator must formulate some "speed-accuracy" strategy in order to perform the single-step tracking task in the shortest amount of time that permits acceptable accuracy (i.e., target acquisition).

Various response strategies are available for single-step tracking. When relatively large, quick control movements are required, the operator may perform a series of separate, but progressively smaller responses. For example, each response might reduce the remaining distance to the target by 90%. This "successive approximation" controls strategy, however, is not very efficient for smaller, fine control movements require to align a cursor with a target. As an alternative method, the response-control strategy may be to monitor the movement continuously. Such a "closed-loop response" strategy requires more processing of additional information so that the movement speed is slower.

When a person is required to be both quick and accurate, that person may respond in a manner that is neither totally preprogrammed (successive approximation) nor completely visually monitored (closed-loop). That individual may start with a quick preprogrammed movement but then slow that movement down as the target is approached. In this situation, a correction may begin before an initial movement is completed, so that the end of the first movement overlaps the beginning of a corrected movement. Finally, the last part of the movement may be visually monitored, and its rate depends upon the degree of accuracy which is necessary.

In many cases, movements in single-step tracking must be both large and accurate will follow this two-phase process. The first phase of the movement is a rapid, open-loop ballistic movement (as though a successive approximation strategy is predominant). However, the second final phase of the operator response is characterized by a slower visually monitored positioning movement (characteristic of the closed-loop strategy).

The purpose of a human operator informatic model is to represent the human operator subsystem in order to then characterize a human-technological system. An appropriate definition of a human operator informatic model will allow the human factors engineer to then perform a human-technological control system analysis. The three primary objectives of human-technological control system analysis are the determination of: (1) the degree of system stability; (2) the system steady-state performance; and (3) the transient response of the system. The method of analysis used by the human factors engineer requires the selection of a model for representing the system (block diagram) and the determination of the transfer function for each subsystem (human operator, technological device, and environment). The system model must then be formulated by appropriately connecting the component subsystems so that the human factors engineer may then proceed to determine the system characteristics. This determination of human-technological control system characteristics begins with the definition of a human operator informatic model.

The above design and analysis objectives may be obtained by invoking a set of suitable simplifying approximations for the desired human operator informatic model. There are three essential approximations:

- **Approximation 1.** The human operator informatic model is physically realizable. In effect, conceptual models or theoretical models that cannot be reduced to real-world physical characterization will not result in mathematically tractable solutions. Negative damping coefficients and denominators that go to zero are but two examples.
- **Approximation 2.** The human operator informatic model is linear and time-invariant (LTI). It is recognized that the human operator informatic model has both nonlinear and time-varying characteristics. For the purposes of applying linear control theory, however, such system characteristics are to be approximated by LTI functions. The human operator is assumed to be trained and motivated.
- **Approximation 3.** The human operator informatic model is a SISO system. This approximation will simplify the writing of transfer functions and result in a parsimonious (most reductionistic) form of the model. It is recognized that a complete description of the human operator informatic model may require a MIMO system. However, an operationally useful model may be approximated by selecting the single input and single output that best characterizes the transient response of the system. This may be at the expense of steady-state characterization.

Preliminary development of the human operator informatic model proceeds by developing a relationship between a first-order time constant (τ_M) and the Fitts' law b coefficient, a reciprocal baud rate. This is accomplished as follows [equations (47) through (64)].

Figure 10.12.b represents a model of the human-machine system in which the output response, $X_0(t)$ is achieved by integrating a display error signal, $e(t)$ in which there is an integration time constant, τ_M. The closed-loop transfer function is:

$$\frac{X_0(t)}{X_d(t)} = \frac{G}{1 + GH} \tag{47}$$

In the case where:

$$G = \frac{1}{\tau_M s} \tag{48}$$

$$H = 1.0 \tag{49}$$

where τ_M = movement time constant (s)
Then substituting equations (48) and (49) into equation (47):

$$\frac{X_0(t)}{X_d(t)} = \left[\frac{\frac{1}{\tau_M}}{s + \frac{1}{\tau_M}}\right] \tag{50}$$

As an alternative expression, where:

$$G = \frac{\alpha_M}{s} \tag{51}$$

and the *movement informatic frequency* (α_M) in units of Hertz is:

$$\alpha_M = \frac{1}{\tau_M} \tag{52}$$

Then substituting equations (49) and (51) into equation (47) results in:

$$\frac{X_0(t)}{X_d(t)} = \frac{\alpha_M}{s + \alpha_M} \tag{53}$$

For a single-step track of amplitude, A at $t = 0$:

$$X_d(t < 0) = 0$$
$$X_d(t \geq 0) = A \tag{54}$$

Rearrange equation (50) to solve for $X_0(t)$:

$$X_0(t) = X_d(t)\left[\frac{\frac{1}{\tau_M}}{s + \frac{1}{\tau_M}}\right] \tag{55}$$

Substituting A (in Laplace notation for a unit step, $\frac{A}{s}$ into equation (55) and simplifying:

$$X_0(t) = \frac{A}{\tau_M}\left[\frac{1}{s\left(s + \frac{1}{\tau_M}\right)}\right] \tag{56}$$

Taking the inverse and simplifying:

$$X - A\left(1 - e^{-\frac{t}{\tau_M}}\right) \tag{57}$$

Figure 10.13 represents the time response of this first-order linear system with respect to the Fitts' law parameters. In proceeding with the mathematical development, we identify the MT of Figure 10.11 as equal to the MT of Figure 10.13. The target is acquired when the response is $A \pm W/2$ *and remains within* $A \pm W/2$ (i.e., steady-state error).

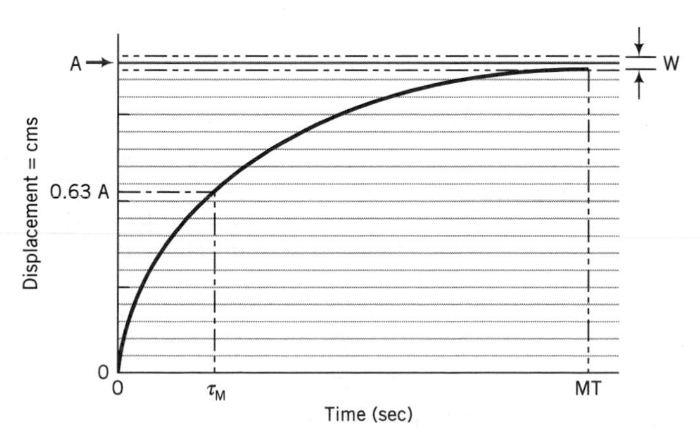

Figure 10.13 Definition of movement time constant (τ_M).

Solving for MT of Figure 10.13:

$$x = A - \frac{W}{2} \tag{58}$$

and substituting equation (58) into equation (57):

$$A - \frac{W}{2} = A\left(1 - e^{-\frac{MT}{\tau_M}}\right) \tag{59}$$

Solving for MT results in:

$$MT = \tau_M \cdot \ln\left(\frac{2A}{W}\right) \tag{60}$$

Substituting equation (35) into equation (60):

$$b \cdot \log_2\left(\frac{2A}{W}\right) = \tau_M \cdot \ln\left(\frac{2A}{W}\right) \tag{61}$$

Rearrange for b:

$$b = \tau_M \left[\frac{\ln\left(\frac{2A}{W}\right)}{\log_2\left(\frac{2A}{W}\right)}\right] \tag{62}$$

Recalling the relationship:

$$\frac{\ln\left[\frac{2A}{W}\right]}{\log_2\left[\frac{2A}{W}\right]} = \ln(2) \tag{63}$$

Then substituting into equation (62):

$$b = \tau_M \ln(2) \tag{64}$$

Figure 10.14 represents the solution of equation (57) [per Figure 10.13] when superimposed upon a typical human operator response as represented in Figure 10.11. The movement time constant, τ_M is indicated with respect to the first-order linear response.

Figure 10.14 allows us to compare the response of a first-order linear system to that of a typical human operator when tracking a single step. The first-order linear response is characteristic of a visco-elastic system that can be uniquely characterized by a system time constant. It is readily apparent that the typical human operator output response is more complex than what is portrayed. First, it appears that there are pronounced inertial effects so that (as human operator initiates the response) the person must overcome the effects of arm mass. Second, as the human operator performs a task requiring speed-accuracy trade-off, there appears to be an initial ballistic phase that merges with a slower visually monitored phase.

Note that in order to satisfy Fitts' law, both systems must acquire the target in exactly the same movement time, MT. The first-order linear response would be characteristic of a hypothetical human operator in which there was zero arm mass

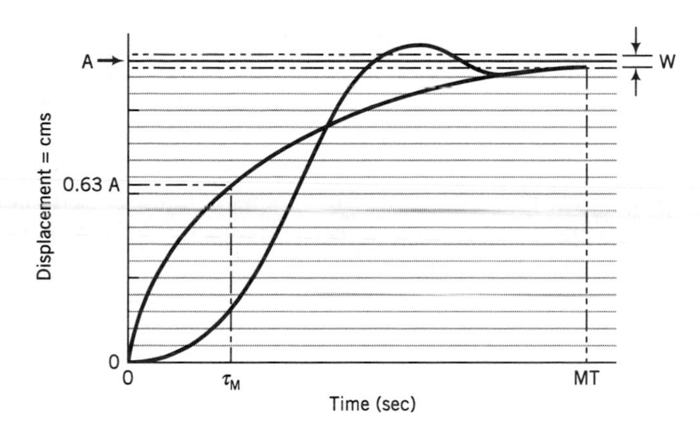

Figure 10.14 First-order step response compared to typical human response.

and the target were acquired by continuous visual monitoring (the "closed-loop" response).

These observations make Fitts' law very useful to the human factors engineer. For certain speed-accuracy trade-off tasks, the model presented in Figure 10.12.b may be applied. The task to be analyzed must be within those boundary conditions for which Fitts' law is applicable. The channel capacity of the human operator for that particular task will then be represented by the movement time constant, τ_M (which has been shown to be equivalent to the information handling capacity parameter, b). It must be emphasized, however, that the original Fitts' law applies a straight-line relationship to the mean responses of the subject population. Therefore, Fitts' law, and consequently the human machine interface model of Figure 10.12.b does not account for subject-to-subject variance.

Also recall that there are various situations in which Fitts' law is not obeyed. These occur as follows: (1) ballistic movements (less than 0.2 s); (2) very narrow targets (less than 0.25 in.); (3) very large targets (greater than 2.0 in.); (4) movement amplitudes less than 2 in.; and (5) movement amplitudes greater than 16 in.

At this particular juncture, only a preliminary form of the human operator informatic model has been developed. However, Figure 10.14 clearly indicates that the preliminary informatic model distinctly oversimplifies the position-time course of the human operator when performing a speed-accuracy trade-off task. The preliminary informatic model represents the response of a first-order linear system to a unit step, while the actual human response is more characteristic of a second or higher order, underdamped linear system. This obviously indicates that second order effects (e.g., mass, inertia) are not characterized by the preliminary informatic model. What we are seeing, in effect, is the position-time trajectory of the human operator response if the human operator were not "encumbered" by mass and inertial effects. Viewed in this perspective, the position-time trajectory of the preliminary informatic model would represent an "ideal" situation. Without mass (inertia), no ballistic movement is required because there is no mass to accelerate (inertial effects to overcome). The entire position-time course of the human operator "speed-accuracy" movement represents the most "economical" trajectory for a unit step "demand" signal.

Economy of movement is of some theoretical interest at the preliminary informatic model phase. Consider a line integral of the position-time path produced by a second order, underdamped linear system. As the mass of this system is progressively reduced, second-order effects are progressively attenuated and the line integral approaches that of a purely visco-elastic (first order) system. The significance of a minimum trajectory path is that it represents a path in which the minimum neuromuscular energy is expended. Deviations of the human trajectory path around the minimum path represent a form of "wasted" neuromuscular energy. In effect, the preliminary informatic model represents an ideal trajectory, which the real human operator attempts to approach through the use of strategy. For example, the student is referred to the "consistent undershoot" strategy of Figure 9.23.b.

iii. Intermediate Model Development

The goal of the intermediate model development phase is to define the human operator informatic model as a transfer function for an LTI (linear and time invariant) system. As a result, the methods of linear control system theory (e.g., frequency-domain analysis and time-domain analysis) can be applied directly (see Section 10.3).

This goal is accomplished by satisfying two objectives. First, we translate the human operator neuromuscular control systems (of Sections 9.3 and 9.4) into a zero-pole-gain transfer function for a SISO system. Second, we relate the resultant zero, pole, and/or gain terms to their equivalent parameter in the human operator informatic model (as developed in Section 10.2.b.2).

Beginning with the human operator *section* of the neuromuscular (isometric) control system (Figure 9.22), let G_p represent the effective gain of the feed-forward elements in the *CNP-B and NMA subsystems:*

$$G_p = G_1 K_p G_2 \tag{65}$$

The block diagram representation of the human operator *section* (of Figure 9.22) is then shown in Figure 10.15.a, and an alternate representation in Figure 10.15.b.

The system may now be reconfigured to identify y_{IN} of the feed-forward path $[y_{IN}(ff)]$ and y_{IN} of the feedback path $[y_{IN}(fb)]$ as indicated in Figure 10.16.a.

Per simplifying approximation 3, we reduce the this to a SISO system by noting that $y_{IN}(fb)$ is the demand signal for the *transient* response, and $y_{IN}(ff)$ is the demand signal for the steady-state response. We invoke Approximation 3 and reduce the

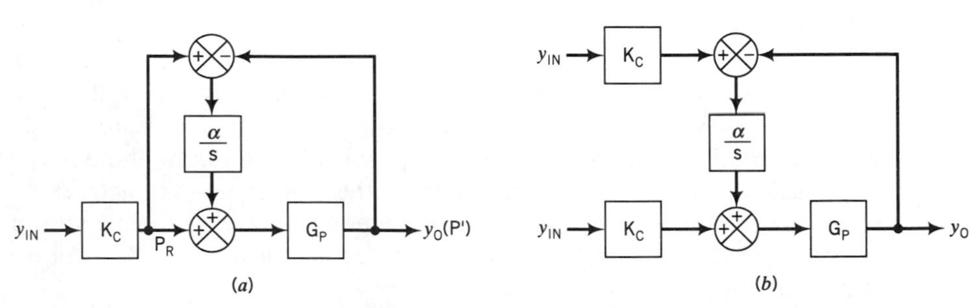

Figure 10.15.a Block diagram representation of the human operator section (of Figure 9.22). **b** Alternative representation of Figure 10.15.a.

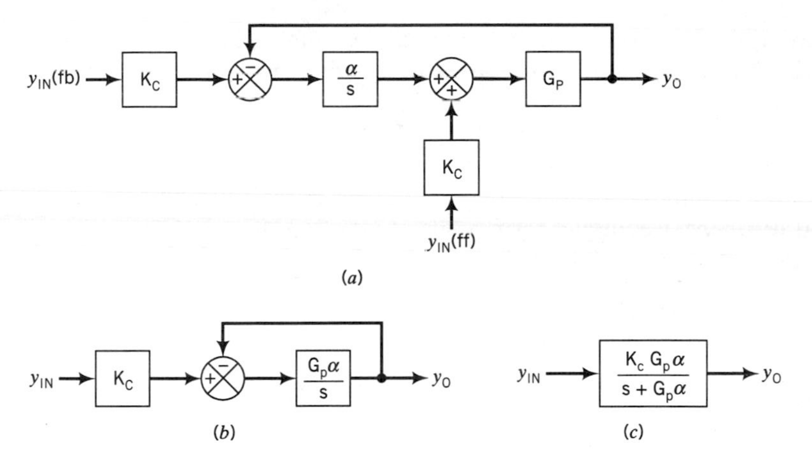

Figure 10.16 Reduction of Figure 10.15.b to a SISO control system.

block diagram to a SISO system that describes the transient response as illustrated by Figure 10.16.b.

Again using block diagram algebra, the overall system transfer function reduces to its most compact form as per Figure 10.16.c.

So that when $K_c = 1$ [of Figure 10.16.c]:

$$\frac{y_0}{y_{IN}} = \frac{G_p\alpha}{s + G_p\alpha} \tag{66}$$

Comparison of equation (52) with equation (64) indicates that:

$$\alpha_M = G_p\alpha \tag{67}$$

for a neuromuscular (isometric) control system.

Continuing with the human operator *section* of the neuromuscular (velocity) control system (Figure 9.26), let G_v represent the effective gain of the feed-forward elements in the *CNP-B* and *NMA* (velocity loop) subsystems:

$$G_v = G_5 K_v G_7 \tag{68}$$

The block diagram representation of the human operator (of Figure 9.26) *section* is then represented in Figure 10.17.a.

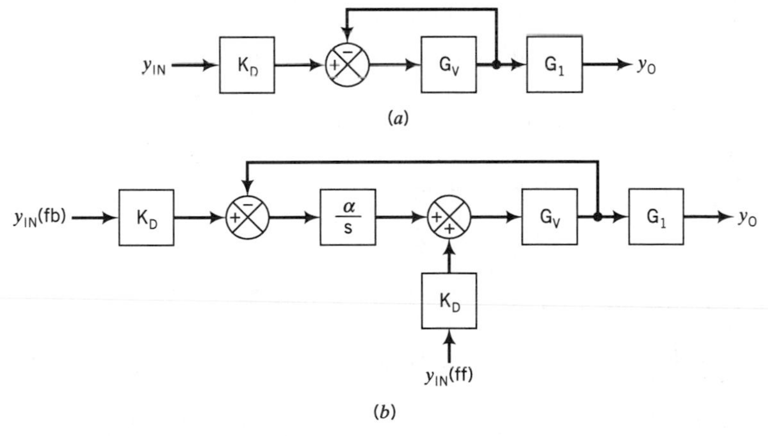

Figure 10.17.a Block diagram of the human operator section (of Figure 9.26). **b** Alternative representation of Figure 10.17.a.

Proceeding in an analogous manner (to the isometric control system), we reconfigure the block diagram with respect to $y_{IN}(fb)$ and $y_{IN}(ff)$ as indicated in Figure 10.17.b.

Reducing the block diagram to a SISO system that describes the transient response is illustrated in Figure 10.18.a.

Again applying block diagram algebra, the overall system transfer function is reduced to its most compact form as per Figure 10.18.b.

So that when $K_D = 1/G_1$ [of Figure 10.18.b]:

$$\frac{y_o}{y_{IN}} = \frac{G_v\alpha}{s + G_v\alpha} \tag{69}$$

Comparison of equation (53) with equation (69) indicates that:

$$\alpha_M = G_v\alpha \tag{70}$$

for the neuromuscular (velocity) control system.

The *general* informatic model may be stated for the case where:

$$G = G_p = G_v \tag{71}$$

So that:

$$\alpha_M = G\alpha \tag{72}$$

and the intermediate informatic model for the human operator (independent of the specific neuromuscular control system) is:

$$\frac{y_0}{y_{IN}} = \frac{\alpha_M}{s + \alpha_M} \tag{73}$$

iv. Final Model Development

The preliminary informatic model has accounted for information processing that represents the movement time (MT) of the human operator. Recall that the human operator response (action time) is the sum of both reaction time and movement time [see equation (26)]. The final development of the human operator informatic model requires us to also account for the additional information processing that occurs during choice reaction time [see equations (27) and (28)].

The final development of the human operator informatic model begins by approximating the information processing associated with choice reaction time as a pure time delay (T_D). Such a representation is reasonably appropriate in an operational setting, and mathematically well defined (and understood) in a control systems setting.

Equation (27) may then be restated as:

$$T_D = a + bH_s \tag{74}$$

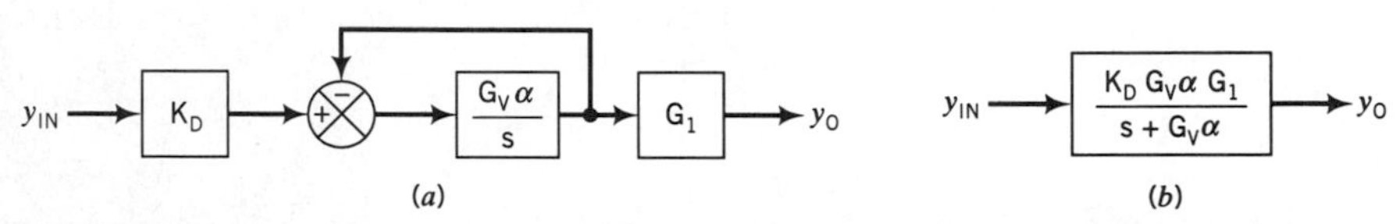

Figure 10.18.a Reduction of block diagram of Figure 10.17.b to an SISO control system.

The human operator has so far been represented as movement time information processing channel. Equation (74) indicates that T_D is distributed along this channel as a noninformation processing (simple reaction) time delay and an information-processing (choice reaction) time delay.

At the anatomical element level, recall that the human operator neuromuscular control system consists of three subsystems in series: CNP-A, CNP-B, and NMA (see Section 9.1). At the human operator subsystem level, T_D is distributed as follows:

$$T_D = T_C + T_N \tag{75}$$

T_C is the higher central nervous time delay. Some representative sources for this fraction of the distributed time delay are: sensory apparatus nerve conduction delays, central (CNP-A subsystem) information processing delays, and other higher CNS (e.g., cerebrum and mid-brain) computational delays.

T_N is the lower central nervous and neuromuscular time delay. T_N is distributed through the CNP-B subsystem and the NMA subsystem. Some representative sources of the CNP-B fraction are: lower CNS (e.g., brainstem and cerebellum) computational and synaptic delays and spinal cord interneuron synaptic and conduction delays. Some representative sources of the NMA fraction are peripheral motor nerve conduction delays and the skeletal muscle electromechanical delay (EMD).

Substituting equation (75) into equation (74) results in:

$$T_N + T_C = a + bH_s \tag{76}$$

As a *very* general approximation, it is then conceptually useful to restate equation (76) as two subsidiary equations:

$$T_N \approx a \tag{77}$$

$$T_C \approx \frac{H_s}{\beta} \tag{78}$$

Recall that a is simple reaction time, which we are now saying is analogous to the human operator reacting at a "midbrain-reflex" level.

In equation (78), β is written directly as baud rate (reciprocal of b) so that T_C is inversely proportional to the baud rate. Specifically, for the same amount of spatial information (H_s) we may state the following. The greater the baud rate, the smaller the effective time delay T_C. Conversely, the lesser the baud rate, the greater the effective time delay T_C.

Alternatively, for the same baud rate: the greater the information content (H_s), the greater the effective (T_C) time delay, and the converse.

This section now concludes by deriving the final human operator informatic model as a transfer function of an LTI systems (SISO).

We first define the reaction time (T_D) information processing block separately from the movement time (first order lag) information processing block with both blocks in cascade. Based on equation (75), distribute the two time delays (T_C and T_N) to represent their spatial (anatomical) arrangement. The final informatic model at this point is represented by Figure 10.19.a.

Combine the two blocks to account for the pure time delay. This represents the reaction time (action time) block for the human operator. The final informatic model at this stage is shown in Figure 10.19.b.

Combine the blocks to define one block, as per Figure 10.19.c.

This results in the transfer function for the human operator informatic model:

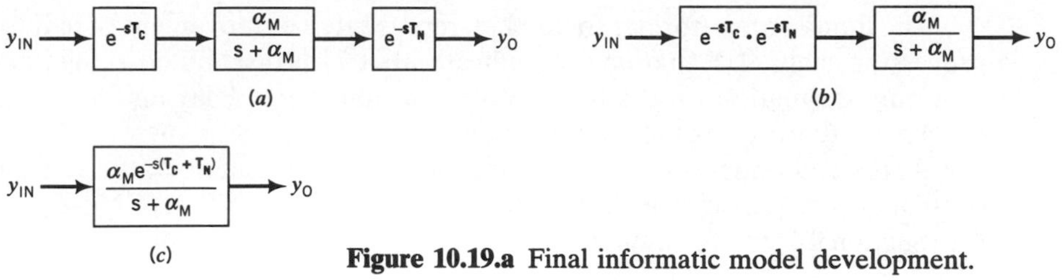

(a) (b) (c)

Figure 10.19.a Final informatic model development.

$$\frac{y_O}{y_{IN}} = \frac{\alpha_M e^{-sT_D}}{s + \alpha_M} \tag{79}$$

10.3 HUMAN CONTROL SYSTEMS ANALYSIS

Consider the block diagram model of a human operator interacting with a technological system as shown in Figure 10.20.

G_c is a central nervous processor gain, and K_p is the technological plant transfer coefficient. Equation (71) is used to represent the human operator informatic model. Recall that K_s is the transfer function for the human operator strategy, K_I is the human-sensory interface transfer function, and G_e is the environmental gain. Figure 10.20 represents the conventional block diagram model that is used for human control systems analysis.

A feedback control system may be reduced to its canonical form as shown in Figure 10.21.a. G is the forward transfer function. For a series of blocks in cascade as per Figure 10.20, the forward transfer function (G) is:

$$G = \frac{G_c K_s K_I K_p \alpha_M e^{-sT_D}}{s + \alpha_M} \tag{80}$$

Referring again to Figure 10.21.a, H is the feedback transfer function. In Figure 10.20 this pathway is represented by two blocks in cascade, so that

$$H = \frac{G_e}{G_c} \tag{81}$$

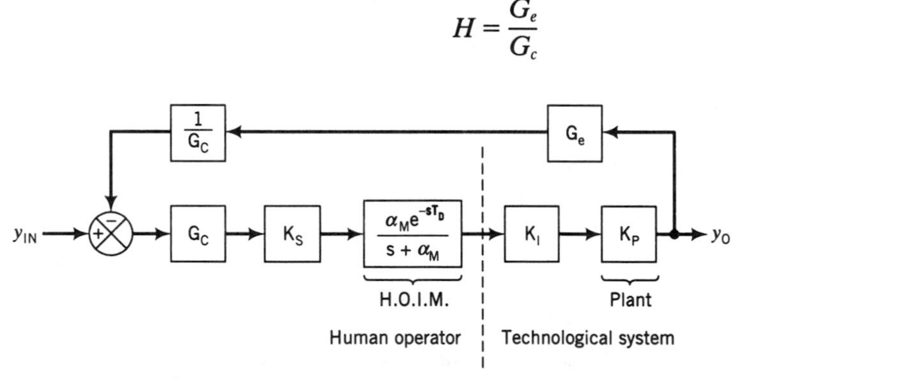

Figure 10.20 General model of human-technological system for control systems analysis.

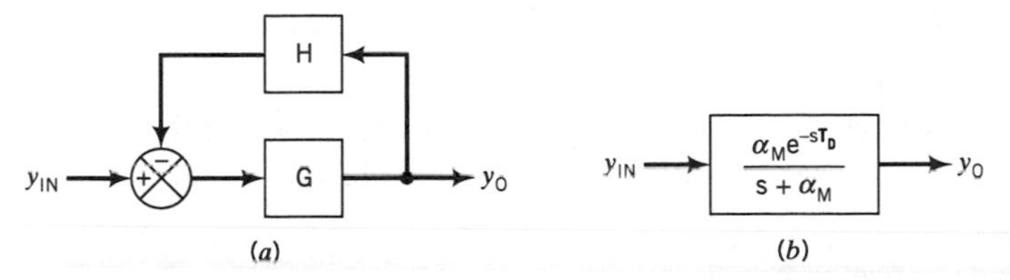

Figure 10.21.a Canonical form of a feedback control system. **b** Transfer function for the human operator informatic model.

The open-loop transfer function is defined as GH (see Figure 10.21.a). For the system shown in Figure 10.20, this is the product of equation (80) with equation (81) and can be written:

$$GH = \frac{K_s K_I K_p G_e \alpha_M e^{-sT_D}}{s + \alpha_M} \tag{82}$$

The closed-loop transfer function (T) for the negative feedback system of Figure 10.21.a is:

$$\frac{y_O}{y_{IN}} = \frac{G}{1 + GH} \tag{83}$$

Substituting equations (80) and (82) into equation (83) and simplifying results in the closed-loop transfer function (T) for Figure 10.20:

$$\frac{y_O}{y_{IN}} = \frac{G_c K_s K_I K_p \alpha_M e^{-sT_D}}{s + \alpha_M + K_s K_I K_p \alpha_M e^{-sT_D}} \tag{84}$$

The primary objective of the human control systems engineer is to satisfy a set of human-system performance specifications. When a mathematical formulation of the human-technological system has been realized, these performance specifications then represent constraints placed upon those mathematical functions which describe the human-system characteristics. This section on human control system analysis will now consider two control system categories: feed-forward control system and feedback control system.

a. Feed-Forward Control System

A feed-forward (or "open-loop") system is one in which the command or control action is independent of the output. The command input or control action is initiated and the subsequent processing by the system elements results in a system output that may or may not be that which is desired depending upon any system disturbances.

"Open-loop" tasks performed by a human operator have the following characteristics.

1. The task performance is *automatic*. The human utilizes experience, training, and judgment to generate a command input or control action that is then executed in a practiced manner resulting in output that is expected based upon that experience, training, and judgment.
2. Since there is no feedback (e.g., visual or auditory), the outcome may or may not be acceptable depending on system disturbances that cannot be directly controlled by the human operator.
3. The degree of accuracy in performing an "open-loop" task is therefore

dependent upon: (a) the human operator's skill level, and (b) the extent to which system disturbances are present.

4. The human operator output does not demonstrate oscillations or other "stability" problems that are inherent in "closed-loop" systems.

Many applications of practical interest to the human factors engineer involve the human operator performing a task that may be described as a stimulus-response paradigm. In these tasks, the stimulus has information content and the human operator must make one or more responses as rapidly as possible and with sufficient accuracy to activate a technological system.

A common example would be operating a motor vehicle, and as the person drives along the road, the driver approaches an intersection (with people at the corners and the traffic light green). Another car suddenly runs the intersection (against a red light). The operator must make a response to this stimulus, and does so by moving the leg and foot (originally resting on the car's floor) to the brake pedal and depressing it. The operator may also blow the car's horn. Many features of this driver-vehicle-environment task might be the focus of an engineering analysis: (a) the unobstructed visibility through the windshield for various driver body sizes; (b) the size, position, and activation force of the brake pedal for drivers of varying biomechanical abilities and physical dimensions; (c) the steering wheel configuration and angle necessary for good grip strength and stabilization of the operator in a rapidly decelerating situation; and (d) the location, size, and activation pressure of the horn switch (on the steering wheel) and the required horn sounding intensity.

Many types of stimulus-response tasks may be reduced to a feed-forward control system in which the human operator is simply modeled as per equation (79) and Figure 10.21.b. This is accomplished in the following manner.

Referring again to Figure 10.20, for the stimulus-response task of interest, the inter-force (K_I) and the plant dynamics (K_p) of the technological system are noncontributory so that the $K_I K_p$ product of Figure 10.20 is unity.

The human operator action (i.e., the combination of reaction and movement (as per Section 10.2.a) is considered as an open-loop action, so that feedback dynamics (H) of Figure 10.21.a are also noncontributory. Finally, the stimulus (required unit-step change) displayed to the human operator (y_{IN}) is of exactly the same magnitude as the desired response. As a result, the central nervous system gain (G_c) of Figure 10.20 is essentially unity, and the human operator is approximated as using a "unity" strategy. Hence, K_s of Figure 10.20 is also unity. Invoking the above approximations reduces the human operator control system to that shown in Figure 10.21.b.

The human operator (informatic model) feed-forward transfer function is then:

$$\frac{y_0}{y_{IN}} = \frac{\alpha_M e^{-sT_D}}{s + \alpha_M} \tag{79}$$

For a unit-step input of magnitude A, the output Laplace transform of this control system is:

$$y_0(s) = \frac{A\alpha_M e^{-sT_D}}{s(s + \alpha_M)} \tag{85}$$

The inverse Laplace transform of equation (85) is a function, $f(t - T_D)$ where $T_D > 0$ and:

For $t \leq T_D$:

$$y_0 = 0 \tag{86}$$

For $t > T_D$:

$$y_0 = A[1 - e^{-\alpha_M(t-T_D)}] \tag{87}$$

The following example illustrates how the human factors engineer might use the human operator informatic model in a feed-forward control systems analysis for a stimulus-response type of task.

EXAMPLE 10.4

A ship's sonar operator is sitting at a console viewing a sonar screen with an information content of $H_s = 5$ bits. The human factors engineer is evaluating two types of operators (operator A and operator B) that are being used in order to evaluate two different sonar station-console switch configurations (State 1 and State 2). Operator A is a well-trained, experienced, and fast-reacting individual ($\beta = 9.2$ bits/s) and Operator B is a recently trained, less experienced, and slower reacting individual ($\beta = 5.0$ bits/s).

When a suspect object suddenly appears on the sonar screen, the operator will immediately alert the computer system (and the watch supervisor) by manually depressing an alert button located on the console panel at a vertical height above the operator's hand. The hand normally rests on the console desk top.

In the first proposed switch configuration (State 1), the center of a square button (of height H_1) is to be located at a vertical height (A_1) of 0.5 M above the operator's hand and will require a coordinated hand-arm-shoulder movement ($\alpha_M = 4.5$/s) to reach and depress the pushbutton switch. In the second proposed switch configuration (State 2), the center of the square button (of height H_2) is to be located at a vertical height (A_2) of 0.1 M and will require only a hand-wrist movement ($\alpha_M = 10$/s) to reach and depress the pushbutton switch.

(a) If the average diameter of an operator's index finger height (h) is .01 M, find H_1 and H_2 so that both switch configurations have an index of difficulty (ID) of 5 bits;

(b) Develop the relevant forward transfer function (Human Operator Informatic Model) for each human operator-switch configurations combination; and

(c) Solve for the action time that the human operator informatic model would predict for each of the four operator-state combinations.

SOLUTION 10.4(a)

Regarding design of the push-button switch, consider the following diagram:

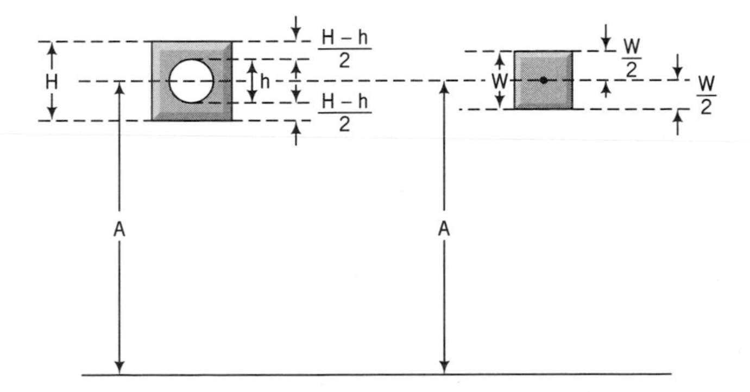

If the *effective* target height (W) is simply defined as:

$$W = H - h \tag{i}$$

then, the design of the pushbutton switch reduces to a Fitt's law task. Consequently, the index of difficulty (ID) is now:

$$\text{ID} = \log_2\left[\frac{A}{\left(\frac{H-h}{2}\right)}\right] \tag{ii}$$

Rearranging for H:

$$2^{(\text{ID})} = \frac{A}{\left(\frac{H-h}{2}\right)}$$

So that:

$$H = h + \frac{2A}{2^{(\text{ID})}} \tag{iii}$$

Substituting the given data into equation (iii):

$$H_1 = .01 + \frac{1.0}{32} = .041\ \text{M} \tag{iv}$$

And solving for H_2:

$$H_2 = .01 + \frac{0.2}{323} = .016\ \text{M} \tag{v}$$

SOLUTION 10.4(b)

Develop the forward transfer function (human operator informatic model):
First, solve for the T_D by substituting equation (28) into equation (74):

$$T_D = a + \frac{H_s}{\beta} \tag{i}$$

Approximate the simple reaction time (a) for both operators as 0.20 s. Then substitute the given values into equation (i):
For operator A:

$$T_D = 0.20 + \frac{5}{9.2} = 0.75\ \text{s} \tag{ii}$$

For operator B:

$$T_D = 0.20 + \frac{5}{5} = 1.20\ \text{s} \tag{iii}$$

Second, substitute the relevant vertical height (A), the relevant movement rate constant (α_M), and the relevant time delay (T_D) into equation (85):
Operator A, State 1:

$$y_0(A1) = \frac{(0.5)(4.5)e^{-s(0.75)}}{s(s+4.5)} \qquad \text{(iv)}$$

Operator A, State 2:

$$y_0(A2) = \frac{(0.1)(10)e^{-s(0.75)}}{s(s+10)} \qquad \text{(v)}$$

Operator B, State 1:

$$y_0(B1) = \frac{(0.5)(4.5)e^{-s(1.2)}}{s(s+4.5)} \qquad \text{(vi)}$$

Operator B, State 2:

$$y_0(B2) = \frac{(0.1)(10)e^{-s(1.2)}}{s(s+10)} \qquad \text{(vii)}$$

SOLUTION 10.4(c)

Solve for the action time (t_a) predicted for each of the four operator-state combinations:
First, solve for the time response of equations 10.4(b)(iv)–10.4(b)(vii):
This is accomplished by appropriate substitution into equation (87):

$$y_0(A1) = 0.5[1 - e^{-4.5(t-0.75)}] \qquad \text{(i)}$$
$$y_0(A2) = 0.1[1 - e^{-10(t-0.75)}] \qquad \text{(ii)}$$
$$y_0(B1) = 0.5[1 - e^{-4.5(t-1.2)}] \qquad \text{(iii)}$$
$$y_0(B2) = 0.1[1 - e^{-10(t-1.2)}] \qquad \text{(iv)}$$

Second, solve for the y_0 that defines the action time, t_a, i.e., the y_0 at which the pushbutton is engaged. This will occur when:

$$y_0 = A - \frac{W}{2} \qquad \text{(v)}$$

For State 1, substituting into equation 10.4(a)(i):

$$W = H - h = .041 - .010 = .031 \text{ M} \qquad \text{(vi)}$$

Recalling that $A = 0.5$ M, and substituting the result of equation (vi) into equation (v):

$$y_0(A1) = y_0(B1) = 0.50 - \frac{.031}{2} = .485 \text{ M} \qquad \text{(vii)}$$

For State 2, substituting into equation 10.4(a)(i):

$$W = H - h = .016 - .010 = .006 \text{ M} \qquad \text{(viii)}$$

Recalling that $A = 0.1$ M, and substituting the result of equation (viii) into equation (v):

$$y_0(A2) = y_0(B2) = 0.10 - \frac{.006}{2} = .097 \text{ M} \qquad \text{(ix)}$$

Third, solve for the specific action times (t_as) by substituting the results of equations (vii) and (ix) into equations (i) through (iv):

$$0.485 = 0.5 - (0.5)e^{-4.5(t_a - 0.75)} \quad \text{(x)}$$
$$0.097 = 0.1 - (0.1)e^{-10(t_a - 0.75)} \quad \text{(xi)}$$
$$0.485 = 0.5 - (0.5)e^{-4.5(t_a - 1.2)} \quad \text{(xii)}$$
$$0.097 = 0.1 - (0.1)e^{-10(t_a - 1.2)} \quad \text{(xiii)}$$

Rearrange equations (x) through (xiii) to obtain the specific t_as:

$$t_a(A,1) = 1.53 \text{ s}$$
$$t_a(A,2) = 1.10 \text{ s}$$
$$t_a(B,1) = 1.98 \text{ s}$$
$$t_a(B,2) = 1.55 \text{ s}$$

For two switch configurations states, each of an *equal* index of difficulty, when the information content of the stimulus (H_s) is 5 bits, the second switch configuration improves operator action time by an average of 25%, and the less experienced operator (using the second switch configuration) has an action time comparable to the more experienced operator (using the first switch configuration). This demonstrates a trade-off between the choice reaction time baud rate (β) and the movement rate constant (α_M).

---

b. Feedback Control System

A feed-back (or "closed-loop") system is one in which there is some comparison between the system output and the command input or control action *and* some operation (within the system) so that some modification of the input-output relation occurs in order that the system output is close to that which is desired. Recall that the feedback for the human operator may be visual, auditory and/or mechanical (somatosensory).

"Closed-loop" tasks performed by the human operator have the following characteristics (as compared to "open-loop" tasks):

1. The task performance requires *continuous* monitoring of the output and its comparison to the command input or control action.
2. The task performance has increased accuracy, as defined (for example) by an ability to more precisely reproduce the command input.
3. The system input-output relationship is less sensitive to uncontrolled disturbances that alter the system output.
4. The human operator output has an increased tendency toward oscillation and instability.

In order to realize the first three characteristics (and to minimize the fourth), the HFE must design the human-technological control system in the appropriate manner so as to satisfy a set of performance requirements.

Performance requirements may be stated with respect to frequency-domain specifications. The desired quantitative parameters would be expressed as functions of frequency. These may be conveniently evaluated by graphical analysis using a Bode plot (see Example 10.7).

Desired human-system performance requirements may also be stated in terms of time-domain specifications. The particular quantitative parameters will be formulated with respect to a specific time-response. For many purposes, it is convenient to evaluate these time-domain specifications with respect to a unit step input (per Example 10.7).

The human factors engineer may use either or both of these above domain specifications in formulating and analyzing the desired human-system characteristics. This section will first review typical frequency-domain specifications and then some typical time-domain specifications.

Of the various frequency-domain specifications, three are of particular interest to the human factors engineer when evaluating human-system control: the gain margin, the phase margin, and the bandwidth.

The gain margin (GM) is one measure of the relative stability of the system. GM represents the magnitude of the reciprocal of the open-loop transfer function (GH) in decibels, when evaluated at the phase crossover frequency (ω_π). ω_π is the frequency at which the argument of the open-loop transfer function is negative π radians per second. This may expressed mathematically as:

$$\text{GM} = 20 \log_{10}\left[\frac{1}{|\text{GH}(j\omega_\pi)|}\right] \tag{88}$$

where GM is in decibels;

The phase crossover frequency (ω_π) is:

$$\omega_\pi = \angle\text{GH}(j\omega_\pi) = -\pi \tag{89}$$

where ω_π is in radians/sec.

The phase margin (PM) is another measure of the relative system stability. PM is defined as 180° plus the phase angle (ϕ_1) of the open-loop transfer function (GH) at the gain crossover frequency (ω_1). ω_1 is that frequency at which the magnitude of the open-loop transfer function is unity. Mathematically this may be expressed as:

$$\text{PM} = 180 + \angle\text{GH}(j\omega_1) \tag{90}$$

where PM is in degrees.

The gain cross-over frequency (ω_1) is:

$$\omega_1 = |\text{GH}(j\omega_1)| \tag{91}$$

where ω_1 is in rad/s.

The bandwidth (BW) of a low pass filter approximation is defined as the range of frequencies over which the magnitude ratio of the closed-loop transfer function ($T(j\omega)$) does not decrease by more than -3 decibels from its specified value (when the frequency is zero). BW is often specified by the cutoff frequency (ω_c), which is the -3 db frequency (in rad/sec) with respect to the zero frequency (DC) value. Recall that BW (as expressed by ω_c) is often used to represent the maximum channel capacity for a continuous-time system (see Section 10.2.a). BW is also a measure of the speed of response of a system. For example, a large bandwidth will correlate with a faster rise time because a higher frequency range is transferred to the system output. Alternatively, for a smaller bandwidth, only the lower frequency range is transferred through the system, and the rise time will consequently be lower.

The resonant peak (M_p) is characteristic of second order, underdamped ($0 < \zeta < 1$) systems and is another measure of relative stability. M_p is defined as the maximum absolute value of the closed-loop frequency response magnitude:

$$M_p = \max|T(j\omega)| \tag{92}$$

where M_p is expressed in decibels.

Note that M_p occurs at the resonant frequency (ω_p) expressed as radians per second:

$$M_p = |T(j\omega_p)| \tag{93}$$

Time-domain specifications are stated in terms of the time response to a specified input function. Unit-step functions, ramp functions and parabolic functions have been used to characterize the system time response. The custom will be followed in this text of using a unit-step function for time-domain specification. Recall that the time-response to a unit-step function will consist of both a steady-state component and a transient component. The steady-state performance is characterized by means of steady-state error, which is a measure of system accuracy (for the particular input applied). The transient performance is customarily specified with respect to a unit-step function, and there are four typical time-response specifications.

Maximum overshoot is a measure of the relative stability of a system. It is defined as the maximum difference between the transient component and the steady-state component ($y_{max} - y_{ss}$) in response to a unit-step forcing function. Percent maximum overshoot (OS) is often reported as a percentage of the final steady-state component (y_{ss}) as follows:

$$\text{OS} = \%\ \text{Maximum Overshoot} = \left(\frac{y_{max} - y_{ss}}{y_{ss}}\right) \times 100 \tag{94}$$

The delay time (T_d) is a measure of the system speed of response. T_d is defined as the time (in seconds) required for the output of the system to reach 50% of its final steady-state value in response to a unit-step function input. The delay time (T_d) should not be confused with the time delay (T_D) as described in Section 10.2.b.4.

The rise time (T_r) is another measure of the system speed of response. T_r is defined as the time required for the system output to rise from 10% to 90% of its final steady-state value (in response to a unit-step input function). T_r is in units of seconds.

The settling time (T_s) is yet another measure of the system speed of response. T_s is the time required for a system output to approach and remain within a specified percentage (usually either 2% or 5%) of the final steady-state value. T_s is in response to a unit-step input function and reported in units of seconds.

The human operator informatic model has been developed in Section 10.2.b as a transfer function of a LTI system, and so may be analyzed using classical control theory techniques. The human operator informatic model of equation (79) has a pure time delay, so that the transfer function contains an exponential delay function in which T_D is the time delay in seconds. Many computational systems that operate on transfer functions will handle only rational functions. One common form is the zero-pole-gain transfer function. For a nth-order linear SISO system this is represented by a numerator, $N(s)$ and a denominator, $D(s)$:

$$G(s) = \frac{N(s)}{D(s)} = \frac{(s + z_1)(s + z_2)\cdots(s + z_n)}{(s + p_1)(s + p_2)\cdots(s + p_n)} \tag{95}$$

For this characteristic transfer function, the pure time delay exponential function may be restated as a Pade approximation of the time delay as follows:

$$e^{-sT_D} \approx -\left[\frac{(s-\tau)}{(s+\tau)}\right] \tag{96}$$

where the *reaction informatic frequency* (τ) in Hertz is:

$$\tau = \frac{1}{\left(\frac{T_D}{2}\right)} \tag{97}$$

Substituting equation (96) into equation (79) results in the zero-pole-gain form of the human operator informatic model transfer function, which we shall term the *approximate informatic model:*

$$\frac{y_O}{y_{IN}} = \frac{-\alpha_M(s-\tau)}{(s+\alpha_M)(s+\tau)} \tag{98}$$

Consider the feedback control system of Figure 10.22, which is a more simplified version of the one presented in Figure 10.20. In this operationally useful version, the human operator is interacting directly with the technological plant so that $K_I = 1$ (of Figure 10.20). There is direct and environmentally neutral feedback such that $G_e = 1$ (of Figure 10.20). Finally, the task is somewhat straightforward (and repetitive) so that a direct transfer ("unity") strategy is employed and so $K_s = 1$ (of Figure 10.20). Invoking these approximations reduces Figure 10.20 to the operationally simplified feedback control system of Figure 10.22. Finally, the human operator time delay transfer function (e^{-sT_D}) is replaced by a first-order Pade's approximation [equation (96)] and the three elements in cascade of the forward transfer function are combined so that the *approximate operational model* of Figure 10.23 is obtained. It is to be strongly emphasized so that the "approximate operational" model is only an initial design phase "first-order" approximation and only applicable when all of the above approximations just invoked are reasonable.

Referring to Figure 10.23, it is apparent that the approximate operational model may be described as follows:
The forward transfer function is:

$$G = \frac{-G_c K_p \alpha_M (s-\tau)}{(s+\alpha_M)(s+\tau)} \tag{99}$$

The feedback transfer function is:

$$H = \frac{1}{G_c} \tag{100}$$

The open-loop transfer function (GH) is the product of equations (99) and (100):

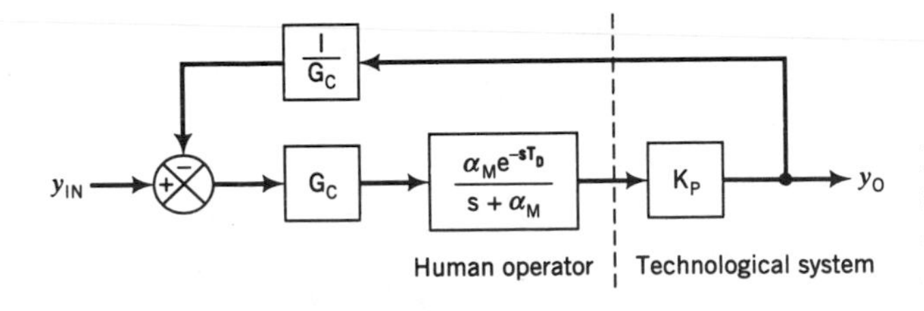

Figure 10.22 Approximate model of human-technological system for control systems analysis.

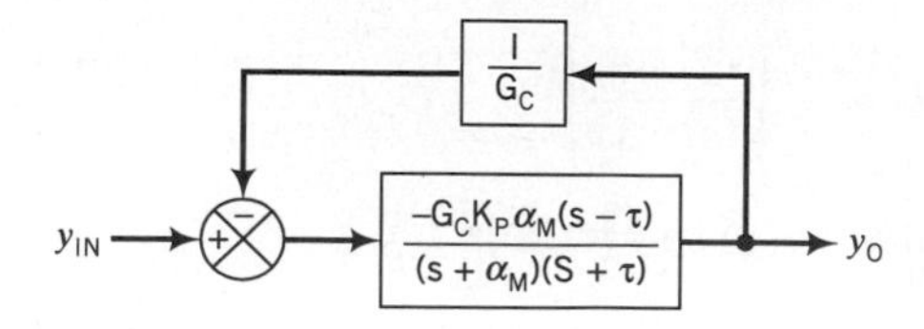

Figure 10.23 Canonical form of the approximate model (see Figure 10.22).

$$\text{GH} = \frac{-K_p\alpha_M(s-\tau)}{(s+\alpha_M)(s+\tau)} \tag{101}$$

The closed-loop transfer function (T) results from substituting equations (99) and (101) into equation (83), and simplifying:

$$T = \frac{y_O}{y_{IN}} = \frac{-G_cK_p\alpha_M(s-\tau)}{s^2+[\tau+\alpha_M(1-K_p)]s+\alpha_M\tau(1+K_p)} \tag{102}$$

EXAMPLE 10.5

For the approximate operational model of equation (102) in Figure 10.23:

(a) Solve for the central nervous system gain (G_c) at which the closed-loop gain (T) at steady state is unity and then derive the unity gain closed-loop transfer function;

(b) Solve for the natural (undamped) frequency (ω_N) and the damping coefficient (ζ).

SOLUTION 10.5(a)

Solve for G_c at $T = 1$:
Recall that in the frequency domain, $s = j\omega$, and that the DC solution ($\omega = 0$) is the steady-state solution ($s = 0$), so that equation (102) reduces to:

$$\frac{y_O}{y_{IN}}\text{(steady-state)} = \frac{G_cK_P\alpha_M\tau}{\alpha_M\tau(1+K_p)} \tag{i}$$

At unity steady-state gain:

$$\frac{G_cK_p\alpha_M\tau}{\alpha_M\tau(1+K_p)} = 1 \tag{ii}$$

And solving for G_c:

$$G_c = \frac{1+K_p}{K_p} \tag{iii}$$

Substituting equation (iii) into equation (102) results in the unity gain closed-loop transfer function ($T = 1$):

$$T = \frac{y_O}{y_{IN}} = \frac{-(1+K_p)\alpha_M(s-\tau)}{s^2+[\tau+\alpha_M(1-K_p)]s+\alpha_M\tau(1+K_p)} \tag{iv}$$

SOLUTION 10.5(b)

Solve for ω_N and ζ:

For a second-order linear system, the denominator of equation 10.5.a.iv may be approximated in the characteristic form of a prototype second-order system:

$$\frac{1}{s^2 + 2\zeta\omega_N s + \omega_N^2} = \frac{1}{s^2 + 2\pi[\tau + \alpha_M(1 - K_p)]s + (2\pi)^2\alpha_M\tau(1 + K_p)} \quad \text{(i)}$$

Note that ω_N is angular frequency in radians per second. Recall that α_M and τ are in seconds [equations (52) and (97), respectively]. In order to relate α_M and τ to ω_N, they are converted to radians per second in equation (i) by multiplying α_M and τ (in Hertz) by 2π radians.
From inspection of equation (i), the undamped natural frequency (ω_N) is:

$$\omega_N = 2\pi\sqrt{\alpha_M\tau(1 + K_p)} \quad \text{(ii)}$$

Also by inspection of equation (i):

$$2\zeta\omega_N = 2\pi[\tau + \alpha_M(1 - K_p)] \quad \text{(iii)}$$

Substituting equation (ii) into equation (iii), and rearranging for the damping coefficient (ζ):

$$\zeta = \frac{1}{2}\left[\frac{\tau + \alpha_M(1 - K_p)}{\sqrt{\alpha_M\tau(1 + K_p)}}\right] \quad \text{(iv)}$$

Two observations are important regarding the approximate operational model of Figure 10.23. Recall that:

$$H = \frac{1}{G_c} \quad (100)$$

Note that the forward transfer function (G) of equation (99) exhibits a CNS gain (G_c), but that the open-loop transfer function (GH) of equation (101) does not.

Equation (100) is a form of feedback gain compensation. The human operator approximate informatic model requires G_c in the forward transfer function in order for the closed-loop transfer function (T) to realize unity gain [see equation (102)]. However, excess gain of the open-loop (GH) transfer function will lead to stability problems with the human operator output (see Example 10.7). Therefore, the feedback (H) transfer function provides compensation (cancellation) of the extra gain (G_c) present in the forward (G) transfer function.

Second, the use of a first-order Pade's approximation for the time delay in the approximate informatic model results in a negative G transfer function [per equation (99)]. By subsequent substitution, there is also a negative T transfer function [see equation (102)]. This apparent anomaly should not unnecessarily disturb the student. The Pade first-order approximation introduces an extra zero [τ] and an extra pole [$-\tau$] into the transfer function. The result is that the human operator output response to a unit-step applied to either the G transfer function alone, or the T transfer function will be negative (reverse) for the initial period of the response *in order to* introduce time delay. The human operator output response so modeled, then crosses zero (initial (time = zero) output condition) and then swings positive (obverse) for the entire time course of the remainder of the response. In practice, the student may simply modify the computational program, so any reverse output values calculated by the model are set and plotted at zero.

We now conclude with three example (numerical and graphical) of human control systems analysis.

EXAMPLE 10.6

A human factors engineer is consulted in the evaluation of a marksmanship training system at a police academy firing range. A specific officer (operator) is being used for the system evaluation, who represents the median skill level (fiftieth percentile score) for the "bull's eye" target shooting.

Two different rifle models are being evaluated (representing the technological plant, K_p). Model A is a rifle (0.80 M in length) with an extended stock so that when the officer is sighting and the end plate is against the officer's shoulder, the extended arm grips the rifle stock at 0.725 M from the end plate. Model B is also 0.80 M in length, but with a shorter stock that is gripped with the extended arm at 0.55 M from the shoulder rest (end plate).

The target display system is at a distance down range and consists of a set of pop-up/knock-down bull's eye targets arranged in a single horizontal row. For any one task event, any one target will randomly "pop-up" and remain so until hit by a shot striking its bull's eye center, at which time it will "knock-down." After a brief pause, while the shooter (operator) holds their position, any target in the set, including the one just knocked down, will then pop up in random order and of equal probability.

The shooter then rotates the end of the rifle barrel by using hand-arm movement ($\alpha_M = 2.5/s$) through the horizontal distance between the prior target (just knocked-down) to the new target (just popped-up). Over the course of one thousand such task events, it is observed that the median movement of the end of the rifle barrel is 0.50 M, linear horizontal distance (you may neglect curvilinear effects). Two target displays are also being evaluated. State 1 consists of 4 pop-up/knock-down bull's-eye targets. State 2 consists of 12 such targets. The horizontal distance between the two end targets (whether in a row of 4 or a row of 12), however, is the same so that the median movement of the rifle barrel end is the same in either state.

(a) Consider the width, W_T (diameter of the target bull's eye center) as it would appear in relation to the width, when W_B (diameter of the far end of the gun barrel) the person is sighting the target along the barrel. If W_B is 0.015 M, calculate W_T for a 2% steady-state error, and calculate the resultant index of difficulty (ID);

(b) Solve for the plant gain, K_p (for Model A and Model B), and also solve for the choice reaction time, T_D and the reaction informatic frequency, τ (for State 1 and State 2); and

(c) Write the symbolic form of the forward transfer function [$G(s)$], the feedback transfer function [$H(s)$], the open-loop transfer function [GH(s)], and the closed-loop transfer function [T(s)].

SOLUTION 10.6(a)

Calculate W_T for a 2% steady-state error and the ID:

A 2% steady-state error means that after the transient response dies out, the steady-state response will be within *plus or minus* 2% of the step amplitude. For Fick's law, this is equivalent to:

$$\frac{W}{2} = (.02)A \tag{i}$$

Since the step amplitude, $A = 0.5$ M:

$$W = .02 \text{ M} \tag{ii}$$

In a manner analogous to equation 10.4.a.i, we may define W as:

$$W = W_T - W_B \tag{iii}$$

Rearranging equation (iii) and substituting for the calculated W [per equation (ii)] and given W_B, the target width (W_T) for a 2% steady-state error is:

$$W_T = W + W_B = .02 + .015 = .035 \text{ M} \tag{iv}$$

In a manner analogous to equation 10.4.a.ii, the index of difficulty (ID) is:

$$\text{ID} = \log_2\left[\frac{A}{\left(\frac{W_T - W_B}{2}\right)}\right] \tag{v}$$

Substituting the result of equation (iv), and the given W_B and given A into equation (v), results in:

$$\text{ID} = \log_2[50] = 5.64 \text{ bits} \tag{vi}$$

SOLUTION 10.6(b)

Solve for K_p and T_D and τ for the four model-state combinations:
We make the approximation that the shoulder rest (or end plate) of the rifle is a stationary point during the 0.50 M horizontal (assumed linear) displacement of the end of the rifle barrel. Remember this is a first-order approximation.
We can then invoke the approximation of similar triangles:

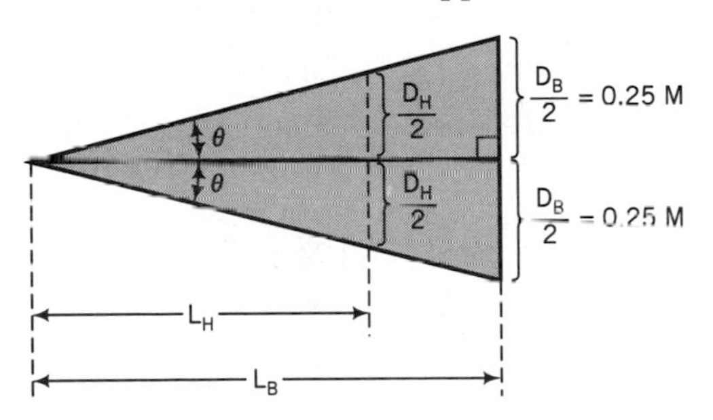

The gain of the plant (K_p) is defined as the *output* displacement of the end of the barrel (D_B) in response to the input displacement of the hand on the rifle stock (D_H).

$$K_p = \frac{D_B}{D_H} \tag{i}$$

Inspection of the diagram of similar triangles indicates that:

$$\tan\theta = \frac{D_B}{L_B} = \frac{D_H}{L_H} \tag{ii}$$

where

L_B = end plate-to-barrel end length (i.e., total rifle length)
L_H = end plate-to-hand grip length

So that rearranging equation (ii) for the plant gain (K_p) defined by equation (i):

$$K_p = \frac{D_B}{D_H} = \frac{L_B}{L_H} \tag{iii}$$

For rifle model A:

$$K_p(A) = \frac{0.80}{0.725} = 1.1 \tag{iv}$$

And for rifle model B:

$$K_p(B) = \frac{0.80}{0.55} = 1.45 \tag{v}$$

For the choice reaction time delay, T_D, we proceed as follows.
In State 1 there are four equiprobable alternatives of choice, so that the stimulus information content (H_s) is:

$$H_s(1) = \log_2(4) = 2 \text{ bits} \tag{vi}$$

In State 2, there are twelve equiprobable alternatives of choice, so that the stimulus information content (H_s) is:

$$H_s(2) = \log_2(12) = 3.59 \text{ bits} \tag{vii}$$

Recalling that the Hick-Hyman law of equation (27) may be restated as equation (74), then the operational Hick-Hyman law of equation (29) may be stated as:

$$T_D = 0.20 + (0.15)H_S \tag{viii}$$

Substitute the result of equation (vi) into equation (viii), and also substitute the result of equation (vii) into equation (viii).
For State 1, the choice reaction time (T_D) is:

$$T_D(1) = 0.2 + (0.15)(2.0) = 0.50 \text{ s} \tag{ix}$$

For State 2, the choice reaction time (T_D) is:

$$T_D(2) = 0.2 + (0.15)(3.59) = 0.74 \text{ s} \tag{x}$$

For the reaction informatic frequency (τ):
Recall equation (97) for τ (in hertz):

$$\tau = \frac{1}{\left(\frac{T_D}{2}\right)} \tag{97}$$

Substitute the calculated $T_D(1)$ of equation (ix) into equation (97):
τ for State 1:

$$\tau(1) = 4/\text{s} \tag{xi}$$

Substituting the calculated $T_D(2)$ of equation (x) into equation (97):
τ for State 2:

$$\tau(2) = 2.7/s \tag{xii}$$

SOLUTION 10.6(c)

Recall that by using block diagram algebra, *any* feedback control system may be reduced to its canonical form (as per Figure 10.21.a). For the system of interest, refer to Figure 10.23.

The forward transfer function (G) is:

$$G(s) = \frac{-G_c K_p \alpha_M (s - \tau)}{(s + \alpha_M)(s + \tau)} \tag{i}$$

The feedback transfer function (H) is:

$$H(s) = \frac{1}{G_c} \tag{ii}$$

So that the open-loop transfer function (GH) is:

$$\mathrm{GH}(s) = \frac{-K_p \alpha_M (s - \tau)}{(s + \alpha_M)(s + \tau)} \tag{iii}$$

The closed-loop transfer function (T) for a negative-feedback system is:

$$T(s) = \frac{y_0}{y_{IN}} = \frac{G}{1 + \mathrm{GH}} \tag{iv}$$

Substituting equations (i) and (iii) into equation (iv) and first simplifying:

$$T(s) = \frac{y_0}{y_{IN}} = \frac{-G_c K_p \alpha_M (s - \tau)}{(s^2 + [\tau + \alpha_M]s + \alpha\tau) - K_p \alpha_M (s - \tau)} \tag{v}$$

So that by further simplifying:

$$T(s) = \frac{y_0}{y_{IN}} = \frac{-G_c K_p \alpha_M (s - \tau)}{s^2 + [\tau + \alpha_M(1 - K_p)]s + \alpha_M \tau (1 + K_p)} \tag{vi}$$

Example 10.6 represents the preliminary (numerical) analysis of a human operator control system. Example 10.7 presents the graphical analysis of a human operator control system.

EXAMPLE 10.7

This example continues the evaluation of a marksmanship training system as described in Example 10.6.

For each of the four model-state combinations (A1, A2, B1, and B2) of Example 10.6:

(a) Write the numerical form of the open-loop transfer function [$\mathrm{GH}(j\omega)$] and the unity-gain closed-loop transfer function [$T(s)$].

(b) Write a computational program for performing a Bode plot analysis on the open-loop transfer function, $[\mathrm{GH}(j\omega)]$, and graphically determine the following: GM, ω_π, PM and ω_1.

(c) Solve for ζ and ω_N of the unity-gain closed-loop transfer function (T).

(d) Write a computational program for performing a 0.5 meter step analysis on the closed-loop transfer function, T, and graphically determine the following: T_r, OS, and T_s (2%).

SOLUTION 10.7(a)

Recall equation 10.6.c.iii, which is the symbolic form of the open-loop transfer function, and expand the denominator:

$$\mathrm{GH}(s) = \frac{-K_p\alpha_M(s-\tau)}{s^2+[\tau+\alpha_M]s+\alpha_M\tau} \tag{i}$$

Since we require the *frequency-domain* open-loop transfer function, $\mathrm{GH}(j\omega)$, recall that:

$$s = j\omega \tag{ii}$$

and define $\mathrm{GH}(j\omega)$:

$$\mathrm{GH}(j\omega) = \frac{-K_p\omega_\alpha(j\omega-\omega_\tau)}{(j\omega)^2+[\omega_\tau+\omega_\alpha](j\omega)+\omega_\alpha\omega_\tau} \tag{iii}$$

where the *movement informatic angular frequency* (ω_α) in radians per second is:

$$\omega_\alpha = \frac{2\pi}{\tau_M} = 2\pi\alpha_M \tag{iv}$$

the *reaction informatic angular frequency* (ω_τ) is

$$\omega_\tau = \frac{2\pi}{\left(\frac{T_D}{2}\right)} = 2\pi\tau \tag{v}$$

and ω_τ is in radians per second.
Substitute G_c from equation 10.5.a.iii into equation 10.6.c.vi to obtain the desired symbolic form of the unity gain closed-loop transfer function $T(s)$. After some simplification:

$$T(s) = \frac{y_0}{y_{IN}} = \frac{-(1+K_p)\alpha_M(s-\tau)}{s^2+[\tau+\alpha_M(1-K_p)]s+\alpha_M\tau(1+K_p)} \tag{vi}$$

Alternatively, the student may simply recall equation 10.5.a.iv.

Table 10.6 First Table of Values

Model State	K_p	α_M (Hz)	ω_a (rad/s)	τ (Hz)	ω_t (rad/s)
A1	1.1	2.5	15.7	4.0	25.1
A2	1.1	2.5	15.7	2.7	17.0
B1	1.45	2.5	15.7	4.0	25.1
B2	1.45	2.5	15.7	2.7	17.0

For the open-loop transfer function make the appropriate numerical substitutions for K_p, ω_α, and ω_τ (from Table 10.6) into equation (iii):

$$\text{GH}(A1) = \frac{-17.3(j\omega - 25.1)}{(j\omega)^2 + 40.8(j\omega) + 394} \quad \text{(vii)}$$

$$\text{GH}(A2) = \frac{-17.3(j\omega - 17.0)}{(j\omega)^2 + 32.7(j\omega) + 267} \quad \text{(viii)}$$

$$\text{GH}(B1) = \frac{-22.8(j\omega - 25.1)}{(j\omega)^2 + 40.8(j\omega) + 394} \quad \text{(ix)}$$

$$\text{GH}(B2) = \frac{-22.8(j\omega - 17.0)}{(j\omega)^2 + 32.7(j\omega) + 267} \quad \text{(x)}$$

For the unity-gain closed-loop transfer function make the appropriate substitutions for K_p, α_M, and τ (from Table 10.6) into equation (vi):

$$T(A1) = \frac{-5.25(s - 4)}{s^2 + 3.75s + 21} \quad \text{(xi)}$$

$$T(A2) = \frac{-5.25(s - 2.7)}{s^2 + 2.45s + 14.2} \quad \text{(xii)}$$

$$T(B1) = \frac{-6.12(s - 4)}{s^2 + 2.88s + 24.5} \quad \text{(xiii)}$$

$$T(B2) = \frac{-6.12(s - 2.7)}{s^2 + 1.58s + 16.5} \quad \text{(xiv)}$$

SOLUTION 10.7(b)

A computational program may be written in MATLAB to perform a Bode plot on the open-loop transfer function, GH($A1$), as represented by equation 10.7.a.vii. The program is as follows:

```
num = −17.3*[1,−25.1];
den = [1,40.8,394];
bode(num,den)
```

Graphical analysis of the resultant plot (shown in Figure E10.7.1) allows us to define the following frequency-domain parameters:

$$\text{GM} = 7.5 \text{ dB}$$

$$\omega_\pi = 38.\text{rad/s}$$

$$\text{PM} = 120°$$

$$\omega_1 = 7.0 \text{ rad/s}$$

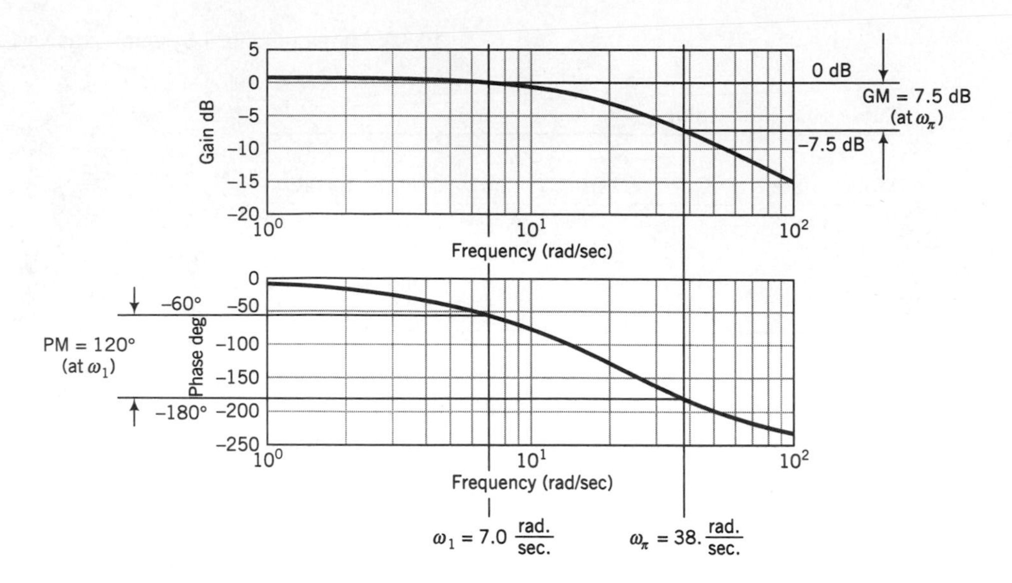

Figure E10.7.1

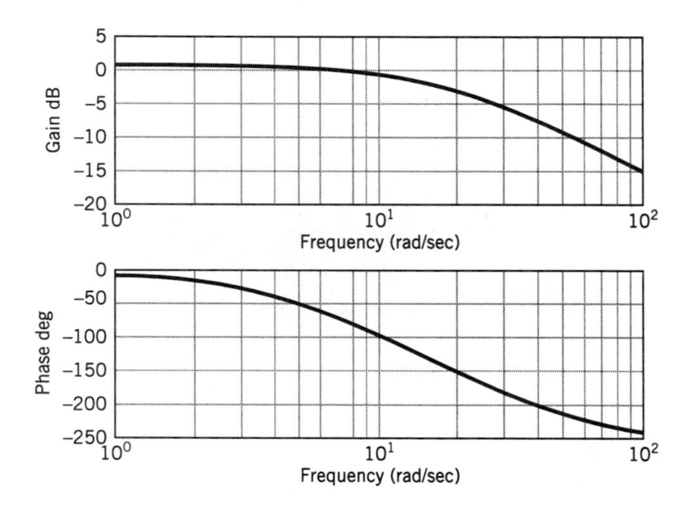

Figure E10.7.2

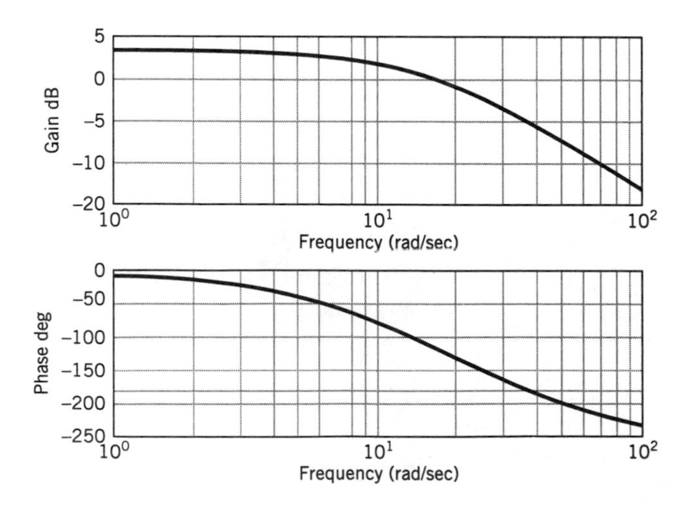

Figure E10.7.3

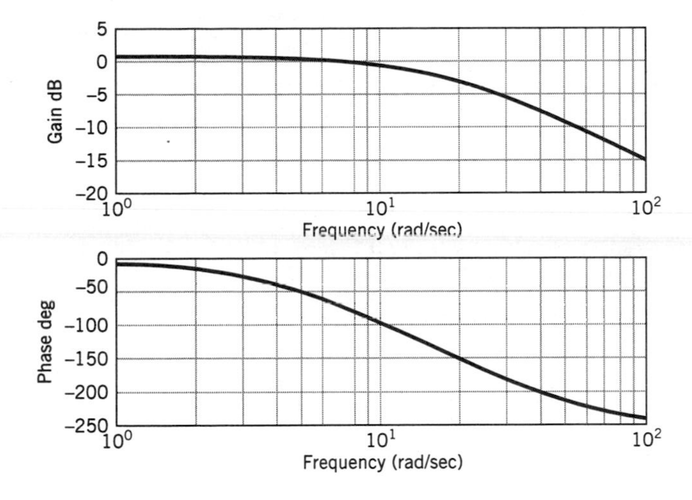

Figure E10.7.4

The above procedure is repeated for: GH($A2$) per equation 10.7.a.viii, and the resultant plot is depicted in Figure E10.7.2.
In a similar manner, GH($B1$) as per equation 10.7.a.ix is plotted as represented by Figure E10.7.3.
Finally, GH(B2) defined by equation 10.7.a.x is plotted as indicated by Figure E10.7.4.
Appropriate graphical analysis is then performed (see Figure E10.7.1) and the results entered into Table 10.7.

SOLUTION 10.7(c)

The damping coefficient (ζ) and the undamped natural frequency (ω_N) are now obtained for $T(A1)$ by appropriate substitution (for K_p, α_M, and τ) from Table 10.6 into equations 10.5.b.iv and 10.5.b.ii, respectively.

$$\zeta = 0.41$$
$$\omega_N = 29 \text{ rad/s}$$

The above procedure is repeated for: $T(A2)$, $T(B1)$, and $T(B2)$. The results are then entered into Table 10.7.

Table 10.7 Second Table of Values

Model-State	K_p	α_M (Hz)	τ (Hz)	ω_1 (rad/s)	PM (degrees)	ω_π (rad/s)	GM (dB)	ζ	ω_N (rad/s)	T_r (s)	OS (%)	T_s (2%) (s)
A1	1.1	2.5	4.0	7.0	120	38	7.5	0.41	29	0.24	33	2.0
A2	1.1	2.5	2.7	7.0	110	28	5.5	0.32	24	0.23	52	3.8
B1	1.45	2.5	4.0	16.	70	38	5.0	0.29	31	0.17	55	3.0
B2	1.45	2.5	2.7	16.	45	28	3.0	0.19	26	0.16	88	5.7

SOLUTION 10.7(d)

A computational program may be written in MATLAB to perform a unit-step response on the closed-loop transfer function, $T(A1)$ as represented by equation 10.7.a.xi:
Recall that:

$$\frac{y_0}{y_{IN}} = T(A1) \tag{i}$$

For the output time response (y_0) to a unit-step (y_{IN}):

$$y_0 = (y_{IN})T(A1) = \left[\frac{1}{s}\right] T(A1) \tag{ii}$$

Substituting equation 10.7.a.xi into equation (ii) results in:

$$y_0(A1) = \frac{-5.25(s-4)}{s[s^2 + 3.75 + 21]} \tag{iii}$$

Since the computational program in MATLAB executes the "step" command by inserting the $1/s$ term (a unit-step!), then the program is as follows:

```
num = −5.25*[1,−4];
den = [1,3.75,21];
step(num,den)
```

Graphical analysis of the resultant plot is shown in Figure E10.7.5 and allows us to determine the following time-domain parameters:

$$T_r = 0.24 \text{ s}$$
$$OS = 33\%$$
$$T_s\,(2\%) = 2.0 \text{ s}$$

This process is then repeated for $T(A2)$ per equation 10.7.a.xii and the resultant plot is illustrated in Figure E10.7.6.

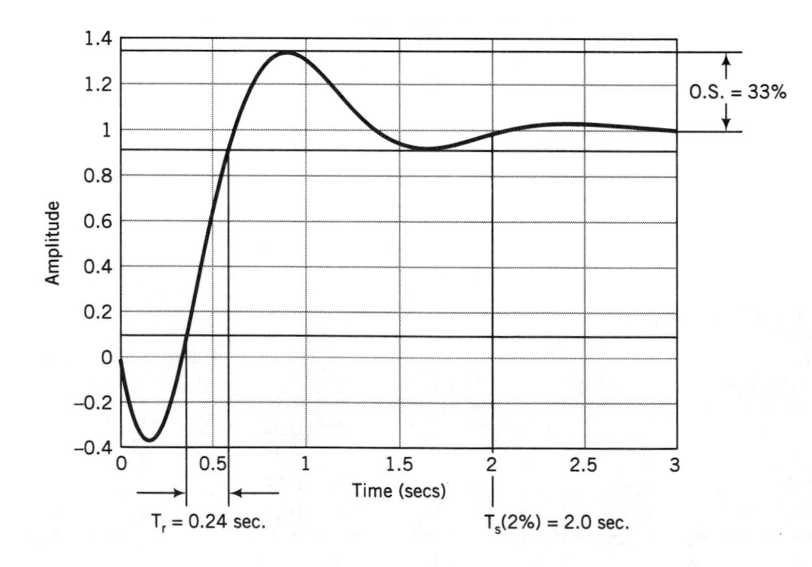

Figure E10.7.5

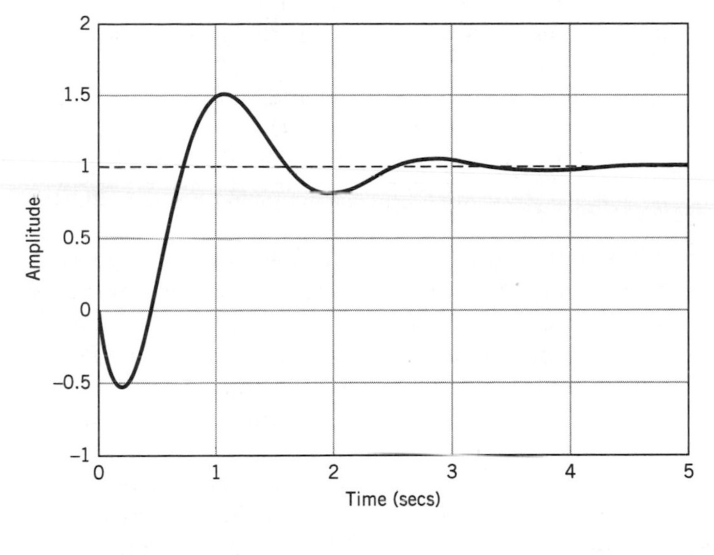

Figure E10.7.6

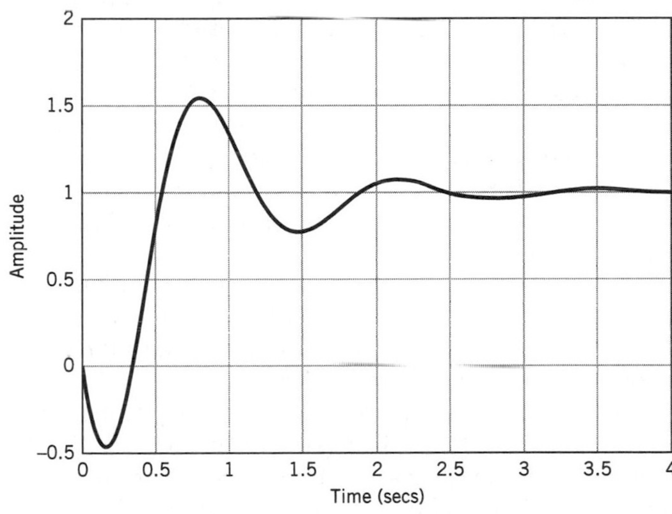

Figure E10.7.7

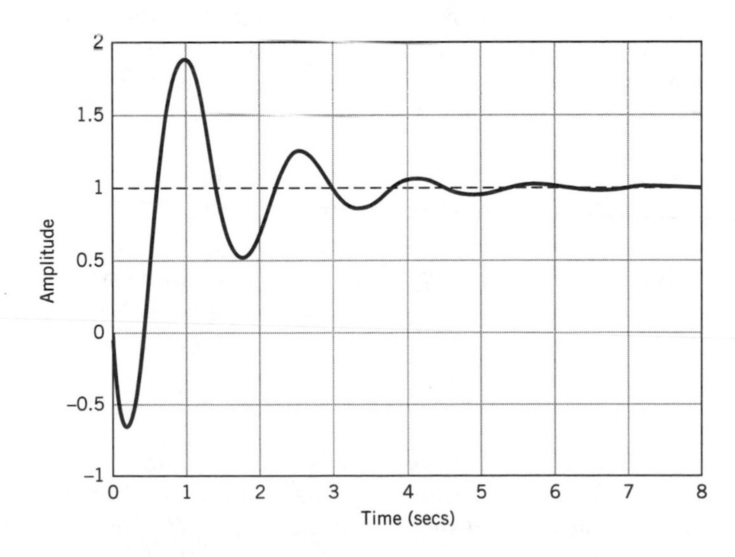

Figure E10.7.8

In a similar manner, $T(B1)$ as defined by equation 10.7.a.xiii is plotted as depicted in Figure E10.7.7.
Finally, $T(B2)$ as represented by equation 10.7.a.xiv is plotted as per Figure E10.7.8. Appropriate graphical analysis is then performed (in a manner analogous to that of Figure E10.7.5) and the results entered into Table 10.7.

Examination of Table 10.7 from Example 10.7 indicates some basic principles of human operator-technological system control as predicted by human control system analysis. These principles apply for a set of system gain and time delay combinations when the human operator movement informatic frequency (α_M) is constant and the task index-of-difficulty (ID) is constant across the various combinations (as per Example 10.6). The relatively most stable human operator-technological system configuration (A1) occurs at the lower technological system gain ($K_p = 1.1$) and the higher reaction informatic frequency ($\tau = 4$ Hz), which indicates a shorter choice reaction time delay ($T_D = 0.50$ s). Human control system analysis characterizes this relatively most stable system by a larger phase margin and a larger gain margin, a larger damping coefficient, and a smaller percent overshoot (Table 10.7). Note that this relatively most stable configuration has the larger rise time, but the smallest settling time of the four combinations (as per Table 10.7).

Two relatively intermediate stable human operator-technological system configurations (B1 and A2) occur when either the system gain is higher ($K_p = 1.45$), as with configuration B1, or *alternatively,* the reaction informatic frequency is lower ($\tau = 2.7$ Hz). This latter situation occurs in configuration A2 and indicates that the choice reaction time delay is longer ($T_D = 0.74$ s). Human control system analysis characterizes these two relatively intermediate stable systems by intermediate values of gain margin, damping coefficient, and percent overshoot (Table 10.7). Note that these two relatively intermediate stable configurations have the intermediate settling times of the four combinations summarized in Table 10.7.

The relatively least stable human operator-technological system configurations (B2) occur at both the higher technological system gain ($K_p = 1.45$) and also the lower reaction informatic frequency ($\tau = 2.7$ Hz). Human control system analysis characterizes this relatively least stable system by the smallest phase margin, and smallest gain margin, the smallest damping coefficient, and the largest percent overshoot of the four combinations evaluated in Table 10.7. Note that while this relatively least stable configuration has the shortest rise time, it also demonstrates the longest settling time of the four combinations that were analyzed (as per Table 10.7).

Numerical analysis and graphical analysis can also be used to calculate the operator-system bandwidth (BW) and cut-off frequency (ω_c). In some operator-system interactions a resonant peak (M_p) and resonant frequency (ω_p) will occur and can be calculated. This is illustrated in the example that follows.

EXAMPLE 10.8

Prior to completing the final consulting report that was requested in the beginning of Example 10.6, the HFE decides to evaluate the operator-system bandwidth (BW) and resonant peak (M_p) for the four model states ($A1$, $A2$, $B1$, and $B2$).

Using the numerical method followed by the graphical method (as per Examples 10.6 and 10.7), determine for the four model states: the bandwidth [BW($A1$), BW($A2$), BW($B1$), and BW($B2$)]; the cut-off frequency [$\omega_c(A1)$, $\omega_c(A2)$, $\omega_c(B1)$, and $\omega(B2)$]; the resonant peak [$M_p(A1)$, $M_p(A2)$, $M_p(B1)$, and $M_p(B2)$]; and the resonant frequency [$\omega_p(A1)$, $\omega_p(A2)$, $\omega_p(B1)$, and $\omega_p(B2)$].

SOLUTION 10.8

Recall equation 10.7.a.vi which is the symbolic form of the unity gain closed-loop transfer function:

$$T(s) = \frac{-(1 + K_p)\alpha_M(s - \tau)}{s^2 + [\tau + \alpha_M(1 - K_p)]s + \alpha_M\tau(1 + K_p)} \quad \text{(i)}$$

Since we require the frequency-domain closed-loop transfer function, $T(j\omega)$, recall that:

$$s = j\omega \quad \text{(ii)}$$

and define $T(j\omega)$:

$$T(j\omega) = \frac{-(1 + K_p)\omega_{\alpha}(j\omega - \omega_\tau)}{(j\omega)^2 + [\omega_\tau + \omega_\alpha(1 - K_p)](j\omega) + \omega_\alpha\omega_\tau(1 + K_p)} \quad \text{(iii)}$$

where the *movement informatic angular frequency* (ω_α) is defined by equation 10.7.b.iv, and the *reaction informatic angular frequency* (ω_τ) is defined by equation 10.7.b.v.

For the unity gain closed loop transfer function make the appropriate numerical substitutions for K_p, ω_α, and ω_τ (per Table 10.6) into equation (iii).

For model state ($A1$):

$$T(A1) = \frac{-(2.1)(15.7)(j\omega - 25.1)}{(j\omega)^2 + [25.1 + (15.7)(-.1)](j\omega) + (15.7)(25.1)(2.1)}$$

which simplifies to:

$$T(A1) = \frac{-33.0(j\omega - 25.1)}{(j\omega)^2 + 23.5(j\omega) + 828} \quad \text{(iv)}$$

For model state ($A2$):

$$T(A2) = \frac{-(2.1)(15.7)(j\omega - 17.0)}{(j\omega)^2 + [17.0 + (15.7)(-.1)](j\omega) + (15.7)(17.0)(2.1)}$$

which simplifies to:

$$T(A2) = \frac{-33.0(j\omega - 17.0)}{(j\omega)^2 + 15.4(j\omega) + 560} \quad \text{(v)}$$

For model state ($B1$):

$$T(B1) = \frac{-(2.45)(15.7)(j\omega - 25.1)}{(j\omega)^2 + [25.1 + (15.7)(-.45)](j\omega) + (15.7)(25.1)(2.45)}$$

which simplifies to:

$$T(B1) = \frac{-38.5(j\omega - 25.1)}{(j\omega)^2 + 18.0(j\omega) + (965)} \quad \text{(vi)}$$

For model state ($B2$):

$$T(B2) = \frac{-(2.45)(15.7)(j\omega - 17.0)}{(j\omega)^2 + [17.0 + (15.7)(-.45)](j\omega) + (15.7)(17.0)(2.45)}$$

which simplifies to:

$$T(B2) = \frac{-38.5(j\omega - 17.0)}{(j\omega)^2 + 9.9(j\omega) + 654} \quad \text{(vii)}$$

A computational program is written in MATLAB to perform a Bode plot on the unity gain closed-loop transfer function, $T(A1)$, as represented by equation (iv). The program is as follows:

```
num = −33.0*[1,-25.1];
den = [1,23.5,828];
bode(num,den)
```

Graphical analysis of the resultant plot (shown in Figure E10.8.1) allows us to define the requested parameters for the human operator-technological system. From the Bode plot, the bandwidth can be defined as the −3 dB point when the transfer function is reduced to 70.7% of the magnitude ratio (M) at the zero frequency (DC) value.

$$BW(A1) = 9.5 \text{ Hz}$$
$$\omega_c(A1) = 60 \text{ rad/s}$$
$$M_p(A1) = 5.5 \text{ dB}$$
$$\omega_p(A1) = 26 \text{ rad/s}$$

The above procedure is repeated for $T(A2)$ per equation 10.8.v, and the resultant plot is depicted in Figure E10.8.2.
In a similar manner, $T(B1)$ per equation 10.8.vi is plotted as shown by Figure E10.8.3. Finally, $T(B2)$ defined by equation 10.8.vii is plotted as indicated by Figure E10.8.4. Appropriate graphical analysis (as per Figure E10.8.1) is then performed on the Bode plots of Figures E10.8.2, E.10.8.3, and E10.8.4. The results are entered into Table 10.8.

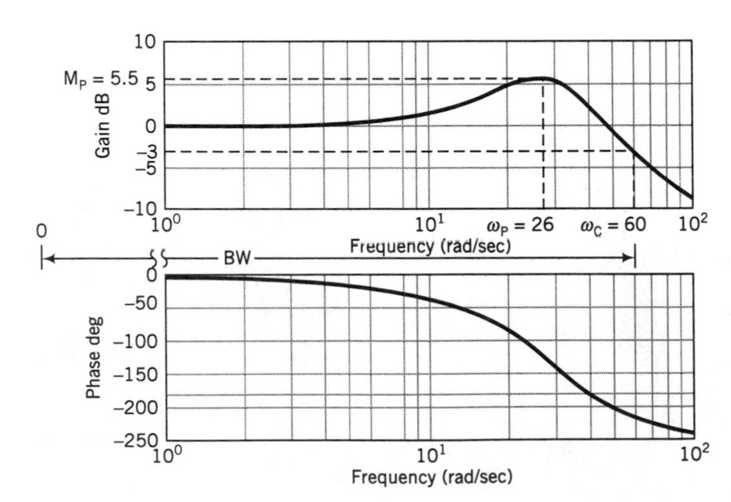

Figure E10.8.1

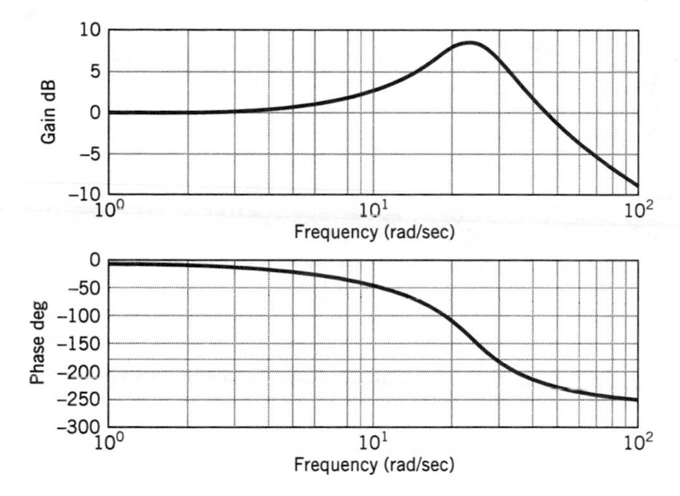

Figure E10.8.2

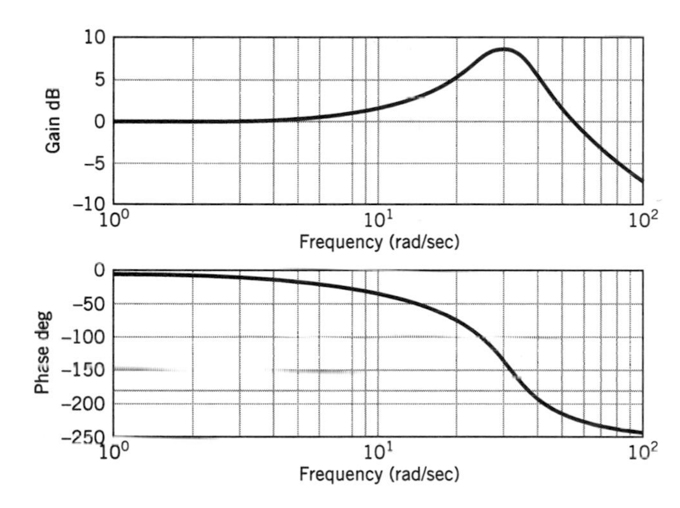

Figure E10.8.3

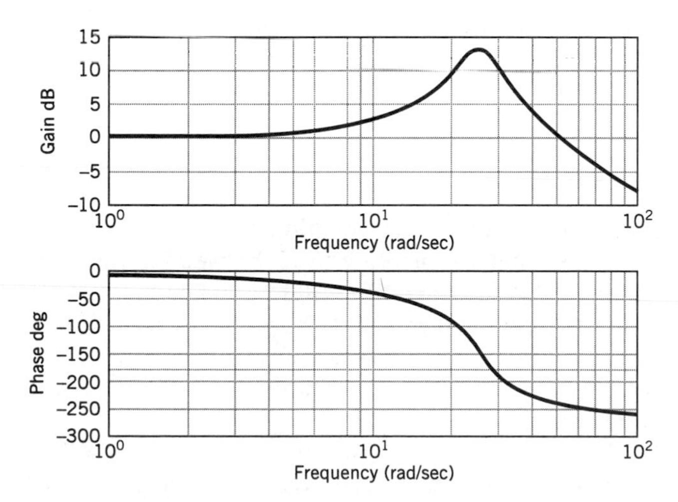

Figure E10.8.4

Table 10.8 Third Table of Values

Model-State	K_p	α_M (Hz)	τ (Hz)	BW (Hz)	ω_c (rad/s)	M_p (dB)	ω_p (rad/s)
A1	1.1	2.5	4.0	9.5	60	5.5	26
A2	1.1	2.5	2.7	9.1	57	8.5	23
B1	1.45	2.5	4.0	11.1	70	9.0	30
B2	1.45	2.5	2.7	10.6	66	13.0	24

Examination of Table 10.8 from Example 10.8 indicates some additional and basic principles of human operator-technological system control beyond those described previously (for Table 10.7 from Example 10.7). A primary feature of the human operator-technological system bandwidth (BW) and cutoff frequency (ω_c) is that they are extended by increasing the plant gain (K_p). Let us refer back to the closed-loop transfer function (T) of equation (102) and rewrite that equation in its zero-pole-gain form.

To find the roots (poles) of the denominator of equation (102), we use the quadratic equation theorem:

$$D(s) = as^2 + bs + c \tag{103}$$

So that:

$$p_1 = \frac{-b + \sqrt{b^2 - 4ac}}{2a} \tag{104}$$

$$p_2 = \frac{-b + \sqrt{b^2 - 4ac}}{2a} \tag{105}$$

where:

$$a = 1$$
$$b = \tau + \alpha_M(1 - K_p)$$
$$c = \alpha_M \tau(1 + K_p)$$

The zero-pole-gain form of equation (102) is then:

$$T(s) = \frac{-K(s - z_1)}{(s + p_1)(s + p_2)} \tag{106}$$

where:

$$z_1 = \tau \tag{107}$$
$$K = G_c K_p \alpha_M \tag{108}$$

and p_1 and p_2 are from equation (104) and equation (105), respectively.

Equation (108) indicates that the increase in K_p between State A and State B increases the gain factor (K) of the operator-system transfer function. Recall that G_c and α_M of equation (108) remained constant. However, a change in any of these three variables will directly affect the gain factor. It is now apparent that K_p has a primary and proportional effect on BW and ω_c through its effect on K. Increasing K_P has limits, however.

Comparing Table 10.7 with Table 10.8 reveals a characteristic effect BW and ω_c have on the rise time (T_r). Recall that the rise time is one measure of the system speed of response. From Table 10.7 it appears that State A has the slower (longer duration) T_r and that State B has the faster (shorter duration) T_r. Reference to Table 10.8 shows that State A has the lower BW and ω_c and that State B has the higher BW and ω_c. The higher BW and ω_c allow higher frequency signals to pass through the system (without attenuation) than is the case for a system with a lower BW and ω_c. It is these higher frequency signal components that contribute to a faster (shorter duration) rise time. It is now apparent that BW and ω_c have a primary and inversely proportional effect on T_r.

The resonant peak (M_p) is a measure of relative stability. Per Table 10.8, it is apparent that the relatively most stable system ($A1$) has the lower M_p, the relatively intermediate stable systems ($A2$ and $B1$) have the intermediate M_p, and the relatively least stable system ($B2$) has the higher M_p. Recall that M_p is characteristic of second-order, underdamped systems. Referring to Table 10.7, it is noted that state $A1$ has the higher damping coefficient ($\zeta = 0.41$), states $A2$ and $B1$ have the intermediate damping coefficients ($\zeta = 0.32$ and ($\zeta = 0.29$) and that state $B2$ has the lower damping coefficient ($\zeta = 0.19$) It is now apparent that the damping coefficient (ζ) has a primary and inversely proportional effect on the resonant peak.

Overshoot (OS) is also a measure of relative stability. It is instructive to compare OS from Table 10.7 with M_p from Table 10.8. For a second-order, underdamped system, it is seen that overshoot in the time-domain is related to the resonant peak in the frequency domain.

FURTHER INFORMATION

J.J. DiStefano, A.R. Stubberud and I.J. Williams: *Theory and Problems of Feedback and Control Systems*. 2nd Edition. McGraw-Hill. New York. 1990.

D.C. Hanselman and B.C. Kuo: *Matlab Tools for Control System Analysis and Design*. 2nd Edition. Prentice Hall. Upper Saddle River, NJ. 1995.

B.C. Kuo: *Automatic Control Systems*. 7th Edition. Prentice Hall. Upper Saddle River, NJ. 1995

C.A. Phillips and D.W. Repperger (eds.): *Human Interaction with Technological Systems* (AUTOMEDICA 16:4). Gordon and Breach. New York. 1998.

E.C. Poulton: *Tracking Skill and Motor Control*. Academic Press. New York. 1974.

PROBLEMS

10.1. For Example 10.1, complete the following for each of the subparts (a)–(h):

i. Identify the specific computational elements of Figure 10.5 for which the two variables have an input-output relationship.
ii. Identify the governing equation for that computational element.
iii. Plot the indicated variables as a function of time.
iv. Explain the result.

The subparts are as follows:

a. e' vs. L' e. $\hat{P}_c$ vs. e'
b. x_0 vs. L' f. $\hat{P}_c$ vs. L'
c. V_R vs. x_e' g. $\hat{P}_c$ vs. V'
d. x_0 vs. V' h. $\hat{P}_c$ vs. x_0

10.2. For Example 10.1, complete the following for each of the subparts (a)–(g):

i. Identify the specific computational elements of Figure 10.5 for which the two variables have an input-output relationship.
ii. Identify the governing equation for that computational element.
iii. Plot the indicated variables as a function of time.
iv. Explain the result.

The subparts are as follows:

a. V' vs. L'
b. $\Delta\overline{\$}$ vs. V'
c. A_v vs. V'
d. V_R vs. V_e
e. $\hat{P}_0\ (L')$ vs. L'
f. $\overline{\$}_{Pc}$ vs. $\hat{P}_0(L')$
g. $\overline{\$}_{Pc}$ vs. $\hat{P}_c$

10.3. Continue Problem 10.2 for each of the subparts (h)–(o):

h. $\hat{V}_0(L')$ vs. L'
i. $\hat{V}_c(L')$ vs. $\hat{V}_0(L')$
j. $\hat{V}_c(L')$ vs. $\hat{P}_0(L')$
k. $\hat{V}_c(l')$ vs. A_v
l. $\overline{\$}_{Pc}$ vs. $\Delta\overline{\$}$
m. $\hat{V}_c(L')$ vs. V'
n. $\overline{\$}_{Pc}$ vs. V'

10.4. A person using the flexor muscles of the fingers is grasping the dish sprayer to clean the dishes. In the case of this elastic stick, the amount of resistance increases as the amount of displacement increases. The more the person compresses down on the handle to activate the flow of water, the harder it is to pull down. This is due to the mechanisms within the area near the nozzle of the sprayer. There is also an added amount of resistance due to the rubber material that surrounds the hose in the region of the handle. The more that this is compressed, the harder it becomes. The amount of displacement required to move a black greasy spot from its original position on a dish to just off the edge of the dish is $dx = 3.0$ cm. The initial muscle length of the finger flexor muscles is $1.25L_0$. The spring constant of the handle is $K_S = 0.7$, and the handle-to-human hand interaction is $K_T = 0.9$. The combination of jet spray, sprayer handle, and connecting hose (under water pressure) has a plant gain of $K_P = 9$. The entire operation occurs in a well-lighted kitchen so that the environmental feedback gain is $G_e = 1$.

a. Write the appropriate computer simulation program (for Figure 10.5) that represents this problem.
b. Plot the following:
   i. V_R vs. V'
   ii. L'
   iii. x' vs. x_e

10.5. Regarding Example 10.4:

a. Consider that State 1 ($A_1 = 0.5$ M) and $\alpha_M = 4.5$/s exist for *both* Operator A and Operator B. If H_{1A} for Operator A is 0.041 M (so that ID = 5 bits):
   i. Find the value of H_{1B} for Operator B so that:

$$t_a(\text{B},1) - t_a(\text{A},1) = 0.33$$

   ii. Calculate the resultant ID for Operator B.

b. Consider that $H_1 = 0.041$ M and $\alpha_M = 4.5$/s exists for *both* Operator A and Operator B. If $A_{1A} = 0.5$ M for Operator A (so that ID = 5 bits):

i. Find the value of A_{1B} for Operator B so that:

$$t_a(\text{B},1) - t_a(\text{A},1) = 0.33$$

ii. Calculate the resultant ID for Operator B.

10.6. Regarding Example 10.4, two new operators are now being evaluated: Operator A′ ($\beta' = 8.0$/s) and Operator B′ ($\beta' = 6.0$/s).

a. For State 1, it is observed that:

$$t_a(\text{B}',1) = t_a(\text{A}',1)$$

If $\alpha_M(\text{A}',1) = 4.5$/s, calculate the $\alpha_M(\text{B}',1)$ required for this to happen.

b. For State 2, it is observed that:

$$t_a(\text{B}',2) = t_a(\text{A}',2)$$

If $\alpha_M(\text{A},'2) = 10$/s, calculate the $\alpha_M(\text{B}',2)$ required for this to happen.

10.7. Regarding Example 10.4, for Operator A in State 1:

i. a $\beta = 10.1$/s (10% rate increase) results in what percent change in t_a?
ii. a $\beta = 8.2$/s (10% rate decrease) results in what percent change in t_a?
iii. a $\alpha_M = 5.0$/s (10% rate increase) results in what percent change in t_a?
iv. a $\alpha_M = 4.0$/s (10% rate decrease) results in what percent change in t_a?
v. a $\beta = 10.1$/s *and simultaneously* a $\alpha_M = 5.0$/s results in what percent change in t_a?
vi. a $\beta = 8.2$/s *and simultaneously* a $\alpha_M = 4.0$/s results in what percent change in t_a?

10.8. Regarding Example 10.4, for Operator B in State 1:

i. a $\beta = 5.5$/s (10% rate increase) results in what percent change in t_a?
ii. a $\beta = 4.5$/s (10% rate decrease) results in what percent change in t_a?
iii. a $\alpha_M = 4.95$/s (10% rate increase) results in what percent change in t_a?
iv. a $\alpha_M = 4.05$/s (10% rate decrease) results in what percent change in t_a?
v. a $\beta = 5.5$/s *and simultaneously* a $\alpha_M = 4.95$/s results in what percent change in t_a?
vi. a $\beta = 4.5$/s *and simultaneously* a $\alpha_M = 4.05$/s results in what percent change in t_a?

10.9. Regarding Example 10.4:

a. For Operator A in State 1:

i. Calculate H_{1A} that results in a 10% faster action time, $t_a(\text{A},1)$. What is the new ID?

ii. Calculate H_{1A} that results in a 10% slower action time, $t_a(\text{A},1)$. What is the new ID?

b. For Operator B in State 1:

i. Calculate H_{1B} that results in a 10% faster action time, $t_a(\text{B},1)$. What is the new ID?

ii. Calculate H_{1B} that results in a 10% slower action time, $t_a(\text{B},1)$. What is the new ID?

10.10. Repeat Example 10.4, keeping all parameters and activities constant, except:

a. The information content of the stimulus, H_s = 2.5 bits.
b. The information content of the stimulus, H_s = 7.5 bits.

10.11. A senior marksman has observed the performance of the human operator as described in Example 10.6. Two recommendations are then made (based upon those observations) in order to improve the human operator performance.

First Obesrvation/Recommendation

The police officer is incorrectly pointing the *barrel* of the rifle at the black "bull's eye" target. This observation leads to the recommendation that the police officer point the *front sight* (located above the barrel muzzle) at the black "bull's eye" target.

The HFE then takes an additional measurement:

$$W_S = .002 \text{ M}$$

where W_S is the horizontal width of the front sight of the rifle. The HFE also recalls that

$$W_T = .037 \text{ M}$$

where W_T is the black "bull's eye" target horizontal width as determined by equation 10.6.a.iv.

Second Observation/Recommendation:

The police officer is pointing and directing the rifle primarily with the hand that firmly grips the rifle stock. The other hand (the trigger hand) is only lightly contacting the area around the trigger guard. This observation leads to the recommendation that the trigger hand also be used to point and direct the rifle by firmly gripping the area around the trigger guard *simultaneously* with the hand that already firmly grips the rifle stock.

The HFE then determines the following. One, the horizontal distance from the rifle end-plate (resting against the officer's shoulder) and the grip point of the trigger hand is 0.38 M. Two, the plant gain (K_p') when pointing and directing the rifle barrel with *two* hands is:

$$K_p' = \frac{D_B}{D_H + D_H'}$$

where:

D_B = horizontal output displacement of the muzzle end of the rifle barrel;
D_H = horizontal input displacement of the hand on the rifle stock area;
D_H' = horizontal input displacement of the hand on the trigger guard area.

K_p' for Model A is then identified as $K_p(A')$:
K_p' for Model B is then identified as $K_p(B')$:

a. With respect to the first observation/recommendation:

i. Define the percent steady-state error that will now result.
ii. Calculate the resultant index of difficulty (I.D.).

b. With respect to the second observation/recommendation:

i. Calculate the plant gain, $K_p(A')$.
ii. Calculate the plant gain, $K_p(B')$.

10.12. Regarding Example 10.6, define the police officer as Operator I (characterized by a specific β and a specific α_M). Inspection of equation 10.6.b.viii will indicate that:

$$\beta = \frac{1}{0.15\ \text{s}} = 6.7/\text{s}$$

Recall that for Operator I:

$$\alpha_M = 2.5/\text{s}$$

The performance of Operator I has been evaluated with respect to $K_p(A)$, $K_p(B)$, $H_S(1)$, and $H_S(2)$ in Examples 10.6, 10.7, and 10.8.

Now define an *Operator I'* (β = 6.7/s, α_M = 2.5/s) for which all activities and parameters of Example 10.6 the same, *except:*
Replace equation 10.6.a.ii with:

$$W = .035\ \text{M}\ (W_T = .037\ \text{M and}\ W_S = .002\ \text{M})$$

Replace equation 10.6.b.iv with:

$$K_p(A') = 0.72$$
$$K_p(B') = 0.86$$

For Operator I' (β = 6.7/s, α_M = 2.5/s):

a. Solve for the choice reaction time, T_D, and the reaction informatic frequency, τ (for State 1 and State 2).
b. Recall the symbolic form of GH(s) per equation 10.7.a.i, GH($j\omega$) per equation 10.7.a.iii, $T(s)$ per equation 10.7.a.vi and $T(j\omega)$ per equation 10.8.iii. *Then for each of the four model-state combinations* ($A'1$, $A'2$, $B'1$, and $B'2$):

    i. Write the numerical form of the open-loop transfer function [GH($j\omega$)], and then write a computational program for performing a Bode plot analysis on GH($j\omega$) and graphically determine the following: GM, ω_π, PM, and ω_1.
    ii. Write the numerical form of the unity-gain closed-loop transfer function [$T(j\omega)$] and then write a computational program for performing a Bode plot analysis on $T(j\omega)$ and graphically determine the following: BW, ω_c, M_P, and ω_P.
    iii. Solve for ζ and ω_N of the unity-gain closed-loop transfer function [$T(s)$].
    iv. Write the numerical form of the unity-gain closed-loop transfer function [$T(s)$] and then write a computational program for performing a unit-step analysisonT(s) and graphically determine the following: T_r, OS, and T_s(3.5%).

c. For Operator I', prepare a first table of values (Problem Table I.1), a second table of values (Problem Table I.2) and a third table of values (Problem Table I.3), that are patterned after Tables 10.6, 10.7, and 10.8 (for Operator I).
d. Interpret the perormance of Operator I' (Problem Tables I.1, I.2, and I.3) with that of Operator I (Tables 10.6, 10.7, and 10.8).

10.13. Now define an *Operator II* (β *= 10/s,* α_M *= 2.5/s)* for which all activities and parameters of Example 10.6 the same, *except:*

Replace equation 10.6.a.ii with:

$$W = 0.35 \text{ M } (W_T = .037 \text{ M and } W_S = .002 \text{ M})$$

Replace equation 10.6.b.iv with:

$$K_p(A') = 0.72$$
$$K_p(B') = 0.86$$

For Operator II ($\beta = 10/s$, $\alpha_M = 2.5/s$):

a. Solve for the choice reaction time, T_D, and the reaction informatic frequency, τ (for State 1 and State 2).

b. Recall the symbolic form of GH(s) per equation 10.7.a.i, GH($j\omega$) per equation 10.7.a.iii, $T(s)$ per equation 10.7.a.vi and $T(j\omega)$ per equation 10.8.iii. *Then for each of the four-model combinations* ($A'1$, $A'2$, $B'1$, *and* $B'2$):

   i. Write the numerical form of the open-loop transfer function [GH($j\omega$)], and then write a computational program for performing a Bode plot analysis on GH($j\omega$) and graphically determine the following: GM, ω_π, PM, and ω_1.
   ii. Write the numerical form of the unity-gain closed-loop transfer function [$T(j\omega)$] and then write a computational program for performing a Bode plot analysis on $T(j\omega)$ and graphically determine the following: BW, ω_c, M_P, and ω_P.
   iii. Solve for ζ and ω_N of the unity-gain closed-loop transfer function [$T(s)$].
   iv. Write the numerical form of the unity-gain closed-loop transfer function [$T(s)$] and then write a computational program for performing a unit-step analysis on $T(s)$ and graphically determine the following: T_r, OS, and T_s (3.5%).

c. For Operator II, prepare a first table of values (Problem Table II.1), a second table of values (Problem Table II.2) and a third table of values (Problem Table II.3), that are patterned after Tables 10.6, 10.7, and 10.8 of the test.

d. Interpret the performance of Operator II (Problem Tables II.1, II.2, and II.3) with that of Operator I′ (Problem Tables I.1, I.2, and I.3).

10.14. Repeat Problem 10.13 for *Operator III* ($\beta = 4/s$, $\alpha_M = 2.5/s$). Note that for part (d) you should interpret the performance of Operator III (Problem Tables III.1, III.2, and III.3) with that of Operator I′ (Problem Tables I.1, I.2, and I.3).

10.15. Repeat Problem 10.13 for *Operator IV* ($\beta = 6.7/s$, $\alpha_M = 5.0/s$). Note that for part (d) you should interpret the performance of Operator IV (Problem Tables IV.1, IV.2, and IV.3) with that of Operator I′ (Problem Tables I.1, I.2, and I.3).

10.16. Repeat Problem 10.13 for *Operator V* ($\beta = 6.7/s$, $\alpha_M = 7.5/s$). Note that for part (d) you should interpret the performance of Operator V (Problem Tables V.1, V.2, and V.3) with that of Operator I′ (Problem Tables I.1, I.2, and I.3).

10.17. Repeat Problem 10.13 for *Operator VI* ($\beta = 10/s$, $\alpha_M = 7.5/s$). Note that for part (d) you should interpret the performance of Operator VI (Problem Tables VI.1, VI.2, and VI.3) with that of Operator I′ (Problem Tables I.1, I.2, and I.3).

Appendix

Various worked examples in the text will require solving a first-order, linear differential equation with constant coefficients of the form:

$$a\dot{y} + by - c = 0 \tag{A1}$$

This will occur in Chapter 5 (i.e., equations 5.0.a.ix, 5.0.c.ii, 5.7.a.i, and 5.7.b.ii) and in Chapter 6 (i.e., equations 6.6.xvi, 6.7.x, 6.8.xiv, and 6.9.x).

These equations will be in a rearranged form of equation (A1) in which $a = 1$:

$$\dot{y} + by = c \tag{A2}$$

The solution of equation (A2) is as follows:
Rearrange equation (A2):

$$\frac{dy}{dt} = c - by$$

Separate variables and integrate:

$$\int \frac{dy}{c - by} = \int dt \tag{A3}$$

Recall that:

$$\int \frac{dx}{x} = \ln x \tag{A4}$$

Then define:

$$x = c - by \tag{A5}$$

So that:

$$dx = -bdy \tag{A6}$$

Substituting equations (A5) and (A6) into equation (A4), and factoring:

$$-b \int \frac{dy}{c - by} = \ln(c - by) \tag{A7}$$

Rearrange:

$$\int \frac{dy}{c - by} = -\frac{1}{b}\ln(c - by) \qquad \text{(A8)}$$

Substitute equation (A8) into equation (A3) and complete the integration:

$$-\frac{1}{b}\ln(c - by) = t + C \qquad \text{(A9)}$$

Recall that C is the constant of integration, evaluated at $t = 0$. Impose the boundary condition:

$$y(t = 0) = y_0 \qquad \text{(A10)}$$

and solve for C by substituting equation (A10) into (A9) at $t = 0$:

$$= -\frac{1}{b}\ln(c - by_0) \qquad \text{(A11)}$$

Now proceed by substituting equation (A11) into equation (A9):

$$-\frac{1}{b}\ln(c - by) = t - \frac{1}{b}\ln(c - by_0) \qquad \text{(A12)}$$

We now proceed to solve equation (A12) for $y = f(t)$. Begin by clearing the denominators:

$$\ln(c - by) = -bt + \ln(c - by_0) \qquad \text{(A13)}$$

Separate variables:

$$\ln(c - by) - \ln(c - by_0) = -bt \qquad \text{(A14)}$$

Invoke the logarithmic identity:

$$\ln\left(\frac{u}{v}\right) = \ln(u) - \ln(v) \qquad \text{(A15)}$$

Define:

$$u = c - by \qquad \text{(A16)}$$

$$v = c - by_0 \qquad \text{(A17)}$$

So that by substituting equations (A16) and (A17) into equation (A15) and that result into equation (A14):

$$\ln\left[\frac{c - by}{c - by_0}\right] = -bt \qquad \text{(A18)}$$

Taking the antilog of both sides of equation (A18):

$$\frac{c - by}{c - by_0} = e^{-bt} \qquad \text{(A19)}$$

Rearrange:

$$c - by = ce^{-bt} - by_0e^{-bt} \tag{A20}$$

Transpose and multiply by -1:

$$by = c - ce^{-bt} + by_0e^{-bt} \tag{A21}$$

Solve for y and factor:

$$y = \frac{c}{b}(1 - e^{-bt}) + y_0e^{-bt} \tag{A22}$$

INDEX